Electric current flow in excitable cells

Electric current flow in excitable cells

J. J. B. JACK

FELLOW OF UNIVERSITY COLLEGE, OXFORD
UNIVERSITY LECTURER IN PHYSIOLOGY

D. NOBLE

FELLOW OF BALLIOL COLLEGE, OXFORD
UNIVERSITY LECTURER IN PHYSIOLOGY

R. W. TSIEN

PROFESSOR OF PHYSIOLOGY, YALE UNIVERSITY
FORMERLY RESEARCH FELLOW OF UNIVERSITY COLLEGE, OXFORD

CLARENDON PRESS · OXFORD

Oxford University Press, Walton Street, Oxford OX2 6DP

London New York Toronto
Delhi Bombay Calcutta Madras Karachi
Kuala Lumpur Singapore Hong Kong Tokyo
Nairobi Dar es Salaam Cape Town
Melbourne Auckland
and associated companies in
Beirut Berlin Ibadan Mexico City Nicosia

Oxford is a trade mark of Oxford University Press

ISBN 0 19 857527 0

First published 1975
First issued in paperback 1983
Reprinted 1985

Printed in Hong Kong

Preface to paperback edition

THIS edition is a reprint of the original hardcover edition published in 1975 and has been produced in response to requests for a cheaper version that would make the work more readily available to graduate students. In the text we have corrected the very few errors that have been detected. So far as we are aware, the equations in the original edition were all correct (apart from a pair of missing brackets in equation (3.60)), which we feel is a tribute to the careful proof reading done by some of our colleagues whom we thank in the Acknowledgements. For each chapter we have added a set of references (in the supplementary bibliography) to relevant papers or books that have appeared over the last six years. Inevitably, these form only a selection of the vast amount of work that has appeared. Our criterion in choosing the papers cited was that they illustrate how electric current theory has developed since this book was written.

Oxford and Yale   J.J.B.J
January 1983   D.N.
           R.W.T.

Preface

The theory of electric current flow in excitable cells has developed extensively since Lord Kelvin first presented the equations for cable transmission a century ago. This development has been particularly rapid during the last 30 years or so, following the first detailed experimental applications of Kelvin's equations to nerve and muscle fibres. As a result, cable theory now plays a central role in many areas of electrophysiology, so that biologists find themselves using mathematical methods of analysis involving techniques considerably more advanced than those with which they are familiar from their undergraduate training. The first aim of this book, therefore, is to give a systematic and explanatory account of the basic mathematical theory that we hope will be of use to research workers in the field as well as in university courses in electrophysiology and in biological mathematics.

When we started writing several years ago this was the sole aim of our work. However, in attempting to write such an account, we encountered a number of areas in which the relevant theory required further development, not only to enable a reasonably systematic account to be given but also to allow us to use the explanatory device of looking at particular problems from different aspects. Where possible, we have attempted the development ourselves, and this second aim has grown as the book was written. As a result, much of the material of the book is new, as a glance at our illustrations will show. If we need to apologize for writing a rather longer and more advanced book than our original aim required, our justification is simply that, in our view, the subject requires such a book. The mathematics of excitation and conduction theory is much more complex, and the experimental work on which it is based is considerably more extensive, than when Bernard Katz wrote his classic review of excitation (*Electric excitation of nerve*) in 1939. Our subject is essentially the same as his, but its content has been greatly transformed and expanded, and we have found it impossible to restrict ourselves to the 100 or so pages that sufficed 35 years ago.

The year 1939 saw the first results of the intracellular recording techniques that were to completely revolutionize the subject after World War II. Katz's book therefore appeared at an ideal moment of time. He was able to review the development of excitation theory (including some of the valuable insights developed during the 1930s) before he and others became so successfully involved in using the new techniques. A watershed had been passed, and for two decades the insights of the 1930s must have paled before the immense power of being able to directly record the events about which the physiologists of the 1930s could only theorize.

We have now reached a somewhat analogous stage of development, although the events about which we now theorize (but wish we could directly record) are molecular rather than cellular. Furthermore, as the new intracellular methods have been applied to progressively more intractable problems in nerve and muscle physiology, so the need to use fairly elaborate, and more highly theoretical, models has returned. Moreover, it is not surprising that some of the insights of the 1930s (for example, Rushton's work on initiation and propagation of the impulse) are proving more useful again. The same problem has returned: even with the new techniques we cannot always record directly all the events we may wish to, and the need to use simplified models of the excitation process itself (not unlike those used in the 1930s) becomes greater when more complex physiological systems are studied. To some extent then the wheel has turned fully round. As a result, we have felt the need to relate some of the older insights to the modern theory of excitation.

We are keenly aware of the fact that the aims of introduction and development co-exist uneasily in the writing of a book. The result is a hybrid. Nonetheless, we have ensured that the introductory and explanatory core is still present as a substantial and identifiable body, although its parts are necessarily interspersed with more advanced development. For the guidance of students interested in our first aim we have indicated in Chapter 1 where the introductory parts are to be found. So far as our second aim is concerned, we cannot say that we are fully satisfied. The developments we have attempted are primarily analytical, largely because we expect the insights gained to be more general than those to be obtained from numerical computer models. However, we have not succeeded in obtaining useful analytical solutions for more than a fraction of the problems that interest electrophysiologists. We present our work as a stimulus to others as much as a record of our own explorations.

Oxford and Yale,   J. J. B. J.
September 1974   D. N.
                                                            R. W. T.

Acknowledgements

THIS work originated in the form of circulated notes for graduate lectures and seminars given at Oxford, Homburg (Saar), and Monash. We have also used some of the more introductory material in lectures and tutorials for final-year undergraduates studying Physiological Sciences at Oxford. We have, therefore, benefited from innumerable discussions with colleagues and students. To acknowledge all of these sources of criticism personally is impossible but we should like to acknowledge valuable discussions with Dr. R. H. Adrian, Mr. D. Attwell, Dr. W. K. Chandler, Dr. I. Cohen, Dr. L. L. Costantin, Dr. J. Daut, Dr. R. S. Eisenberg, Professor A. L. Hodgkin, Dr. D. Kernell, Professor D. G. Lampard, Professor K. Morztyn, Dr. K. G. Pearson, Dr. D. Perkel, Dr. W. Rall, Dr. S. J. Redman, Mr. H. Sackin, Dr. M. Schneider, Dr. P. G. Sokolove, and Dr. R. B. Stein. We are also grateful to some of our colleagues for allowing us to see unpublished work and, in some cases, for kindly reading various chapters. Needless to say, the responsibility for remaining errors is ours.

We have benefited from various sources of financial support during the course of writing the book. J. J. B. J. would like to acknowledge the support of a Foulerton Gift Research Fellowship of the Royal Society; D. N. would like to thank Professor M. Schachter and the University of Alberta for a Visiting Professorship, and the Canadian Medical Research Council for a research grant, part of which was used to provide secretarial help and to pay for travel to enable the authors to meet in North America; R. W. T. would like to acknowledge the support of a Rhodes Scholarship and of a Weir Junior Research Fellowship at University College, Oxford.

For permission to reproduce text figures from a number of publications we should like to thank the authors, who are mentioned in the figure captions, and the following: Academic Press, Inc. (*Experimental Neurology*) for Figs 7.3 and 7.13; the American Association for the Advancement of Science (*Science*, © 1964) for Fig. 11.39. The American Physiological Society (*Physiological Reviews*) for Figs 9.1, 9.2, and 10.2; *The Bulletin of Mathematical Biology* for Fig. 11.36; the Company of Biologists, Ltd. for Figs 7.43 (*Journal of Cell Science*) and 11.17 (*J. Experimental Biology*); *The Journal of Physiology* for Figs 3.11(b), 3.19, 3.20, 6.6, 6.7, 6.9, 7.14, 7.15, 7.16, 7.18, 7.19, 9.4, 9.5, 9.6, 9.7, 9.8, 10.6(a), 10.8, 11.8(b), 11.16, 11.22, 11.23, 11.24 11.27, 12.19, 12.21, and 12.22; the New York Academy of Sciences for Figs 10.6, 10.7, and 11.25; Pergamon Press Ltd. for Fig. 8.6; the Rockefeller University Press for Figs 6.1 (*Journal of Cell Biology*), 6.5, 11.15, 11.28, 11.29 (*J. General Physiology*); the Royal Society for Figs 6.3(a), 6.4, and 11.14; Springer-Verlag for Figs 11.21 (*Pflügers Archiv für die gesamte Physiologie des Menschen und der Tiere*) and 11.37 (*Kybernetik*); the Regents of the University of California for Fig. 2.10(a); and the Wistar Press for Figs 10.12 and 10.13.

The great majority of the illustrations are, however, new and were prepared for us by Mr. F. Loeffler at Alberta and by Miss A. G. Smith, Miss A. Goodwin, and Miss H. Cripps at Oxford.

Finally, we should like to thank the staff of the Clarendon Press for guidance on preparation for publication. Theirs has also been the thankless task of waiting so patiently for the manuscript during the many occasions on which we delayed final submission to incorporate new material and revisions.

J. J. B. J.
D. N.
R. W. T.

Contents

8. Nonlinear properties of excitable membranes

9. Nonlinear cable theory: excitation

10. Nonlinear cable theory: conduction

11. Repetitive activity in excitable cells

12. Nonlinear cable theory: analytical approaches using polynomial models

13. Mathematical appendix

List of notation and definitions

E_m — transmembrane potential expressed as potential of intracellular fluid with respect to that of extracellular fluid (mV).

E_r — resting value of E_m (mV).

V — transmembrane potential expressed as deviation of intracellular potential from resting potential, $V = E_m - E_r$ (mV).

R_i — intracellular resistivity (kΩ cm).

R_m — membrane resistance (kΩ cm^2).

C_m — membrane capacitance (μF cm^{-2}).

Z_m — membrane impedance (kΩ cm^2).

X_m — membrane reactance (kΩ cm^2).

f — frequency (Hz).

ω — radial frequency $= 2\pi f$ (rad).

I_i — membrane ionic (resistance) current (μA cm^{-2}).

I_c — membrane capacity current (μA cm^{-2}).

I_m — total membrane current, usually $I_i + I_c$ (μA cm^{-2}).

I — applied current (μA).

a — fibre radius (one-dimensional theory) (cm).

b — preparation thickness (two-dimensional theory) (cm).

i_i — membrane ionic current per unit length of fibre $(= 2\pi a I_i)$ (μA cm^{-1}).

i_c — membrane capacity current per unit length fibre $(= 2\pi a I_c)$. (μA cm^{-1}).

i_m — membrane current per unit length fibre $(= 2\pi a I_m)$ (μA cm^{-1}).

r_m — membrane resistance per unit length fibre $(= R_m/2\pi a)$ (kΩ cm).

r_a — intracellular resistance to axial flow of current along fibre $(= R_i/\pi a^2)$ (kΩ cm^{-1}).

c_m — membrane capacitance per unit length fibre $(= 2\pi a C_m)$(μF cm^{-1}).

τ_m — membrane time constant $(= R_m C_m)$ (ms).

λ — fibre space constant $(= \sqrt{(R_m a/2R_i)} = \sqrt{(r_m/r_a)})$ (cm).

x — distance along fibre (unidimensional theory) (cm).

X — $= x/\lambda$.

r — distance from point electrode (two-dimensional theory) (cm).

λ_2 — two-dimensional space constant (see Chapter 5).

R — $= r/\lambda_2$.

t — time (ms).

T — $= t/\tau_m$.

i_a — intracellular axial current (μA).

R_{in} — input resistance (recorded potential/applied current) (Ω).

Q — charge (nC).

G_m membrane chord conductance (ms cm^{-2}).

g_m membrane chord conductance in unit length of fibre (ms cm^{-1}).

θ conduction velocity (ms^{-1}).

K $= 2R_i\theta^2 C_m/a$ (ms^{-1} or s^{-1}).

$\bar{V}, \bar{I}$, etc.   Laplace transforms of V, I, etc.

s Laplace transform variable.

l length of fibre.

L l/λ.

Other symbols are defined as they are introduced.

1. Introductory remarks

THE evolution of electrically excitable membranes in living systems was an essential step in the development of those forms of life which display the complex kinds of behaviour which we associate with the possession of a nervous system. The chemical basis of this excitability is still largely unknown. However, the physical aspects are now very well understood and the theory of current flow in excitable cells is well developed. Unfortunately, many of the important results are still to be found only in the original papers or in fairly specialist reviews and, although some excellent elementary textbooks now exist, there is no systematic account of the more mathematical aspects. We hope that this book will fill this gap.

Virtually all living cells maintain an electrical potential difference between their interiors and the environment, and this potential is one of the factors determining the energy barriers encountered by charged substances entering or leaving the cell. The special characteristic of excitable cells is that the potential may change in response to variations in the chemical environment or in response to current flow. These potential changes (receptor potentials, synaptic potentials, pacemaker potentials, and action potentials) underly the ability of nervous systems to process and to transmit information. They also serve as the triggers of mechanical activity in the case of effector cells such as muscles.

The transmission of information over long distances is carried out by thin projections of nerve cells called nerve axons. Moreover, for anatomical reasons, muscle cells are also often arranged in long fibres. Thus two of the most important kinds of excitable cell have a geometry resembling that of an electric cable. The theory of current flow in electric cables, initially developed for submarine cables by Lord Kelvin (1855, 1856, 1872), was first used in work related to excitable cells towards the end of the nineteenth century by Weber (1873, 1884), Cremer (1899, 1909), and Hermann (1879, 1899, 1905). One of the most important results of this early work was Hermann's suggestion that current flow of the kind described by cable theory may be adequate to maintain nerve impulse propagation. Since then, the theory has frequently been used in the study of nerve and muscle, particularly in work on the responses to electrical stimuli which are small enough to neglect the gross nonlinearities which appear in response to strong stimuli. The theory has also been successfully applied to the mechanism of impulse propagation. This work developed rapidly in the 1930s and 1940s (see Rashevsky 1931; Rushton, 1934, 1937; Monnier 1934; Cole and Curtis 1936, 1939, 1941; Rosenberg 1937a, b; Hodgkin 1937; Cole and Hodgkin 1939; Katz 1939; Offner, Weinberg, and

Young 1940; Weinberg, 1941; Hodgkin and Rushton 1946; Lorente de Nó 1947) so that by 1946, when Hodgkin and Rushton published their experimental and theoretical analysis of the subthreshold responses of nerve axons to locally applied currents, the nature of the purely passive ('electrotonic') flow of current in nerve axons was largely clarified and some important clues to the nature of the nonlinear properties had emerged. More recently, the most exciting developments in this field have concerned the analysis of the nonlinear properties of excitable cells using the voltage control ('voltage-clamp') technique introduced by Cole and Marmont in 1949. The theoretical interpretation of this work has been based largely on the semi-empirical equations, given by Hodgkin and Huxley in 1952, for describing the time and voltage dependence of the membrane current. However, these developments have continued to require the use of cable theory and of extensions to it that were designed to deal with nonlinear systems and with more complex geometries. The result has been a steady but considerable development of the theory, the importance of which does not depend directly on any particular theory concerning the mechanism of the membrane nonlinearities. Moreover, since Fatt and Katz's (1951) quantitative analysis of the end-plate potential, the theory has also been used in the study of synaptic mechanisms. Therefore it has become an important part of many branches of biophysics and neurophysiology, and it may be useful to graduate students and research workers in these fields to have a more comprehensive introduction to the theory than is at present available.

In this book we attempt to give a systematic account of the theory and its applications. Some of the results are well known and may be found in the physiological literature or in some of the standard mathematical texts (e.g. Carslaw and Jaeger 1959; Jaeger 1951; Luikov 1968). Some of the more recent work may not be so well known. Moreover, the mathematical methods and the notation used have not always been uniform and, largely because the theory has been developed for particular applications, some results of more general importance have not always been obtained or given the attention they may deserve. In view of this situation, we give fairly complete derivations for some of the important and widely-used results, together with references to sources containing the original derivations. Where no references are given we believe the results to be new, but we apologize if we have inadvertently neglected any previous work of importance.

Wherever possible, we have tried to be simple and explanatory rather than complete and general. Some of the consequences of this policy are worth mentioning here, since we may in this way warn our more sophisticated readers where they may expect to find limitations which are largely of our own choosing. First, the derivations given are not always the most general since we believe that most biologists find particular derivations easier to follow. Second, we do not attempt to give a complete review of the physiological applications of importance, but some examples of applications are referred to in order to illustrate the theory. Third, although most of the early development of cable theory was directed towards work with extracellular electrodes, more recent work has often used intracellular electrodes. Since, in many cases, this allows the equations to be simplified, we use this simplification wherever possible. In keeping with this approach, the theory of extracellular fields (see Lorente de Nó 1947; Plonsey 1964; Clark and Plonsey 1966, 1968; Rall and Shepherd 1968; Rosenfalck 1969; Nicholson and Llinas 1971) is omitted. We also omit the theory of three dimensional fields in the vicinity of current sources (see Eisenberg and Johnson 1970).

It may be helpful to readers to have some guidance on how the book might best be used. First, it should be emphasized that we have not written a general introduction to cellular electrophysiology. On the contrary, some familiarity with the subject is assumed, and readers who have no previous knowledge would be well advised to first read an introductory account such as Katz's *Nerve, muscle, and synapse* (1966) or Aidley's *The physiology of excitable cells* (1971). Some parts of the present work will then be found fairly easy to follow. In particular, Chapter 2, many sections of Chapter 3, and Chapters 8, 9, 10, and most of 11 are intended as introductions to the basic principles of linear and nonlinear cable and excitation theory, and some parts of each of these chapters will be found to be relatively elementary. It should be noted that the introductory sections of the book do not necessarily appear at its beginning. We have deliberately deferred some of the introductory material to later chapters dealing with nonlinear cable theory since it is in these chapters that the appropriate applications occur.

Chapters 6 and 7 are concerned with particular applications of cable theory to problems in muscle excitation and the theory of nerve cells. These chapters are written largely as reviews of the present state of the field, and they may well become out of date more quickly than other chapters. We feel, however, that these chapters will give a useful indication of the way in which the theory is used in problems of current interest, and we hope that they will also serve as introductions to these two fields for those who do not have the time to adequately study the complex, and sometimes rather inaccessible, literature. Similarly, Chapter 11 is, to a considerable extent, a survey of the analysis of repetitive firing; as such it is likely to become incomplete as new work takes the analysis further.

Chapters 4, 5, and 12 contain fairly advanced or specialist material, and we suggest that they should be omitted on a first reading. These chapters should prove more useful to those already familiar with the basic principles and to those who need equations for particular problems.

Finally, it will be obvious on perusing the book that some parts assume a fair degree of mathematical knowledge. However, we hope that this will not deter non-mathematical physiologists. In the introductory chapters mentioned above we have tried to explain the derivations in fairly easy stages and, if the reader confines himself to these chapters initially, he should find that little more than a general knowledge of calculus is required. The 'little more' knowledge required involves the use of the Laplace transformation in solving differential equations. We have included a short introduction to this method and some of the other mathematical techniques employed in the book in Chapter 13. Even if the reader does no more than simply accept that the Laplace transformation works without understanding how it does so, we suspect that a fairly good understanding of the basic principles and major applications of cable theory may be achieved, and we hope that this book may help many physiologists to do this more easily and more quickly than we did ourselves. More than anything else, this book is motivated by the desire to make what is admittedly one of the most difficult areas of physiological theory less forbidding and more accessible than it usually appears.

2. Linear electrical properties of excitable cells

EXCITABLE tissues are formed of cells. Electric current may flow therefore either within the cells as intracellular current, or it may flow between the intracellular and extracellular fluids by entering or leaving the cells as membrane current. The impedance to current flow within the cells will be referred to as intracellular impedance, even though in some cases this may involve cell-to-cell impedance in addition to that of the intracellular medium. The impedance to current entering or leaving the cells will be referred to as membrane impedance. The theory of current flow which we shall discuss in this book is based on two assumptions concerning the impedances. These are that the intracellular medium acts as an ohmic resistance and that the surface membrane acts as a leaky capacitance. In this chapter we shall review the evidence for these assumptions. Chapters 3–5 and 7 are concerned with the theory of current flow when the assumptions are unmodified. Later chapters will be concerned with the complexities which arise when the assumptions must be modified, either because the membrane capacitance is only partially located at the cell surface (Chapter 6; see also pp. 18–21 of this chapter) or because the leak is not a simple linear resistance (Chapters 8–12).

Forms of excitation used in studying living cells

In order to investigate the electrical properties of cells, we may apply various forms of current stimulus (excitation) and investigate the way in which the cell responds. A major aim of a theory of current flow is to obtain equations for these responses. We shall begin by briefly describing the forms of excitation which are most frequently used in electrophysiology. The four types of stimulus to be applied are shown in Fig. 2.1.

Very brief current pulse (delta function)

A very simple form of excitation is the application of a known quantity of charge Q in a very brief period of time. Provided that the time taken to apply the charge may be regarded as infinitesimal the current input is called a delta function. This form of excitation is useful partly because the mathematical theory of delta functions is well developed (see Chapter 13) and partly because some physiological events (e.g. brief synaptic or receptor currents) may be approximated by delta functions (see Chapter 3, p. 46). As shown in Chapter 13, the delta function is important in current-flow theory since it may be used theoretically to obtain the response to any other form of excitation. However, its use is limited in practice by the fact that in some experimental

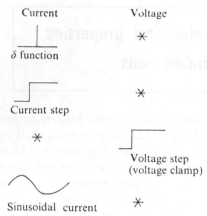

FIG. 2.1. Forms of excitation most frequently used in the analysis of excitable cells. The delta function is an infinitesimally brief pulse of current characterized by the amount of charge applied. (Ideally, the *magnitude* of the current pulse would be infinite.) The current or voltage step is a sudden change in current or voltage from one constant value to another. These steps are characterized by their magnitude and duration. Sinusoidal currents are characterized by amplitude and frequency. The aim of current flow theory is to obtain equations for the voltage (or current if voltage steps are used) at the point of excitation and for the voltage and current at other points in response to various forms of excitation.

situations, particularly those involving the use of high-resistance current electrodes, the current intensity may not be large enough to apply charge sufficiently quickly. Moreover, since at the point of charge application the current density may be very large, the simplest equations for current flow may be invalid (Eisenberg and Johnson 1970). Despite the theoretical advantages of delta function inputs, other forms of excitation are therefore more frequently used experimentally.

Step current waveform

Much of the early work on excitable cells was done using step current changes (i.e. rectangular current pulses of various durations). The theory of current flow in response to this form of excitation therefore has been extensively used in electrophysiology, and we shall consider this case in some detail. A step current is often achieved experimentally by applying a step voltage to a resistance connected via an electrode to the cell (see Fig. 2.2). Provided that the series resistance is very large compared to the impedance of the cell the current flow will be practically independent of any changes in the cell impedance. More recently, constant currents have been achieved using operational amplifier circuits (Moore 1963). These circuits are more satisfactory than the conventional one illustrated in Fig. 2.2 since the magnitude of the current flow is not limited by a large series resistance. The principle of the circuit is that the current flow is measured and compared at the summing junction of an operational amplifier with a standard current step, which is

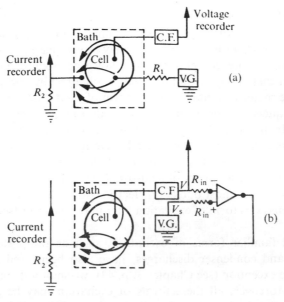

Fɪɢ. 2.2. (a) Simple circuit for applying controlled currents to excitable cells. A voltage. generator feeds current to cell via a large resistance so that impedance of the cell is negligible. As mentioned in the text, operational amplifier circuits which do not require large resistance in series with the cell may also be used. (b) Simplified circuit for controlling cell potential (voltage clamp). Mode of operation is described in text.

generated separately. The output of the amplifier is then used to adjust the current flow through the preparation so that it remains constant and equal in magnitude to the standard current step.

Voltage step (voltage clamp)

One of the major advances in electrophysiological technique was the invention of a circuit which allows the control of voltage rather than current (Cole 1949; Marmont 1949). This is achieved by forcing the cell potential to follow a reference potential V_s, which is generated electronically. A simplified circuit for doing this is shown in Fig. 2.2. The cell potential is compared with the reference potential at the summing junction of an operational amplifier. The input to the amplifier is therefore proportional to $(V-V_s)$. The output is given by $-A(V-V_s)$, where A is a large amplification factor. This output is then fed back to the cell via a current-passing electrode. Hence a large current will flow whenever $(V-V_s)$ does not approximate to zero. Moreover, since the output is proportional to $-(V-V_s)$, the current flows in such a direction as to minimize the magnitude of $(V-V_s)$. Therefore, provided that A is sufficiently large to keep $(V-V_s)$ near zero, the value of V is forced to follow V_s. This technique is now widely used in studying the nonlinear properties of living cells, and we shall refer to it frequently in the later chapters of this book.

Sinusoidal current

Provided that the number of electrical components in a system is not too great, the use of delta functions or steps may be adequate to fully analyse the cell impedance. More complex circuits, however, require much more information for their analysis, and this may be obtained by applying sinusoidal currents. The major advantages of this form of excitation are the following.

1. The frequency of the sinusoidal excitation may be varied over a wide range. In practice, this gives a more accurate characterization of the cell impedance than do individual wave forms such as a step or delta function which contain components of all frequencies (see Chapter 13).
2. The mathematical theory of sinusoidal excitation is well developed (see Chapter 13 and most books on electric circuit theory).

We shall discuss this form of excitation later in this chapter (see p. 14) and in Chapter 6.

Other less familiar forms of excitation, such as ramp steps, sawtooth waveforms, and condenser discharges, have also been used in physiology but, with one exception (see Chapter 3, p. 43), we shall not consider them in this book. Moreover, all these forms of excitation may be applied either uniformly or nonuniformly to a cell membrane. The differences between responses to uniform and nonuniform excitation are usually very large and, since most physiological forms of excitation are nonuniform, a major aim of current-flow theory in physiology is to analyse the effects of nonuniform excitation. In order to do this it is sometimes necessary and always desirable to know the behaviour of the system in response to uniform excitation. We shall discuss therefore the effects of uniform excitation before obtaining equations for nonuniform excitation. In the case of linear systems, we shall do this later in this chapter (*Membrane impedance*, p. 11). The uniform responses of nonlinear systems will be discussed in Chapter 8.

Intracellular impedance

The intracellular impedance of nerve and muscle cells is usually represented by an ohmic resistance to *axial* flow of current. The radius of the cylinder formed by the intracellular fluid is assumed to be small enough for the resistance to *radial* flow of current to be negligible compared to the surface membrane impedance. Unless high-frequency responses are being considered (when the membrane impedance becomes very low—see *Membrane impedance* p. 15) the second assumption is generally valid, since the membrane resistance is extremely high (see p. 23). However, in more accurate work and at high frequencies, radial fields may not be negligible in the immediate vicinity of current sources, and equations allowing for these fields must then be used (see Falk and Fatt 1964; Eisenberg 1967; Eisenberg and Johnson 1970; Rall 1969b). In the equations used in this book, radial fields will be assumed to be negligible.

Nerve axons

In the case of nerve axons the assumption that the intracellular medium acts as an ohmic resistance is now well established. The axoplasm forming the intracellular fluid is a relatively structureless aqueous gel and, provided that most of the ions in the axoplasm are freely mobile, it should act as a simple low-resistance medium. However, this assumption has been questioned and, in particular, it is sometimes held that the ions inside cells are not freely mobile but may be largely bound to slowly moving or immobile macromolecules (see e.g. Ling 1962; Troshin 1966). Although it is now clear that the relatively small number of multivalent cations (calcium, magnesium) are largely bound to intracellular macromolecules this is not the case for the much more numerous monovalent cations. In particular, potassium ions, which are the most numerous cations inside the cell, have about the same activity coefficient as in free solution outside the cell (Lev 1964). Apart from attempts to measure internal ion activities directly, which are still subject to considerable uncertainties, there are three other approaches which support the view that most of the internal cations are freely mobile.

First, Cole and Hodgkin (1939), using cable-theory equations and measurements of the resistance of a squid axon between external electrodes at various distances apart, obtained an electrical measurement of the internal resistivity which was only about 40 per cent greater than that of sea water. The ohmic behaviour of axoplasm was confirmed by Hodgkin and Rushton (1946) and by Rashbass and Rushton (1949). Similar electrical estimates have now been made for a large number of excitable cells with similar results (see, however, Carpenter, Hovey, and Bak (1973)). Although Cole and Hodgkin's experiments provided the first d.c. estimates of intracellular resistance, previous estimates had been made using high-frequency sinusoidal currents to reduce the membrane impedance (see the section *Membrane impedance*, p. 15). The theoretical stimulus for these investigations was Bernstein's (1902) assumption that intracellular fluids behave as simple electrolytes, which formed an important part of his 'membrane' hypothesis. These high-frequency estimates of the intracellular resistivity also gave values not very much greater than that of the extracellular fluids (see Cole 1941*a*, 1968).

A second approach to this problem was employed by Hodgkin and Keynes (1953), who measured the rate of diffusion and the ionic mobility of radioactive potassium ions in nerve axons. They obtained values similar to those in free solution. By contrast, calcium ions were found to have extremely low mobility in axoplasm (Hodgkin and Keynes 1957; see also Kushmerick and Podolsky 1969).

The third approach is based on the measurement of the kinetics of transfer of ions between the inside and outside of a cell. If the ions are freely mixed inside the cell, the only rate-limiting step in this process should be the transfer of the ions across the cell membrane. If this is a relatively thin layer (see p. 23) and the transfer process is one of simple diffusion, or involves any other mechanism in which the net membrane flux is proportional to the difference in activity across the membrane, the kinetics governing the loss of ions from the cell should be of first order (see Chapter 8). Thus the concentration of radioactive ions inside a cell previously loaded with radioactive ions and exposed to a solution free of radioactive ions should decay exponentially. By contrast, a non-exponential decay would be expected if ion movements were restricted inside the cell (cf. Harris and Sjodin 1961). The best available evidence on single cells for potassium ions shows that the major part of the decay of concentration is exponential (Hodgkin and Keynes 1955). It should be emphasized, however, that this approach must be used with caution. A simple result may be good evidence that the internal ions are freely mobile, but the converse does not necessarily apply. Thus, if the membrane transfer process is more complex (e.g. the rate may be proportional to a power of the activity difference) non-exponential decay curves might be obtained even in cases where all the internal ions are freely mobile. Moreover, the internal ions may be compartmentalized and relatively free within each compartment. Whether this may impede the longitudinal flow of current will depend on the internal organization of the cell.

Muscle fibres

In the case of most muscle cells, the intracellular phase is highly structured (see Smith 1966). The sarcoplasmic reticulum forms an elaborately organized system of membranes which is involved in excitation–contraction coupling (see Chapter 6). Moreover, in cardiac and smooth muscle, the tissues are traversed by membranes (the intercalated discs in cardiac muscle) formed by the apposition of the individual cells forming the tissue. Surprisingly perhaps, these internal membrane structures do not necessarily have a large influence on the intracellular impedance of the fibres or tissue. In skeletal muscle large spaces exist between the reticulum structures so that current can flow very easily between them. The internal resistance is probably increased by the presence of these structures (Kushmerick and Podolsky 1969), but it is still largely an ohmic resistance (Katz 1948; Fatt and Katz 1951, 1953; Fatt and Ginsborg 1958; Hencek and Zachar 1965). In fact the sarcoplasmic reticulum influences the impedance between the inside and outside of the muscle cells far more than the intracellular impedance (see the section *Membrane impedance*, p. 18 and Chapter 6).

In cardiac muscle the question whether the intercalated disc membranes impede the internal flow of ions between cells has been highly controversial (see Kavaler 1959; Tarr and Sperelakis 1964; Woodbury and Crill 1961; Barr, Dewey, and Berger 1965; Weidmann 1952, 1966; Woodbury and Gordon 1965; Johnson and Sommer 1967; Freygang and Trautwein 1970). Much of the evidence in favour of the view that current may flow easily between cells has been excellently reviewed by Weidmann (1967, 1969). When internal current flow is restricted to one dimension, either by using the cable-like Purkinje strands (Weidmann 1952) or by applying current to dissected strips of ventricle muscle via an external suction electrode (Woodbury and Gordon 1965), the longitudinal spread of current is much more extensive than would be expected if the discs provided a high resistance to current flow. Moreover, the resistance to movement of radioactive potassium ions along a ventricle strip is relatively low (Weidmann 1966). On the basis of diffusion of radioactive potassium ions along ventricle strips, Weidmann estimates that the intercalated discs may approximately double the effective internal longitudinal resistance, but this would still mean that the internal resistance is very low compared to the surface membrane resistance. Freygang and Trautwein (1970) have shown that the axial impedance of Purkinje fibres is not entirely resistive. At frequencies which may be important in impulse propagation, some current flow between cells is capacitive.

The decay of voltage in response to current from a small intracellular source is sometimes (Woodbury and Crill 1961), but not always (Tille 1966), a simple function of distance. At the microscopic level, therefore, there may be preferred pathways from one region to another (Sommer and Johnson 1968), which are nevertheless too numerous to greatly influence the behaviour of the system in response to current applied to a large number of cells. Evidence that heart cells do not always form good electrical junctions has been obtained by Sperelakis (1972), who has shown that individual cells in heart tissue cultures are often electrically isolated. The fact that heart tissue cultures may also show synchronous spontaneous electrical activity does suggest however that good electrical junctions are formed in cultured tissue under some circumstances. For a general review of electrical connections between cells the reader is referred to Furshpan and Potter (1969).

Smooth muscle tissues have been studied recently under conditions in which the current flow is restricted to one dimension and, here also, the tissue may behave as a simple cable (Tomita 1966, 1967; Abe and Tomita 1968) with a fairly low resistance to internal current flow between the cells.

Membrane impedance

Bernstein (1902, 1912) and Nernst (1889) first developed 'membrane' theories to explain electrophysiological phenomena. It is largely to Bernstein that we owe the theory that cells behave electrically as simple electrolytes surrounded by thin polarizable membranes which are very sparingly permeable to ions. Höber (1910, 1912), Lapicque (1907), and others produced some of the early experimental evidence in favour of this view. More quantitative evidence concerning the nature of the membrane impedance came from the work of Fricke (1925) and Cole (1928). Cole has written a valuable review of this early work in his book *Membranes, ions, and impulses* (1968). The reader should refer to this book and to the papers of Fricke, Cole, and Schwan (references given in Cole 1968) for more advanced treatments than the one we shall give here.

Fricke used sinusoidal current analysis of red blood cell suspensions to obtain an electrical equivalent circuit for the cell membrane and showed that the membrane could be represented by a capacitance. He could not obtain an estimate of the membrane resistance, although his results showed that it must be very high compared to the resistance of the fluids between the cells. His estimate of the membrane capacity was close to 1 μF cm^{-2}. Using a nominal value of 3 for the dielectric constant of the membrane substance (thought to be a lipid) he also estimated the membrane thickness to be 33 Å. This value is close to 50 Å, the modern estimate for the lipid part of the membrane, which is based on electron microscope and X-ray measurements (see Robertson 1960; Finean 1966; Levine 1972). The membrane capacitance of nerve cells has also been studied with a.c. techniques (see e.g. Curtis and Cole 1938; Schwan 1965; Taylor 1965). Another approach has been to record the capacitive transients in the current response to step voltage changes using the voltage-clamp technique (Cole 1949; Hodgkin, Huxley, and Katz 1952). Both of these methods give estimates of about 1 μF cm^{-2} for the nerve membrane capacitance.

The capacitance of muscle cells is more complicated than that of nerve cells (Schwan 1954; Falk and Fatt 1964; Fozzard 1966; Eisenberg 1967; Freygang, Rapoport, and Peachey 1967; Freygang and Trautwein 1970; Schneider 1970). Recent measurements on muscle have also used a.c. techniques to obtain electrical equivalent circuits for the membrane impedance. It may help in understanding the results first to explain how a simple imped-ance is analysed before discussing how the more complicated equivalent circuits have been obtained.

Step current analysis

We will assume that the membrane impedance may be represented by the circuit shown in Fig. 2.3. Then the membrane current I_m will be given by the sum of two components,

$$I_m = I_c + I_i, \tag{2.1}$$

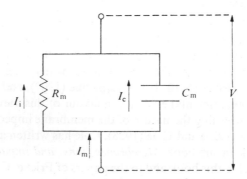

FIG. 2.3. Parallel resistance and capacitance circuit used to represent cell membrane impedance.

I_i is the ionic current which is carried by the flow of ions across the membrane. This component will be given by V/R_m, where R_m is the membrane resistance. In the early chapters of this book we will assume that R_m is constant, that is, that I_i is a linear function of V. The resulting theory is called linear cable theory. The consequences of relaxing this assumption will be discussed in the chapters on nonlinear cable theory.

I_c is the capacitive current flow which results in a change in the amount of charge separated by the membrane. (It should be noted that this current flow does not involve charge actually crossing the cell membrane—cf. Chapter 10, p. 279.) The voltage established across the membrane is proportional to the amount of charge separation which has occurred, so that the capacitive current flow determines the *rate* at which the voltage changes, that is, $\partial V/\partial t$. The larger the capacity to be charged, the more current flow is required to reach a given voltage. Hence

$$I_m = C_m \cdot \partial V/\partial t + V/R_m. \tag{2.2}$$

If a sudden step change in I_m from zero to some constant value is applied to the membrane, the solution to eqn (2.2) is a simple exponential equation:

$$V = I_m R_m \{1 - \exp(-t/\tau_m)\}, \tag{2.3}$$

where $\tau_m = R_m C_m$. τ_m is called the membrane *time constant*. According to eqn (2.3), τ_m is given by the time at which the voltage rises to $(1 - e^{-1})$ (approximately two-thirds) of its steady state value $I_m R_m$. Hence, under conditions which allow current to be applied uniformly to a cell membrane, the determination of τ_m from the time course of the voltage change in response to a step current change gives a value for $R_m C_m$. Since the value of R_m may be calculated from the steady state voltage reached (see Fig. 2.4), it is possible to obtain a value for C_m from the time-constant measurement. Unfortunately, the conditions required for uniform polarization of a cell membrane are often difficult to achieve experimentally, and it is more usually the case that current is applied nonuniformly to the membrane. It should be emphasized that eqn (2.3) does not then apply and the time taken to reach two-thirds of the final voltage may be very different from the membrane time constant. In these cases more complicated equations are required (see Chapters 3, 4, 5, and 7).

Fig. 2.4 shows the voltage and currents during a step change in membrane current applied uniformly to a membrane whose equivalent circuit is given by Fig. 2.3. Following the application of the current step, the capacitive current falls exponentially to zero as more charge leaks through the membrane resistance to form resistive current. After the termination of the rectangular current pulse the charge on the membrane capacitance dissipates by flowing through the membrane resistance. The resistive and capacitive currents are then equal and opposite to give a net membrane current equal to zero.

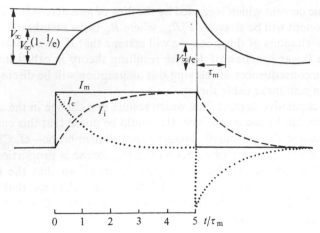

F_{IG}. 2.4. Voltage and currents during and following application of a step current pulse to the circuit shown in Fig. 2.3.

Sinusoidal current (a.c.) analysis

Although the use of step current analysis gives all the required information in the case of the simple circuit shown in Fig. 2.3, it usually does not give sufficient information about more complicated circuits. Moreover, in cases where some of the 'reactive' properties of the membrane are attributable to nonlinearities (see p. 21) the step-current method may give misleading estimates of the membrane capacitance. In general, considerably more information about the membrane impedance may be obtained by studying the response to sinusoidal currents over a wide range of frequencies.

Since the current flow through a pure capacitance is proportional to $\partial V/\partial t$, the voltage response to a sinusoidal current will also be sinusoidal but will 'lag' in time behind the current by a quarter cycle (90°) (see Fig. 2.5). The existence of a phase lag in the response of a capacitance means that we cannot compare its impedance directly with that of a resistance, which has no phase lag. This difficulty is overcome by representing the impedance of any complex system as two separate components, which are the resistive and reactive components of an equivalent *in series* circuit. The resistive component R_s is plotted along the abscissa and the reactive component X_s is plotted at 90° to the resistance axis and so forms the ordinate. A particular pair of values R_s and X_s will specify a point in the plane. The distance from the origin to this point is the magnitude of the impedance (this is the ratio of the maximum voltage to the maximum current and has the dimensions of resistance) while the angle formed with the R-axis is the phase angle (see Fig. 2.6). A pure resistance would have a zero phase angle and would be a line on the abscissa. A pure capacitance would be represented as a line on the ordinate. The impedance of a circuit with both resistive and capacitive properties will be

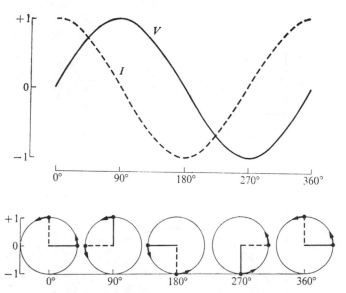

FIG. 2.5. Voltage and current flowing through pure capacitance during one cycle of sinusoidal waveform. The current 'leads' the voltage by a quarter cycle (90°). In a pure resistance the current and voltage are exactly in phase.

represented by a line with a phase angle less than 90°. It is important to note that, although this method of representing the impedance of a circuit gives a pair of values of R_s and X_s for each frequency, these values are not those of the actual circuit elements unless the circuit being investigated consists of a resistance and capacitance in series. In general, the R_s and X_s values given by the coordinates of the point Z_s are those of an *equivalent* series circuit. Thus the value of the equivalent R_s may vary with frequency although the value of each circuit resistance element in the real circuit would be independent of frequency. However, although the values of the circuit components may not be obtained directly from the coordinates of the point Z_s, they may be obtained indirectly by investigating the locus of Z_s as the frequency of the applied current is varied and then fitting the locus with equations based on possible equivalent circuits. We will illustrate this process by deriving the equations for the locus of Z_s in the case of the simple parallel resistance and capacity circuit.

Equations for impedance locus diagrams. The impedance of a pure resistance is, of course, independent of the frequency of the applied current. The impedance of a capacity, however, is highly frequency-dependent, since the ability of the capacity to pass current is determined by the rate of change of voltage, which is greater at high frequencies that at low frequencies. If the radial frequency is ω, where $\omega = 2\pi f$, the reactance of the capacity is given by

$$X_c = 1/j\omega C_m, \qquad (2.4)$$

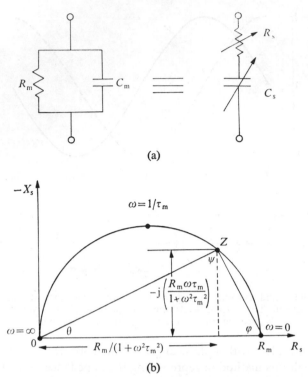

(a)

(b)

F𝗂𝗀. 2.6. (a) The equivalent series representation of the circuit in Fig. 2.3. (b) Impedance locus diagram for the circuit shown in Fig. 2.3. The locus is given by eqn (2.6). The horizontal coordinate is given by the first term $R_s = R_m/(1+\omega^2\tau_m^2)$. The vertical coordinate is given by $X_s = -jR_m\omega\tau_m/(1+\omega^2\tau_m^2)$. Note that both R_s and X_s are functions of R_m, C_m, and ω. The components of the equivalent series circuit are not therefore constants (see (a)) but are functions of the frequency of the applied current. As shown in text the locus is a semicircle of radius $R_m/2$.

where j indicates that the magnitude of the reactance must be plotted at 90° to the R-axis. j is $\sqrt{-1}$, but this term is sometimes puzzling to those who are unfamiliar with the theory of complex numbers and in the relatively simple equations used here the reader will find it sufficient to regard j as a means of indicating that the quantity which it precedes must be plotted on the ordinate of an X–R plot.

Since R_m and C_m are assumed to be in parallel, the total impedance Z_m will be given by

$$1/Z_m = 1/R_m + 1/X_c = 1/R_m + j\omega C_m$$

or

$$Z_m = R_m/(1+j\omega\tau_m). \tag{2.5}$$

Eqn (2.5) does not give the coordinates of Z_m directly since it is not of the form $x+jy$. (In this form x is called the real part and y the imaginary part.) We may separate the real and imaginary parts of Z_m by noting that

$$(1+jA)(1-jA) = 1+A^2,$$

since j^2 is -1. Hence

$$Z_m = \frac{R_m}{1+\omega^2\tau_m^2} - j\frac{R_m\omega\tau_m}{1+\omega^2\tau_m^2}. \tag{2.6}$$

Thus the imaginary part is always negative. This results from the fact that we have assumed the reactance to be purely capacitive. (Inductive reactance gives positive values for the imaginary part of an impedance, cf. Fig. 2.10.) The usual physiological convention is to plot capacitive reactance above the R-axis. We will follow this convention and label the reactance ordinate $-X$ to emphasize that the ordinate has been inverted. We may now plot Z_m on an X–R plot; the first term in eqn (2.6) giving the R_s value and the second term the X_s value of the coordinates of Z_m. When $\omega = 0$ the reactance is zero and $Z_m = R_m$. When $\omega = \infty$, $Z_m = 0$. At intermediate values of ω, Z_m is given by the top point of the triangle OZR (see Fig. 2.6) and the phase angle is θ. Now from Fig. 2.6 it is evident that

$$\tan\theta = \omega\tau_m,$$

and

$$\tan\phi = 1/\omega\tau_m.$$

Hence

$$\tan\theta\tan\phi = 1$$

or

$$\theta + \phi = 90°.$$

This means that OZR is a right-angled triangle for all values of ω. Hence the locus of Z_m as ω is varied from 0 to ∞ must be a semicircle of diameter R_m whose centre lies on the R_s-axis at $R_m/2$. The maximum reactance must occur when $\theta = 45°$ or when $\tan\theta = 1$. Hence the maximum reactance will occur at $\omega = 1/\tau_m$ (see Fig. 2.7).

In practice, although the impedance locus diagrams obtained for many cell membranes in response to small currents are semicircular, they may differ from that shown in Fig. 2.7 in three important respects.

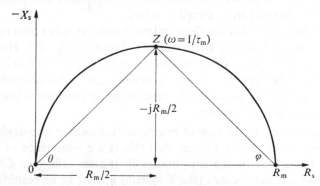

FIG. 2.7. Coordinates of Z_m when $\omega = 1/\tau_m$. $X_s = -jR_m/2$. $R_s = R_m/2$. X_s is maximal at this frequency.

1. The impedance tends to a finite value of R_s as $\omega \to \infty$. This represents the fact that a small resistance exists in series with the membrane capacitance and this resistance is at least partly attributable to the fluids surrounding the cell membrane between the current passing electrodes. This resistance presents no serious difficulty in the analysis since it simply shifts the semicircular locus along the R_s-axis.

2. In many cases the centre of the semicircle lies below the R_s-axis so that the curve forms an angle at its intersection with the R_s-axis which is less than 90°. This deviation from the behaviour of a pure capacitance is still unexplained. Palti and Adelman (1969) have shown that, in squid nerve, the capacitance is independent of frequency when the current is passed entirely across the cell membrane by using an intracellular electrode. The impedance locus diagram would then form an angle of 90° to the R-axis. Since lower estimates of this angle were obtained using external electrodes, Palti and Adelman suggest that the lower angles may be attributable to a surface impedance, perhaps caused by counter-ion movement (Schwan 1965). Whether an explanation of this kind is possible for all cases in which frequency-dependent capacities have been observed is not yet clear. For further discussion, particularly of early work in this field, the reader is referred to Cole's book (1968).

3. In some cases, the impedance locus is not a simple semicircle over the whole range of frequencies. The deviations which have been observed may be classified into two kinds.

(a) Some curves, such as those for squid axon (Cole and Baker 1941), actually cross the resistance axis at low frequencies to enter the inductive half of the $X–R$ plane. This effect will be discussed further below (see p. 21).

(b) The impedance locus may follow one semicircle at high frequencies and another semicircle at low frequencies. This phenomenon has been studied in detail in skeletal muscle and we shall discuss it below as an illustration of the way in which more complicated equivalent circuits may be obtained from impedance loci.

It is clear, therefore, that the electrical properties are often too complicated to be represented accurately by the circuit shown in Fig. 2.3. However, in many cases, the deviations from the behaviour of the parallel R–C circuit are small enough to be neglected when investigating the cable properties of the cells. In particular, the capacity of nerve axons approximates fairly closely to that of a pure parallel capacity.

Muscle impedance. In the case of muscle cells, however, the deviations are much larger, and it is now evident that this is a consequence of the more complex arrangement of the membranes in muscle cells (see Chapter 6). Some intracellular membranes (the T system) appear to be continuous with the surface membrane, forming an array of very thin tubular invaginations, and there is good evidence that some form of electrical signal spreads along

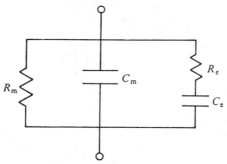

FIG. 2.8. Four-component circuit representing muscle impedance used by Falk and Fatt (1964). R_e and C_e are attributable to the transverse tubular system. This circuit is one of the simplest which may be used for muscle and is not the most accurate. More realistic and accurate circuits will be discussed in Chapter 6.

these tubules in the activation of muscle contraction (Huxley and Taylor 1958; see also Chapter 6). Fatt (1964) and Falk and Fatt (1964) have studied the impedance of frog striated muscle fibres, and a reasonable fit was obtained with a circuit which has an additional current pathway, a resistance and capacitance in series (see Fig. 2.8). C_e represents the capacitance of the tubular walls; the simplest interpretation of R_e is that it corresponds to the resistance of the tubular lumen (cf. Falk and Fatt 1964; see Chapter 6 for further discussion). Assuming this circuit, the impedance Z_e of the T system will be given by

$$Z_e = R_e + 1/j\omega C_e$$

and the total impedance will be given by

$$1/Z = 1/R_m + j\omega C_m + 1/(R_e + 1/j\omega C_e). \tag{2.7}$$

The impedance locus diagram for eqn (2.7) is more difficult to analyse than for eqn (2.5). However, the method is similar. Eqn (2.7) is rearranged to give an equation for Z in which the real and imaginary parts are separated. When this is done it is found that the real part of the equation may be approximated fairly closely at high frequencies by an equation for a two-component system and at low frequencies by an equation for a three-component system (see Fig. 2.9). The algebraic proof is cumbersome and we shall simply give the important results here.

First, the two semicircles formed by the two- and three-component systems intercept the resistance axis at a common point R_{int} given by

$$R_{int} = R_m \left(\frac{\tau_e^2 + \tau_e \tau_3}{\tau_{sum}^2 - 2\tau_m \tau_3} \right), \tag{2.8}$$

where $\tau_m = R_m C_m$, $\tau_e = R_e C_e$, $\tau_3 = R_m C_e$, and $\tau_{sum} = \tau_m + \tau_e + \tau_3$. Second, the high-frequency semicircle is given by this resistance in parallel with a capacitance. The time constant of this circuit (which also gives the value of

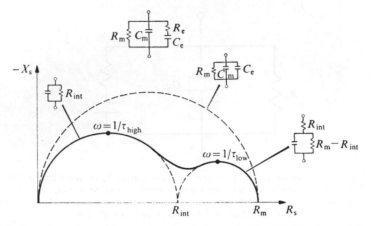

FIG. 2.9. Impedance locus diagram for circuit shown in Fig. 2.8. The locus follows one semicircle at high frequencies and a second semicircle at low frequencies. Explanation in text.

ω at which the maximum reactance is obtained) is given by

$$\tau_{\text{high}} = \tau_m \tau_e/(\tau_{\text{sum}}^2 - 2\tau_m \tau_e)^{\frac{1}{2}}, \tag{2.9}$$

Hence the 'capacitance' of the equivalent high frequency circuit will be given by $C = \tau_{\text{high}}/R_{\text{int}}$.

Finally, the low-frequency semicircle is given by a circuit formed by the resistance R_{int} in series with a parallel R–C circuit. The resistance of the parallel circuit is $R_m - R_{\text{int}}$, and its time constant is given by

$$\tau_{\text{low}} = (\tau_{\text{sum}}^2 - 2\tau_m \tau_e)^{\frac{1}{2}}. \tag{2.10}$$

The 'capacitance' of the equivalent low frequency circuit therefore is equal to $\tau_{\text{low}}/(R_m - R_{\text{int}})$. From eqns (2.9) and (2.10) it can be seen that the product $\tau_{\text{low}}\tau_{\text{high}} = \tau_m \tau_e$. Since R_m is given by the zero-frequency intercept, eqns (2.8) (2.9), and the product $\tau_{\text{low}}\tau_{\text{high}}$ give three equations which may be solved for the three unknowns τ_m, τ_e, and τ_3 from which, together with the value for R_m, we may calculate the remaining circuit components C_m, C_e, and R_e. The actual values obtained for the circuit components in the case of skeletal muscle fibres will be discussed in Chapter 6. It is usually found that R_e is fairly small compared to R_m and that C_e is of the same order of magnitude as (although larger than) C_m. τ_e therefore is usually small compared to τ_m so that, at fairly low frequencies, the charging time constant τ_e may be neglected and the system is approximated by a simple parallel R–C circuit given by R_m in parallel with $C_m + C_e$. For low frequencies, therefore, this enables the muscle impedance to be represented by an equation of the same form as that required for nerve membrane impedance (eqn 2.5) and the same equations for spread of electric current may be used. However, this approximation will not be valid for high frequency signals and, in the case of the muscle action potential the charging time constant τ_e may not be neglected (see Chapter 6, p. 117).

The discovery that most of the muscle fibre capacity is not located at the surface membrane is reassuring since, quite apart from the important implications in connection with excitation–contraction coupling, we no longer have to suppose that the specific capacity of muscle membrane is very much greater than that of nerve. This would be a very surprising disagreement. For example, in the case of crab muscle the total capacity is as high as $40\mu F$ cm² of fibre surface area. As in frog muscle, however, most of this capacity is attributable to the T system and to infoldings of the sarcolemma itself (Eisenberg 1967). We shall discuss the precise allocation of the total capacitance to surface and internal membranes in Chapter 6.

Fozzard (1966) has shown that the high capacitance of cardiac tissue may also be explained by supposing that a large fraction of the total membrane area is internal. In this case, however, it is not possible to attribute the extra capacitance to the T system. Fozzard worked on Purkinje strands and, under the electron microscope, these preparations show little or no T system (Sommer and Johnson 1968). However, it is possible that some of the spaces between the cells forming the strands may be continuous with the external solution and that the extra capacitance is located in the cell membranes in the deeper part of the strands (Freygang and Trautwein 1970; Mobley and Page 1972).

In the case of linear systems, the sinusoidal and transient methods of analysis must be interrelated so that, given the response to one form of stimulus, it should be possible to calculate the response to the other form of stimulus. Thus Teorell (1946) has described how impedance locus diagrams may be constructed from the response to a single-step current. However, it should be emphasized that although the response to all frequencies is 'contained' in the step response, only the response to those frequencies around $1/\tau_m$ may be accurately extracted. The responses to higher frequencies are best studied with the sinusoidal method.

Inductive properties of cell membranes

We have so far assumed that the membrane exhibits resistive and capacitive properties only. In some cases, however, the impedance locus may enter the inductive half of the X–R plane. An important example of this behaviour was discovered by Cole and Baker (1941). Their result for squid nerve is shown in Fig. 2.10. It can be seen that at low frequencies, instead of following a simple semicircle and terminating at the R-axis the locus swings down below the R-axis. At these frequencies, the voltage leads the current as in an inductance.

This effect has been shown to arise from time- and voltage-dependent changes in membrane *conductance* (Hodgkin and Huxley 1952). It has already

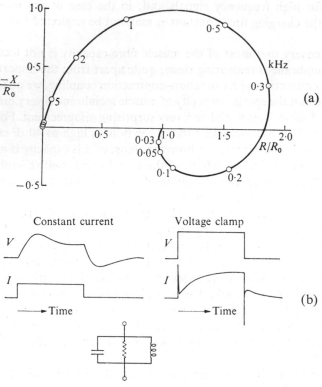

FIG. 2.10. (a) Impedance locus diagram for squid nerve, obtained by Cole and Baker. This diagram enters the inductive half of the X–R plane at low frequencies. This effect is attributable to time-dependent membrane conductance changes. (From Cole 1968.) (b) Time-dependent voltage and current changes of a form corresponding to the behaviour of an inductive circuit. *Left.* Voltage response to constant current pulse. *Right.* Current response to constant voltage pulse.

been noted that outside a small range of potentials the membrane conductance is not constant. Thus, in most excitable cells, depolarization increases both the sodium and potassium conductance of the membrane. These changes will be discussed in Chapter 8. The way in which they may produce an effect which is similar to that of an inductance may be shown by considering the potassium conductance change only. An increase in potassium conductance with time following a change in current will result in the voltage change in response to the current becoming smaller with time. Therefore the voltage will rise first to a peak as the membrane capacitance is charged and then decline to a lower value as the membrane conductance is increased. Conversely, if the voltage is held constant, the current will initially be small and then slowly increase with time towards a steady state value. Since an inductance may be regarded as an element which impedes sudden changes of current flow, the membrane may be said to behave like an inductance. In fact, for small voltage changes the membrane impedance may be represented fairly accurately by including a constant inductance in the equivalent circuit. One of the important properties of circuits containing both capacitances and inductances is that they tend to oscillate (see Chapter 10, p. 303; Chandler, FitzHugh, and Cole 1962; Sabah and Liebovic 1969; Mauro, Conti, Dodge, and Schor 1970). Outside a small range of voltages, the system is best treated as a nonlinear one, and it is no longer convenient to represent the time-dependent conductance changes by an inductance. The oscillatory activity which may occur in nonlinear systems will be discussed in Chapter 11.

Validity of assumptions used in current flow theory: summary

One of the important assumptions of the theory we shall discuss in sub-sequent chapters is that the membrane impedance is sufficiently high for the resistance of the intracellular fluid to current entering or leaving the cell to be negligible. The membrane current may then be represented by eqn (2.2) (with V/R_m replaced with a suitable function if the ionic current is a nonlinear function of voltage). In general, this assumption can now be seen to be valid except at very high frequencies when the membrane impedance tends towards a short-circuit through the membrane capacity. At lower frequencies the membrane resistance is not short-circuited in this way. Moreover, under normal circumstances the low-frequency membrane impedance is remarkably high. Although the membrane is only 50–100 Å thick, its resistance to steady current flow is of the order of 1–10 kΩ cm^2 or even higher (e.g. the estimate of 40 kΩ cm^2 for smooth muscle (Abe and Tomita 1968)). This means that the specific resistivity of the membrane substance is of the order of 10^{10} Ω cm, which is about 10^8 times larger than that of the intracellular fluid, whose resistivity is about 100 Ω cm. The origin of the very large specific resistance of the membrane is that ions seem to be able to cross only at very sparsely distributed sites (see Hille 1970). Thus, an estimate of the number of 'channels' which may be available for carrying sodium ions in nerve membrane is only about 50 per square micrometre of membrane surface (Moore, Narahashi, and Shaw 1967; Keynes, Ritchie, and Rojas 1971).

Although the intracellular resistivity R_i is very low compared to the membrane resistance R_m the very small size of excitable cells ensures that the corresponding resistances per unit length (r_a and r_m) become more comparable in magnitude. In fact, r_m is usually smaller than r_a. Thus, in a fibre about 10 μm in radius, the resistances become

$$r_a = (0{\cdot}1 \text{ k}\Omega \text{ cm})/(\pi \times 10^{-6} \text{ cm}^2) \simeq 30 \text{ M}\Omega \text{ cm}^{-1},$$

$$r_m = (10 \text{ k}\Omega \text{ cm}^2)/(2\pi \times 10^{-3} \text{ cm}) \simeq 1{\cdot}6 \text{ M}\Omega \text{ cm}.$$

Hence, in a centimetre length of fibre, current injected at one end will largely flow out of the fibre through r_m rather than along the fibre through r_a. Also, there will be some length, less than 1cm, at which the resistances are equal. For a length x, the axial resistance will be $r_a x$ and the membrane resistance will be r_m/x. When these values are equal we obtain

$$r_a x = r_m/x \quad \text{or} \quad x = (r_m/r_a)^{\frac{1}{2}}.$$

This value of x is very important in cable theory and is known as the space constant λ (see Chapter 3, p. 27). For the values of resistances given above, we obtain

$$\lambda = \left(\frac{1 \cdot 6}{30}\right)^{\frac{1}{2}} \simeq 0 \cdot 23 \text{ cm},$$

that is, the space constant would be about 2 mm. As will be shown in Chapter 3, it is the value of the space constant which determines how far current may spread along the fibre.

The assumption that the intracellular impedance to axial current flow is a simple resistance is valid in nerve and in skeletal muscle. In cardiac muscle (Freygang and Trautwein 1970) and, perhaps, in smooth muscle (Jones and Tomita 1967) some of the intracellular impedance is also capacitive. For the sake of simplicity we shall neglect this complication in subsequent chapters. However, it should not be neglected in quantitative studies of conduction in cardiac and smooth muscle.

Finally, the assumption that the membrane acts as a simple leaky capacitance is completely valid only for some excitable cells under certain conditions For nerve axons subjected to small voltage changes, the assumption is obeyed very accurately, except when the inductive effects due to time-dependent conductance changes are appreciable. In muscle cells, the assumption is rarely accurate. Nevertheless, even in these cases the theory based on making the simplest assumptions is a useful approximation. The way in which the theory must be modified for muscle will be discussed in Chapter 6.

3. Linear cable theory

IN this chapter we shall use the assumptions discussed at the end of Chapter 2 to develop equations for current flow in cylindrical nerve and muscle fibres subjected to voltage changes which are small enough for the membrane properties to be linear.† The axial current is assumed to flow along an un-branched cylinder of uniform cross-section. The axial current is then one-dimensional, and the current density changes *only* if current enters or leaves through the cell membrane or via an intracellular electrode. We shall assume also that the cable is of unlimited length. The equations for terminated cables will be obtained and discussed in Chapter 4.

These assumptions give rise to the simplest equations, and the theory may be called linear one-dimensional cable theory. Nerve axons and skeletal muscle fibres are the most obvious excitable tissues which may be treated as one-dimensional systems. However, it is also possible to apply current to cardiac and smooth muscle tissues in a way which allows one-dimensional theory to be used (Weidmann 1952; Woodbury and Gordon 1965; Tomita 1967). One-dimensional theory is very well developed, and it is the theory which is most frequently used in applications of cable theory to excitable tissues. This chapter will, therefore, be fairly long. To help the reader we have divided it into four parts. Part I is a short section in which the basic differential equation is derived. Part II obtains solutions for maintained current or voltage changes. Part III is concerned with the equations for brief changes in current and voltage. Part IV deals with the responses to a change in membrane conductance.

Part I: Derivation of general differential equation

The one-dimensional cable may be visualized as a cylindrical membrane surrounding an intracellular fluid phase of constant cross-sectional area (see Fig. 3.1). We assume that the intracellular fluid impedance may be represented by an ohmic resistance and that the non-axial voltage gradients may be neglected. In the case where no leakage of current occurs through the cell membrane the relation between intracellular axial current i_a and the intracellular voltage V_i is given by Ohm's law,

$$\Delta V_i / \Delta x = -r_a i_a, \tag{3.1}$$

where x is distance along the cable and r_a is the intracellular resistance per unit length of cable. Eqn (3.1) applies in the steady state to a cable with zero membrane conductance. In fact, however, the membrane conductance is not

† A careful discussion of these and other assumptions may be found in Scott (1971, 1972).

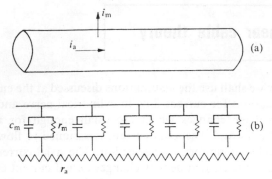

FIG. 3.1. The one-dimensional cable model (a) and its electrical equivalent circuit (b). In keeping with most of the theory discussed in this chapter, the extracellular resistance has been neglected.

negligible (see p. 23) so that, even in the steady state, current leaks across the membrane. Hence i_a will not be constant over any finite length. We therefore consider an infinitesimal distance, ∂x, across which a voltage difference ∂V occurs. The differential form of eqn (3.1) may then be used,

$$(\partial V_i/\partial x)_x = -r_a(i_a)_x, \tag{3.2}$$

where the suffixes indicate that $\partial V/\partial x$ and i_a are measured at the same point x. In future, however, we will omit these suffixes. The negative signs in eqns (3.1) and (3.2) arise from the convention that flow of positive charge from low to high values of x is regarded as a positive current. In this case the voltage must decrease as x increases so that $\partial V/\partial x$ must be negative when a positive current flows.

We will also assume that the fibre is immersed in a large volume of extracellular fluid. The extracellular resistance may then be neglected and the transmembrane voltage V may be identified with V_i, since the extracellular potential V_0 will be constant and is conveniently assumed to be zero. Many experimental situations correspond to this state of affairs, but it is not always valid to make this simplification. When nerve and muscle cells are surrounded by rather small volumes of extracellular fluid the extracellular fluid resistance must be taken into account (Cole and Curtis 1936; Hodgkin 1939; Hodgkin and Rushton 1946). However, provided that the extracellular resistance is uniform, the form of the equations which are obtained for the intracellular potential and current flow is not greatly altered. We shall indicate later how the presence of extracellular resistance may be taken into account (see p. 27).

The axial current may change either as a result of current crossing the cell membrane or as a result of current applied through an internal electrode. Hence at all points apart from those at which current is applied from an electrode, the rate of change in i_a with distance along the cable must be equal and opposite to the density of membrane current i_m,

$$\partial i_a / \partial x = -i_m. \tag{3.3}$$

We may combine eqns (3.2) and (3.3) by differentiating eqn (3.2) to give

$$\partial^2 V / \partial x^2 = -r_a \, \partial i_a / \partial x. \tag{3.4}$$

Substituting eqn (3.3) in eqn (3.4),

$$(1/r_a)(\partial^2 V / \partial x^2) = i_m, \tag{3.5}$$

which gives an equation for the membrane current. From the assumptions made about the membrane impedance we already have another equation for the membrane current (eqn (2.2)). Eqn (2.2) may be rewritten in terms of quantities expressed per unit length of cable rather than per square centimetre of membrane area,

$$i_m = c_m \cdot \partial V / \partial t + V / r_m. \tag{3.6}$$

Combining eqns (3.5) and (3.6) we obtain

$$(1/r_a) \, \partial^2 V / \partial x^2 = c_m \cdot \partial V / \partial t + V / r_m, \tag{3.7}$$

c_m and r_m are the capacitance and resistance, respectively, of the membrane enclosed by a unit length of cable. This is the basic differential equation of linear cable theory. The constants r_m, c_m, and r_a are related to C_m, R_m, and R_i by the equations

$$r_m = R_m / 2\pi a,$$

$$r_a = R_i / \pi a^2,$$

and

$$c_m = C_m \cdot 2\pi a,$$

where a is the cable radius. Hence eqn (3.7) may also be written

$$(a/2R_i) \, \partial^2 V / \partial x^2 = C_m \cdot \partial V / \partial t + V / R_m. \tag{3.8}$$

Also, by rearranging the constants in eqn 3.7 we may obtain

$$\lambda^2 \, \partial^2 V / \partial x^2 = \tau_m \cdot \partial V / \partial t + V, \tag{3.9}$$

where $\lambda = \sqrt{(r_m / r_a)}$ and $\tau_m = r_m c_m$. τ_m is the membrane *time constant*, which was discussed in the previous chapter (p. 13). λ has the dimensions of distance and is known as the *space constant*. Since it depends on the ratio of the membrane resistance to intracellular resistance, this constant determines the tendency for current to spread along the cable (cf. Chapter 2, p. 24; see Fig. 3.2). The extent of current spread also depends on the extracellular resistance when this may not be neglected. λ is then defined by

$$\lambda = \sqrt{\{r_m / (r_a + r_0)\}}, \tag{3.10}$$

where r_0 is the external resistance per unit length of cable (see Cole and Curtis 1936; Hodgkin and Rushton 1946). Otherwise, the equations are unchanged.

However, it should be noted that eqn (3.10) applies only when the extracellular space forms a uniform resistance and when it is small enough for radial fields to be negligible. In practice, these restrictions will be met in a rather limited range of experimental conditions, e.g. when the extracellular fluid is restricted to a cylindrical shell surrounding the fibre by immersing it in a non-conducting medium such as oil. When the extracellular space is effectively unbounded, it may then be necessary to take account of extracellular fields (see Clark and Plonsey 1968). Moreover, in the case of nerve trunks and whole muscles it may be necessary to include other fibres in the 'extracellular' resistance for a particular fibre.

Eqns (3.7), (3.8), and (3.9) apply to *linear* cables in which r_m, c_m, and r_a are independent of V, x, and t. Outside a small range of potentials, however, r_m is usually strongly dependent on V and t (see Chapter 8). We must then use eqn (3.7) in the form

$$(1/r_a)\ \partial^2 V/\partial x^2 = c_m(\partial V/\partial t) + i_i(V, t) \tag{3.11}$$

or

$$(a/2R_i)\ \partial^2 V/\partial x^2 = C_m(\partial V/\partial t) + I_i(V, t), \tag{3.12}$$

where i_i is the membrane ionic current density referred to the membrane area enclosing a unit length of cable and I_i is the ionic current density referred to a unit area of membrane. $I_i(V, t)$ is a nonlinear function of voltage and time (see Chapter 8). We shall use the equations in these forms when discussing the applications of cable theory to nonlinear systems (Chapters 9–12).

In this chapter we shall derive and discuss solutions of the linear equations for cases of physiological interest. In order to do this it will be convenient to transform the equations into a form which allows solutions for various cases to be obtained relatively easily. The best-known method for doing this uses the Laplace transform, which offers the advantage that very simple relations exist between the transform of the function and the transforms of its derivatives. This method is explained in Chapter 13. In the case of partial differential equations involving two independent variables differentiation with respect to only one of the variables remains after applying the transformation. The result is an ordinary differential equation which may be solved for the transformed function by usual methods.

We first redefine the independent variables in terms of the dimensionless quantities,

$$X = x/\lambda$$

and

$$T = t/\tau_m.$$

Eqn (3.9) then becomes

$$\partial^2 V/\partial X^2 - \partial V/\partial T - V = 0. \tag{3.13}$$

Let $\bar{V}$ be the Laplace transform of V as a function of T. Then

$$\mathrm{d}\bar{V}/\mathrm{d}T = s\bar{V} - V(0, X), \qquad (3.14)$$

where s is the Laplace transform variable (see Chapter 13) and $V(0, X)$ gives the initial values of V at all X when $T = 0$. In an initially quiescent cable, $V(0, X) = 0$ so that eqn (3.13) transforms to

$$\mathrm{d}^2\bar{V}/\mathrm{d}X^2 - (s+1)\bar{V} = 0. \qquad (3.15)$$

A general solution to this equation is

$$\bar{V} = \bar{A}\exp\{-X\sqrt{(s+1)}\} + \bar{B}\exp\{X\sqrt{(s+1)}\}, \qquad (3.16)$$

where $\bar{A}$ and $\bar{B}$ are constants or functions of s. This is the general transformed equation for a quiescent cable. In order to obtain particular solutions, $\bar{A}$ and $\bar{B}$ must be determined from the boundary conditions for each case.

Part II: Maintained changes

Response of infinite cable to prolonged step current change

This is an important case since the step current change remains one of the most commonly used excitation waveforms for investigating the properties of excitable cells. It is also convenient to describe this case in detail as an illustration of the mathematical methods used. Subsequent examples will be dealt with more concisely.

The conditions to be satisfied are:
(1) $V \to 0$ as $X \to \infty$,
(2) a current I of constant amplitude I_0 is applied internally at the point $X = 0$ at times $T \geqslant 0$.

Condition (1) requires that $\bar{B} = 0$. Hence

$$\bar{V} = \bar{A}\exp\{-X\sqrt{(s+1)}\}. \qquad (3.17)$$

From eqn (3.2)

$$i_{\mathrm{a}} = -(1/r_{\mathrm{a}}\lambda)\,\partial V/\partial X$$

so that

$$\bar{i}_{\mathrm{a}} = -(1/r_{\mathrm{a}}\lambda)\,\mathrm{d}\bar{V}/\mathrm{d}X. \qquad (3.18)$$

where $\bar{i}_{\mathrm{a}}$ is the Laplace transform of i_{a}. Differentiating eqn (3.17),

$$\mathrm{d}\bar{V}/\mathrm{d}X = -\bar{A}\sqrt{(s+1)}\exp\{-X\sqrt{(s+1)}\}. \qquad (3.19)$$

Hence

$$\bar{i}_{\mathrm{a}} = [\{\bar{A}\sqrt{(s+1)}\}/r_{\mathrm{a}}\lambda]\exp\{-X\sqrt{(s+1)}\}. \qquad (3.20)$$

Now, at $X = 0$, $i_{\mathrm{a}} = I_0/2$, since the applied current flows equally into both halves of the infinite cable. Hence

$$I = 2\bar{A}\sqrt{(s+1)}/r_{\mathrm{a}}\lambda$$

or

$$\bar{A} = r_\mathrm{a}\bar{I}\lambda/2\sqrt{(s+1)}.$$

Now I is a step function of T so that $\bar{I} = I_0/s$ (see Chapter 13). Hence

$$\bar{A} = r_\mathrm{a}I_0\lambda/2s\sqrt{(s+1)} \tag{3.21}$$

and eqn (3.17) becomes

$$\bar{V} = \frac{r_\mathrm{a}I_0\lambda}{2s\sqrt{(s+1)}}\exp\{-X\sqrt{(s+1)}\}. \tag{3.22}$$

This gives the Laplace transform of V. In order to obtain V itself eqn (3.22) must be arranged into a form for which the inverse transform may be found. This procedure usually involves finding a way of fractionating the expression containing s to obtain a sum of expressions which are known transforms. In this case, a suitable fractionation is

$$\frac{2}{s\sqrt{(s+1)}} = \frac{1}{s+1-\sqrt{(s+1)}} - \frac{1}{s+1+\sqrt{(s+1)}},$$

so that eqn (3.22) may be rewritten

$$\bar{V} = \frac{r_\mathrm{a}I_0\lambda}{4}\left\{\frac{\exp\{-X\sqrt{(s+1)}\}}{s+1-\sqrt{(s+1)}} - \frac{\exp\{-X\sqrt{(s+1)}\}}{s+1+\sqrt{(s+1)}}\right\}. \tag{3.23}$$

From Roberts and Kaufman (1966, p. 248, pair 3.2.28),

$$\frac{\exp(-a\sqrt{s})}{s+b\sqrt{s}} \leftrightarrow \exp(b^2T+ab)\mathrm{erfc}\left(\frac{a}{2\sqrt{T}}+b\sqrt{T}\right),$$

where $\mathrm{erfc}(y) = 1-\mathrm{erf}(y)$. (For further information on the error function $\mathrm{erf}(y)$ see Chapter 13.) In our case, $a = X$, $b = \pm 1$ and, in place of s, we have $s+1$ which has the effect of multiplying the solution for V by $\exp(-T)$. Hence

$$V = \frac{r_\mathrm{a}I_0\lambda}{4}\left\{\exp(-X)\mathrm{erfc}\left(\frac{X}{2\sqrt{T}}-\sqrt{T}\right) - \exp(X)\mathrm{erfc}\left(\frac{X}{2\sqrt{T}}+\sqrt{T}\right)\right\}. \tag{3.24}$$

This is the solution obtained by Hodgkin and Rushton (1946, eqn 4.1), using a less well-known method, and the solution is tabulated by them. This solution was also obtained by Davis and Lorente de No (1947, eqn 36). Hodgkin and Rushton give the solution for the responses to the termination of an infinitely long current, which may be obtained from eqn (3.24) by use of the superposition principle (see Chapter 13). The termination of the current pulse may be regarded as the superposition of a second current step, equal in magnitude but opposite in sign to the first, applied at a long time after the first step. The general properties of these solutions may become clear by considering some limiting cases.

Steady state. When $T \to \infty$, erfc(T) $\to 0$, erfc($-T$) $\to 2$, and eqn (3.24) becomes
$$V = (r_a I_0 \lambda / 2) \exp(-x/\lambda). \tag{3.25}$$

This result may also be obtained by putting $\partial V/\partial T = 0$ in eqn (3.13) and solving the resulting ordinary differential equation. The properties of eqn (3.25) are illustrated in Fig. 3.2. V decays exponentially towards zero, falling to e^{-1} of its initial value in one space constant. On a semilogarithmic plot the relation is a straight line whose slope is $-1/\lambda$. Measurement of this slope

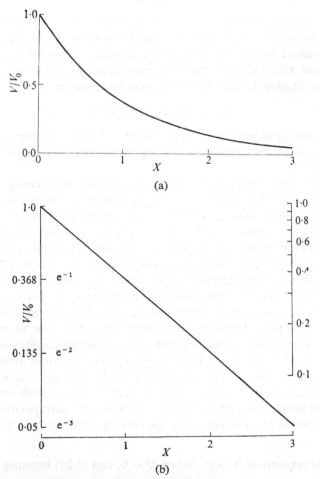

FIG. 3.2. Steady state spatial decay of potential. (a) Curve plotted on linear coordinates. The ordinate is normalized in terms of V_0, the voltage at $X = 0$. (b) Same curve plotted on semilogarithmic coordinates. At left, the values of e^{-1}, e^{-2}, and e^{-3} are shown; at right a conventional logarithmic scale is shown. Note that voltage decays to e^{-1} of initial value in one space constant.

is the usual method for determining λ and hence for obtaining r_m/r_a. λ is frequently of the order of one or two millimetres, in the large nerve and muscle fibres usually used in cable analysis, but very small fibres tend to have much smaller space constants. Since $\sqrt{(r_m/r_a)} = \sqrt{(aR_m/2R_i)}$, λ will be proportional to $\sqrt{a}$ if R_m and R_i are constant.

The *input resistance* of the cable is defined as

$$R_{in} = V(x = 0, t = \infty)/I_0$$

and may be obtained from eqn (3.25) by putting $x = 0$ and rearranging to give

$$R_{in} = \{(r_m r_a)^{\frac{1}{2}}/2\}, \tag{3.26}$$

that is, the resistance of each half of the cable is $(r_m r_a)^{\frac{1}{2}}$. Therefore we may obtain an estimate for $r_m r_a$ from the input resistance. Since the ratio r_m/r_a may be obtained from the space constant, separate values for r_m and r_a may be calculated. Using an estimate of the fibre diameter, the values of R_i and R_m may be calculated, since eqn (3.26) may be rewritten

$$R_{in} = (R_m R_i/2)^{\frac{1}{2}}/(2\pi a^{\frac{3}{2}}) \tag{3.27}$$

Note that the input resistance varies as the $-\frac{3}{2}$ power of the fibre radius. This result is of importance in dealing with branching cable structures (see Chapter 7).

The cable input resistance is usually determined by inserting two intracellular electrodes (for voltage recording and current passing) at a distance apart which is very small compared to λ. It should be noted that the one-dimensional approximation used in this chapter is least valid in the immediate vicinity of the microelectrode tip, which is effectively a point source of current. Near the tip, a significantly large radial component of current may flow and in such cases a full three-dimensional treatment is required. Eisenberg and Johnson (1970) have obtained equations for this situation. Their results show that the errors arising from neglecting radial fields are minimal when the voltage and current electrodes are inserted into the fibre at an angular separation of about 60° in the transverse plane. A further check on such errors may be obtained by measuring the voltage at several distances away from the current electrode. On a semilogarithmic plot the voltage–distance relation should be linear (see Fig. 3.2) so that, even if there is a small error arising from radial fields, this may be largely eliminated by extrapolating the line formed by points at a distance from the current electrode back to the point $X = 0$.

Transient response at $X = 0$. When $X = 0$, eqn (3.24) becomes

$$V = (r_a I_0 \lambda/2) \, \text{erf}\{(t/\tau_m)^{\frac{1}{2}}\} \tag{3.28}$$

Since erf(1) = 0·84, V rises to 84 per cent of its steady state value in one time constant. This is considerably faster than a simple exponential, as shown in

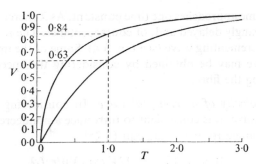

FIG. 3.3. Comparison of time courses of responses to step current pulse applied uniformly and at one point. The response to current applied at one point increases much more quickly than the response to uniformly applied current. Thus, in one time constant, the response reaches 84 per cent of the steady state value, compared to 63 per cent for the uniform response.

Fig. 3.3. The 84 per cent time thus gives a measure of $r_m c_m$. Since r_m is already known from estimates of the space constant and input resistance, c_m may also be calculated.

Responses at various distances. When X is not zero, there will be a period of time (which increases as X increases) when X is large compared to T. The complementary error function terms in eqn (3.24) are then dominated by $X/(2\sqrt{T})$ and, since T is small, these terms will be nearly zero. Eqn (3.24) as a whole will, therefore, be nearly zero for a certain period of time.† The physical basis for this result is that the membrane capacitance near the current source must be charged first to develop the axial voltage gradient required to drive current through the series resistance r_a. This current may then charge the capacitance of more distant areas of membrane. Thus at one space constant, V remains near zero for about a quarter time constant. At three space

† It should be noted, however, that apart from $T = 0$, eqn (3.24) is not strictly zero for any values of X and T. Hence, the beginning of the wave actually propagates virtually instantaneously (at the speed of electromagnetic radiation). Only in cables involving inductances (across which the current may not be changed instantaneously) may waves occur which propagate more slowly than the electromagnetic velocity.

Nevertheless, it is useful in electrophysiological applications of cable theory to use the term 'propagation' in a rather different sense from its strict meaning in electrical engineering. As noted in Chapter 2, there are no linear inductive elements in physiological systems so that, in a strict sense, all disturbances travel virtually instantaneously. However, physiological systems characteristically require a fairly large 'threshold' magnitude of voltage or current to be achieved. The functionally important parameter, therefore, is not the propagation velocity of the beginning of the wave but rather the time taken for a certain magnitude to be reached. We shall use the term 'propagation time' to refer to this delay and the term 'conduction velocity' to refer to the speed at which a well-defined point (e.g. the half-amplitude point, inflection point, etc.) travels along the cable.

This usage has several advantages, of which two may be mentioned here. First, the definition of conduction velocity becomes a very natural one when the propagation of action potentials in nonlinear cables is treated (see Chapter 10). Second, the usage adopted by physiologists is also helpful in treating some of the properties of terminated cables (see Chapter 4, p. 69, footnote).

constants this time approaches one time constant. As X increases, the response becomes increasingly delayed and, of course, smaller. The response therefore appears as a decrementing wave (see Fig. 3.4). Some of the important properties of this wave may be obtained by considering the speed with which it propagates along the fibre.

Conduction velocity of decremental wave. In considering the conduction velocity of the wave, it is convenient to introduce a nondecremental function W which is obtained from V using eqn (3.25),

$$W = V/V_{t=\infty} = \{2V \exp(X)\}/r_a I_0 \lambda. \tag{3.29}$$

From eqns (3.24) and (3.29),

$$W = \tfrac{1}{2}[\mathrm{erfc}\{(X/2\sqrt{T}) - \sqrt{T}\} - \exp(2X)\mathrm{erfc}\{(X/2\sqrt{T}) + \sqrt{T}\}]. \tag{3.30}$$

The useful property of this function is that for each value of X it has the same shape as V when plotted as a function of T but, instead of decrementing with X, it varies from 0 to 1 for all values of X as T varies from 0 to ∞. This function is plotted in Fig. 3.5 (cf. Davis and Lorente de Nó 1947, Fig. 6), and it can be seen that as the wave propagates along the fibre it assumes a fairly constant sigmoid shape and that the time delay between the waves tends towards a constant, which is $\tfrac{1}{2}\tau_m$ per space constant. Most points on the wave, therefore, propagate with a speed of the order of $2\lambda/\tau_m$.

In the case of the half-amplitude point $W = 0.5$, it may be shown analytically that the conduction velocity converges onto the value $2\lambda/\tau_m$ as T increases and that the equation for this point for all but small values of X

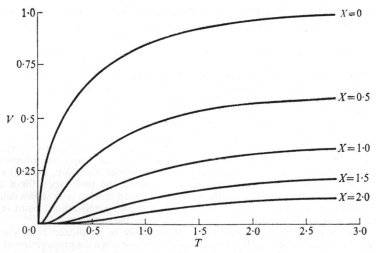

FIG. 3.4. Effect of propagation on shape and magnitude of response. The curves are solutions of eqn (3.24) for $X = 0, 0.5, 1.0, 1.5$, and 2.0. Note that the response becomes smaller and shows an increasing delay as X increases.

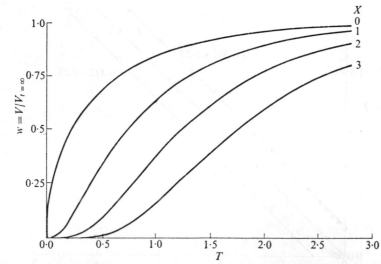

FIG. 3.5. Effect of propagation on the shape of the normalized response. All curves are normalized in terms of their own steady state values. This method of representation illustrates the increasing delay more clearly and allows the speed of propagation to be studied. The curves are solutions of eqn (3.30) for $X = 0, 1, 2,$ and 3.

and T is

$$X = 2T+C, \tag{3.31}$$

where C is a constant. This result may be obtained by showing that a constant value for C satisfies the equation obtained by substituting eqn (3.31) into eqn (3.30) for $W = 0.5$. $W = 0.5$ when

$$\text{erfc}\{(X/2\sqrt{T})-\sqrt{T}\}-\exp(2X)\text{erfc}\{(X/2\sqrt{T})+\sqrt{T}\} = 1.$$

Assuming $X = 2T+C$, we obtain

$$\text{erfc}(C/2\sqrt{T})-\exp(4T+2C)\,\text{erfc}\{2\sqrt{T}+(C/2\sqrt{T})\} = 1. \tag{3.32}$$

Using a series expansion for erfc (see Chapter 13), it can be shown that for large values of T compared to C

$$\text{erfc}(C/2\sqrt{T}) \rightarrow 1-C/\sqrt{(\pi T)}$$

and

$$\text{erfc}\{2\sqrt{T}+(C/2\sqrt{T})\} \rightarrow \{\exp(-4T-2C)\}/2\sqrt{(\pi T)}.$$

Hence, for large values of T compared to C, eqn(3.32) becomes

$$(C+0.5)/\sqrt{(\pi T)} = 0. \tag{3.33}$$

Hence eqn (3.31) will be satisfied at large values of T when $C = -0.5$. Hence the equation for the point $W = 0.5$ is

$$X = 2T-0.5$$

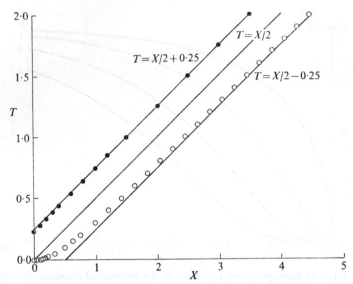

FIG. 3.6. Propagation of half-amplitude (●) and inflection (○) points. At $X = 0$, the half-amplitude point occurs at nearly $0 \cdot 25\tau_{m}$ and then propagates at a velocity that is almost exactly $2\lambda/\tau_{m}$ (compare points with upper line). The inflection point always occurs earlier. As the wave propagates the inflection point approaches a line $0 \cdot 5\tau_{m}$ earlier than the half-amplitude line and which also propagates at a velocity of $2\lambda/\tau_{m}$. The centre line corresponds to conduction at $2\lambda/\tau_{m}$ with no advance or delay and is included for comparison with the other lines and points. Note that in this diagram and in Fig. 3.9, T is plotted as a function of X and equations such as eqn (3.34) have been transformed accordingly.

or

$$x = (2\lambda/\tau_{m})t - 0 \cdot 5\lambda. \tag{3.34}$$

In Fig. 3.6 the values of X and T at which $W = 0 \cdot 5$ are plotted and compared with eqn (3.34). It can be seen that the deviation from eqn (3.34) is extremely small even at very small values of X and T. Thus, at $X = 0$, $W = 0 \cdot 5$ when T is $0 \cdot 23$, which is quite close to the value of $0 \cdot 25$ given by eqn (3.34). Since the half-amplitude time is easily measured experimentally a plot of this kind is useful for obtaining a further estimate of the cable parameters and has been used for this purpose (e.g. Hodgkin and Rushton 1946; Shaw 1972). The value of τ may be estimated from the slope of the line (if λ is known) or from the t intercept, which is equal to $0 \cdot 25\tau_{m}$. The latter may be estimated more accurately than the slope of the line and does not require a value for λ.

Another important point for which a simple propagation equation may be obtained is the inflection point. Differentiation of eqn (3.30) gives

$$\frac{\partial^2 W}{\partial T^2} = \frac{X^2 - 2T - 4T^2}{4\sqrt{(\pi)}T^{\frac{5}{2}}} \exp\left(\frac{X - T - X^2}{4T}\right)$$

and this equation is equal to zero when

$$X^2 = 4T^2 + 2T. \tag{3.35}$$

This equation may be rewritten

$$\sqrt{(X^2+0.25)} = 2T+0.5, \qquad (3.35a)$$

from which it is clear that for all but small values of X the inflection point will be given fairly accurately by the equation

$$X = 2T+0.5$$

or

$$x = (2\lambda/\tau_m)t+0.5\lambda. \qquad (3.36)$$

In this case also, therefore, the propagation velocity tends towards $2\lambda/\tau_m$ although, as can be seen in Fig. 3.6, the velocity approaches this asymptotic value more slowly than does the half-amplitude point. Moreover, the inflection point is less easily measured than is the half-amplitude point. However, as will be shown later (see p. 47), eqn (3.36) is important in relation to junctional mechanisms since the inflection point of the response to an infinite step current corresponds to the peak of the response to a very short current.

These results enable us to characterize the asymptotic form of the wave. As X increases, the wave becomes sigmoid, the inflection point occurs at $T = 0.5X-0.25$, and the half-amplitude points occurs at $T = 0.5X+0.25$, so that the inflection point will occur half a time constant before the half-amplitude point.

Influence of fibre size on propagation speed. Since $\lambda = (R_m a/2R_i)^{\frac{1}{2}}$ and $\tau_m = R_m C_m$,

$$\frac{2\lambda}{\tau_m} = \left(\frac{2a}{R_m R_i C_m^2}\right)^{\frac{1}{2}},$$

the conduction velocity will also be proportional to the square root of the fibre diameter, provided that R_i, R_m, and C_m do not vary with a (see, however, Hencek and Zachar (1965) for an example in which these conditions do not hold). As will be shown later (see p. 292) this result is also applicable to the conduction velocity in nonlinear cables when the response is sufficiently regenerative to show no decrement.

In addition to the possibility that R_i, R_m, and C_m may vary with a, there are other reasons for being cautious in applying this result to physiological problems. Thus, in determining the time for the voltage to reach a certain value (e.g. when calculating synaptic delays in the initiation of firing) it is necessary to take account of at least two other factors.

1. The spatial decay of voltage in the *steady state* also scales with $\sqrt{a}$. This property is omitted by the nondecremental function W and for the purpose of calculating times for reaching a threshold value of potential it is best to use eqn (3.24) directly.

2. The value of V at $x = 0$ depends on the current and on the input resistance, and this was shown above to be proportional to $a^{-\frac{3}{2}}$ (eqn (3.27)). Hence it may be important to know whether fibres of different

sizes are being excited by similar *currents* (in which case the steady state value of V at $x = 0$ will fall as a increases) or whether a constant *voltage* waveform is being applied. Thus, in considering the spread of synaptic potentials, the degree to which the potential approaches the reversal potential will be important since this effectively determines whether the synaptic input corresponds to a constant current waveform or a constant voltage waveform (see p. 62, Fig. 3.25).

Time course of total membrane charge. The time taken for the voltage to reach a certain threshold value is of great importance in electrophysiology. Much of the early work on the initiation of action potentials in excitable cells was concerned with how this time varies with the shape and duration of the excitatory current. In particular, the relation between the magnitude and duration of a threshold rectangular current pulse was found to be a simple one involving an exponential term (see Chapter 9, eqn (9.12)).

In a uniformly polarized membrane, the rise of voltage in response to a constant current step is exponential (eqn (2.3)), and this fact has frequently been used in theories of excitation which account for the exponential form of the strength–duration equation since this equation (eqn (9.12)) then follows from the assumption that the voltage threshold is a constant. In the case of a cable polarized at one point, the voltage does not change exponentially (see Fig. 3.3) and the usual strength–duration equation would not be obeyed if it were sufficient to raise the voltage at $x = 0$ to some constant threshold value for excitation to occur. However, it is possible to define a variable which does change exponentially in this case. This is the total charge on the cable Q, where

$$Q = c_m \int_{-\infty}^{+\infty} V \, dx = 2c_m \int_0^\infty V \, dx. \tag{3.37}$$

The time course of Q may be obtained as follows. Integration of eqn (3.9) gives

$$\lambda^2 \int_0^\infty (\partial^2 V/\partial x^2) \, dx - \tau_m \int_0^\infty (\partial V/\partial t) \, dx = Q/2c_m.$$

Now

$$\int_0^\infty \left(\frac{\partial V}{\partial t}\right) dx = \frac{\partial}{\partial t}\left(\int_0^\infty V \, dx\right) = \frac{1}{2c_m} \frac{\partial Q}{\partial t}$$

and

$$\int_0^\infty (\partial^2 V/\partial x^2) \, dx = [\partial V/\partial x]_0^\infty = \tfrac{1}{2} r_a I_0.$$

Hence

$$\tau_m I_0 - \tau_m(\partial Q/\partial t) - Q = 0. \tag{3.38}$$

The solution to this equation is

$$Q = \tau_m I_0\{1 - \exp(-t/\tau_m)\} \tag{3.39}$$

when a step change in I is applied to the quiescent cable at $t = 0$, and

$$Q = \tau_m I_0 \exp(-t/\tau_m) \tag{3.40}$$

when the current is switched off after a long period of time.

It is possible, therefore, that the explanation of the strength–duration curve for a cable might still be relatively simple provided that the conditions for excitation may be stated in terms of a constant charge threshold for excitation. This problem will be considered further in Chapter 9 (see p. 267) and in Chapter 12 (see p. 416).

Response of infinite cable to prolonged step voltage change applied at $X = 0$

In many cases of physiological interest, it is more appropriate to obtain the response to a step voltage change than to a step current change. The application of a voltage step to a cable has become possible through the development of voltage-clamp techniques using microelectrodes (see Adrian and Freygang 1962; Deck, Kern, and Trautwein 1964; Adrian, Chandler and Hodgkin 1966, 1970). Although the experimental situation usually uses a termination, i.e. a short-cable configuration, the response at short times will approximate to that of an infinite cable (see Chapter 4).

Up to eqn (3.17), $\bar{V} = \bar{A} \exp\{-X \sqrt{(s+1)}\}$,

the derivation is identical with that given previously (p. 29). Now at $X = 0$, V is a step function. Hence $\bar{V}_{x=0} = V_0/s$,

where V_0 is the amplitude of the voltage step. Hence

$$\bar{V} = (V_0/s) \exp\{-X \sqrt{(s+1)}\}.$$

From Roberts and Kaufman (p. 85, pair 16.2.5) the inverse transform may be obtained:

$$V = \frac{V_0}{2}\left\{\exp(-X)\mathrm{erfc}\left(\frac{X}{2\sqrt{T}}-\sqrt{T}\right)+\exp(X)\mathrm{erfc}\left(\frac{X}{2\sqrt{T}}+\sqrt{T}\right)\right\}. \tag{3.41}$$

This solution strongly resembles the voltage response to a step current excitation (eqn (3.24)). In this equation the braces contain the sum of two terms rather than their difference. An analogous diffusion problem is treated in Crank (1956, p. 130, eqn 8.46).

Eqn (3.41) has been evaluated for various values of X and is plotted in Fig. 3.7. In the steady state, the voltage decrements along the cable in as exponential fashion, just as it does for a constant current excitation. However, the transient solutions to a voltage step will clearly differ. Since there is on delay in charging the capacitance at $X = 0$, it is clear that at any X the response to a voltage step (vs) will have a briefer latency than the response to a

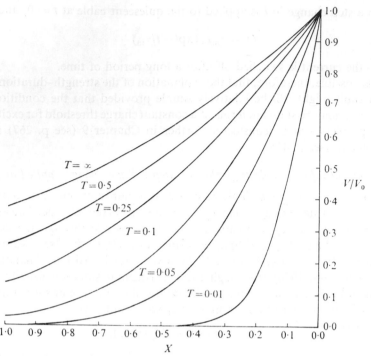

FIG. 3.7. Voltage response to a step voltage change at $X = 0$ in an infinite one-dimensional cable. Curves are labelled by the parameter T for the values $0.01, 0.05, 0.1, 0.25, 0.5$, and ∞. Note that the voltage distribution approaches the steady state (exponential decrement, labelled ∞) more quickly than in the case of a point injection of a step current (Hodgkin and Rushton 1946, Fig. 2; and also Fig. 3.4, noting the difference in time scale).

The abscissa has been arranged with X decreasing from left to right to allow the comparison of this case with the plot of the radially inward spread of depolarization within the tubular lattice of muscle fibres (Fig. 6.7).

current step. This comparison may be clarified by considering a non-decrementing function

$$W' = V/V_{t=\infty}.$$

The prime notation distinguishes this function from the W defined for the response to a step current excitation (eqn (3.29)). W' is plotted in Fig. 3.8.

Let us now consider the half-amplitude point $W' = 0.5$. From eqn (3.41), $W' = 0.5$ when

$$\operatorname{erfc}\left(\frac{X}{2\sqrt{T}} - \sqrt{T}\right) + \exp(2X)\operatorname{erfc}\left(\frac{X}{2\sqrt{T}} + \sqrt{T}\right) = 1. \qquad (3.41a)$$

By analogy with the treatment above (eqns (3.32) and (3.33)), one might expect that the $W' = 0.5$ point will move with velocity $2\lambda/\tau_{\mathrm{m}}$, for large values of T. Assuming, then, that $X = 2T + C'$,

$$\operatorname{erfc}(C'/2\sqrt{T}) + \exp(4T + 2C')\operatorname{erfc}(2\sqrt{T} + C'/2\sqrt{T}) = 1$$

Using an approximation for $\mathrm{erfc}(y)$ when y is small (see p. 460), we obtain

$$(-C'+0\cdot5)/\sqrt{(\pi T)} = 0.$$

Hence, eqn (3.41a) will be satisfied for large values of T when $C' = 0\cdot5$. $W' = 0\cdot5$ when

$$X = 2T+0\cdot5. \tag{3.42}$$

This confirms the assumption that $W' = 0\cdot5$ moves with constant velocity $2\lambda/\tau_{\mathrm{m}}$, when X is not small. This may also be seen in Fig. 3.9, which compares eqn (3.42) with values of X and T, where $W' = 0\cdot5$. The figure also shows a dashed line for eqn (3.34). For large T, the W' wave is a full space constant ahead of the W wave for a step-current excitation. At a particular location (for X not small), $W' = 0\cdot5$ precedes $W = 0\cdot5$ by $\frac{1}{2}\tau_{\mathrm{m}}$.

It may also be useful to consider the propagation of W' for small values of X and T. This problem is relevant to applications of the voltage-clamp technique where current is applied at one point, since the behaviour of a terminated cable resembles that of an infinite cable for a brief period after the application of the step voltage change (see Chapter 4).

For small values of T, the dominant process is the longitudinal spread of charge. Since dissipation of charge through r_{m} is negligible, the voltage spread can be described by allowing $r_{\mathrm{m}} \to \infty$ in eqn (3.41). As $r_{\mathrm{m}} \to \infty$,

$$\exp(\pm X) = \exp\left\{\pm\frac{x}{\sqrt{(r_{\mathrm{m}}/r_{\mathrm{a}})}}\right\} \to 1$$

and

$$\frac{X}{2\sqrt{T}}\pm\sqrt{T} \to \frac{X}{2\sqrt{T}},$$

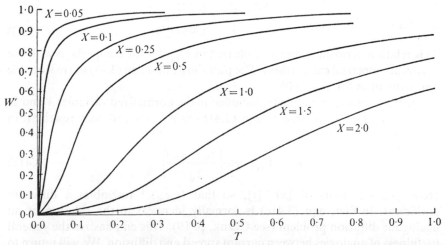

W'

FIG. 3.8. Propagation of response to step voltage change. The curves are solutions of eqn (3.41) normalized at each distance in terms of the steady state voltage:
$$W' = V_{\mathrm{vs}}/V_{t=\infty}.$$

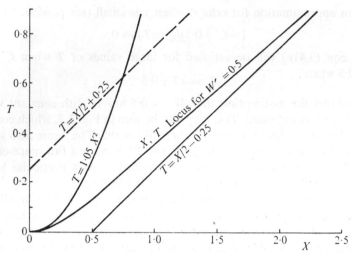

FIG. 3.9. Propagation of half-amplitude point for response to step voltage change. At large values of T the curve approaches the line $T = X/2 - 0.25$. At short times, the curve follows $T = 1.05X^2$. The dashed line shows $T = X/2 + 0.25$ which is the asymptote for the response to a current step (see Fig. 3.6). The half-amplitude point for the step voltage response leads that for the step current response by one space constant at large values of T.

and thus eqn (3.41) can be simplified to give

$$V = V_{x=0} \, \text{erfc}(X/2\sqrt{T}). \tag{3.41b}$$

Imposing the condition that $W'_{x=0} = V/V_{x=0} \big|_{t=\infty} = 0.5$, we find that

$$\text{erfc}(X/2\sqrt{T}) = 0.5$$

or

$$X/2\sqrt{T} = 0.477$$

and thus

$$X = 0.954 \sqrt{T}. \tag{3.43}$$

This relation between X and T is plotted in Fig. 3.9 for comparison with the results of numerical calculations. The plot shows that eqn (3.43) is a reasonable approximation for $T < 0.05$.

The parameter r_m is implicitly included in the normalized variables X and T. To give more physical sense to eqn (3.41b) as $r_m \to \infty$, we may rewrite it in terms of x and t, giving

$$V = V_{x=0} \, \text{erfc}\left\{\frac{x\sqrt{(r_a c_m)}}{2\sqrt{t}}\right\}. \tag{3.41c}$$

Here $r_a c_m$ has units of $[x]^{-2} [t]$, so that $(r_a c_m)^{-1}$ resembles a diffusion coefficient. In fact, eqn (3.41c) is formally identical to the solution of an analogous diffusion problem (see Crank, p. 19). This emphasizes the general usefulness of analogies between current spread and diffusion. We will return to this analogy when treating the spread of excitation in muscle (see Chapter 6, *Velocity of radial spread of excitation*), and nerve (see Chapter 9, p. 268).

The current which must be injected into the cable to produce the step voltage response (that is, the current which would be supplied by a voltage-clamp circuit) may be obtained by differentiating eqn (3.41) with respect to X to obtain $\partial V/\partial X$ (using the method described on p. 460 in Chapter 13). This is related to the axial current by

$$i_a = -\frac{\partial V/\partial X}{\sqrt{(r_m r_a)}}.$$

At $X = 0$, $i_a = I/2$, where I is the applied current.

It may be shown that this gives eqn (3.44) for the applied current,

$$I = \frac{2V_{x=0}}{\sqrt{(r_m r_a)}}\left\{\mathrm{erf}(\sqrt{T}) + \frac{\exp(-T)}{\sqrt{(\pi T)}}\right\} = V_{x=0}I_{\mathrm{step}}(T). \tag{3.44}$$

where I_{step} (T) is the response for a voltage step of unit amplitude. This equation is plotted in Fig. 3.10(b). Note that the current is infinitely large at $T = 0$. This is required in order to instantaneously change the voltage across the capacitance in the immediate vicinity of the current source. Step voltage changes are achieved experimentally by passing a very intense current for a short time in order to change the voltage as quickly as possible.

Response to non-instantaneous voltage change

Since the charging current in a voltage-clamp system cannot be infinite, and may in fact be very severely limited if high-resistance microelectrodes are used as current sources, the charging current required in actual voltage-clamp situations may be approximated better by calculating the response to a non-instantaneous change in voltage. However, it is important to distinguish between two cases, namely, uncontrolled and controlled voltage changes.

1. *Uncontrolled voltage change.* If the charging current required to change the membrane potential at the speed of the 'reference potential' change (see Chapter 2, p. 7) is greater than the voltage-clamp system may provide without losing control, then the actual voltage change will not follow the reference potential change and, during this time, the voltage and current flow will be uncontrolled. In order to treat this case, the equation for the current flow must include the properties of the voltage-clamp apparatus, the electrodes, and the fibre itself. In general, the equations will be very cumbersome, and it is difficult to imagine situations in which they would be useful in interpreting experimental results.

2. *Controlled voltage change.* If the 'reference potential' is also changed slowly, then the membrane voltage may remain controlled. Thus, the reference voltage may be a ramp function in which dV/dt is constant during the voltage change. In the case of a uniformly polarized membrane, the capacity current is then also constant. The ionic current will increase linearly with time so long as the membrane remains linear.

In the case of a cable polarized at one point, the membrane-current time course will be more complex. We shall derive the current response for this case from eqn (3.44) and show that it approaches eqn (3.44) as $T \to \infty$ or as ΔT (the time taken to change the voltage) $\to 0$.

The voltage change may be represented as the sum of two equal and opposite ramp voltage waveforms, one of which is delayed by ΔT. Thus,

$$V(T) = (V_0/\Delta T) \{ur(T) - ur(T - \Delta T)\},$$

where V_0 and ΔT are, respectively, the amplitude and duration of the voltage change. The unit ramp waveform $ur(T)$ is defined as

$$\left. \begin{array}{ll} ur(T) = 0 & T \leqslant 0 \\ ur(T) = T & T > 0 \end{array} \right\}.$$

According to the superposition principle (Chapter 13) the response to the non-instantaneous voltage change may be expressed in terms of the current response to a unit ramp voltage $I_{ur}(t)$. Thus

$$I(T) = (V_0/\Delta T)\{I_{ur}(T) - I_{ur}(T - \Delta T)\},$$

where $I_{ur}(t)$ is given by the integral of $I_{step}(T)$, the response to a unit step voltage (eqn (3.44) gives $V_0 I_{step}(T)$); namely

$$I_{ur}(T) = \int_0^T I_{step}(T') \, dT'.$$

For times before the voltage change is complete ($0 < T < \Delta T$),

$$I(T) = \frac{V_0}{\Delta T} I_{ur}(t) = \frac{V_0}{\Delta T} \int_0^T I_{step}(T') \, dT' \tag{3.45}$$

and for $t \geqslant \Delta T$,

$$I(T) = \frac{V_0}{\Delta T} \left(\int_0^T I_{step}(T') \, dT' - \int_0^{T - \Delta T} I_{step}(T') \, dT' \right)$$

$$= \frac{V_0}{\Delta T} \int_{T - \Delta T}^T I_{step}(T') \, dT'. \tag{3.45a}$$

This equation may be best interpreted graphically. The integral corresponds to the area under the unit step response $I_{step}(T)$, between $T - \Delta T$ and T (see Fig. 3.10). As $\Delta T/T \to 0$, i.e. for long times or fast voltage changes, expression (3.45a) approaches expression (3.44). Also, an approximation can be given for (3.45). Since $T' < \Delta T$, and if $\Delta T \ll 1$ (voltage change is fast compared to membrane time constant), $\text{erf}\{(T')^{\frac{1}{2}}\} \to 0$, and $\exp(-T') \to 1$.

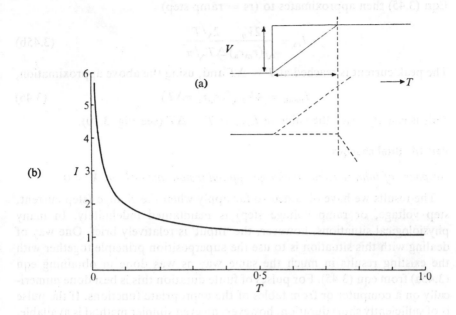

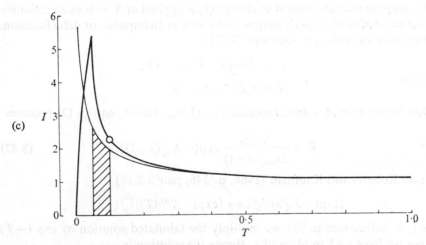

FIG. 3.10. Responses to step or ramp voltage changes. (a) Voltage changes imposed at $X = 0$. The V step is replaced by a ramp function (top) which is obtained by adding the responses to a rising ramp and a delayed falling ramp (bottom) where the delay is equal to the duration of the required ramp function. (b) Current flow required to produce step voltage change. The curve is calculated from eqn (3.44) and is plotted in units of $V_0/R_{in} = 2V_0/\sqrt{(r_m r_a)}$. (c) Illustration showing how response to ramp voltage change may be obtained by integration of response to instantaneous voltage change. For each point T the current is obtained by integrating between $T-\Delta T$ and T, where ΔT is duration of ramp.

Eqn (3.45) then approximates to (rs = ramp step)

$$I_{rs} = \frac{2V_0}{\sqrt{(r_m r_a)}} \frac{2\sqrt{T}}{\Delta T \sqrt{\pi}}. \qquad (3.45b)$$

The peak current is reached at $T = \Delta T$ and, using the above approximation,

$$I_{peak} = 4V_0/\sqrt{(r_m r_a \pi \Delta T)}. \qquad (3.46)$$

This is exactly twice the value of I_{step} at $T = \Delta T$ (see Fig. 3.10).

Part III: Brief changes

Response of infinite cable to charge applied instantaneously at $X = 0$

The results we have obtained so far apply when the change (step current, step voltage, or ramp voltage step) is maintained indefinitely. In many physiological situations, however, the 'input' is relatively brief. One way of dealing with this situation is to use the superposition principle together with the existing results in much the same way as was done in obtaining eqn (3.45a) from eqn (3.45). For pulses of finite duration this is best done numerically on a computer or from tables of the appropriate functions. If the pulse is of sufficiently short duration, however, an even simpler method is available. We suppose that an amount of charge Q_0 is applied at $X = 0$ as an infinitely brief pulse of current. Such an input is known as an impulse, or delta function. We derive the solution from eqn (3.17):

$$\bar{V} = \bar{A} \exp\{-X\sqrt{(s+1)}\},$$

where

$$\bar{A} = r_a \bar{I} \lambda / 2 \sqrt{(s+1)}.$$

Now in the case of a delta function $\bar{I} = Q_0/\tau_m$. Hence, eqn (3.17) becomes

$$\bar{V} = \frac{r_a \lambda Q_0}{2\tau_m \sqrt{(s+1)}} \exp\{-X\sqrt{(s+1)}\}. \qquad (3.47)$$

From Roberts and Kaufman (1966, p. 246, pair 3.2.16)

$$\{\exp(-X\sqrt{s})\}/\sqrt{s} \leftrightarrow \{\exp(-X^2/4T)\}/\sqrt{(\pi T)}$$

and, as before (see p. 30), we multiply the tabulated solution by $\exp(-T)$ since we have $s+1$ in place of s. Hence the solution is

$$V = \frac{Q_0}{2c_m \lambda \sqrt{(\pi T)}} \exp\{(-X^2 - 4T^2)/4T\}. \qquad (3.48)$$

This equation has been given by Hodgkin (in Fatt and Katz 1951) and its derivation in a slightly different form is given by Noble and Stein (1966).

Another way of obtaining eqn (3.48) also illustrates one of its important properties. This uses the superposition principle and the solution to an infinitely long current pulse (eqn (3.24)). If V_1 is the response to a step current of unit intensity, then the response to a short square pulse applying Q_0 in time δt is given by

$$V = \frac{Q_0}{\delta t}\{V_1(t) - V_1(t - \delta t)\},$$

as $\delta t \to 0$

$$V = Q_0 \frac{\partial V_1}{\partial t}. \tag{3.49}$$

Eqn (3.48) may be obtained therefore by differentiating eqn (3.24).

We shall use eqn (3.48) extensively in this section and the ones which follow it in order to illustrate the applications of cable theory to synaptic potentials. At this stage, however, it is worth noting the main limitation of eqn (3.48): the pulse is assumed to be infinitely brief, whereas synaptic potentials in fact are produced by short but finite pulses of current. Where this deficiency influences the results we shall compare the delta function response with one of two other possible ways of simulating synaptic potentials. The first method simply uses the square pulse (as above) but keeps δt finite. The result is obtained from eqn (3.24) by numerical analysis. Although the square pulse representation is an improvement on the delta function, an even more realistic approximation to the synaptic current may be obtained by using the equation

$$I(t) = \alpha^2 t e^{-\alpha t}. \tag{3.50}$$

We shall refer to this function as the 'alpha function'. The forms of membrane current given by these various methods are illustrated in Fig. 3.11. The time course of the voltage response to current of the form given by eqn (3.50) may be obtained by convolution (see Chapter 13) with eqn (3.48). This has been done by Jack and Redman (1971a) and we shall simply use their results here rather than give the full derivation.

Despite its limitations, eqn (3.48) has a major advantage over the other two methods since it readily allows further analytic treatment. Thus, the time of the peak response at various distances may be obtained directly by differentiating eqn (3.48) and setting the resulting expression to zero:

$$\frac{Q_0}{2\lambda c_m \sqrt{\pi}} \left(\frac{X^2 - 4T^2 - 2T}{4T^{\frac{5}{2}}} \right) \exp\left(\frac{-X^2 - 4T^2}{4T} \right) = 0.$$

Hence

$$X^2 = 4T^2 + 2T, \tag{3.35}$$

which is identical with the equation for the propagation of the inflection point in the response to an infinitely long current step (see p. 37). This result is not surprising in view of the relations between the delta function and prolonged step responses given by eqn (3.49). As shown previously (eqn (3.36)) the speed of propagation of the peak approaches $2\lambda/\tau_m$ as the wave propagates away from the current source.

The propagation of the two inflection points may be obtained by setting the

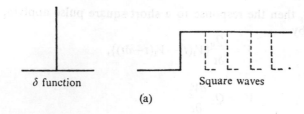

δ function   Square waves

(a)

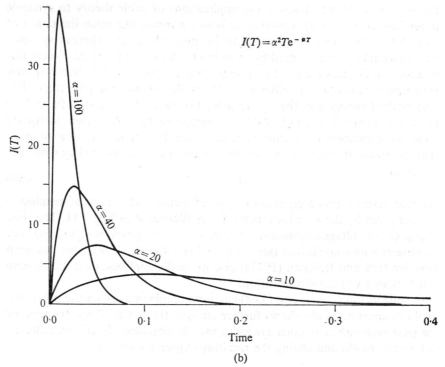

$$I(T) = \alpha^2 T e^{-\alpha T}$$

FIG. 3.11. Forms of current used to simulate synaptic events. (a) The delta function (left) and the short step current (right). (b) Smooth time course of current given by the alpha function (eqn (3.50)) for $\alpha = 10, 20, 40,$ and 100. (From Jack and Redman 1971a.) The alpha function is often the most realistic simulation but is more difficult to treat mathematically. From the mathematical point of view the simplest simulation is the delta function.

second time differential of eqn (3.48) to zero. It may then be shown that the inflection points are given by the positive roots of $\{4T^2 + 6T \pm 2T\sqrt{(8T+6)}\}^{\frac{1}{2}}$.

From these equations it is possible to obtain the maximum rates of rise and fall of the potential. It is also a relatively simple matter to evaluate the 'latency' (here defined as the time taken to reach 10 per cent of peak response), 10–90 per cent rise time, time to half decay, and the 'half-width' (defined by Rall (1967) as the time between the points at which the potential has half its peak amplitude). These measurements have been increasingly useful in experimental work. We have, therefore, calculated the various parameters for a delta function excitation (Fig. 3.12).

Another important property is the relation between the peak amplitude of the response and distance from the current source. This may be obtained by inserting eqn (3.35), rearranged as

$$T = 0.25 \{\sqrt{(4X^2+1)}-1\})$$

into eqn (3.48). We then obtain

$$V_{peak} = \frac{Q_0}{c_m\lambda\sqrt{\pi}} \frac{\exp\{-\frac{1}{2}\sqrt{(4X^2+1)}\}}{\{\sqrt{(4X^2+1)}-1\}^{\frac{1}{2}}} \tag{3.51}$$

Although this equation is useful in obtaining the decrement of peak voltage at some distance from the current source, it will become increasingly inaccurate as the current source is approached. Near $X = 0$, serious errors will arise from the assumption that the charge is applied instantaneously. A more realistic representation of the spatial decay may be obtained therefore by calculating the responses to finite current pulses. This was done by applying the superposition principle to eqn (3.24) for square waves of 0.01, 0.02, 0.05, 0.1, and 0.2T duration. The results are shown in Fig. 3.13 and Fig. 3.14.

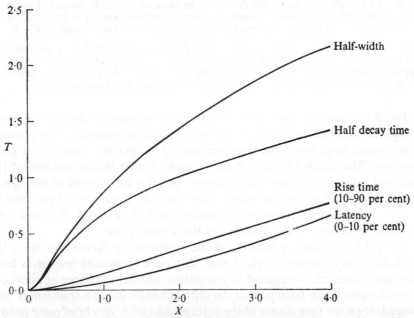

FIG. 3.12. Time taken for response to delta function current to arrive at various points on cable (measured as time taken for voltage to rise to 10 per cent of peak value at each point) and to decay to half-amplitude. The rise time (10 per cent to 90 per cent) and half-width (duration of potential change measured at half-amplitude) are also plotted.

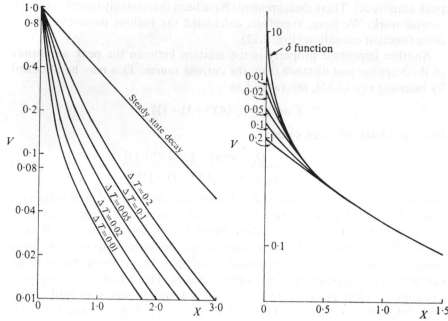

FIG. 3.13 (left). Influence of current duration on the decrement of potential along the cable. The straight line shows the decrement in response to steady current. The curves show decrements in response to square wave currents of duration $\Delta T = 0.01, 0.02, 0.05, 0.1,$ and 0.2. The curves have been normalized in terms of peak voltage at $X = 0$.

FIG. 3.14 (right). Comparison between voltage decrements for delta function current and for square wave currents of same durations as in Fig. 3.13. The same amount of charge is applied to the cable in each case. Note that beyond about $X = 0.5$ the responses converge. Beyond this distance, therefore, the response is insensitive to current duration.

Fig. 3.13 shows the decrement of the peak response to the square waves compared with the decrement of the response to an infinitely long current. The various responses have been normalized to give the same initial peak response. The striking feature of these results is that the attenuation of the transient responses is very much greater than the attenuation of the steady state response. Whereas the steady state response decays to 37 per cent at $X = 1$, the transient potentials decrement to 2.95, 4.2, 6.7, 9.6, and 13.8 per cent depending on the duration of the current pulse.

Fig. 3.14 shows a comparison of the square-wave decrements with the delta-function decrement (eqn (3.51)). The peak amplitude responses have now been scaled to correspond to a constant amount of charge application. It can be seen that a large part of the initial attenuation of a transient is very dependent on the time course of the voltage change. A very brief pulse produces a high local accumulation of charge, and hence a large voltage, which rapidly redistributes to other regions of membrane. When the charge is applied more slowly, some of the redistribution occurs during the application of the charge, so that the peak density at $X = 0$ is smaller and attenuates less rapidly with distance. However, each of the attenuation curves eventually joins the curve for the delta-function response. As the waves propagate, they become smoother and slower (see Fig. 3.17) and the attenuation curves then converge onto that given by eqn (3.51). At larger distances, therefore, the attenuation depends on the distance propagated rather than on the pulse duration.

In contrast to the steep decay of the peak amplitude of the transients at increasing distances along the cable, there is much less attenuation of the total area of the waveform. This may be demonstrated by integrating eqn (3.48) with respect to time.

$$\int_0^\infty \frac{Q_0 \exp\{-(X^2+4T^2)/4T\}}{2\lambda c_m \sqrt{(\pi T)}} \, dt = \frac{Q_0 \tau_m}{2\lambda c_m} \exp(-X). \tag{3.52}$$

This simple exponential decline of the area of the waveform with distance holds generally, not only for the delta function but also for any arbitrary waveform of current injection (see Chapter 7, p. 187). Note also that the relative decline is independent of the time constant of the membrane and depends only on the electrotonic distance propagated.

The time course of the membrane current at each distance may be obtained by differentiating eqn (3.48) twice with respect to X:

$$i_m = \frac{1}{r_a \lambda^2} \frac{\partial^2 V}{\partial X^2} = \frac{Q_0}{8 r_a \lambda c_m \sqrt{\pi}} \frac{(X^2 - 2T)}{T^{\frac{5}{2}}} \exp\left\{\frac{-(X^2+4T^2)}{4T}\right\}. \tag{3.53}$$

Inspection of this equation shows that at $X = 0$ the membrane current will always be negative beyond $T = 0$. Thus, for a depolarizing transient, the current at $X = 0$ will be an outward current only while the charge is being applied (in eqn (3.53) it is, of course, applied instantaneously). After this time the charge will redistribute to other areas of membrane, and this region will be subject to an inward current. (For hyperpolarizing transients, these current directions will be reversed.) At increasing values of X, there will be an increasing time of initial positive current followed by a period of negative membrane current. The exact point of reversal is given by $X^2 = 2T$. These results are illustrated in Fig. 3.15 which shows the membrane currents at $X = 0.2$, 0.5, 1.0, and 2.0. The time course of these curves will be used in the next section to explain some important features of the time course of the voltage at various distances. The time of reversal of the membrane current is plotted in Fig. 3.16 together with the time of the peak potential for comparison.

An important result may also be obtained from the third differential of eqn (3.48) with respect to X:

$$\frac{\partial^3 V}{\partial X^3} = \frac{Q_0}{16 c_m \lambda \sqrt{\pi}} \frac{(6TX - X^3)}{T^{\frac{7}{2}}} \exp\left\{\frac{-(X^2+4T^2)}{4T}\right\} \tag{3.54}$$

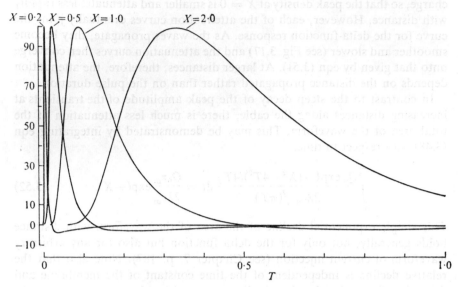

FIG. 3.15. Time course of membrane current in response to delta function input. The curves were computed from eqn (3.53) for $X = 0.2$, 0.5, 1.0, and 2.0. Each curve is normalized in terms of the peak current.

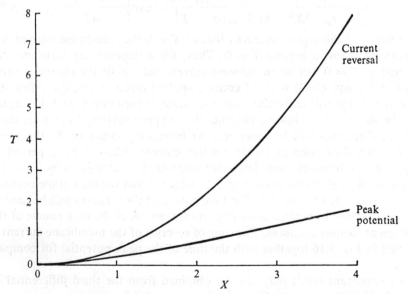

FIG. 3.16. Times of occurrence of peak potential and reversal of membrane current during propagation of response to delta function.

Setting this equation to zero gives

$$X^2 = 6T \quad \text{or} \quad X = \sqrt{(6T)}. \tag{3.55}$$

This equation gives the propagation of the peak positive membrane current with distance and is therefore relevant to the interpretation of extracellularly recorded potentials, as has been pointed out by Stevens (1966).

Applications to junctional mechanisms

In the remaining parts of this chapter, we shall discuss the application of cable theory to junctional potentials. In many of the cases which have been studied experimentally, these potentials may be attributed to a transient change in the ionic conductance of the postjunctional membrane (Fatt and Katz 1951; Takeuchi and Takeuchi 1959; Eccles 1964; Ginsborg 1967). The direction and magnitude of the ionic current which flows as a result of the conductance change depends not only on the conductance change itself but also on two other factors: the reversal potential of the conductance mechanism and the magnitude of the variation in potential during the conductance change. In general, the reversal potential for excitatory junctional changes is very positive to the resting potential and is usually near $E = 0$ (mV). The reversal potential for inhibitory junctional changes usually lies close to or negative to the resting potential. When the reversal potential is very different from the resting potential then, for small synaptic potentials, the driving force $(E-E_{\text{Rev}})$ remains nearly constant and the synaptic current flow $i = g(E-E_{\text{Rev}})$ will be nearly proportional to the conductance change. In this case, therefore, we may use the equations for the responses to brief currents. When the reversal potential is close to the resting potential or when the synaptic potential is large in amplitude, this approximation will not be valid and it will then be necessary to use equations for the responses to a conductance change. These equations will be discussed later in this chapter (see p. 59).

In addition to the possibility that equations for conductance changes rather than current changes may be required, there are other conditions which must be satisfied for the infinite-cable equations to be applied.

1. The postjunctional cell must be cable-like and sufficiently long for the effects of terminations to be negligible. If the latter condition is not satisfied, the equations for terminated cables given in Chapter 4 will be required. Moreover, if the postjunctional cell is not a simple cable, the one-dimensional equations may be invalid or require modifications. This problem will be discussed further in Chapter 7 in connection with the properties of nerve cells.

2. The conductance change must be localized to a region of the cell which is small compared to the space constant if the equations for point application of current are to be used. However, this restriction applies only to individual responses. The combined effects of several responses

occurring in the same cell may be treated even when they are spatially separated. In some circumstances (p. 64) this may be done by super-posing the responses to individual synaptic currents, assuming that the nonlinearity introduced by each junctional change has a negligible effect on the component of response to other conductance changes. Since one of the most important functions of junctions is that they allow summation and other operations to be performed on several inputs, the problem of the interaction of junctional potentials which are temporally and spati-ally dispersed is an important one.

It is not possible to choose a structure which satisfies all these conditions. In the first place, very few examples of junctional transmission have been submitted to a rigorous quantitative analysis. The best available example is the analysis of the skeletal muscle end-plate potential by Fatt and Katz (1951), and we shall refer to some of their results in this section. However, this case is of limited interest from the theoretical point of view since there is no spatial interaction of synaptic potentials to be considered.

Time course of the voltage transient. In order to illustrate the time course of the transient and the changes which occur as it propagates along the fibre, we have calculated the responses to a square wave of current injection of duration $0{\cdot}1T$. These are plotted in Fig. 3.17 and are very similar to the records shown in Fig. 8 of Fatt and Katz (1951). It is clear, as X increases, that there is an increasing delay in the onset of the wave, and the rise and fall of potential become considerably slower. Close to the site of current genera-tion, the response waveform will be very dependent on the shape of the current pulse, but as the wave propagates the response becomes relatively independ-ent of the time course of current injection, as expected. This is illustrated by comparing the rise and decay of potentials in response to current pulses of various durations with those in response to a delta function. Fig. 3.18 shows the influence of pulse duration on the 10–90 per cent rise time and on the half-width time. Fig. 3.19 summarizes the information in the form of a rise-time versus half-width plot. Provided that the time course of the synaptic current is brief and has no prolonged residual phase, significant deviations from the delta-function response appear only at near distances and at early times (see also Fatt and Katz 1951, p. 332). Notice also that the rise time is more sensitive to current duration than is the half-width.

Since the time course of the synaptic potential is to some extent dependent on the shape and duration of the junctional current there have been attempts to deduce the time course of the membrane current from the time course of the junctional potential. This may be done relatively easily in the case of a response in a uniformly polarized membrane provided that the membrane properties are not grossly nonlinear. Once the synaptic current has ceased to flow, the potential should decline exponentially with the time constant of the membrane. Any deviation from this decay time course may be attributed,

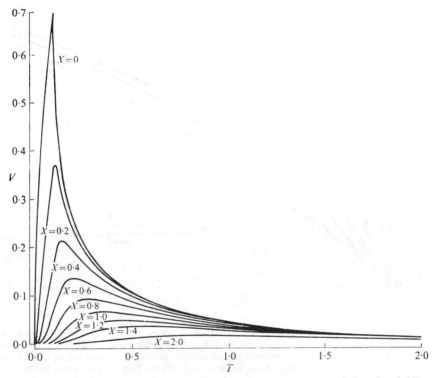

FIG. 3.17. Potential changes in response to square wave current of duration $0.1T$ at $X = 0, 0.2, 0.4, 0.6, 0.8, 1.0, 1.2, 1.4$, and 2.0. Current applied at $X = 0$.

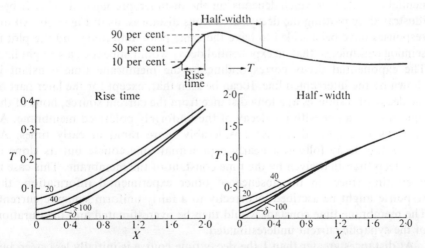

FIG. 3.18. Influence of duration of synaptic current on rise time and half-width of response. The synaptic current was simulated using the alpha function (eqn (3.50)) and current duration is increased by reducing the value of α (cf. Fig. 3.11).

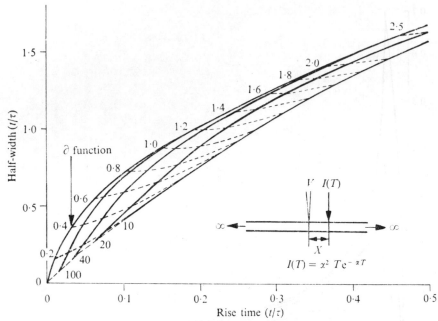

FIG. 3.19. The results of Fig. 3.18 are plotted as a series of half-width versus rise-time curves. If the synaptic current is well simulated by an alpha function, a pair of experimental values for half-width and rise time enables the current duration (determined by α) and the distance from the recording site to be estimated. (From Jack and Redman 1971a.)

therefore, to a residual flow of synaptic current. In the case of a propagating potential, however, the time course of decay deviates from a simple exponential in a manner which depends on the distance propagated. This is best illustrated by plotting the decays at various distances, as in Fig. 3.20. All the responses have been scaled to have the same peak amplitude and the plot is semilogarithmic, so that an exponential decay would appear as a straight line. The exponential decay corresponding to the membrane time constant is shown by the interrupted line. It can be seen that, except for the later part of the decay of responses at a long distance from the current source, none of the time courses agree with the decay of the uniformly polarized membrane. At near distances, the decay is considerably more rapid at early times. At $X = 1$ the decay follows a nearly exponential time course but its slope is less steep than that given by the time constant of the membrane. This case is interesting since, in the absence of other experimental information, the response might be ascribed incorrectly to a fairly uniform synaptic current. The membrane time constant would then be overestimated and the duration of the synaptic current underestimated.

At distances greater than 1 the decay time course is initially less steep but eventually approaches the slope of the membrane time constant. On the assumption that the synaptic current is brief and localized some indication of the distance propagated may therefore be obtained from the shape of the decay curve (see below).

In the absence of axial current flow, the charge on each area of membrane would decay exponentially with the time constant of the membrane: deviations from this time course of decay (as in Fig. 3.20) may be attributed to net membrane current flowing as a consequence of the nonuniformity of the polarization. It should be possible, therefore, to explain the decay curves in terms of the membrane currents. The latter were plotted for the delta-function response in Fig. 3.15. At short distances, the membrane current soon becomes negative, and the potential therefore decays very quickly as charge spreads to other, less polarized, areas of membrane. At longer distances away from the current source, the time course of positive current is prolonged and the transition to negative current more gradual. Moreover, there is an increasing lag between the time of peak voltage (given by $X^2 = 4T^2 + 2T$) and the time of reversal of membrane current (given by $X^2 = 2T$). This was shown in Fig. 3.16 from which it is clear that, beyond about $X = 0.5$, the initial decay of the potential occurs while the membrane current is still positive. Hence, it is not surprising that the rate of decay of voltage is slower than that given by the membrane time constant and will remain so for the time interval between the two curves in Fig. 3.16.

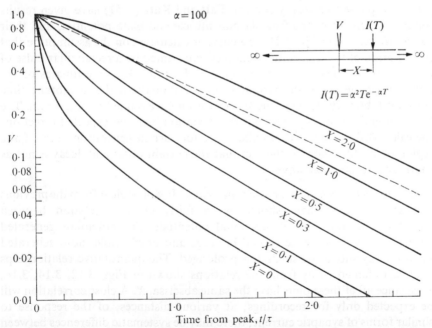

FIG. 3.20. Logarithmic plot of the decay of response generated by synaptic current simulated by an alpha function ($\alpha = 100$) at distances of 0, 0·1, 0·3, 0·5, 1·0, and 2·0 from the recording electrode. Each response is normalized in terms of the peak potential and the abscissa is time from the peak potential. (From Jack and Redman 1971.)

These results are of considerable importance in the interpretation of synaptic potentials. First, it is clear that the geometry of the excitable cell is crucial to the interpretation of recordings. If the synaptic current flow is localized rather than uniform, then use of measures such as the half-decay time to calculate the membrane time constant will be inaccurate, and will give either overestimates or underestimates depending on whether the synaptic current is generated further away or nearer to the recording electrode. Alternatively, if the membrane time constant is already known (but see Chapter 7 for a discussion of some of the uncertainties in measuring membrane time constants), it is in general not valid to infer the time course of the synaptic current by the use of Hill's local potential equation (cf. Curtis and Eccles 1959). Secondly, if the geometry is that of a simple cable, the total *shape* of the decay curve, particularly if plotted semilogarithmically, gives more information than single measures such as the half-width or half-decay. Together, the rise time, decay time, and time course of decay may allow an estimate of both the electrotonic distance propagated and the membrane time constant. Further discussion of this problem will be found in Chapter 7.

Amplitude of the voltage transient. As shown above (Fig. 3.13) the decrement of amplitude with distance depends to some extent on the duration of the synaptic current; in general the decrement is considerably steeper than it is for the response to a steady current. Fatt and Katz (1951) have given results for the decrement of the frog skeletal muscle end-plate potential, and their values are plotted in Fig. 3.21. The curve is calculated on the assumption that the duration of the transmitter action is represented by a rectangular pulse of duration $0 \cdot 1 T$. This value corresponds to Fatt and Katz's estimates of the duration of the end-plate current (about 2–3 ms) and the membrane time constant (about 20–30 ms). The space constant is assumed to be 2·4 mm. The fit is not entirely satisfactory, but the deviation is in the expected direction; the calculated curve makes no allowance for current generation over a finite region of the fibre nor for the fact that the current onset and decay is not as sharp as a rectangular wave.

Relation between amplitude and time course. It will be clear from the previous two sections that there should be a fairly close correlation between amplitude and duration of junctional potentials. The potentials generated near a recording electrode should be large and brief, while those generated further away should be small and prolonged. The quantitative relationships may be obtained easily from the relations shown in Figs. 3.12, 3.14, 3.16, 3.18, since all of these plots have the same abscissa, X. A close correlation will be expected only for recordings, at various distances, of the response to similar forms of synaptic current; there may be systematic differences between the time courses and initial amplitudes of the responses owing to variations in time course and magnitude of transmitter action. Such effects will be particularly important when the synaptic currents are generated close to the

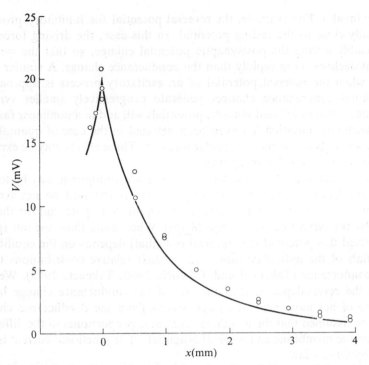

F_{IG}. 3.21. Amplitude of synaptic potential as function of distance. Points show peak values of end-plate potential given by Fatt and Katz (1951, Fig. 7). The curve shows expected decrement of voltage if synaptic current is represented by a square wave of duration 0·1T.

recording site, since it is at near distances that amplitude attenuation and time course are most dependent on the exact form of the current generating the response. This is the situation where experimental measurements have been made (see, e.g. Del Castillo and Katz 1956; Ginsborg, 1960; Burke 1967). The lack of correlation between amplitude and time course of response presumably reflects variability of the magnitude and/or time course of the transmitter action (Gage and McBurney 1972).

Part IV: Response of infinite cable to step change in membrane conductance

It has already been noted that synaptic effects are usually generated by a brief localized membrane conductance change (see p. 53). Provided that the reversal potential for the ionic current flowing as a consequence of this conductance change is very different from the resting potential, the driving force $(E-E_{\mathrm{Rev}})$ will be virtually constant for relatively small postsynaptic potentials so that the synaptic current will have the same time course as the conductance change. The preceding analysis should then apply.

However, there are other circumstances where the assumption that the ionic current has the same time course as the conductance change will be clearly invalid. For example, the reversal potential for inhibitory processes is usually close to the resting potential. In this case, the driving force falls appreciably during the postsynaptic potential change, so that the synaptic current declines more rapidly than the conductance change. A similar effect arises when the reversal potential of an excitatory process is approached. Additional conductance changes generate progressively smaller synaptic currents so that individual synaptic potentials will add in a nonlinear fashion. Such nonlinear addition has even been detected in the case of quantal end-plate potentials at the neuromuscular junction. These effects may be explored quantitatively in the following way.

When an additional conductance occurs in a membrane it may be to only one ion species or to several. If only one ion is involved no net synaptic current flows across the membrane at the equilibrium potential for the ion, given by the Nernst equation (eqn (8.15)). When more than one ion species is involved this potential (the reversal potential) depends on the equilibrium potentials of the individual ions, and on their relative contributions to the total conductance (Takeuchi and Takeuchi 1960; Takeuchi 1963). We may define the reversal potential at the site of the conductance change by the absence of net movement of charge arising from the conductance change. It will be assumed that the junctional current is proportional to the difference between the membrane and reversal potentials. The junctional current is then given by Ohm's law.

$$I_j(t) = G_j \{(E_r - V(t)) - E_{Rev}\}, \tag{3.56}$$

where $I_j(t)$ is the current at time t, E_r is the resting membrane potential, E_{Rev} is the reversal potential for the current flowing through the junctional conductance G_j, and $V(t)$ is the voltage displacement, from the resting potential, generated by the flow of junctional current.

We wish to determine the time course of $V(t)$. Before considering the infinite-cable case it may be helpful to derive the response for a simpler situation, where spatial spread of current can be ignored (e.g. a spherical cell). The equivalent circuit is then represented by Fig. 3.22.

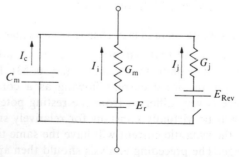

FIG. 3.22. Equivalent circuit used to represent effect of junctional conductance on a uniformly polarized cell. The arrows attached to each element indicate the convention for positive current flow (and not the expected behaviour).

The junctional current which flows when the switch closes flows through the membrane capacitance C_m, and the resting membrane conductance G_m. The equation is

$$I_j = VG_m + C_m \, dV/dt.$$

Hence

$$VG_m + C_m \, dV/dt = G_j\{(E_r - V) - E_{Rev}\}.$$

This gives the simple differential equation

$$\frac{dV}{dt} + V\left(\frac{G_m + G_j}{C_m}\right) = \frac{G_j}{C_m}(E_r - E_{Rev}), \tag{3.57}$$

which has the solution

$$V = \frac{G_j(E_r - E_{Rev})}{(G_m + G_j)}\left\{1 - \exp\left(-\frac{G_m + G_j}{C_m}t\right)\right\}. \tag{3.58}$$

This solution is formally analogous to the response of a parallel R–C circuit to a step of *current* (eqn (2.3)). The steady state solution is

$$V = \frac{G_j(E_r - E_{Rev})}{(G_m + G_j)}. \tag{3.59}$$

These equations differ notably from the step-current response, since the time constant of the exponential is a function of G_j. For small values of G_j relative to G_m, the time constant approximates to C_m/G_m (that is, $R_m C_m$, the resting membrane time constant). For larger values of G_j the time constant is briefer than the membrane time constant.

The response of an infinite cable to a step conductance change is more complicated. By using the Laplace-transform method it can be shown that the voltage as a function of distance and time is given by

$$V = \frac{G_j(E_m - E_{Rev})}{G_{in}\{1 - (G_j/G_{in})^2\}}\left[\frac{(1 - G_j/G_{in})}{2}\exp(-X)\text{erfc}\left(\frac{X - 2T}{2\sqrt{T}}\right) - \right.$$
$$\frac{(1 + G_j/G_{in})}{2}\exp(X)\text{erfc}\left(\frac{X + 2T}{2\sqrt{T}}\right) +$$
$$\left. + \frac{G_j}{G_{in}}\exp\left[\frac{G_j}{G_{in}}X - \{1 - (G_j/G_{in})^2\}T\right]\text{erfc}\left(\frac{X + 2G_jT/G_{in}}{2\sqrt{T}}\right)\right], \tag{3.60}$$

where X is the electrotonic distance (x/λ) away from the site of the conductance increase at which V is recorded, G_{in} represents the input conductance of the cable (that is, $2/\sqrt{(r_m r_a)}$—see eqn (3.26)), and T is the time expressed in units of the resting membrane time constant (t/τ_m). For the special case $X = 0$ this equation simplifies to

$$V = \frac{(E_r - E_{Rev})}{\{1 - (G_{in}/G_j)^2\}}\left\{1 - \frac{G_{in}}{G_j}\text{erf}\sqrt{T} - \exp(-T)\psi\left(\frac{G_j\sqrt{T}}{G_{in}}\right)\right\}, \tag{3.61}$$

where $\psi(x) = \exp(x^2)\mathrm{erfc}(x)$, the exerfc function which is tabulated in Carslaw and Jaeger (Appendix II). Eqn (3.61) has been given in slightly different forms by E. J. Harris (in Fatt and Katz 1951) and by B. Ginsborg (in Burke 1957).

Fig. 3.23 shows the time course of development of the voltage displacement at $X = 0$ for various magnitudes of junctional conductance. The values plotted have been selected for comparison with the measurements made by Fatt and Katz (1951, Fig. 32) on an artificial transmission line. Note that the initial rate of rise of voltage is very rapid for the large conductance changes plotted here. Thus, when the junctional conductance is 5 times larger than the input conductance of the cable, the voltage displacement takes approximately 0.15τ to reach 84 per cent of its steady state value, compared to 1.0τ for the voltage response to a current step. The reason for this large difference is that as the voltage displacement develops, the 'driving voltage' $(E_{\mathrm{m}} - V - E_{\mathrm{Rev}})$ falls. Hence the current flow through the junctional conductance also falls and the voltage response rapidly 'levels off'.

The time course of the junctional current may be calculated using eqn (3.56) and is plotted in Fig. 3.24. When the conductance change (expressed as the ratio $G_{\mathrm{j}}/G_{\mathrm{in}}$) is relatively small, the current flowing does not differ significantly from a step. Thus the time course of the voltage displacement will be similar to the time course of voltage in response to a step of current, that is, $\mathrm{erf}\,(\sqrt{T})$. This is shown in Fig. 3.25, where comparison of the time courses has been made easier by scaling the steady state voltages to the same value.

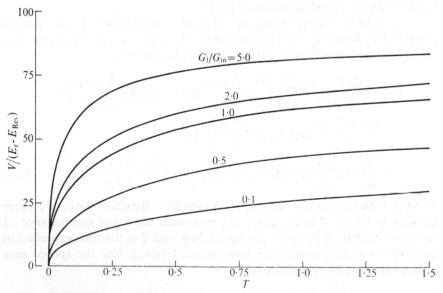

FIG. 3.23. Time course of voltage change at $X = 0$ in an infinite cable in response to various magnitudes of conductance change. The responses have been computed from eqn (3.61) for $X = 0$. The figure on each curve is the value of $G_{\mathrm{j}}/G_{\mathrm{in}}$.

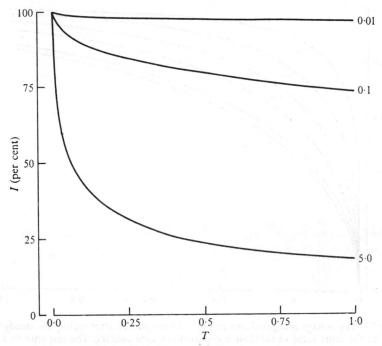

FIG. 3.24. Time course of junctional current flowing in response to conductance changes
cf. Fig. 3.23.

The same result can be shown to follow mathematically from eqn (3.60).
When $G_j/G_{in} \to 0$ this equation becomes

$$V(X, T)_{G_j/G_{in} \to 0} = \frac{G_j(E_m - E_{Rev})}{2G_{in}} \left\{ \exp(-X)\mathrm{erfc}\left(\frac{X-2T}{2\sqrt{T}}\right) - \right.$$
$$\left. - \exp(-X)\mathrm{erfc}\left(\frac{X+2T}{2\sqrt{T}}\right) \right\},$$

which is of the same form as eqn (3.24) for the response to a current step.

If, on the other hand, the ratio G_j/G_{in} is very large eqn (3.60) becomes

$$V(X, T)_{G_j/G_{in} \to \infty} = \frac{(E_m - E_{Rev})}{2} \left\{ \exp(-X)\mathrm{erfc}\left(\frac{X-2T}{2\sqrt{T}}\right) + \right.$$
$$\left. + \exp(X)\mathrm{erfc}\left(\frac{X+2T}{2\sqrt{T}}\right) \right\},$$

which has the same form as eqn (3.41) for the response of a cable to a step
change of voltage at $X = 0$. A large conductance increase at $X = 0$ therefore
has the same effect as a step voltage change applied at this point on the cable.

These equations only describe the time course of voltage displacement
during the conductance and not subsequent to its cessation. If the full time

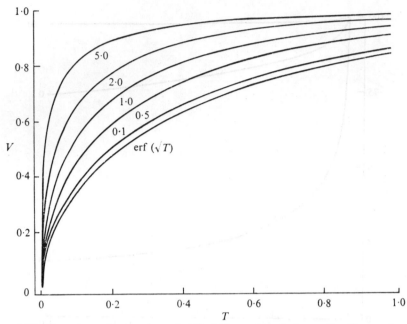

FIG. 3.25. The voltage changes shown in Fig. 3.23 are plotted after scaling the steady state values to the same value to facilitate comparison of time courses. The response to a step current change is also shown labelled erf ($\sqrt{T}$).

course of the voltage response to a rectangular pulse of conductance is required it is not valid to use the superposition principle (see Chapter 13), as was done previously to derive the cable response to a rectangular pulse of current. The conductance changes disturb the membrane constants of the cable; superposition in the time domain is invalid therefore because the differential equation itself has been altered. Similarly, superposition of responses produced by conductance changes at different points on the cable is not valid. For synaptic inhibition, in particular, the effect on cable parameters, such as the input conductance, may be especially large.

Linear superposition of the voltage responses to two *synchronous* brief conductances at different points on a cable might be expected in the circumstance when the finite latency of propagation of voltage along the cable is greater than the duration of the conductance change. In other circumstances numerical solution of the differential equation would be required. Discussion which includes the results of such calculations and which illustrate the form and degree of interaction between conductances will be found in the papers of Rall (1962a, 1962b, 1964, 1967).

One advantage of the relatively simple mathematical description presented here is that it gives an estimate of the degree of nonlinear addition of conductance changes. Providing that the conductance changes occur synchronously at a localized site, it is possible (using eqn (3.61)) to calculate the peak amplitude of the voltage response to different magnitudes of conductance. This can only be done at $X = 0$ since this is the only place on the cable at which the time of peak amplitude of the voltage response occurs at the termination of the pulse. At other points it is necessary to compute the voltage responses at later times which cannot be obtained from eqn (3.61). Despite these restrictions, the simple calculation may serve as a model for the addition of quantal conductance changes which have been studied at the neuro-muscular junctions of frog and mammal (del Castillo and Katz 1954; Martin 1955, 1966).

In presenting these calculations it is necessary to illustrate the dual effect of varying the magnitude and duration of the conductance increase. Fig. 3.26 shows a set of calculations in which the value of voltage has been normalized (V_N) to compare the degree of nonlinear addition of conductances of different duration but similar magnitude. Linear addition is represented by the straight line relation between V_N and G_j/G_{in}. The maximum degree of non-linearity will of course be displayed by the curve for the steady state ($\Delta T \to \infty$). The intermediate curves show the relative degree of nonlinearity produced by various durations of conductance change. The briefer the duration of the conductance change, the more linearly it adds. This is because the membrane capacitance delays the development of the voltage displacement so that the reduction in driving force responsible for the nonlinear addition takes time to occur.

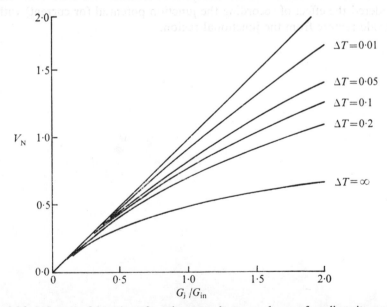

FIG. 3.26. Influence of duration of conductance change on degree of nonlinearity expected on addition of conductance changes. For very brief conductance changes, the addition of responses is almost linear. The nonlinearity becomes most marked when the conductance change is of long duration.

These graphs are analogous to comparing an end-plate potential with the 'end-plate current', which is directly proportional to the end-plate conductance change. Takeuchi and Takeuchi (1959) measured the end-plate current under voltage-clamp conditions at the frog neuromuscular junction. They found that for conductance changes of normal duration, the relation between peak end-plate current and peak end-plate potential was roughly linear over a greater range than for end-plate conductances prolonged by eserine.

In 1955 Martin gave a formula for the correction of the estimate of the number of quanta responsible for an end-plate potential. This formula assumed a steady state condition (eqn (3.59)). Since the duration of the conductance change at many junctions is usually of the order of 0.05 to 0.2τ, so that much less nonlinear addition occurs, it is apparent that the formula will give an overestimate of the number of quanta (cf. Takeuchi and Takeuchi 1959). It is easy to derive more accurate, though still approximate, formulae by expanding the exponential and error functions in eqns (3.58) and (3.61) as power series in t (providing t remains small). After algebraic manipulation polynomial formulae for m (quantal number) can be obtained (Jack, unpublished). However, Hubbard, Llinas, and Quastel (1969, pp. 67–8) have pointed out that the effect of extracellular resistivity may be important. Quastel (quoted in Hubbard, Llinas, and Quastel 1969) has calculated that Martin's formula may not, in these circumstances, over-correct. It would be useful to see full details of his analysis. Auerbach and Betz (1971) have considered the effect of recording the junction potential (or current) with an electrode remote from the junctional region.

4. Properties of finite cables

In many physiological and experimental situations, the length of the preparation is not large enough compared to the space constant λ for the cable to be regarded as infinite. The equations discussed in the previous chapter will not apply since one of the boundary conditions ($V \rightarrow 0$ as $X \rightarrow \infty$) will not hold. It then becomes necessary to consider the behaviour of cables of finite length, treating the effects of terminations at a finite distance from the excitatory source by making appropriate modifications of the boundary conditions. The basic assumptions and equations, apart from the boundary conditions, are the same as for the infinite-cable case, and it will be shown that the finite-cable solutions approach those derived previously when the length l of the cable is sufficiently large.

Semi-infinite cable

We shall consider the general case in which current is injected at one point and then spreads in both directions along the cable. The cable may have different or identical terminations on each side at arbitrary distances from the current source. In Chapter 7 (*Mathematical models of the nerve cell*) we shall discuss the case of a cable terminated at one end in an R–C circuit. In this chapter we will consider two simpler terminations which may be good approximations in many biological situations. The more typical termination is an *open-circuit*. This condition will occur when a fibre terminates naturally and is enclosed by a high-resistance membrane (e.g. the ends of skeletal muscle fibres and some nerve endings) or when shortened preparations acquire a high resistance (as in the case of cardiac muscle in which cut ends acquire a high resistance, possibly as a result of a change in the resistance of exposed intercalated disc membranes (Weidmann 1952; Délèze 1971)). The open-circuit condition may apply also to the peripheral end of a dendrite, particularly if there is a very rapid taper, or if branching abruptly decreases the total dendritic cross-section (see Rall 1959; and Chapter 7).

The other simple termination is a *short-circuit*. This would be a good approximation for the unhealed ends of a nerve or muscle fibre. It also might apply to certain nerve cell geometries; for example, where a single dendritic trunk terminates centrally in a large soma or where the peripheral branching in a dendritic tree is so profuse that it can be treated as a very low impedance beyond the branching point.

These termination conditions may be combined in various ways, e.g. we might consider a dendrite as ending in a short-circuit at the soma and in an open-circuit at its peripheral end. For simplicity we will consider first the case when one side of the cable is terminated while the other side is infinite.

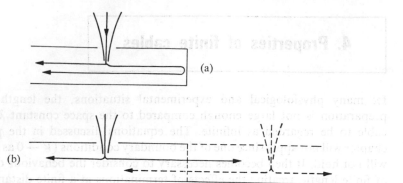

FIG. 4.1. The semi-infinite cable (a) and its infinite cable equivalent (b). In the semi-infinite cable there are two current or voltage terms to be considered, one of which is a 'reflection' term arising from the termination. This situation may be simulated in an equivalent infinite cable by using two identical current sources, one acting as the 'image' of the real source at a distance $2L$ away from it. Hence the response of semi-infinite cable may be obtained from superposition of two infinite cable responses.

The solutions for this *semi-infinite cable* may be obtained readily from the infinite cable solutions by using a method which gives some helpful insight into the properties of short cables. We will illustrate this method by treating the semi-infinite cable with an *open-circuit* termination at $X = L$ (see Fig. 4.1).

The open-circuit boundary condition requires that no axial current flows beyond the site of termination. One way of simulating this condition is to use an imaginary infinite cable in which an axial current of identical time course but opposite (longitudinal) direction is superimposed upon the original axial current at the point where the boundary condition is to apply. This may be achieved mathematically by using an 'image' current source, that is, by placing a hypothetical current source identical to the first at an equal distance away from the boundary point along the hypothetical infinite cable at $X = 2L$. The individual axial currents at $X = L$ must then be equal and opposite so that the total axial current is zero. The voltage response at all points will be described simply by superposition of the responses to the individual current sources. Thus, we may use the infinite cable response to a delta function (eqn (3.48)) to derive the corresponding semi-infinite cable response. If the termination is at $X = L$ and the infinite half of the cable is in the direction of $-\infty$, we obtain

$$V = \frac{Q_0}{2c_m\lambda\sqrt{(\pi T)}}\left[\exp\left\{\frac{(-X^2-4T^2)}{4T}\right\} + \exp\left\{\frac{-(2L-X)^2-4T^2}{4T}\right\}\right]. \quad (4.1)$$

In this case there are two exponential terms. The first term is the infinite-cable term. The second term is that resulting from the mathematical 'device' of injecting current at $X = 2L$. The second term may be called a 'reflection' term since the effect of the device is equivalent to 'reflecting' the wave at the open-circuit (see Fig. 4.1). In this case, there is one reflection term only since there is only one termination at which reflection may occur.†

The boundary condition for a semi-infinite short-circuited cable requires that the value of V should be zero at $X = L$. Clearly this condition may be achieved in a similar way by reversing the polarity of the 'image' current source at $X = 2L$. This is equivalent to 'reflecting' and *changing the sign* of the wave at $X = L$. Hence the solution will be identical with eqns (4.1) and (4.2) except that the exponential terms must now be subtracted instead of being added. The semi-infinite-cable responses to other waveforms for which the infinite-cable responses are known (see Chapter 3) may be obtained very easily by this method.

Cables terminated at both ends: method using 'reflections'

If both sides of the cable are terminated, both of the waves will be reflected back and forth an infinite number of times as they become progressively smaller. Therefore there must be an infinite number of terms in the equation obtained from the infinite cable response (see eqn (4.7a) below). Using the previous analogy, the response is equivalent to the response of an infinite cable with an infinite number of current sources at $X = 0, 2L, 4L$, etc. and $X = -2L, -4L$, etc. In practice, however, it is usually sufficient to consider only the first few reflection terms in order to obtain an approximate solution. Thus, in a cable which is long enough for a substantial decrement to occur between each reflection, very few terms will be significant. This is illustrated in Fig. 4.3. The cable is terminated at $X = L$ and at $X = -L$ by open-circuits. The voltage is recorded at $X = 0.2$. The lowest curve shows the infinite-cable response to a delta function. The other curves show the effect of adding the first, second, third, and fourth reflection terms. It is apparent in

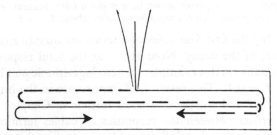

FIG. 4.2. Cable terminated with open-circuits at both ends. In this case there are an infinite number of 'reflection' terms which become progressively smaller. The series solutions obtained for this case therefore converge after a certain number of terms (see Fig. 4.3).

† The use of the term 'reflection' is an extension of the usage relating to 'propagation' discussed in Chapter 3 (see p. 33, footnote). The wave is regarded as propagating to the termination at which it is reflected and propagated back. However, this usage differs from the usual terminology in electrical engineering where the second source is described as an 'image' source and the waves (in systems with no inductance) are regarded as propagating virtually instantaneously.

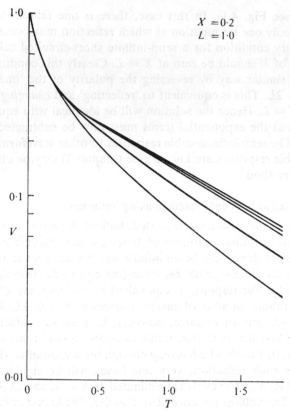

$X = 0.2$
$L = 1.0$

FIG. 4.3. Convergence of series solution for open-circuit short cable. The figure shows voltage curves in response to delta function excitation recorded at $X = 0.2$ in a cable of length $L = 1$. The series converges fairly rapidly and after the addition of four 'reflection' terms the early part of the response shown here is given fairly accurately. Note also that response becomes exponential after about $T = 0.5$.

this case that only the first four reflection terms are quantitatively significant in the early part of the decay. Note also that the total response eventually decays exponentially (in this example, the semilogarithmic plot becomes linear beyond about $T = 0.4$). This result is of importance in the interpretation of synaptic potentials and will be discussed further in Chapter 7. (For more detailed comparison between the responses of cables terminated in open-circuits and the response of an infinite cable, see Jack and Redman (1971a).)

The short-circuit condition may be satisfied, as previously noted, by reversing the sign of the wave at each reflection. As indicated in Fig. 4.4, this procedure will generate an infinite series of positive and negative terms (see eqn 4.14a). Lux (1967) has illustrated the delta function response of short-circuit cables of various lengths when voltage and current are recorded at the same point.

In general, therefore, the response of cables terminated at both ends may be

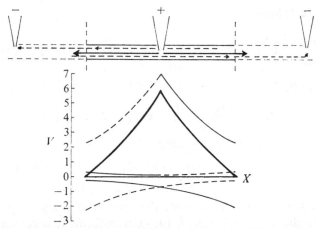

F IG. 4.4. Cable terminated with short-circuits at both ends. This case also generates an infinite number of terms but their signs alternate. Further explanation in text.

expressed by the sum of an infinite number of terms, each of the form of an infinite-cable response. The signs of the terms depend on the nature of the terminations. As we shall also show below, delta function excitation gives rise to series which are theta functions. These functions are defined on p. 27 of Roberts and Kaufman (1966) and their Laplace transforms are tabulated on p. 82–5 of their book.

Cables terminated at both ends: Laplace-transform method

The method described in the previous sections may be used whenever the infinite-cable response is known. Another approach which does not require the infinite-cable solution is to use the Laplace transformation directly. As already noted above many of the required Laplace transforms are tabulated. More-over, this method may more conveniently illustrate some of the other prop-erties of short cables. Therefore we will give a fairly full derivation of the response of an open-circuit short cable to a delta function current input. Since the general equations are the same as in the case of infinite cable, we may simply refer to eqn (3.16):

$$\bar{V} = \bar{A}\exp\{-X\sqrt{(s+1)}\} + \bar{B}\exp\{X\sqrt{(s+1)}\}.$$

Now, instead of setting $\bar{B} = 0$, as in the infinite-cable case, we set the axial current, or its transform $\bar{i}_a$, to zero at $X = L$. From eqns (3.16) and (3.18) we obtain the following equation for $\bar{i}_a$,

$$\bar{i}_a = \frac{\sqrt{(s+1)}}{r_a\lambda}[\bar{A}\exp\{-X\sqrt{(s+1)}\} - \bar{B}\exp\{X\sqrt{(s+1)}\}]. \qquad (4.2)$$

Setting $\bar{i}_a$ at $X = L$ to zero and solving for $\bar{B}$,

$$\bar{B} = \bar{A}\exp\{-2L\sqrt{(s+1)}\}. \qquad (4.3)$$

Hence eqn (4.2) becomes

$$\bar{i}_a = \frac{\bar{A}\sqrt{(s+1)}}{r_a\lambda}[\exp\{-X\sqrt{(s+1)}\}-\exp\{-2L\sqrt{(s+1)}\}\exp\{X\sqrt{(s+1)}\}].$$

(4.4)

When $X = 0$,

$$\bar{i}_a(0, s) = \frac{\bar{A}\sqrt{(s+1)}}{r_a\lambda}[1-\exp\{-2L\sqrt{(s+1)}\}].$$

Hence $\bar{A}$ may be eliminated by writing

$$\bar{i}_a = \bar{i}_a(0, s)\frac{\exp\{-X\sqrt{(s+1)}\}-\exp\{-2L\sqrt{(s+1)}\}\exp\{X\sqrt{(s+1)}\}}{1-\exp\{-2L\sqrt{(s+1)}\}}.$$

Now, for a delta function excitation, the amount of charge applied instantaneously to the cable at $X = 0$, $\bar{i}_a(0, s)$, transforms to a constant $Q_0/2\tau_m$ if the charge is applied to the centre of a cable, of length $2l$. Hence

$$\bar{i}_a = \frac{Q_0}{2\tau_m}\frac{\exp\{(L-X)\sqrt{(s+1)}\}-\exp\{-(L-X)\sqrt{(s+1)}\}}{\exp\{L\sqrt{(s+1)}\}-\exp\{-L\sqrt{(s+1)}\}}$$

$$= \frac{Q_0}{2\tau_m}\frac{\sinh\{(L-X)\sqrt{(s+1)}\}}{\sinh\{L\sqrt{(s+1)}\}}.$$

(4.5)

The transform $\bar{V}$ follows directly since $i_a = -(1/r_a)\partial V/\partial x$ and $\sinh(ay) = d\{(1/a)\cosh(ay)\}/dy$,

$$\bar{V} = \frac{r_a\lambda Q_0}{2\tau_m\sqrt{(s+1)}}\frac{\cosh\{(L-X)\sqrt{(s+1)}\}}{\sinh\{L\sqrt{(s+1)}\}}.$$

(4.6)

Using the transform (Roberts and Kaufman, p. 82, pair 15.9),

$$\frac{a}{\sqrt{s}}\cosh(v\sqrt{s})\mathrm{cosech}(a\sqrt{s}) \leftrightarrow \theta_4\left(\frac{v}{2a}\Big|\frac{T}{a^2}\right)$$

and putting $s = s+1$, $v = L-X$, and $a = L$, we obtain

$$V = \frac{r_a\lambda Q_0\exp(-T)}{2\tau_m}\frac{L}{L}\theta_4\left(\frac{L-X}{2L}\Big|\frac{T}{L^2}\right),$$

(4.7)

where θ_4 is a theta function (Roberts and Kaufman, p. xxvii). As expected from the nature of the short-cable response discussed in the previous section, this function is an infinite series of exponential terms. For computations it is useful to have eqn (4.7) in series form:

$$V = \frac{r_a\lambda Q_0\exp(-T)}{2\tau_m}\frac{1}{\sqrt{(\pi T)}}\sum_{n=-\infty}^{\infty}\exp\left\{-\left(-\frac{X}{2}+nL\right)^2\Big|T\right\}.$$

(4.7a)

The sum may be interpreted as follows. Since

$$\sum_{-\infty}^{\infty}\exp\{-(\tfrac{1}{2}X+nL)^2/T\} = \sum_{-\infty}^{\infty}\exp\left\{-\left(n-\frac{X}{2L}\right)^2\frac{L^2}{T}\right\},$$

each of the terms in the sum corresponds to a point on a Gaussian curve whose standard deviation is $\sqrt{T/L}\sqrt{2}$. (see Fig. 4.5). For most values of T and L this sum is best evaluated explicitly. However, there are two extreme cases which may be treated analytically.

Response of long cable at relatively short times

As the variance of the Gaussian decreases, the sum becomes dominated by the $n = 0$ term so that eqn (4.7a) approximates to

$$V = \frac{r_a \lambda Q_0}{2\tau_m \sqrt{(\pi T)}} \exp\left(-\frac{X^2}{4T} - T\right), \qquad (4.8)$$

which is identical with eqn (3.48) obtained previously for the response of an infinite cable. Hence, at short times a terminated cable behaves as an infinite cable. The physical reason for this result is that it takes an appreciable time for distant parts of the cable to respond (see Chapter 3; *Conduction velocity of decremental wave*), so that initially the presence or absence of a termination will not be evident. However, as the amplitude of the reflection terms become significant the responses of short and infinite cables will begin to diverge.

Response of short cable at relatively long times: steady state in short cables

As the variance of the Gaussian increases, that is $\sqrt{T/L}\sqrt{2} \gg 1$, the sum in eqn (4.7a) approaches the integral or area under the Gaussian so that we obtain

$$V = (r_a \lambda Q_0 / 2\tau_m L)\exp(-T) = (Q_0 / 2\tau_m l) \exp(-T). \qquad (4.9)$$

This is the voltage response of a simple parallel R–C circuit formed by the area of membrane in a cable of length $2l$ so that, in this case, the decline of voltage with distance may be neglected, and the axial resistance becomes negligible. For this result to apply at all times, particularly at very short times, the termination must be close. However, provided that the nature of the response at short times does not matter, a similar result will apply to cables which are surprisingly long (see below). This becomes apparent when the steady state behaviour of the open-circuited short cable is obtained. This may be done directly by integrating the cable differential equations with $\partial V/\partial T = 0$ or by setting $s = 0$ (corresponding to $\partial/\partial T = 0$) in the short-cable equations for $\bar{V}$. In either case eqn (4.10) is obtained:

$$V = V_{x=0} \frac{\cosh\{(l-x)/\lambda\}}{\cosh(l/\lambda)}, \qquad (4.10)$$

where $V_{x=0}$ is the steady state voltage at $x = 0$. This equation is also given in Weidmann (1952).

Eqn (4.10) is plotted for various values of L in Fig. 4.6. It can be seen that, provided that L is not too large, the decrement of V is very much smaller than in an infinite cable. Thus, even in a cable which is one space

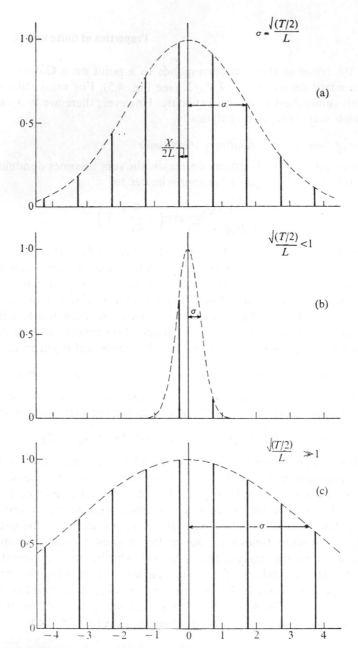

FIG. 4.5. Graphical interpretation of short-cable solution (eqn 4.7a). (a) The individual terms in the summation

$$\sum_{n=-\infty}^{n=\infty} \exp\left\{-\left(n-\frac{X}{2L}\right)^2 \frac{L^2}{T}\right\}$$

are represented by the vertical bars. The envelope of the bars (dashed) is a Gaussian curve, $\exp(-X^2/2\sigma^2)$, where $\sigma^2 = T/2L^2$. In this example, $X/L = \frac{1}{2}$. (b) The limiting case of relatively short times. The $n = 0$ term (large vertical bar) dominates the $n = +1$ term (small bar) and the other terms are negligible. (c) The limiting case of long times. Here, the variance σ is large in comparison to the spacing between the bars. The summation is well-approximated by the area under the Gaussian curve (see text).

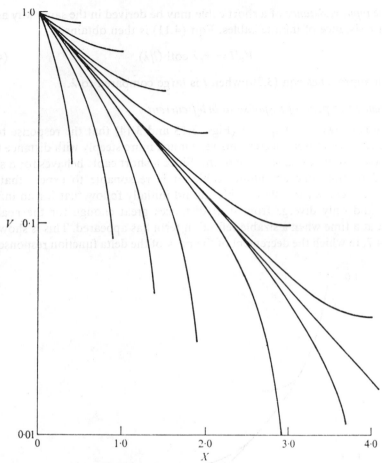

FIG. 4.6. Steady state decay of voltage with distance in terminated cables of various lengths. The linear relation (corresponding to an exponential decay since the plot is semi-logarithmic) is the infinite-cable case. The curves above this line are those for open-circuited cables. The curves below the line are those for short-circuited cables.

constant long ($L = 1$), the voltage at the end of the cable is $0.66\,V_{x=0}$, whereas in an infinite cable the voltage at this distance would be $0.37\,V_{x=0}$. This is an important result since it forms the basis of some voltage-clamp techniques in muscle fibres in which fairly uniform polarization of a fairly large area of membrane is achieved by using short-cable properties (Adrian and Freygang 1962; Deck, Kern, and Trautwein 1964; Hecht, Hutter and Lywood 1964; Rougier, Vassort, and Stämpfli 1968; Brown and Noble 1969). It should be noted, however, that approximately uniform polarization is realized at fairly long times only. Thus, in cables as short as one space constant, the voltage-clamp step will produce significantly longer-lasting initial capacitive transients than in the case of strictly uniform polarization.

The *input resistance* of a short cable may be derived in the same way as the input resistance of infinite cables. Eqn (4.11) is then obtained:

$$V_0/I = r_a\lambda \coth(l/\lambda), \tag{4.11}$$

which approaches eqn (3.26) when l is large compared to λ.

Decrement of peak of response to brief current

It was shown in Chapter 3 (Figs. 3.13 and 3.14) that the response to an infinitely brief current decrements very much more steeply with distance than does the response to a steady current. Since a short cable behaves for a short period of time like an infinite cable, it is reasonable to expect that the decrement curve for a short cable should initially follow that for an infinite cable and only diverge from it at distances great enough for the peak to occur at a time when a sizable reflection term has appeared. This is shown in Fig. 4.7, in which the decrement of the peak of the delta function response has

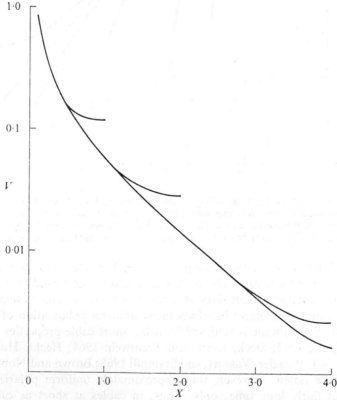

FIG. 4.7. Decrement with distance of peak of response to delta function input for various short cables. Note that decrement is much greater than for corresponding steady state curves (cf. Fig. 4.6).

been plotted for an infinite cable and for cables terminated symmetrically, with $L = 1$, 2, and 4. Comparison with Fig. 4.6 shows that the decrement of the delta function response is very much steeper than the corresponding spatial decay in response to a steady current.

Time course of response to brief current

The time courses of the voltage response to a delta-function input in a cable with the terminations one space constant from the centre, where the charge is applied, are plotted in Figs. 4.8 and 4.9. Fig. 4.8 shows the responses at various distances at early times (up to $T = 0.34$). As in the case of the response of an infinite cable there is a progressive increase in the

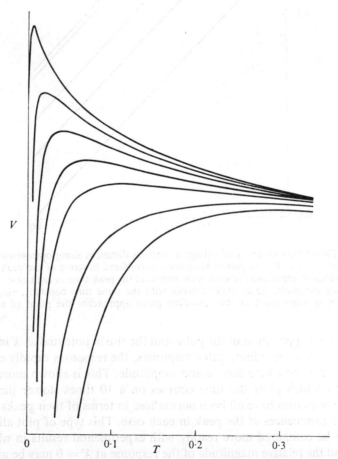

FIG. 4.8. Convergence of time course of voltages in response to delta-function input recorded at various distances along open-circuited short cable where $L = 1.0$. After $T = 0.3$ the cable becomes fairly uniformly polarized and the decay of voltage with time is then exponential (cf. Fig. 4.9).

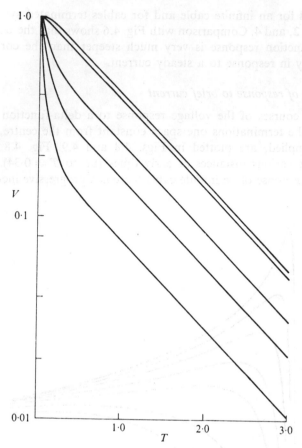

FIG. 4.9. Time course of decay of voltage at various distances along open-circuited short cable where $L = 1 \cdot 0$. The responses have been normalized in terms of the peak response to a delta function input and the time scale begins at the peak time in each case. Note that all responses eventually decay exponentially with the same time constant, τ_m, but that initial decay is more rapid as the recording point approaches the point of excitation.

delay between application of the pulse and the rise in potential as X increases. However, unlike the infinite-cable responses, the responses rapidly converge towards a common time course and amplitude. This is shown more clearly in Fig. 4.9 which plots the time courses on a 10 times slower time scale. Here, the responses have all been normalized in terms of their peaks and the time scale commences at the peak in each case. This type of plot allows the results to be compared more readily with experimental results in which the timing and the relative magnitude of the response at $X = 0$ may be unknown (see Chapter 7). It is clear that the time courses all become exponential after about $T = 0 \cdot 3$. Moreover, the time constant of the exponential decay is the membrane time constant $r_m c_m$. The reason for this is that at later times the voltage responses at various distances converge (Fig. 4.8) so that the cable becomes nearly uniformly polarized. The decay of potential then follows the equation for a uniformly polarized membrane (see Fig. 2.4). The importance of this result in relation to synaptic mechanisms will be discussed later (Chapter 7).

Response to long-lasting current pulse

In Chapter 3 (eqn (3.49)) we used a simple property of linear systems: the response to a step input may be obtained by integration of the response to a delta-function input (see Chapter 13). In a short cable, this procedure requires the integration of an infinite series of terms (eqn (4.7a)). The response therefore will be the sum of an infinite series of terms, each containing error functions, as in the infinite-cable response (eqn (3.24)). The appropriate equation may also be derived by the Laplace-transform method. For a step-current excitation eqn (4.6) becomes

$$\bar{V} = \frac{r_a \lambda I}{2s\sqrt{(s+1)}} \frac{\cosh\{(L-X)\sqrt{(s+1)}\}}{\sinh\{L\sqrt{(s+1)}\}}.$$

Here Q is replaced by I/s, the transform of a current step of magnitude I. Since

$$\frac{\cosh\{(L-X)\sqrt{(s+1)}\}}{\sinh\{L\sqrt{(s+1)}\}}$$

$$= \sum_{n=0}^{\infty} \exp\{-(2nL+X)\sqrt{(s+1)}\} + \sum_{n=1}^{\infty} \exp\{-(2nL-X)\sqrt{(s+1)}\}$$

and

$$\frac{1}{s\sqrt{(s+1)}} = \frac{1}{2}\left(\frac{1}{s+1-\sqrt{(s+1)}} - \frac{1}{s+1+\sqrt{(s+1)}}\right),$$

we obtain

$$\bar{V} = \frac{r_a \lambda I}{4}\left[\sum_{n=0}^{\infty}\frac{\exp\{-(2nL+X)\sqrt{(s+1)}\}}{s+1-\sqrt{(s+1)}} - \sum_{n=0}^{\infty}\frac{\exp\{-(2nL+X)\sqrt{(s+1)}\}}{s+1+\sqrt{(s+1)}} + \right.$$

$$\left. + \sum_{n=1}^{\infty}\frac{\exp\{-(2nL-X)\sqrt{(s+1)}\}}{s+1-\sqrt{(s+1)}} - \sum_{n=1}^{\infty}\frac{\exp\{-(2nL-X)\sqrt{(s+1)}\}}{s+1+\sqrt{(s+1)}}\right].$$

From Roberts and Kaufman (p. 248, pair 28),

$$\frac{\exp(-a\sqrt{s})}{s+b\sqrt{s}} \leftrightarrow \exp(b^2 T + ab)\,\mathrm{erfc}\left(\frac{a+2bT}{2\sqrt{T}}\right).$$

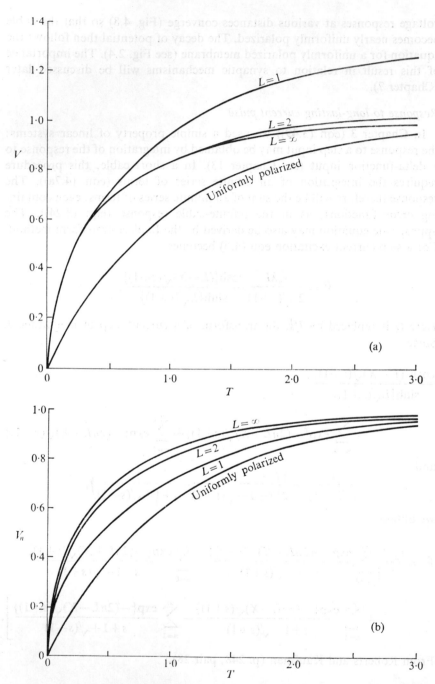

FIG. 4.10. Time courses of response to step current inputs of cables of different lengths. The response of an infinitely short (or uniformly polarized) cable is an exponential and that of an infinitely long cable is an error function (cf. Fig. 3.3). (a) shows the non-normalized responses. In (b), all responses have been normalized in terms of the steady state voltage.

Setting $a = 2nL \pm X$ and $b = \pm 1$ we obtain

$$V = \frac{r_a \lambda I}{4} \left[\sum_{n=0}^{\infty} \left\{ \exp(-2nL-X) \operatorname{erfc}\left(\frac{2nL+X-2T}{2\sqrt{T}}\right) - \right. \right.$$

$$\left. -\exp(2nL+X)\operatorname{erfc}\left(\frac{2nL+X+2T}{2\sqrt{T}}\right) \right\} +$$

$$+ \sum_{n=1}^{\infty} \left\{ \exp(-2nL+X)\operatorname{erfc}\left(\frac{2nL-X-2T}{2\sqrt{T}}\right) - \right.$$

$$\left. \left. -\exp(2nL-X)\operatorname{erfc}\left(\frac{2nL-X+2T}{2\sqrt{T}}\right) \right\} \right]. \quad (4.12)$$

As expected, the solution contains an infinite series of error-function terms. The resulting calculations are, of course, very tedious unless performed on a computer. Fig. 4.10 shows four responses to a long-lasting step current input: infinite cable, short cables with lengths $L = 1, 2$, and a uniformly polarized membrane (or extremely short cable). The responses are all calculated at $X = 0$ and are normalized in terms of the steady state voltage. It can be seen that the short-cable responses follow the infinite-cable response for a period of time which increases as the cable length increases; the short-cable responses then diverge towards the exponential response of the uniformly polarized membrane.

Response of cable terminated in short-circuit

In the case of cables terminated in a short-circuit (unhealed end or rapidly branching load) the boundary condition requires that the voltage at the termination should be zero. The solutions for this case may be obtained in a similar way and we shall simply quote the important results. The equation giving $\bar{V}$ for a delta function input is

$$\bar{V} = \frac{r_a \lambda Q_0}{2\tau_m(\sqrt{s+1})} \sinh\{(L-X)\sqrt{(s+1)}\} \operatorname{sech}\{L\sqrt{(s+1)}\}. \quad (4.13)$$

Using Roberts and Kaufman (p. 82, pair 15.1):

$$V = -\frac{r_a \lambda Q_0 \exp(-T)}{2\tau_m L} \theta_1\left(\frac{L-X}{2L} \bigg| \frac{T}{L^2}\right) \quad (4.14)$$

or, in series form,

$$V = -\frac{r_a \lambda Q_0 \exp(-T)}{2\tau_m\sqrt{(\pi T)}} \sum_{n=-\infty}^{\infty} (-1)^n \exp\left\{\frac{(-\frac{1}{2}X+nL)^2}{T}\right\}. \quad (4.14a)$$

As for the open-circuit termination, the response to long-lasting currents may be obtained by integration. The steady state responses decrement much more rapidly than in the case of an infinite cable. These responses have been plotted in Fig. 4.6 where they may be compared with the corresponding responses for open-circuit terminated cables.

Response of short cable to step voltage change

The solution to this case has been obtained, using the Laplace-transform method, by Waltman (1966, eqns 1.8 and 1.10).

Summary of applications of finite-cable theory

The use of equations for terminated cables is called for in a number of situations, some of which will be discussed in greater detail in later chapters. At this stage, we will simply summarize the most important applications.

1. *Nerve endings and short muscle fibres.* These are fairly obvious cases where finite-cable theory is required. Thus in some work on cardiac muscle (e.g. Weidmann 1952) it may not be possible to obtain a preparation which is long enough for the infinite-cable equations to be used. As we have noted already in this chapter (p. 75) some voltage-clamp techniques make use of the fact that a short preparation ending in an open-circuit is nearly uniformly polarized in the steady state, provided that the fibre length is not large and the space constant is not too short.

2. *The T system of muscle.* The steady state spread of current in the T tubular system of muscle in response to surface depolarization has been found to be too great to allow the T system to be regarded as infinite. However, in this case, the current spread is two-dimensional. It is necessary therefore to take account both of termination effects and of multi-dimensional current spread of the kind described by the equations given in the next chapter. We will discuss this problem in Chapter 6.

3. *Nerve-cell dendrites.* The question whether nerve-cell dendrites act as short cables and, if so, whether they are open-circuit or short-circuit cables has been controversial. The application of terminated-cable theory to this case will be discussed in Chapter 7.

4. Finally, the presence and nature of terminations may have large effects on current spread and current–voltage relations in nonlinear cables. These effects will be discussed in Chapter 12.

5. Current flow in multi-dimensional systems

In certain biological situations the current spread may take place in more than one spatial dimension. Thus, dendritic trees, syncytial tissues (such as cardiac and smooth muscle), and the internal membrane system of skeletal muscle (see Chapter 6) may be better described by two- or three-dimensional cable models. Such multi-dimensional systems require a more complicated mathematical treatment than is adequate for the unidimensional model. Moreover, even in cases where the current flow is for the most part unidimensional, it may still be necessary, for some purposes, to consider the three-dimensional spread of current near current sources (Eisenberg and Johnson 1970). Various different approaches have been used to deal with these problems. One method is to use a network of electrical components (George 1961; Tomita 1966) which resembles the microscopic structure of the preparation. Another approach that has been applied to dendritic systems (see Chapter 7) uses 'lumped' elements to represent mathematically the electrical properties of the dendrites. These approaches have the advantage of adaptability; with sufficient refinement, a useful model may be constructed for particular cases. However, such models often have little application in other cases and may also require considerable computing time.

An alternative and more general approach is to use a rather idealized geometry so that analytical solutions may be obtained. Thus if the branching is frequent compared to the space constant and if the membrane area increases as a simple function (e.g. as r^2 or r^3) of the distance r from a point current source, the system may be described by a continuous geometry. In the case of a preparation in which the branching is largely in one plane and in which the network is microscopically homogeneous the membrane area will increase as r^2 from a given point. Although there are a number of experimental situations which satisfy these conditions the simplest model is that of a pair of parallel sheet membranes. We will consider this case in some detail, in order to illustrate some of the properties of multi-dimensional current flow.

Current flow in infinite plane cell

The derivation of the equations for this case is analogous to that of the one-dimensional cable and the same physical laws (Ohm's and Kirchhoff's) concerning current flow are used. The model is illustrated in Fig. 5.1. Although current is allowed to flow in two dimensions, only one distance variable r is needed since radial symmetry about the point of current injection is assumed. This is, of course, a convenient simplification which may

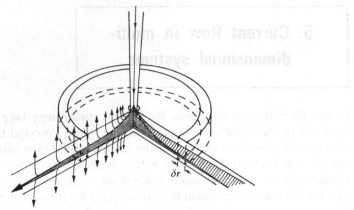

Fig. 5.1. Infinite plane cell model. Current is applied intracellularly at one point $r = 0$, and flows between and across two plane membranes which represent the properties of a microscopic network of membranes. The differential equations are obtained by considering a small circular segment of width δr. Current flow from the source decrements partly as a consequence of current flow across the membranes (shown on left-hand part of cut-away) and partly as a consequence of 'dilution' of current as it spreads out from the source (shown on right-hand part of cut-away). The second effect is absent in the one-dimensional cable (see Fig. 3.1) and is responsible for the differences between current flow in single and multi-dimensional systems.

not always apply in experimental situations. Thus Woodbury and Crill (1961) studied the two-dimensional spread of potential in the sheet-like structure formed by the atrium and found that current spreads more easily in the direction of the fibre bundles in the preparation than at right angles to the bundles. This finding is not surprising since it is likely that there are more connections between cells forming an individual bundle than between cells in neighbouring bundles. In this case the infinite plane cell equations may still be applied by assuming a simple transformation of the coordinates to produce radial symmetry. This may be done by allowing the distance scale to be a function of the angle formed with the fibre axis. Provided that there are no gross geometrical discontinuities, similar transformations would allow a variety of possible physical situations to be analysed in this way using the infinite plane cell equations.

In addition to the variables used in the unidimensional case, we require a variable for the *total* radial intracellular current at any distance from the point cource. We will use I_r for this quantity. It should be noted that I_r is not equal to the intracellular current density, which varies as I_r/r. Changes in I_r due to current crossing the cell membrane will be given by Kirchhoff's law:

$$\partial I_r = -2I_m . 2\pi r\, \partial r, \tag{5.1}$$

where ∂I_r is the change in total radial intracellular current between r and $r + \partial r$ (see Fig. 5.1). The radial current will also be related to the change in voltage according to Ohm's law:

$$\partial V = -I_r R_i \, \partial r / 2\pi r b, \tag{5.2}$$

where b is the distance between the parallel membranes of the plane cell and the expression $R_i \, \partial r / 2\pi r b$ gives the radial resistance of a cylindrical shell of intracellular fluid of radial thickness ∂r.

We may rearrange and differentiate eqn (5.2) to give

$$r\frac{\partial^2 V}{\partial r^2} + \frac{\partial V}{\partial r} = -\frac{R_i}{2\pi b}\frac{\partial I_r}{\partial r}. \tag{5.3}$$

The membrane current is given once again by

$$I_m = C_m\frac{\partial V}{\partial t} + \frac{V}{R_m}. \tag{2.2}$$

Combining eqns (5.1), (5.2), and (2.2) we obtain

$$\frac{\partial^2 V}{\partial r^2} + \frac{1}{r}\frac{\partial V}{\partial r} - \frac{2R_i}{R_m b}\left(\tau_m\frac{\partial V}{\partial t} + V\right) = 0, \tag{5.4}$$

where $\tau_m = R_m C_m$. We will replace t/τ_m by T as before. Eqn (5.4) may be transformed term by term, using the property

$$\mathrm{d}f(T)/\mathrm{d}T \leftrightarrow s\bar{f}(s)$$

to give

$$\frac{\mathrm{d}^2\bar{V}}{\mathrm{d}r^2} + \frac{1}{r}\frac{\mathrm{d}\bar{V}}{\mathrm{d}r} - \frac{2R_i(s+1)}{R_m b}\bar{V} = 0$$

or

$$-(\lambda_2)^2\frac{\mathrm{d}^2\bar{V}}{\mathrm{d}r^2} - \frac{(\lambda_2)^2}{r}\frac{\mathrm{d}\bar{V}}{\mathrm{d}r} + (s+1)\bar{V} = 0, \tag{5.5}$$

where $\lambda_2 = \sqrt{(R_m b/2R_i)}$. Putting $Z = \sqrt{(s+1)}jr/\lambda_2$ we obtain

$$\frac{\mathrm{d}^2\bar{V}}{\mathrm{d}Z^2} + \frac{1}{Z}\frac{\mathrm{d}\bar{V}}{\mathrm{d}Z} + \bar{V} = 0. \tag{5.6}$$

This differential equation gives rise to a Bessel function of order zero (see Watson 1966, Chapter 11) and the form of the solution which approaches zero as $Z \to \infty$ is

$$\bar{V} = CK_0\{R\sqrt{(s+1)}\} = CK_0\{r\sqrt{(s+1)}/\lambda_2\}, \tag{5.7}$$

where K_0 is a zero-order Bessel function (see Chapter 13) and $R = r/\lambda_2$ $(= Z/j\sqrt{(s+1)})$. This is the general solution for the infinite plane cell. C is a constant or a function of s and is determined by the particular form of excitation which is applied at $r = 0$.

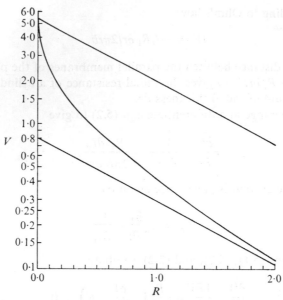

FIG. 5.2. Steady state spatial decay of voltage. The curve is a solution of eqn (5.8). At $R = 0$, the voltage is infinite. It falls very steeply up to about $R = 1$ and then begins to approach an exponential decay. The straight lines correspond to exponential decays which would be obtained in a one-dimensional cable. The upper line starts at the same voltage as the curve at a point very close to the current source. The lower line is an approximate asymptote to the curve at large values of R.

Distribution of voltage and current in steady state

In the steady state, $\partial/\partial t = 0$ so that $s = 0$ and eqn (5.7) becomes

$$V = CK_0(r/\lambda_2). \tag{5.8}$$

This equation has been plotted in Fig. 5.2, where it is compared with the corresponding equation (eqn (3.25)) for the infinite one-dimensional cable. There are several important differences.

1. When $r \to 0$, that is, in the immediate vicinity of the current electrode, eqn (5.8) simplifies to

$$V = C \ln(2\lambda_2/r). \tag{5.9}$$

Hence, V is a very steep function of r and becomes infinite when $r = 0$. This result arises from the fact that it is assumed that the current originates from a perpendicular line source (that is, at one point in the r-plane), so that at $r = 0$ the current density is infinite. In any real situation, of course, this will not be the case since the current must be injected from a finite source. However, provided that the electrode is very small compared to the space constant, eqn (5.9) will be fairly accurate for all but very small values of r. At very small values of r, the assumption that a perpendicular line source is used will be violated and the three-dimensional spread of current around a point source must then be considered (Eisenberg and Johnson 1970). A further complication is that the assumption that the geometry of the preparation is continuous will also be invalid for very small values of r comparable to the dimensions of the elements of the network (see Chapter 7, pp. 157–60). All three problems make it difficult to draw a direct comparison with the one-dimensional equations. In Fig. 5.2 we have matched one of the exponentials (corresponding to the one-dimensional case) to the voltage given by eqn (5.8) at a very small distance away from the current source. This procedure is, of course, arbitrary, but it is useful for showing how the voltage decrements very much more steeply with distance in the infinite plane cell case than it does in the one-dimensional case.

2. When r is large, eqn (5.8) approximates to

$$V = C \sqrt{(\pi \lambda_2 / 2r)} \exp(-r/\lambda_2).$$ (5.10)

This result is obtained by neglecting all but the first term in the asymptotic expansion of K_0 given by Watson (1966, p. 202). At large enough values of r the exponential term in eqn (5.10) varies more steeply than $1/\sqrt{r}$ so that, when plotted semilogarithmically, the slope of the curve approaches that of the corresponding one-dimensional decay. To illustrate this point in Fig. 5.2 we have included a second exponential which forms an approximate 'asymptote' to eqn (5.10). It should be noted, however, that this curve is not a strict asymptote since the voltage given by eqn (5.10) always decrements faster than an exponential. Moreover, this approximation is valid only at values of r at which V is already very small. Thus, at $r = 2\lambda_2$, V is only 2 per cent of its value at $r = 0\cdot01\lambda_2$. Most experimental records are likely to be obtained at values of r at which the differences between K_0 and a simple exponential function are large.

3. The steady state distribution of current is also non-exponential. Using methods analogous to those used in previous chapters, we obtain eqn (5.11) for I_r as a function of r:

$$I_r = -(rI_0/\lambda_2) \, K_1 \, (r/\lambda_2),$$ (5.11)

where I_0 is the total current applied at $r = 0$ and K_1 is a first-order Bessel function (see Chapter 13). The intracellular current density may be obtained by dividing eqn (5.11) by $2\pi rb$.

Transient response to current step at r = 0

To obtain the transient response to a current step we must first obtain C in eqn (5.7). From eqn (5.2)

$$\partial V / \partial r = -I_r \, R_i / 2\pi rb.$$

Transforming to the s-domain this becomes

$$d\bar{V}/dr = -\bar{I}_r \, R_i / 2\pi rb.$$ (5.12)

Now I is a step function of amplitude I_0. Hence

$$\bar{I}_r(r=0) = I_0/s$$

From eqns (5.7) and (5.12), and noting that

$$dK_0(kx)/dx = -kK_1(kx),$$

we may evaluate C at the line source of current, i.e. by letting $r \to 0$, thus

$$C = \frac{R_i I_0}{2\pi b s} \frac{\lambda_2}{\sqrt{(s+1)}} \lim_{r \to 0} \left[\frac{1}{rK_1\{r\sqrt{(s+1)}/\lambda_2\}} \right].$$

Since $K_1(x) \to 1/x$ as $x \to 0$, the expression in square brackets tends to $\sqrt{(s+1)}/\lambda_2$ as $r \to 0$, hence

$$C = I_0 R_i / 2\pi b s$$

and

$$\bar{V} = \frac{I_0 R_i}{2\pi b s} K_0 \left\{ \frac{r}{\lambda_2} \sqrt{(s+1)} \right\}. \tag{5.13}$$

Using the transform (Roberts and Kaufman, p. 304, pair 13.1.1)

$$K_0(a\sqrt{s}) \leftrightarrow (1/2t)\exp(-a^2/4t)$$

and the properties

$$\bar{f}(s+1) \leftrightarrow \exp(-t)f(t);$$

$$\bar{f}(s)/s \leftrightarrow \int_0^t f(t')\,dt',$$

we obtain

$$V = \frac{R_i I_0}{2\pi b} \int_0^T \frac{1}{2t'} \exp\left\{ -\frac{(r/\lambda_2)^2}{4t'} - t' \right\} dt'. \tag{5.14}$$

Eqn (5.14) may also be expressed in terms of a gamma function which may be more useful for some purposes. Letting $R_i I_0/4\pi b = A$ and $R^2/4t' = y$ we obtain

$$V = A \int_0^t \frac{1}{t'} \exp(-y-t')\,dt'.$$

Since $dy = -(R^2/4t'^2)\,dt'$, $dt' = -(R^2/4y^2)\,dy$ and therefore

$$V = A \int_\infty^{(R^2/4t)} \frac{4y}{R^2} \exp(-y-R^2/4y)\frac{-R^2}{4y^2}\,dy.$$

Replacing $\exp(-R^2/4y)$ by its series expansion and manipulating the resulting expression we obtain

$$V = A \sum_{n=0}^\infty \frac{(-1)^n}{n!} \left(\frac{R^2}{4} \right)^n \int_{(R^2/4t)}^\infty \left(\frac{1}{y} \right)^{n+1} \exp(-y)\,dy.$$

The integral in this equation is the incomplete gamma function. Hence

$$V = \frac{R_i I_0}{4\pi b} \sum_{n=0}^{\infty} \frac{(-1)^n}{n!}\left(\frac{R^2}{4}\right)^n \Gamma\left(-n; \frac{R^2}{4t}\right), \tag{5.14a}$$

where $R^2/4t > 0$.

This equation is best evaluated numerically. This has been done and the results plotted for various values of r in Fig. 5.3. It can be seen that, as in the one-dimensional case (see Fig. 3.4), the response becomes progressively more delayed as r increases. However, there are some important differences which are characteristic of responses in systems with multi-dimensional current spread.

1. At small values of r, V rises very rapidly indeed towards its steady state value. Moreover, the time taken to rise to a given fraction of the steady state value is very dependent on r. Hence, the time course of the response will give a very unreliable indication of the membrane time constant. In fact, the true membrane time constant may be as much as one or two orders of magnitude greater than the 'apparent time constant' obtained from the best exponential fit to the response.

2. As we have already noted above, the steady state value becomes very small as r increases. Most experimental records are therefore likely to be obtained in the region where the voltage changes are fast compared to the membrane time constant.

From eqns (5.12) and (5.13) we may also obtain an equation for the current flow

$$I_r = \frac{I_0 r \sqrt{(s+1)}}{s\lambda_2} K_1\left\{\frac{r}{\lambda_2}\sqrt{(s+1)}\right\}. \tag{5.15}$$

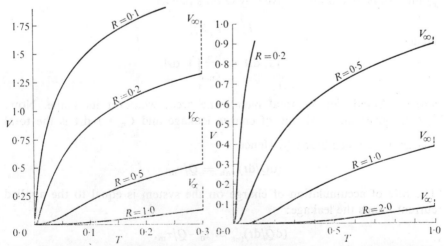

FIG. 5.3. Responses to maintained step change in current. (a) Responses at $R = 0.1, 0.2$, $0.5, 1.0$. (b) Responses at $R = 0.2, 0.5, 1.0$, and 2.0 plotted on more compressed time scale. The interrupted lines at the end of each curve indicate the final steady state values.

Using the transform (Roberts and Kaufman, p. 304, pair 13.2.2)

$$\sqrt{s}\, K_1(a\sqrt{s}) \leftrightarrow \{a \exp(-a^2/4t)\}/4t^2,$$

we obtain

$$I_r = I_0 \int_0^t \frac{(r/\lambda_2)^2}{4t'^2} \exp\{-t' -(r/\lambda_2)^2/4t'\}\, \mathrm{d}t', \tag{5.16}$$

and the current density is given by dividing eqn (5.16) by $2\pi r b$.

Time course of total membrane charge in multi-dimensional systems

Integration of eqn (5.4) gives

$$\lambda^2 \int_0^\infty \left(r \frac{\partial^2 V}{\partial r^2} + \frac{\partial V}{\partial r} \right) \mathrm{d}r - \tau_m \int_0^\infty r \frac{\partial V}{\partial t} \, \mathrm{d}r - \int_0^\infty rV \, \mathrm{d}r = 0.$$

Since $Q = C_m \int_0^\infty rV \, \mathrm{d}r$ we may obtain

$$\tau_m I_0 - \tau_m (\mathrm{d}Q/\mathrm{d}t) - Q = 0.$$

This equation is identical with eqn (3.38) obtained previously for the one-dimensional cable, and the same exponential solutions (eqns (3.39) and (3.40)) apply. This result may appear surprising since it shows that the time course of total charge on the membrane is not affected by the large difference in the extent of spatial spread of current. In fact, this result applies quite generally to any geometrical system provided that the membrane properties are linear and uniform. Thus, for any geometry, the leakage of charge at each point on the membrane is given by Ohm's law as

$$I_i = V/R_m.$$

Hence

$$\int_A I_i \, \mathrm{d}A = \frac{C_m}{\tau_m} \int_A V \, \mathrm{d}A,$$

where A stands for the total membrane area, whatever its shape. Now, $\int_A I_i \, \mathrm{d}A$ gives the total rate of charge leakage and $C_m \int_A V \, \mathrm{d}A$ is the total charge on the membrane Q. Hence

$$(\mathrm{d}Q/\mathrm{d}t)_{\text{leak}} = Q/\tau_m.$$

The rate of accumulation of charge on the system is equal to the applied current minus the leakage:

$$(\mathrm{d}Q/\mathrm{d}t)_{\text{tot}} = I_0 - Q/\tau_m,$$

which is also identical with eqn (3.38). Hence the equations for the total charge are independent of the geometry of the system.

Input resistance

Since the voltage at $r = 0$ is infinite (Fig. 5.2), the input resistance is also, strictly speaking, infinite. However, experimental measurements are made using voltage and current electrodes of finite dimensions at a finite distance apart. Hence, we may obtain an expression for the effective 'input resistance' as a function of the electrode separation. In the steady state

$$\frac{V}{I_0} = \frac{R_i}{2\pi b} K_0 \left(\frac{r}{\lambda_2} \right).$$ (5.17)

For small values of r this becomes

$$\frac{V}{I_0} = \frac{R_i}{2\pi b} \ln \frac{2\lambda_2}{r}.$$ (5.18)

Thus, when r is small (as is usually the case when intracellular microelectrodes are used), the input resistance is extremely insensitive to changes in membrane resistance. For example, when $r = 0 \cdot 1\lambda_2$, a four-fold change in R_m (corresponding to a two-fold change in λ) changes the input resistance by only 30 per cent. This means that it is nearly impossible to determine membrane resistance changes from changes in input resistance in multidimensional systems (see also p. 96). This conclusion also applies in multi-dimensional nonlinear systems (see Noble 1962b).

Response to short current pulse

The effects of multi-dimensional current spread on synaptic processes may be evaluated by considering the response to a short current pulse. As in the one-dimensional case (see p. 46) we may represent the current pulses by two maintained pulses of opposite sign, one delayed with respect to the other by an interval ΔT. We then obtain

$$V = \frac{R_i I_0}{2\pi b} \int_{T-\Delta T}^{T} f(t')\, \mathrm{d}t'$$ (5.19)

for $t' \geq \Delta T$, where

$$f(t') = \frac{1}{2t'} \exp\left\{ -\frac{(r/\lambda_2)^2}{4t'} - t' \right\}.$$

This response is therefore obtained by integrating the function $f(t')$ over the appropriate interval. Of particular interest is the response to an extremely short pulse or delta function. Then $\Delta T \to 0$ or $I_0 \Delta T \to Q_0$ and

$$V = \frac{R_i Q_0}{2\pi b} f(t)$$ (5.20)

that is, the response is proportional to $f(t)$ itself. Eqn (5.20) has been plotted for various values of r in Fig. 5.4. One of the most striking features of these

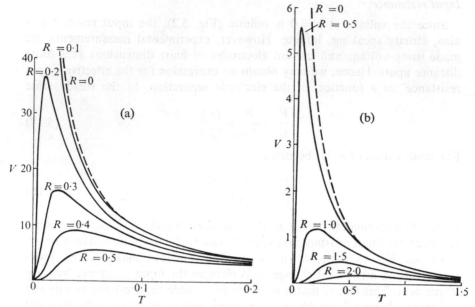

FIG. 5.4. Responses to delta function input. (a) Responses at $R = 0$ (interrupted line), 0·1, 0·2, 0·3, 0·4, and 0·5. (b) Responses at $R = 0$ (interrupted line), 0·5, 1·0, 1·5, 2·0 plotted on more compressed time scale. On this voltage scale, the peak response at $R = 0·1$ is about 150. The peak response at $R = 0$ is, of course, infinite.

curves is that the decay of voltage is extremely rapid. Thus, at $r = 0·2\lambda_2$, the 'time constant' which would be obtained by measuring the time taken to decay to e^{-1} of the peak value is only about $0·05\tau_m$.

Current flow in three-dimensional systems

We close this chapter by considering the spread of current in a three-dimensional cable structure. Extending the comparison between one- and two-dimensional systems, one might expect that the divergence of current will play an even more dominant role in generating steep spatial decrements with distance from a point source of current. This will be shown mathematically by means of equations that are quite similar to the previous treatment. We will therefore present a more concise account of the three-dimensional results.

It may be useful to explain just what is meant by a three-dimensional structure in the present context. In any multi-dimensional structure, the critical feature is the relation between membrane area and the spatial coordinates. Thus, in the one-dimensional cable, increments of membrane area are directly proportional to incremental distance along the fibre axis; i.e.

$$\delta A \propto \delta x. \tag{5.21}$$

In the two-dimensional case, and in particular the infinite plane cell geometry that has been discussed in this chapter,

$$\delta A \propto r \, \delta r. \tag{5.22}$$

We will now consider current spread in a structure where

$$\delta A \propto r^2 \, \delta r. \tag{5.23}$$

This mathematical condition may serve as a definition of a 'three-dimensional' system; clearly, this is a quite specific description which implies more than simply that intracellular current spreads in three dimensions. The condition (5.23) would hold if the membrane area in a thin spherical shell (say, between r and $r+\delta r$) scaled according to the volume of the shell. This assumption might be useful to describe the spread of current from a localized source within an electrically syncytial tissue. If the individual cells are small and relatively uniform in size, the total membrane area per unit volume will be approximately constant.

Examples of three-dimensional cable structures may be found in cardiac tissue or smooth muscle, where there are low-resistance pathways for current flow between adjacent cells. It is also possible that the relationship between membrane area and distance will fall somewhere between the particular cases described above.

The idealized three-dimensional network is relatively difficult to describe pictorially, since the extracellular and intracellular regions must share the over-all volume on a relatively microscopic scale. One might imagine a loose mesh structure, with an easily contiguous extracellular space. For the sake of simplicity, we will ignore the presence of potential gradients in the extracellular volume; the resistivity of the extracellular conductor can be incorporated easily at a later stage, as in Chapter 3 (p. 27). In the three-dimensional cable, we may replace the specific intracellular resistivity R_i by the expression

$$R_i'/f + R_0/(1-f), \tag{5.24}$$

where R_i' and R_0 are the resistivities per unit volume of intracellular and extracellular space respectively, and f is the fraction of tissue volume which is intracellular.

It is convenient to consider only the voltage decrements in the radial direction (that is, to assume spherical symmetry). This simplifies the mathematics, since it leaves only the distance from the current source, r as a spatial variable.

With these simplifying assumptions, we may proceed to derive the differential equation for the three-dimensional cable. By analogy to eqn (5.1), the increment in the total radial intracellular current will be produced by current escaping across the membrane (Kirchhoff's current law). For a spherical shell between r and $r+dr$,

$$dI_r = -I_m (4\pi r^2 \, dr)\chi, \tag{5.25}$$

where I_m is the membrane current per unit area, and χ is the surface-to-volume ratio. The radial current I_r produces an ohmic radial voltage decrement,

$$dV = -I_r(R_i\, dr/4\pi r^2),\tag{5.26}$$

where $(R_i\, dr/4\pi r^2)$ is simply the resistance of a spherical shell to radial current.

We proceed by rearranging eqn (5.26) and differentiating,

$$\frac{\partial}{\partial r}\left(r^2\frac{\partial V}{\partial r}\right) = -\frac{R_i}{4\pi}\frac{\partial I_r}{\partial r}.\tag{5.27}$$

Solving eqn (5.25) for $\partial I_r/\partial r$ and substituting in eqn (5.27) gives

$$\frac{\partial}{\partial r}\left(r^2\frac{\partial V}{\partial r}\right) = R_i r^2 I_m \chi$$

or

$$\frac{1}{r^2}\frac{\partial}{\partial r}\left(r^2\frac{\partial V}{\partial r}\right) - R_i\chi I_m = 0$$

and, making the customary assumption about the membrane current (eqn (2.2)),

$$\frac{1}{r^2}\frac{\partial}{\partial r}\left(r^2\frac{\partial V}{\partial r}\right) - R_i\chi\left(C_m\frac{\partial V}{\partial t} + \frac{V}{R_m}\right) = 0.\tag{5.28}$$

This differential equation is analogous to eqn (3.8) for the unidimensional case and eqn (5.4) for the infinite plane cell. Letting $R_m/\chi R_i = \lambda^2$ and $R_m C_m = \tau_m$, this equation may be rewritten

$$\lambda^2\frac{1}{r^2}\frac{\partial}{\partial r}\left(r^2\frac{\partial V}{\partial r}\right) - \tau_m\frac{\partial V}{\partial t} - V = 0$$

or

$$\frac{\partial^2 V}{\partial r^2} + \frac{2}{r}\frac{\partial V}{\partial r} - R_i\chi\left(C_m\frac{\partial V}{\partial t} + \frac{V}{R_m}\right) = 0.$$

This differential equation is analogous to eqn (3.8) for the one-dimensional case and eqn (5.4) for the infinite plane cell. Letting $R_m/\chi R_i = \lambda^2$, and $R_m C_m = \tau_m$, the equation may be rewritten

$$\lambda^2\left(\frac{\partial^2 V}{\partial r^2} + \frac{2}{r}\frac{\partial V}{\partial r}\right) - \tau_m\frac{\partial V}{\partial t} - V = 0$$

or

$$\frac{\partial^2 V}{\partial R^2} + \frac{2}{R}\frac{\partial V}{\partial R} - \frac{\partial V}{\partial T} - V = 0,\tag{5.29}$$

where $R = r/\lambda$ and $T = t/\tau_m$.

In the Laplace transform domain, eqn (5.29) becomes

$$\frac{d^2\bar{V}}{dR^2} + \frac{2}{R}\frac{d\bar{V}}{dR} - (s+1)\bar{V} = 0,\tag{5.30}$$

which may be solved by making the substitution $\bar{U} = \bar{V}R$, to give

$$\frac{d^2\bar{U}}{dR^2} - (s+1)\bar{U} = 0. \tag{5.31}$$

The general solution is

$$\bar{U} = \bar{A}\exp\{-R\sqrt{(s+1)}\} + \bar{B}\exp\{R\sqrt{(s+1)}\},$$

so that

$$\bar{V} = \frac{1}{R}[\bar{A}\exp\{-R\sqrt{(s+1)}\} + \bar{B}\exp\{R\sqrt{(s+1)}\}]. \tag{5.32}$$

Since $\bar{V} \to 0$ as $R \to \infty$, the solution becomes

$$\bar{V} = \frac{\bar{A}}{R}\exp\{-R\sqrt{(s+1)}\}. \tag{5.33}$$

The value of $\bar{A}$ may be determined by using eqn (5.26) and the boundary condition that $\bar{I}(R) = \bar{I}$ (the transform of the applied current) when $R = 0$, giving

$$\bar{V} = \frac{\bar{I}R_i}{4\pi\lambda R}\exp\{-R\sqrt{(s+1)}\}. \tag{5.34}$$

For a delta-function excitation, $\bar{I} = Q_0/\tau_m$ and eqn (5.34) can be inverted by use of a transform pair given in Roberts and Kaufman (p. 246, pair 3.2.14) and the following relationship from Table 13.1(a) (p. 449):

$$e^{-at}f(t) \longleftrightarrow \bar{F}(s+a).$$

The result is

$$V(R, T) = \frac{Q_0 R_i}{8\lambda\tau_m(\pi T)^{\frac{3}{2}}}\exp(-T - R^2/4T). \tag{5.35}$$

In a similar manner the response to a step of current (where $\bar{I} = I_0/s$) may be obtained by using the same inversion relation that was employed in obtaining the response of a one-dimensional cable to a step of voltage (Chapter 3, p. 39)

$$V_{step}(R, T) = \frac{I_0 R_i}{8\pi\lambda R}\left\{e^{-R}\,\mathrm{erfc}\left(\frac{R}{2\sqrt{T}} - \sqrt{T}\right) + e^R\,\mathrm{erfc}\left(\frac{R}{2\sqrt{T}} + \sqrt{T}\right)\right\}. \tag{5.36}$$

This expression also bears a considerable similarity to the response to a step of current in the one-dimensional cable (eqn (3.24)); the major differences being the presence of R in the denominator and the fact that the solution is the sum of terms rather than the difference.

The steady state voltage distribution follows directly from eqn (5.36)

$$V(T = \infty) = \frac{I_0 R_i}{4\pi\lambda} \cdot \frac{e^{-R}}{R}. \tag{5.37}$$

As in the two-dimensional sheet (eqn (5.8)) there is a voltage singularity at the origin, resulting from the infinite current density at the (point) current source. This singularity does not occur in the one-dimensional cable because an idealized plane source is assumed in that treatment (cf. Eisenberg and Johnson 1970). There is the same difficulty, therefore, for the three-dimensional as well as the two-dimensional case in giving a realistic expression for the input resistance. It is at short distances that the assumptions made in formulating the differential equations are most likely to lead to errors (see Chapter 7, pp. 157–160). It is possible to show that for a small separation between current passing and recording electrodes the 'input resistance' is virtually insensitive to R_m. Thus

$$\frac{V}{I_0} = \frac{R_i}{4\pi r}e^{-r/\lambda} = \frac{R_i}{4\pi r}\left(1 - \frac{r\sqrt{(R_i\lambda)}}{\sqrt{R_m}} + \cdots\right)$$

For example, if $r = 0.05$, a four-fold increase in R_m increases the input resistance by 2·5 per cent and a four-fold decrease in R_m leads to a decrease in input resistance of less than 5 per cent. (The corresponding figures for the two-dimensional case are both about 20 per cent).

Eqn (5.37) may be compared with the steady state voltage distribution in the unidimensional cable

$$V \propto e^{-X}$$

and the two-dimensional sheet where R is not small,

$$V \propto e^{-R}/R^{\frac{1}{2}}.$$

It is evident that the three-dimensional voltage distribution forms an extension of this progression, where voltage declines even more steeply as function of radial distance from the source of current.

Summary of applications of multi-dimensional theory

1. *Spatial spread of current in cardiac muscle.* As noted at the beginning of this chapter, several studies on spatial spread in cardiac and smooth muscle have involved the use of lumped component networks which are designed to represent the electrical properties of particular tissues, whereas other studies have used the alternative approach of assuming a continuous geometry. Woodbury and Crill (1961) were the first to use the infinite plane cell equations to describe the spatial decay of voltage in atrial muscle polarized intracellularly at one point. As expected, they found that the decay is steeper than exponential and is better fitted by a Bessel function. Although they used the two-dimensional equations, it is possible that a three-dimensional model might be more correct. Noble (1962b) has used the two-dimensional equations to study the input resistance and current–voltage relations which may be expected in microelectrode experiments on cardiac muscle. As expected from eqn (5.18), the input resistance is insensitive to changes in membrane resistance and the current–voltage relations are insensitive to nonlinearities in the membrane current–voltage relation.

2. *Junctional potentials in syncytial tissues.* There are very few studies of junctional mechanisms in multidimensional preparations. One of the most interesting examples was provided by Tomita (1967) who measured the temporal and spatial decrement of junctional potentials in smooth muscle. The preparation allowed him to study the effects of nearly uniform synaptic current (produced by whole nerve stimulation), one-dimensional current spread (produced by stimulating with large external electrodes), and multidimensional spread occurring during spontaneous junctional potentials. As expected, the spontaneous junctional potential was found to decay very rapidly indeed by comparison with the other two responses.

3. The T tubular system in skeletal muscle forms a three-dimensional network. Whether two- or three-dimensional equations are required depends on the experimental circumstances. When the muscle surface is depolarized by an action potential or an intracellular electrode the T system is polarized circumferentially and the inward spread of current then requires two-dimensional equations. Unlike the case considered in this chapter, the current is not injected at one point and spreads to regions of *decreasing* membrane area rather than to regions of increasing membrane area. The same differential equations are used, but the different boundary conditions give rise to different solutions which will be discussed in Chapter 6. When the T system is polarized at one point (e.g. the focal stimulation experiments of Huxley and Taylor (1958)) the current will spread out in two or three dimensions. It should then be quite difficult to excite the whole T system, which corresponds well with Huxley and Taylor's observation that only the surface myofibrils were made to contract and that the spread of contraction along the fibrils was limited to the regions served by the T tubule being polarized (see Chapter 6).

Finally, it may be important to note that the applications of three- or two-dimensional equations discussed here differ from some of the theory and applications discussed by Eisenberg and Johnson (1970). We have restricted the discussion to cases where multi-dimensional current spread arises because the membrane being polarized does not form a simple cylindrical cable. Eisenberg and Johnson also consider cases where equations for multi-dimensional spread of current are required, even in simple cable-like structures, because the assumptions made in one-dimensional cable theory are inaccurate. As noted in Chapter 1, we have omitted the theory of these cases, and the reader is referred to Eisenberg and Johnson's article for a detailed discussion of the problems which arise.

6. Spread of excitation in muscle

THE electrical events in nerve and muscle fibres have different functions. Nerves act as transmission lines conveying information between receptors, spinal cord, brain, muscles, and glands. In muscles, the electrical events trigger the mechanical response. Early theories of this triggering process assumed that the electrical events in muscle are very similar to those in nerve. Activation of contraction was thought to be due either to the local circuit current flow which accompanies the action potential as it propagates along the fibre (see Chapter 10) or to the release of an activating substance at the surface membrane. In either case, no electrical events additional to the action potential itself were required. The former theory (the 'window-field' theory) was shown to be incorrect since electric current flow along the myoplasm does not activate contraction (Sten-Knudsen 1954), whereas uniform surface membrane depolarization without local circuit current flow is adequate to cause activation in normal muscles (see e.g. Hodgkin and Horowicz 1960). The second theory was abandoned in its simplest form when A. V. Hill (1949) showed that the onset of the active state of contraction in fast skeletal muscles is too rapid to be accounted for by diffusion of an activating substance (e.g. one of the ions crossing the membrane during the action potential) to the centre of the muscle fibres from the surface.

Nevertheless, until recently, work on the electrical properties of muscle developed in close conjunction with, and largely relied on, similar work on nerve. This is hardly surprising in view of the obvious similarities between the electrical properties of nerve and skeletal muscle fibres. Subthreshold excitations (produced by chemically transmitted conductance changes or by application of current pulses) produce responses that obey the equations of one-dimensional cable theory to a reasonably close approximation. Suprathreshold stimuli, on the other hand, trigger all-or-nothing action potentials whose propagation depends on a common mechanism: local circuit currents flowing between excited and resting regions of the cell (see Chapter 10). Finally, the ionic currents underlying the action potential, as revealed by the voltage-clamp technique (Adrian, Chandler, and Hodgkin 1970; Ildefonse and Rougier 1972), are controlled by the membrane potential in much the same way as in nerve (see Chapter 8).

Clearly, similar mechanisms have developed to initiate and safeguard the spread of excitation in both nerve and muscle fibres. However, it is now known that depolarization of the surface membrane of muscle fibres during the action potential is only one step in a sequence of electrical events, eventually leading to the rise in intracellular calcium concentration which is responsible for activating the contractile machinery (see Ebashi and Endo (1968) and

Ebashi, Endo and Ohtsuki (1969) for reviews of the role of calcium ions in excitation–contraction coupling). This sequence of events involves inward spread of excitation along internal membranes. However, the precise mechanism by which this inward spread leads to a release of calcium ions is still incompletely understood.

One of the main difficulties in analysing the current flow in the internal membrane systems of muscle arises from the complex morphology of these systems. We may distinguish several membranous structures within the fibre interior. First, at regular intervals, both longitudinally and radially, the surface membrane invaginates to form transverse tubules (the T system) which run radially across the muscle fibre. These tubes run between the myofibrils, which consist of regular arrays of the myofilaments forming the contractile machinery. As they do so, they make close contact with the lateral cisternae of the sarcoplasmic reticulum which run longitudinally along the surface of each myofibril. At the junctions with the T tubules the lateral cisternae enlarge to form the terminal cisternae. These structures are closely apposed to the T tubules, and since there is one on each side of the T tubule, the microscopic appearance is that of three adjacent spaces, known as a triad.

A full analysis of the spread of excitation in muscle would therefore be difficult since it involves treating a three-dimensional nonlinear system.

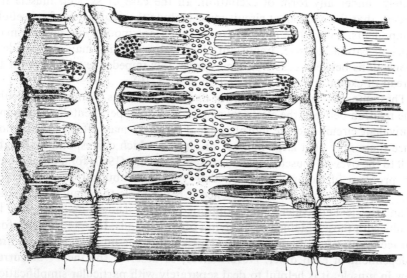

FIG. 6.1. Drawing depicting the internal membrane structures of skeletal muscle (frog sartorius). The surface membrane (sarcolemma) invaginates at the middle of each I band to form a tubule which runs transversely across the fibre between the myofibrils. The terminal cisternae of the sarcoplasmic reticulum make close contact with the T tubules to form a 'triad'. Although the T tubules are small (the diameter is of the order of 10^{-7} m), the total area of T tubule membrane in a fibre of typical size is about 5 times the surface membrane area. (From Peachey 1965.)

Further difficulties arise from our incomplete knowledge of the electrical properties of the subcellular components of muscle. It is not clear, for example, what fraction of the total sodium conductance lies in the membranes of the transverse tubular system, to assist the radial spread of excitation into the fibre centre (Costantin 1970; Peachey and Adrian 1973). Nor do we know the full nature of the relationship between the T tubules and the adjoining terminal cisternae.

In the face of this complexity, it is necessary to make various approximations which emphasize different aspects of the electrical properties of the muscle fibre. This may be achieved experimentally in a number of ways, such as restricting the pattern of current flow, varying the geometry, amplitude, or time course of the current source, or by specific morphological alterations. The most dramatic example of the latter approach is a technique which isolates the transverse tubular system from the surface membrane using osmotic shock (Fujino, Yamaguchi, and Suzuki 1961; Howell and Jenden 1967). After this treatment the sarcolemmal membrane properties can be studied largely in isolation from those of the transverse tubular system (Gage and Eisenberg 1969a,b; Eisenberg and Gage 1969).

One of the aims of these various approaches is to obtain an electrical equivalent circuit for the morphological elements of the muscle. This circuit may then be used to determine the pattern of current flow, at least approximately, under any form of excitation. In the case of skeletal muscle it is evident that the equivalent circuit must include the properties of subcellular components. For other muscle tissues, such as smooth muscle and many types of cardiac muscle, important electrical properties may also reside in connections between cells which may form low-impedance pathways for the spread of excitatory current (see Chapter 2, p. 10).

These morphological complexities underlie a serious theoretical difficulty. The relation between the electrical elements of an experimentally determined equivalent circuit and the morphological elements may often be ambiguous since there may be a large number of circuits which correspond reasonably well to the over-all muscle impedance (see Eisenberg (1967) for a discussion of the problem of canonical forms; also Van Valkenburg 1964; Falk 1968; Freygang and Trautwein 1970). Hence it is desirable to obtain confirmation of an electrical equivalent circuit obtained by one method by comparing the results with those obtained by applying other methods. In practice, therefore, the techniques for obtaining equivalent circuits for muscle are not entirely independent of each other. Nevertheless, to describe the theory of current flow in muscle, it is helpful to deal separately with particular simplifications of the general description. In the case of skeletal muscle it is convenient to make these simplifications correspond to the sequence of events involved in initiating contraction:

(1) the effects of subthreshold excitation;
(2) the propagation of the action potential;

(3) the radial transmission of depolarization from the surface membrane to the interior of the cell.

Since no comprehensive review of the development of electrical equivalents for muscle exists, we shall include a fairly large number of references to recent work. A full coverage of this area is not intended, however, and for a more complete treatment of the impedance theory and further references the reader is referred to Falk and Fatt (1964). An excellent outline of progress in elucidating the mechanism of electrical–mechanical coupling is included in Peachey's review (1968) and also in that of Ebashi, Endo, and Ohtsuki (1970). The electrical properties of the transverse tubular system are treated mathematically in a recent article by Peachey and Adrian (1973), which includes numerical results for a model T system with nonlinear membrane characteristics.

Part I : Subthreshold excitation

Step current analysis

A considerable simplification is achieved by considering the effects of current stimuli that are small enough to allow the use of linear-cable theory. If the effect of the three-dimensional spread of current near the source of current is also neglected (see Eisenberg and Johnson (1970) for a detailed discussion and criticism of the validity of this approximation), the response of the fibre can be treated analytically following the theory developed in Chapter 3.

The simple-cable model (Fig. 3.1) assumes that current crosses the surface of the cylindrical cable by either of two pathways: a simple resistive (i.e. ionic) path or a simple capacitive path. Although this assumption is justified in the case of nerve fibres, it is not valid for muscle where the pathway cannot be treated accurately as a parallel R–C circuit (see Chapter 2, p. 18). Nevertheless, the simple-cable model has been usefully applied in several types of muscle. For example, Fatt and Katz (1951) found that the response of frog skeletal muscle fibres to step current excitation may be fitted reasonably well by the simple-cable equations. However, the value of C_m (about 6 μF cm^{-2}) was considerably larger than that expected for a layer of plasma membrane, which in nerve has a capacitance of only 1 μF cm^{-2}. It is now clear that this difference is largely due to an underestimate of the total membrane area in a unit length of muscle fibre. Electron-microscope studies (Peachey 1965) show that in a typical frog fibre the surface area of the transverse tubular membranes (T system) is roughly 4 or 5 times as large as the area of the sarcolemma, which is generally used for the calculation of specific capacitance (cf. also Hodgkin and Nakajima 1972a).

Large values of capacitance have also been obtained from step current analysis in cardiac Purkinje fibres, which have a capacitance referred to the surface area of a strand which is about 10 μF cm^{-2} (Weidman 1952; Fozzard

1966). The explanation in this case cannot be based on intracellular membrane systems, which are largely absent (Sommer and Johnson 1968), but may be accounted for by including the surface membranes of all the cells in the strands (Mobley and Page 1972). If the cells within the strand are well connected electrically (Weidmann 1966), and if the resistance of the interstitial space is low, 'deep' membranes will also contribute to the total bundle capacitance. Since there are about 5–10 cells in the cross-section of the sheep Purkinje fibre bundles used in cable analysis, the calculated capacitance per unit cell membrane area may be much closer to the expected range of 1–2 μF cm^{-2}. Another factor which may contribute to the large apparent capacitance is suggested by electron microscope studies (Page *et al.* 1970) which show extensive evaginations of the sarcolemmal membrane.

The components of the total muscle capacitance. In the case of skeletal muscle, the use of simple-cable theory for responses to synaptic excitation or subthreshold current injection requires that both the surface membrane capacitance and the transverse tubular capacitance be charged by axial currents. The capacitance calculated from the simple-cable equations may be called C_{tot} since it is the total capacitance which is charged per unit area of fibre surface. Following the terminology used in Chapter 2 we may divide the total capacity into two components. C_m will refer to the sarcolemmal† capacitance only, and C_e will refer to the capacitance of the internal membranes. It is important to note that both of these quantities are expressed in units of μF cm^{-2}, where the area refers to the cylindrical surface of the fibre. This is obviously appropriate in the case of C_m. In the case of C_e this convention is justified on the grounds that C_e lies in parallel with C_m, and must be included in an equivalent circuit for the 'inside to outside' impedance. The total capacity is given by the equation $C_{tot} = C_m + C_e$. Note that the units of C_e do not allow the specific capacitance of the internal membranes to be obtained directly since, as already noted above, the internal membrane area may be much larger than the surface membrane area. Moreover, without further information, it cannot be assumed that the internal membranes are charged to the same potential as the surface membrane. In general, therefore, there are no 'expected' values of C_e which do not rely on additional information of a kind which is still particularly difficult to obtain (see Adrian and Almers 1973, 1974).

In fact, morphological studies on fibres of different sizes show that the area of the tubular membrane system scales in proportion to the volume of the fibre and not the surface (see Peachey (1968) for review and references). Hence the fractional contribution of tubular capacitance to total capacitance

---

† Later in this chapter (p. 113) we will compare this 'lumped' model with a distributed equivalent circuit for the transverse tubular system. Strictly speaking, C_m is defined by its lack of significant series resistance. In this sense, there may be an outer zone of the tubular system which has relatively little series resistance, and which will contribute to the lumped element C_m along with the sarcolemmal membrane *per se*.

should be larger in larger fibres. Step current analysis of the cable properties of isolated frog single muscle fibres shows that large diameter fibres do have a larger C_{tot}, as expected (Nakajima and Hodgkin 1970; Hodgkin and Nakajima 1972a, b; Adrian and Almers, 1974). With these qualifications in mind, it is still convenient to refer the capacitances to the cylindrical surface area, preferably for a standard diameter fibre (see Falk 1968).

Series resistance of interior structures. One of the simplest ways of modifying the cable equivalent circuit for muscle is to place a resistance R_e in series with C_e (cf. Chapter 2, p. 19), as shown in Fig. 6.2(a). Since C_e is expressed with respect to the surface fibre area, the series resistance R_e must be referred to surface fibre area. This parameter must also be converted into other units if specific resistances of internal structures are required.

The structural basis for R_e is still relatively uncertain. Several possibilities have been considered. *First*, the lumen of the tubules must contribute at least some resistance to the lumped element E_e, although it is unclear to what extent the luminal space differs from Ringer's fluid in its specific resistivity (Schneider 1970; Hodgkin and Nakajima 1972b; Peachey and Adrian 1973). *Second*, there may be a significant resistance at the mouths of the tubules (cf. Pugsley 1966) if the openings are constricted in some way, or if there are relatively few structural connections between the T tubular mesh and the surface membrane (cf. McCallister and Hadek 1970). The effect of such a resistance (an 'access resistance') has been treated by Peachey and Adrian (1973) in their mathematical analysis of current spread within the T system. *Third*, some of the series resistance could lie on the *intracellular* side of the capacitance C_e, if there were some additional structure which restricted current flow between the T tubular membrane and the myoplasm (cf. Freygang, Rapoport, and Peachey 1967; see Fig. 6.2(c)).

We will discuss the possible origin or origins of R_e in more detail (see p. 113). However it is worth pointing out at this stage that the elements C_e and R_e are lumped elements which deliberately neglect the possibility of decrements in the radial direction from the fibre surface, a simplification which may be justified for the purposes of describing the longitudinal spread of excitation. In order to take radial spread into account, more complex circuits are required (see p. 121).

The effect of the series resistance R_e will be to make the path of current flow depend on the frequency of the excitation. As the frequency of the excitation increases, an increasingly large fraction of the current will escape from the myoplasm to the external solution via the surface membrane (Chapter 2, p. 18). This will become critically important at frequencies higher than $\omega_e = 1/\tau_e = 1/R_eC_e$ (see Chapter 2 p. 20). Thus the spread of end-plate potentials, which are relatively low-frequency signals, will require the charging of the entire membrane capacitance C_{tot}. On the other hand, the local circulating currents that herald the arrival of propagating action

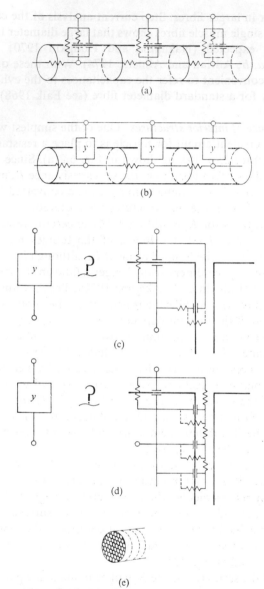

(a)

(b)

(c)

(d)

(e)

FIG. 6.2. Electrical models for passive properties of muscle fibre. (a) The four-component model: in this model the properties of the T system are represented by a series resistance and capacitance pathway which is placed in parallel with the surface membrane capacitance and resistance. (b) A more general model in which the membranes (surface and deep) are represented by an admittance y. This model is of use when it is sufficient to specify y graphically (e.g. by an X–R or phase–frequency plot) without using any particular equivalent circuit. (c) A physical interpretation of the four-component circuit. The capacitance C_e is identified with that of the T tubule membrane and the series resistance R_e with the resistance of a structure which is distributed in the same way as the T membranes, e.g. connections to the lateral cisternae. It is also possible to include the resistance of the T

potentials are relatively high-frequency signals, which will charge the surface capacitance C_m but produce very little charging of C_e. As already shown in Chapter 2, one of the most important methods for analysing more complex circuits than the simple parallel R–C circuit is to use a.c. currents of different frequencies. We shall discuss the application of this method to a distributed cable structure in a later section (*A.c. impedance analysis of distributed structures*).

Early response to step currents. A step current excitation may also be analysed in terms of pure frequency components and, as shown in Chapter 13, in this case some contribution is present at all frequencies. The high-frequency components determine the response at short times and the low-frequency components determine the response at longer times after the onset of the step. Therefore the response of a linear cable will reveal its high-frequency properties only at early times after the application of the current step, when $t < \tau_e$. In the analysis of Fatt and Katz (1951) the early voltage response was neglected on the grounds that it was likely to be distorted by a residual capacitive artifact. The remaining voltage response was well fitted by the simple cable equations (Chapter 3, eqn (3.24)), where τ corresponds to the low-frequency time constant $\tau = R_m(C_m + C_e)$. However, more precise measurement of the response to a step current have shown deviations from the simple-cable response at early times, which is consistent with the two time-constant model obtained by a.c. techniques (see p. 110).

The early response to step currents was used by Adrian and Peachey (1965) to characterize the high-frequency capacitance of slow muscle fibres. If the admittance of the T-system pathway can be neglected at early times, the equivalent circuit simplifies to that of a simple cable (Fig. 3.1) and the voltage response near the point of current injection is simply given by eqn (3.28):

$$V_0 = \tfrac{1}{2}I_0\, r_a\, \lambda\, \mathrm{erf}\{\sqrt{(t/\tau_m)}\}$$
$$= \tfrac{1}{2}I_0\, \sqrt{(r_a r_m)}\, \mathrm{erf}\{\sqrt{(t/r_m c_m)}\}, \qquad (3.28)$$

where c_m is the surface membrane capacitance per unit length of fibre. Since $\mathrm{erf}(x) = x$ for small values of x,

$$V_0 = \tfrac{1}{2}I_0,\, r_a\, \sqrt{(t/r_a c_m)},$$

FIG. 6.2. (*contd.*)
tubule membrane to form a five-component model. In both cases, the resistance of the tubular lumen is neglected. (d) A model in which the T tubules are represented as cables along which radial voltage differences may occur. This kind of model is required to study the propagation of excitation along the T system. (e) an elaboration of the cable model for the T system in which the cables are assumed to interconnect to form a uniform two-dimensional lattice. If the lattice is sufficiently fine, this model resembles the two-dimensional model for current spread discussed in Chapter 5 and the same differential equations apply. Model (e) is the most complete model used so far, but it leads to more complex equations than does the four-component model. As discussed in the text, it is sometimes more convenient and sufficiently accurate to use the four-component model when it is not necessary to describe the radial spread of current in the T system.

where r_m has been eliminated. The product of the axial resistance and the surface capacitance may be interpreted as a time constant for high-frequency axial currents, which charge the surface membrane capacitance in preference to the other current pathways. The product $r_a c_m$ is also useful in the description of local circulating currents preceding the propagating action potential (see p. 117).

Use of models representing radial spread and of generalized admittance models

The four-component model used to represent the membrane impedance in Fig. 6.2(a) does not allow representation of the radial spread of current in the T system since C_e is a lumped equivalent capacitance representing the behaviour of the capacitance of the entire T system. A model which is more useful when radial spread is important is shown in Fig. 6.2(d) where the T tubules are represented as cables with the tubular membrane impedance given by the simple parallel R–C circuit. This model may be further developed by allowing the T system cables to interconnect to form a branching network or lattice. This model will be referred to in later sections of this chapter as the lattice cable model.

The question whether the four-component model or the lattice cable model most accurately represents the over-all impedance of muscle is still unsettled (see p. 113). For some purposes, however, it is not necessary to specify the details of an equivalent circuit, and it is then sufficient to represent the membrane properties per unit length of fibre by a generalized admittance y. Thus, the equations for the 'input impedance' (see below) may be developed without specifying y in any way at all. For other purposes, it may be sufficient to specify y by means of an X–R plot or similar impedance diagram.

A.c. impedance analysis of distributed structures

As already discussed in connection with step current analysis, the pathway of current depends on the frequency of excitation. Measurement of the voltage responses to pure frequencies (that is, sinusoidal currents) therefore should be a valuable method for obtaining an equivalent circuit, since the way in which the pattern of current flow in the fibre changes as the frequency changes must depend on the details of the equivalent circuit.

Although the technique is often experimentally difficult, the principle of a.c. impedance analysis is relatively simple and may be presented as an extension of linear cable theory. Various aspects of a.c. cable theory have been developed and discussed by Tasaki and Hagiwara (1957), Falk and Fatt (1964), Eisenberg (1967), Freygang *et al.* (1967), and Schneider (1970). We will outline the basic principles of the analysis and leave matters of less general interest to the original papers and appropriate mathematical texts (e.g. King 1965; Desoer 1969).

In studying the properties of linear electrical networks, the use of sinusoidal excitation has the major advantage that a steady sinusoidal excitation will produce a steady sinusoidal response. The general relation between a voltage at any point in the network and the applied current may then be described by an impedance term. As already discussed in Chapter 2, it is convenient to describe a sinusoidal signal as the real part (Re) of a complex vector. The impedance Z is defined as a complex quantity which relates a voltage response to a current excitation. For an excitation $I(t) = \text{Re}(Ie^{j\omega t})$,

$$V(t) = \text{Re}(Ve^{j\omega t}) = \text{Re}(ZIe^{j\omega t}). \tag{6.1}$$

As a complex quantity Z may be expressed either in terms of real and imaginary parts, the resistance R and the reactance X,

$$Z = R + jX, \tag{6.2}$$

or in terms of magnitude and phase, $|Z|$ and ϕ,

$$Z = |Z|\, e^{j\phi}. \tag{6.3}$$

The analysis of the frequency dependence of two simple networks (parallel R–C circuit and the four-component circuit) was illustrated in Chapter 2 by use of the X–R plot. In that discussion, we assumed the current to be applied uniformly. We will now describe the theory for nonuniform current.

The customary method of studying the a.c. properties of muscle fibres is to use two intracellular electrodes, measuring the voltage response at a small distance from the point of current application. Physiologists have used the term 'impedance' to describe the relation between the recorded voltage and the applied current although, strictly speaking, the term 'input impedance' would be preferable (cf. the term 'input resistance' discussed in Chapter 3, p. 32 and Chapter 4, p. 76). The sinusoidal voltage response at some distance from the current source can also be described by means of a complex quantity, termed the 'transfer impedance'. Schneider (1970) has measured the magnitude and phase of the transfer impedance of frog sartorius fibres at various inter-electrode separations. His results support the usual assumption that the internal impedance is purely resistive and that the fibre is well described by a one-dimensional cable model under the conditions used for a.c. analysis. With this experimental justification for the one-dimensional cable model, it is possible to restrict the discussion to the input impedance, since measurement of this alone is sufficient to characterize all the parameters with the exception of the axial resistivity. The mathematics of transfer impedance analysis is similar, although more complicated (see Falk and Fatt 1964; Schneider 1970).

The derivation of the input impedance of an infinite cable is analogous to the development of the input resistance in Chapter 3 (see Lux (1967), and Williams, Johnson, and Dainty (1964) for development of a.c. behaviour of terminated cables). We will continue to refer to the admittance per unit length between the fibre interior and the external solution as y, where

$$y = g + jb \text{ and } y^{-1} = z.$$

None of the physical conditions assumed in Chapter 3 in the development of the equations for the simple parallel R–C cable in the steady state depend on frequency. In fact, the previous parameter for the d.c. conductance of the membrane, $g_m = 1/r_m$, corresponds to the value of y as the frequency approaches zero. Generalizing the d.c. equations by substituting $1/y$ for r_m in eqn (3.26)

$$V(x = 0)/I(x = 0) = Z_0 = \tfrac{1}{2}(r_a/y)^{\frac{1}{2}}. \tag{6.4}$$

Various methods have been used to display the properties of Z_0 as a function of frequency. Cole and Baker (1941) chose to extract the properties of y by squaring the complex quantity Z_0.
Thus

$$Z_0^2 = r_a/4y = r_a z/4. \tag{6.5}$$

Hence Z_0 is proportional to the square root of z just as the input resistance for d.c. current is proportional to the square root of the membrane resistance (see Chapter 3, p. 32).

Note on relation between cable and uniform properties. The results given by eqns (3.26) and (6.5) depend on a very general property of the one-dimensional cable, which is that the input characteristics for current applied at one point are always proportional to the square root of a function of the characteristic for uniformly applied current. This property may be obtained directly from two relations: one mathematical, the other physical. The mathematical relation is a general one between first and second derivatives which we shall obtain and discuss in more detail in Chapter 12:

$$\left(\frac{\mathrm{d}V}{\mathrm{d}x}\right)^2 = 2\int_{V_1}^{V_2} \frac{\mathrm{d}^2 V}{\mathrm{d}x^2}\,\mathrm{d}V + C, \tag{12.4}$$

where C is a constant of integration and V_1 and V_2 are voltages at each end of the cable. The physical property is that, in a one-dimensional cable, the input and membrane currents are directly proportional to the first and second derivatives of membrane voltage respectively (see eqns (3.2) and (3.5)). Hence, in general,

$$I \propto \left(\int_{V_1}^{V_2} i_m\,\mathrm{d}V + C\right)^{\frac{1}{2}}. \tag{6.6}$$

The limits in this equation are determined by the boundary conditions and in the case of an infinite cable, the integration is between $V = 0$ and $V = V_{x=0}$. The constant of integration is zero (see Chapter 12, p. 381). If i_m is a linear function of V, eqn (6.6) simplifies to

$$I \propto \left\{(1/r_m)\int_0^{V_{x=0}} V\,\mathrm{d}V\right\}^{\frac{1}{2}} \propto V(1/r_m)^{\frac{1}{2}}. \tag{6.6a}$$

For impedance work, substitute z for r_m and Z_0 for V/I to obtain

$$Z_0 \propto \sqrt{z}, \tag{6.6b}$$

which is the same as the result given in eqn (6.5).

When i_m is a nonlinear function of V, it is generally not possible to simplify eqn (6.6). It may be used however to derive the Cole theorem for the relation between cable input properties and uniform properties (see Chapters 9 and 12). It is not surprising, therefore, that there is some analogy between the use of eqn (6.5) for impedance work on cables and the use of the Cole theorem for work on nonlinear cables. In both cases, the properties for uniform polarization are distributed along a one-dimensional structure but nevertheless may be extracted from the properties of the cable as a whole. The difficulties involved in doing this are similar in both cases, since the distributed structure will generally reduce the degree of nonlinearity or frequency dependence. These difficulties make it desirable to obtain the properties for uniform polarization directly whenever possible. In large nerve fibres this may be done by using axial current electrodes. Apart from certain giant muscle fibres (e.g. barnacle), it is not feasible to use this technique in muscle.

An alternative approach which also avoids the distributed nature of the muscle fibre is the measurement of the transverse impedance of whole muscle (Fatt 1964). Although this method is unsuited to obtaining certain parameters of the equivalent circuit, it does provide a valuable complement to measurements with intracellular electrodes. Apart from the shunting effect of the external solution between the fibres, the impedance locus which is obtained shows the two semicircles that are characteristic of an equivalent circuit with two time constants, such as the four-component circuit (Chapter 2, p. 20.) For further discussion of the transverse impedance method, the reader is referred to Fatt's (1964) original paper and to Cole (1968).

X–R plots for muscle. Although the method involving calculating Z_0^2 is direct, it has the disadvantage of compressing impedance information obtained at higher frequencies. This may be avoided by the use of X–R plots (see Chapter 2, p. 14 for a discussion of these plots).

It may be helpful to present a simplified explanation of the use of X–R plots in distributed structures (following the treatment of Falk and Fatt 1964) to illustrate the transformation between the frequency dependence of uniform and distributed networks.

Fig. 6.3 shows the transformation (eqn (6.5)) for both the single time-constant and two time-constant cases. The high-frequency asymptotes (corresponding to purely capacitive impedance) are indicated by dashed lines, while the low-frequency (that is, purely resistive) point is shown by the open circle. The cross marks the impedance at the lower 'characteristic frequency', which corresponds to the time constant $r_m (c_m + c_e)$.

These diagrams show that the behaviour of a distributed two-time-constant equivalent circuit is quite analogous to that of the uniform two-time-constant

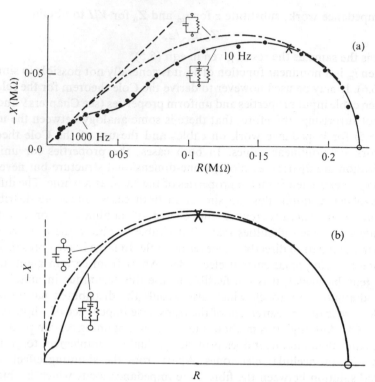

FIG. 6.3. (a) X–R plot of input impedance data for frog sartorius fibre obtained by Falk and Fatt. Each point represents the magnitude and phase of the impedance at a particular frequency (10 Hz and 1000 Hz are labelled) plotted as described in Chapter 2 (Fig. 2.6). The points are fitted by a solid curve computed from the model shown in Fig. 6.2(a). The impedance locus for the simple R–C cable (dot–dash line) is shown for comparison. Note that the simple cable equations are adequate only at lower frequencies. Note also that the phase angle of the input impedance tends to a limit of 45° at high frequencies (interrupted line). (From Falk and Fatt 1964, Fig. 8.) (b) impedance locus for *uniform* polarization of fibre obtained from Fig. 6.3(a) using the transformation equations described in text. The locus closely approximates to a semicircle at low frequencies with a characteristic time-constant (marked by the cross) equal to $r_m(c_m+c_e)$. The value obtained is 32·4 ms. At high frequencies the locus shifts to a second, smaller semicircle. This behaviour is characteristic of the four-component circuit (cf. Fig. 2.9) but is also obtained with the more complex circuits illustrated in Fig. 6.2. As in Fig. 6.3(a), the locus for the simple R–C circuit is shown as a dot–dash line and the high-frequency asymptote as an interrupted line. The latter now has a phase angle of 90°. The scales of the ordinate and abscissa depend on the value assumed for the axial resistance (see eqn (6.5)). For the sake of simplicity arbitrary scales have been used in this illustration.

circuit. Similarly, modifications of the unit admittance produce corresponding changes in the X–R plot of Z_0 (see Falk and Fatt 1964, Figs. 3 and 5; Eisenberg 1967).

Evaluation of the components of the equivalent circuit for uniform polarization follows directly, once an equivalent circuit model is chosen. In particular, the parallel capacitance in a two-time-constant model C_m can be obtained directly from the high-frequency asymptote in the X–R plot. This particular calculation may be facilitated by plotting the impedance magnitude $|Z|$ versus $f^{-\frac{1}{2}}$. Since at high frequencies

$$|Z| \rightarrow (8\pi C_m/r_a)^{-\frac{1}{2}} f^{-\frac{1}{2}}, \qquad (6.7)$$

the slope of the $|Z|$ versus $f^{-\frac{1}{2}}$ plot should lead directly to a value for C_m (Tasaki and Hagiwara 1957).

Phase angle–frequency plots. Both the $|Z_0|^2$ plot and the X–R plot require knowledge of the value of r_a for the evaluation of other circuit parameters. The axial impedance may be determined directly by considering the more general problem of transfer impedances at various interelectrode spacings (Schneider 1970). Assuming that the axial impedance is purely resistive (as Schneider's results indicate), it is also possible to eliminate the parameter r_a by measuring the *phase* of the input impedance ϕ. This technique may be used directly with the aid of sensitive phase-detection techniques (Freygang *et al.* 1967; Freygang and Trautwein 1970).

The phase ϕ of the cable input impedance is related to the phase of the unit admittance y as follows. Starting from eqn (6.5), that is,

$$Z_0^2 = r_a/4y = r_a z/4. \qquad (6.5)$$

Taking the phase of both sides,

$$2\phi = -\phi_y \qquad (6.8)$$

and thus

$$-\tan 2\phi = \tan \phi_y = b/g, \qquad (6.8a)$$

where $y = g + jb$. Freygang *et al.* (1967) give a similar derivation and also show a plot of b/g as a function of the cable-impedance phase angle ϕ.

The relations between $|Z_0|$ and frequency, and phase and frequency obtained by Falk and Fatt (1964) for frog skeletal muscle are shown in Fig. 6.4. It can be seen that, as the frequency increases, Z_0 decreases and the phase angle (plotted as $-\phi$) increases. The theoretical relations expected for the simple parallel R–C circuit and the four-component circuit for the membrane impedance are also shown. It can be seen that the results differ greatly from those expected for the simple parallel R–C cable but are much better fitted by the four-component model cable. It can also be seen that the difference between the two models is much greater in the phase frequency plot than it is in the impedance frequency plot. The phase angle is therefore a more sensitive test between certain kinds of equivalent circuit. As we shall see later (p. 114), this is also true when the four-component circuit is compared with the lattice cable model.

In any network composed only of resistances and capacitances, the phase angle of z cannot exceed 90° (see Chapter 2, p. 14). Since Z_0 is proportional to the square root of z, the maximum phase angle of Z_0 can never exceed 45° (eqn (6.8)). This is illustrated in Fig. 6.4, where the models both converge

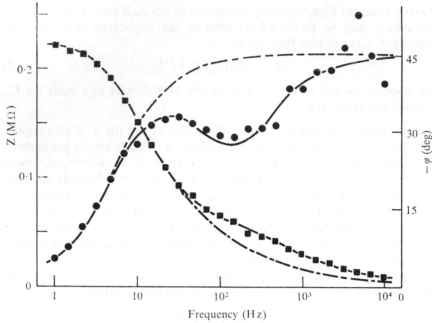

FIG. 6.4. An alternative method for presenting impedance data. The impedance magnitude Z and the phase $-\phi$ are plotted versus log (frequency). The experimental data is the same as that used in Fig. 6.3(a). The dot–dash lines represent the relations expected for a simple R–C model membrane. The continuous lines give the predictions of the four-component model. It can be seen that the deviation from the behaviour of the simple R–C cable is most pronounced in the phase–frequency plot. (From Falk and Fatt 1964, Fig. 9.)

onto a phase angle of 45° at high frequencies, as do the experimental points, and in Fig. 6.3(a), where the 45° line in the $X–R$ plot is a limiting boundary which is reached, but not exceeded, when the impedance is purely capacitive at very high frequencies.

Exceptions to the 45° limit occur in impedance analysis when the assumptions made in deriving the equations we have given here are invalid. For example, if the *axial* impedance to current flow is not purely resistive, as assumed, but also includes a parallel capacitance, the phase angle ϕ of the input impedance may exceed 45°. Such behaviour has been observed in impedance studies of cardiac Purkinje fibres (Freygang and Trautwein 1970) where axial current may flow between cells by capacitive pathways (possibly the apposition between adjacent cells) as well as by resistive pathways (nexuses). This may also be true in the case of smooth muscle bundles (see Jones and Tomita 1967).

The other exception arises from the necessarily finite separation of the current- and voltage-measuring electrodes. If the space constant of the fibre is small compared to the electrode separation, it is possible that the 'input' impedance phase angle may exceed 45°. Falk and Fatt (1964) determined the effect of various electrode separations on theoretical impedance loci, and showed that for $x/\lambda \simeq 0.05$ the predicted deviation from the 45° limit is not severe. Within this limit, there is some advantage to maintaining some small separation between electrodes, since this minimizes the importance of the three-dimensional spread of current immediately surrounding the current source. For further discussion of these problems, the reader is referred to Falk and Fatt (1964, Appendix 1), Eisenberg (1967), Eisenberg and Johnson (1970), Schneider (1970), and Pugsley (1966).

Interpretation of impedance measurements in terms of muscle structure. We may now return to the original question of how the observed electrical properties of muscle fibres are to be correlated with morphological elements of the muscle structure. In the previous section we discussed various lines of evidence suggesting that a large fraction of the fibre capacitance lies in series with a significant resistance—but the precise nature of the resistance was left unsettled. Two alternative equivalent circuits have already been put forward. The choice between them is significant because the equivalent circuits imply quite different physical origins of the series resistance.

The simpler circuit is the *four-component model* (Fig. 6.2(a)), which characterizes the tubular system by a simple lumped resistance R_e and capacitance C_e in series. This assumes that R_e has the same geometrical distribution as C_e, and implicitly, that the resistivity of the tubular lumen is negligible. Falk and Fatt (1964) have suggested that R_e may be assigned to the coupling between the T tubules and the lateral cisternae (as indicated in Fig. 6.2(c)).

The *lattice model* assumes that the impedance of the T system corresponds to that of a lattice-like cable structure, where the series resistance arises from the bulk resistivity of the tubular lumen. The spatial distribution of the resistance causes the tubular capacitance near the fibre centre to be charged through a larger series resistance than the capacitance nearer the fibre's surface. In contrast to the four-component model (which has a single tubular time-constant $R_e C_e$) the *lattice model* has a dispersion of time constants (see Adrian, Chandler, and Hodgkin 1970).

These two equivalent circuits lie at the extremes of a range of possible models (see Falk (1968) for the analysis of intermediate cases). The lattice and four-component models, however, serve to define the issue of the *degree* of distribution of the series resistance, which is important to describing the inward spread of excitation (see Part III). The physical nature of the series resistance may also be relevant to the mechanism of the excitation–contraction trigger.

Low-frequency signals do not distinguish between the four-component and lattice models, since they simply charge the total capacitance. At high frequencies, however, the transverse distribution of the series resistance will influence the pattern of current spread, and hence the input impedance of the

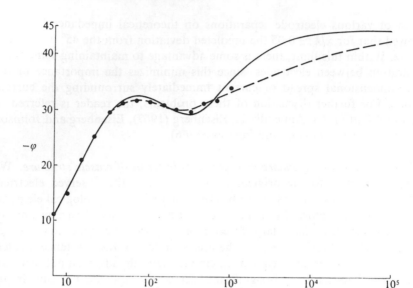

FIG. 6.5. Phase angle versus frequency plots for skeletal muscle. The points are experimental values. The continuous line shows the relation predicted by the four-component model. The interrupted line shows the relation predicted by the lattice model. Note that the major difference between the predicted relations occurs at frequencies beyond those at which it is possible to obtain reliable experimental measurements. At lower frequencies the models predict very similar relations, although the lattice model seems to fit the points slightly better than the four-component model. (From Schneider 1970.)

whole muscle fibre. As we have discussed already, the high-frequency data can be readily presented by plotting the impedance phase angle $(-\phi)$ as a function of frequency. In principle at least, the shape of the $-\phi$–f locus should indicate which model is to be preferred.

Fig. 6.5 has been taken from Schneider's (1970) study of the impedance of frog sartorius fibres. This phase–frequency diagram includes one set of experimental points as well as theoretical curves which have been generated from the two different equivalent circuits. The parameters of each circuit have been adjusted by an iterative procedure to give the best possible fit of the experimental points. Note that the lattice-model curve is less humped and gives a somewhat better fit to the data, which extends to 1 kHz. On the basis of the experimental measurements up to 1–2 kHz, the lattice model seems to be marginally preferable (see recent papers by Valdiosera, Clausen, and Eisenberg (1974a,b,c)).

It is fair to add, however, that the theoretical predictions of impedance phase angle are within a degree or so of each other up to 1 kHz. This frequency corresponds to a time constant $(= (2\pi f)^{-1})$ of 0·16 ms, which is approximately equal to the exponential time-constant of the action potential foot in normal Ringer (Freygang et al. 1967; Hodgkin and Nakajima 1972b).

Thus the local circulating current preceding the foot does not contain high enough frequencies to provide a useful test between the models. Conversely, the relative closeness of the impedance loci at 1 kHz is convenient, since it allows the simpler four-component model to be adequate for reconstruction of the propagating action potential (Adrian, Chandler, and Hodgkin 1970). In this respect, the four-component model is a useful simplification of the lattice model. The value of C_m will then correspond to the strictly sarcolemmal capacitance, *plus* a superficial annulus of tubular system capacitance, which will experience a negligible series resistance at the predominant frequency of the action potential foot (Hodgkin and Nakajima 1972*b*).

Fig. 6.5 also shows that the theoretical curves do diverge considerably at higher frequencies in the range 1–100 kHz. Here the lattice model predicts a more gradual variation in phase angle, corresponding to the dispersion of tubular time constants. Unfortunately, the only published data in this frequency range (Falk and Fatt 1964; see Fig. 6.4) show considerable scatter. With the help of techniques which minimize the capacitive coupling between the microelectrodes (Freygang *et al.* 1967) it might be possible to use the divergence between theoretical curves at high frequencies to produce relatively clear-cut evidence for the more appropriate electrical equivalent. Using considerably improved techniques, Valdiosera, Clausen, and Eisenberg (1973) have carried out phase measurements at frequencies up to 10 kHz. Their data does not support the four-component model as such. The authors suggest (see also Peachey and Adrian 1973) that the most realistic model is probably a hybrid between the four-component and distributed circuits. Such a hybrid model would include both a lumped resistance (perhaps an access resistance at the tubule mouths) and a distributed resistance (a property of the luminal fluid).

This problem provides a good example of the advantages and limitations of different methods of displaying impedance data. For example, the X–R plot in Fig. 6.3 so severely compresses the impedance behaviour at high frequencies that the data appear to support the existence of a single tubular time-constant between 300 Hz and 100 kHz. On the other hand, the $-\phi$–f plot of the same data (Fig. 6.4) gives a much better indication of the uncertainties at higher frequencies.

Although the experimental data are not yet adequate to decide conclusively between the various models, the lattice cable model has the advantage of including explicitly the spread of excitation in the radial direction. Therefore we shall use this model in Part III of this chapter.

Part II: Propagation of muscle action potential

The propagation of an action potential in either nerve or muscle fibres requires the regenerative activation of inward ionic current. Once the threshold for this current is exceeded the assumptions of linear cable theory are no longer even approximately valid, and the cable equations must then be generalized to include both the time and voltage dependence of the ionic current (see Chapter 8). We shall discuss the resulting nonlinear equations in Chapters 9–12. However, it will be convenient to deal with one particular aspect of action potential propagation in this chapter, partly because it concerns that part of the action potential (the 'foot') which may be described by linear cable theory (see discussion in Chapter 10, p. 281) and partly because the results provide a useful and largely independent means of determining some of the passive properties of muscle fibres.

Analytical solutions using linear cable theory may be obtained for the 'foot' of the action potential because this part of the wave represents the depolarization of the membrane by local circuit currents flowing from other areas of the fibre which have already been excited. The inward sodium current responsible for the nonlinear behaviour once the threshold is exceeded takes an appreciable time to be activated (see Chapter 10, p. 281 and Fig. 10.2) and, during this time, the area being invaded by the local circuit current flow behaves as a linear cable. Since the voltage change is fairly fast, most of the depolarizing current flow is capacitive. The magnitude of the capacitive current is determined by the magnitude of the inward ionic current flowing in the activated areas of the fibre, and the resistance through which it flows is determined by the axial resistance r_a. The conduction velocity of the action potential is determined therefore by these three factors: c_m, r_a, and i_i. We shall discuss this problem in more detail in Chapters 10 and 12. For the present we may note that, once the action potential has propagated a sufficient distance in a uniform cable, the conduction velocity becomes a constant and the time and space derivatives of the voltage are then simply related:

$$\frac{\partial^2 V}{\partial x^2} = \frac{1}{\theta^2} \frac{\partial^2 V}{\partial t^2}. \tag{6.9}$$

Substituting the right-hand expression in the cable equation (eqn (3.7)) gives a differential equation in the time domain only:

$$\frac{1}{r_a \theta^2} \frac{d^2 V}{dt^2} = c_m \frac{dV}{dt} + \frac{V}{r_m}, \tag{6.10}$$

which is also given in terms of R_i, C_m, and I_i in Chapter 10 (eqn(10.6)). Putting $\lambda^2 = r_m/r_a$, $\tau_m = r_m c_m$, and rearranging:

$$(\lambda/\theta)^2 \, d^2V/dt^2 - \tau_m \, dV/dt - V = 0. \tag{6.11}$$

The solution of this equation gives the voltage as a function of time at any one point on the cable. In general, the solution may be written as the sum of two exponentials,

$$V(t) = C_1 \exp(\alpha_1 t) + C_2 \exp(\alpha_2 t), \tag{6.12}$$

where α_1 and α_2 are roots of the characteristic equation

$$(\lambda/\theta)^2 \alpha^2 - \tau_m \alpha - 1 = 0.$$

Hence
$$\alpha = \frac{\tau_m \pm \sqrt{\{\tau_m^2 + 4(\lambda/\theta)^2\}}}{2(\lambda/\theta)^2}.$$

Only the positive root has a physical interpretation in this phase of the action potential. Therefore the time constant of the rising foot of the action potential (Cole and Curtis 1938; Hodgkin and Huxley 1952d; Tasaki and Hagiwara 1957) will be

$$\tau_{foot} = 1/\alpha_1 = \frac{2\lambda^2}{\theta^2[\tau_m + \sqrt{\{\tau_m^2 + 4(\lambda/\theta)^2\}}]}, \tag{6.13}$$

where α_1 is the positive root.

The conduction velocity θ is usually large compared to the electrotonic velocity $2\lambda/\tau_m$. Hence $4(\lambda/\theta)^2 \ll \tau_m$, and the expression for the time constant of the action potential foot simplifies to

$$\tau_{foot} = \frac{(\lambda/\theta)^2}{\tau_m}. \tag{6.14}$$

In other words, the ratio between the 'active' and 'passive' time-constants is inversely proportional to the square of the ratio of 'active' and 'passive' velocities,

$$\tau_{foot}/\tau_m = \frac{(\lambda/\tau_m)^2}{\theta^2}. \tag{6.15}$$

These equations may also be written in terms of the parameters r_a and c_m,

$$\tau_{foot} = 1/\theta^2 r_a c_m = a/2\theta^2 R_i C_m. \tag{6.16}$$

Using experimental measurements of the foot of the propagating action potential the product $R_i C_m$ has been evaluated in a variety of muscle preparations, including skeletal muscle (Tasaki and Hagiwara 1957; Freygang, Rapoport, and Peachey 1967; Hodgkin and Nakajima 1972b), Purkinje fibres (Fozzard 1966), ventricular heart muscle (Weidman 1970), and smooth muscle (Abe and Tomita 1968). Provided that the longitudinal resistivity is known, this method provides an estimate of the effective membrane capacitance at the frequencies of the action potential foot. This result may then be compared with a.c. step current or step voltage measurements of the capacitance. In the case of skeletal muscle, the value of capacitance obtained by the 'foot' method is in reasonable agreement with the high-frequency capacitance obtained by other techniques, that is, about 2 μF cm^{-2}. The results for cardiac and smooth muscle however must be interpreted with caution since it is uncertain to what extent the impedance to axial flow of current is reduced at high frequencies by cell-to-cell capacitance in parallel with purely resistive connections. Measurements of longitudinal impedance (Jones and Tomita 1967; Freygang and Trautwein 1970) suggest that the axial capacitance may have an appreciable shunting effect for the circulating currents preceding the action potential.

The equations we have discussed so far in this section assume that the capacitance is a simple capacitance in parallel with the membrane resistance. It is of some importance therefore to ask to what extent the equations and results are modified if more complicated impedances are assumed. Freygang, Rapoport, and Peachey (1967) have treated the more general case of a one-dimensional cable whose membrane impedance is described by the four-component equivalent circuit. The time-constant of the foot of the action potential is then obtained as a solution of a cubic characteristic equation corresponding to the quadratic equation used above. The circuit parameters obtained by a.c. measurements were used to predict values for the time constant of the action potential foot. By including the series R–C pathway corresponding to the T-system impedance they improved the fit between experimental and predicted values, although only slightly. This result is consistent with the idea that the local circulating currents charge up the lumped capacitance C_m in preference to the series capacitance C_e.

Hodgkin and Nakajima (1972b) have analysed the capacitance during the foot of the action potential C_t in terms of the distributed model for the tubular system. In their interpretation, C_t incorporates the capacity of a superficial annulus of T system, as well as the surface membrane per se. This is not surprising, since an outer zone of T-tubular membrane in a distributed model would contribute to the element C_m in a four-component model (see footnote, p. 102).

In conclusion, measurements of the action-potential foot and of propagation velocity may be used to give a useful indication of the magnitude of the capacitance charged at frequencies dominant in the action-potential foot, but the results may not be used to distinguish clearly between different models of muscle impedance. It should be noted also that the method does not distinguish between the frequency dependence of the membrane impedance to current leaving the cells and frequency dependence of the longitudinal impedance to axial currents since eqn (6.16) includes the product $r_a c_m$.

Part III: Radial spread of excitation

The coupling between the action potential and activation of contraction in skeletal muscle involves the radial spread of electrical activity along the internal membrane systems. In this section we shall discuss some of the evidence for this view and the methods which have been employed to study the radial spread of excitation. At the present time, this is a rapidly developing field of research, and it is not yet possible to give a full account of the subject without relying on information which is relatively new or uncertain. The major uncertainties arise from the fact that the cable properties of the internal membrane systems have not been directly accessible. The T tubules and the lateral cisternae are generally too small to be penetrated by micro-electrodes (cf. del Castillo, Houk, and Morales 1967), and information on their electrical properties must be obtained by rather more indirect methods than can be employed in the case of the cable properties of whole cells or fibres. Much of the evidence we shall discuss depends on optical measurements of the inward spread of contraction.

Methods for studying properties of tubular system

In a striking series of experiments, Huxley and Taylor (1958) and Huxley and Straub (1958) showed that it is possible to produce graded contractions of individual sarcomeres in skeletal muscle by applying focal depolarizations to sensitive locations on the surface of the fibres. These locations almost certainly correspond to points at which the T-tubule membranes join with the surface membrane. The focal depolarization experiments therefore provided very good evidence that the signal for calcium release is mediated by the inward spread of depolarization via the T system.

This conclusion has been further strengthened by the use of the glycerol shock technique (Howell and Jenden 1967) which disrupts the continuity between the sarcolemma and T system (Eisenberg and Eisenberg 1968; Nakajima, Nakajima, and Peachey 1968; Howell 1969). In fibres treated in this way, the action potential still occurs at the surface membrane but there is little or no mechanical activity. It is likely that this treatment does not seriously affect the later stages in excitation–contraction coupling since agents such as caffeine are still capable of eliciting contraction, presumably by direct effects on the internal membrane system.

Sugi and Ochi (1967*a, b*) and Sugi (1968) have extended the original experiments of Huxley, Taylor, and Straub, using much larger micropipettes (tip diameter $\sim 10~\mu$m) to produce a less localized depolarization of the surface membrane. As they point out, however, the usefulness of this technique is restricted by the radial asymmetry of the depolarization. The spread of excitation is therefore difficult to describe quantitatively. It is simpler from the theoretical point of view, and also more realistic physiologically, to study the inward spread of excitation under conditions where the surface membrane is depolarized uniformly around the circumference. Adrian, Costantin, and Peachey (1969) have done this by using an intracellular microelectrode to voltage-clamp a muscle fibre at one point. Since the fibre space constant is much larger than the fibre circumference, this achieves a depolarization which is nearly uniform around the circumference, although not along the axis of the fibre.

Using this technique in fibres treated with tetrodotoxin, Adrain, Costantin, and Peachey were able to impose a controlled depolarization around the circumference, while observing the onset of contraction within an optical section through the centre of the fibre. An action-potential waveform at the surface was barely sufficient to produce mechanical activation of the central myofibrils. This does not exclude the possibility that, under normal conditions, the inward spread of depolarization may be assisted by sodium current flow in the tubules, since presumably any tubule sodium current was blocked by using tetrodotoxin. Recently, this view has been supported by similar experiments in 50% sodium Ringer *without* tetrodotoxin (Costantin 1970). Under these conditions, it is sometimes possible to observe a contraction in the axial myofibrils with depolarizations that are too small to produce general activation. It may be, therefore, that regenerative sodium current within the tubules allows an 'action potential' of some kind at the centre of the fibre to escape from the voltage control imposed at the fibre surface. The reconciliation between this finding and the localized surface contraction produced by focal depolarization may lie in the contrasting geometry for current spread in the two cases (see the section *Equations for radial spread of excitation*). Uniform surface depolarization leads to a convergence of depolarizing current within the tubule mesh, which would greatly favour inward spread whereas in the focal depolarization experiments the current must first diverge from the electrode region. In a two- or three-dimensional lattice this divergence would greatly raise the excitation threshold (cf. Noble 1962*b*).

Another approach to this problem is to record the delay between mechanical activation at the surface and at various radii from the fibre centre during normal activation. This is difficult to do successfully since the adjacent myofibrils are linked mechanically so that it is not easy to distinguish between active shortening and passive compression. Gonzalez-Serratos (1966, 1971) has circumvented this problem by an ingenious experimental technique. This involves embedding the fibre in gelatin and compressing it longitudinally until the myofibrils show a wavy appearance. After an action potential has been initiated, activation is signalled by the straightening of the myofibrils as they actively shorten. Using this method, Gonzalez-Serratos estimated that at 20 °C the contraction of the central myofibrils lags behind that of the surface by about 1 ms in a fibre with a 90 μm radius. His result has been compared with the lags predicted by various equivalent circuits for the fibre cross-section (Falk 1968), but it is not yet possible to use the result to decide between different models. In particular, the temperature dependence of the lag (Q_{10} about 2) can be interpreted by either a passive or active mode of propagation in the tubular system so that, although Gonzalez-Serratos' result is important, it does not by itself decide between these two possibilities.

The experiments we have described all involve visual observation of contraction. This means that obtaining an equivalent circuit is at least partly dependent on a knowledge of the final electrical step in the excitation–contraction coupling process. The nature of this step is still not certain, but recent voltage-clamp experiments have clarified the question whether mechanical activation is directly related to the flow of ionic currents underlying the action potential itself. One such suggestion was that the calcium release might be linked causally to the onset of delayed potassium current. Although this idea gained plausibility from the close correspondence of the voltage threshold for the two phenomena and their sensitivity to agents such as nitrate (Kao and Stanfield 1968) or calcium ions (Costantin 1968), the similarity now appears to be fortuitous. Under a variety of conditions, mechanical activation and delayed potassium current may be dissociated (Kao and Stanfield 1968; Heistracher and Hunt 1969; Adrian, Chandler and Hodgkin 1969). In any case, the mechanism of calcium release could not be explained as a simple stoichiometric displacement by other ions. The flow of ionic current during an action potential is simply insufficient to match the calcium requirement for mechanical activation (Freygang 1965). Some sort of 'amplification' therefore must take place at a stage preceding the actual release of calcium itself.

One type of amplification would occur if the liberation of calcium ions from an internal store were dependent on a potential change across some element of the internal membrane system. The relation between current flow and calcium release would then be indirect. Current flow within the muscle would lead to activation by altering the charge on an internal capacitance. Schneider and Chandler (1973) have recently detected a voltage-sensitive displacement current, corresponding to the movement of charges trapped within or near the muscle-fibre membrane. The properties of the charge displacement suggest that a voltage-sensitive structural change (presumably in the T-tubule membranes) may be a link in the sequence of events between electrical depolarization and the release of calcium ions from the sarcoplasmic reticulum.

Equations for radial spread of current

Since the depolarization of the tubular membrane appears to be a necessary step in the normal coupling between excitation and contraction, we need equations to describe the spread of current within the T system. To obtain these equations we may use some useful analogies between the mathematical description of current flow in the T system and in the one- and two-dimensional cable models discussed in Chapters 3–5. First, we may note that the electrical roles of the intracellular and extracellular media are reversed. In the case of the T tubules, the resistance of the myoplasm may be neglected, since it surrounds the T system and has many times its volume. Voltage decrements occur in the 'extracellular' fluid in the lumen of the tubule. Spread of depolarization is, in effect, a wave of negativity in the lumen.

Falk and Fatt (1964, Appendix D), Falk (1968), Adrian, Chandler, and Hodgkin (1969, Appendix), and Schneider (1970) have described the radial spread of depolarization using lattice cable models for the T system. Mathematically, these models are very similar to the two-dimensional model described in Chapter 5. Because the T system ramifies considerably, and probably has uniform properties throughout the fibre, it is reasonable to assume the mathematical property of circular symmetry and that the admittance between myoplasm and tubular lumen scales with fibre volume, i.e. as the square of the distance from the centre of the cylinder. These geometrical assumptions correspond precisely to those made for the two-dimensional model described in Chapter 5.

It follows from the assumption of circular symmetry that only radially directed decrements occur within the network of tubules. Voltage decrement in the tubular lumen is assumed to occur across an ohmic resistivity R_L. Current escapes from the T-system mesh via the tubular walls which are represented by purely resistive and capacitive pathways in parallel (the series R-C case is also considered in Falk and Fatt (1964), Appendix C).

These assumptions lead to a general differential equation (see eqn (6.18) below) which is identical to eqn (5.4) (Chapter 5) apart from various geometrical scaling factors. The major difference between the muscle-fibre model and the two-dimensional model in Chapter 5 lies in the boundary conditions that must be applied. In the case of the T system, the surface membrane potential imposes a boundary condition at the fibre surface $r = a$. At $r = 0$ (the centre of the fibre) the radial current flow must be zero. The latter condition is equivalent to saying that the muscle-fibre model resembles an open-circuited short cable. In this case, however, the mathematical device of placing a second 'image' source equidistant from the termination (see Chapter 4, p. 68) corresponds to reality since the source is, in fact, circularly symmetric. (The assumption of circular symmetry is not strictly correct if the cross-section of the fibre is ellipsoidal or triangular (see Blinks 1965; Hodgkin and Nakajima 1972a; Dulhunty and Gage 1973), but it should be a good approximation.)

Spatial decrement of voltage in steady state. The steady state solution of the general differential equation is obtained in the usual way, by setting the time derivative of eqn (5.4) to zero and applying the boundary conditions. If V_a is the potential applied at $r = a$, the voltage distribution $V(r)$ is given by

$$V(r) = V_a \frac{I_0(r/\lambda_T)}{I_0(a/\lambda_T)}, \tag{6.17}$$

where λ_T is the tubular space constant. I_0 is the hyperbolic Bessel function of order zero, which complements the K_0 function obtained in Chapter 5 (eqn (5.8)). These Bessel functions are cylindrical analogues of the exponential solutions of the one-dimensional cable equation (see Chapter 13).

One of the striking properties of the two-dimensional model described in Chapter 5 is that the steady state potential decays very steeply with distance from a point source of current as a result of the increasing area for the escape of current as it diverges from the source. The opposite effect occurs in the case of an excitation applied at the circumference of a circle since the current then converges as it flows towards the centre. This convergence of current, and the effect of the 'open-circuit' at $r = 0$, will both minimize the radial decrement from the surface of the cylinder to the centre. Adrian, Costantin, and Peachey (1969) have computed solutions to eqn (6.17) for a fibre of radius 50 μm and space constants equal to 60 μm and 120 μm. Their results are shown in Fig. 6.6. For a surface depolarization of 35 mV (from

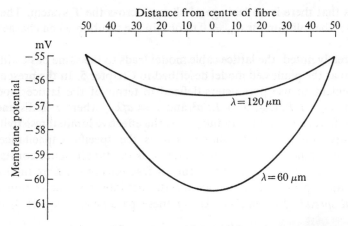

Distance from centre of fibre

FIG. 6.6. The steady state potential across the wall of the tubular system (eqn (6.17)) when the potential difference across the surface membrane of the fibre is altered from -90 mV to -55 mV ($V_a = 35$ mV). The potential distribution is shown for two tubular space constants (120 μm and 60 μm). The diameter of the fibre is 100 μm. (From Adrian, Costantin, and Peachey 1969, Fig. 1.)

-90 mV to -55 mV) the decrement of potential at the centre of the fibre is calculated to be about 5·5 mV for $\lambda_T = 60$ μm and only about 1·5 mV for $\lambda_T = 120$ μm. On the basis of morphological data provided by Peachey (1965) and Page (1965), the space constant expected theoretically is about 100 μm. Adrian, Costantin, and Peachey compared these theoretical calculations with experimental estimates of the decrement of potential in the T system. The estimates were obtained as the difference between the surface depolarization which produces activation of surface myofibrils only, and that depolarization which must be applied to also activate central myofibrils. The experimental decrements were unexpectedly larger than the theoretical predictions in all but the largest fibres used. Peachey and Adrian (1973) have attributed the extra voltage decrement to an ohmic voltage drop at the fibre surface, due to current flowing across an 'access resistance' which lies in series with the tubular system.

In conclusion, the steady state results suggest that the decrement of potential is fairly small but not insignificant.

Response to step voltage change at fibre surface. The normal physiological signal for activation is an action potential, which is too brief to allow the steady state equations to be used to calculate the potential across the T-system membranes (see p. 128). In order to develop the theory for responses to brief changes in potential we will first describe how the response to a step change in potential at the fibre surface may be obtained. This corresponds to the situation when a voltage-clamp step is applied using an intracellular microelectrode. The derivation may be performed in two stages. The first stage ignores the d.c. conductance of the tubule membranes G_W, that is, it

assumes that there is no leakage of charge across the T system. The second
stage uses the solution for $G_W = 0$ to obtain the solution when charge leakage
occurs.

As already noted, the lattice cable model leads to the same basic differential
equation as the plane cell model described in Chapter 5. In the present case it
is convenient to use parameters defined in terms of the lattice model. Let
$R = r/a$, $K = 1/\bar{R}_L \bar{C}_W$, $T = Kt/a^2$ and $v = a/\lambda_T$, where r is distance from
centre of fibre, a is the fibre radius, $\bar{R}_L$ is the effective luminal resistivity in the
radial direction (per unit volume), $\bar{C}_W$ is the specific capacitance of the
tubule membrane per unit volume, and λ_T is the tubule space constant. R is
therefore a distance measured in terms of fractions of the fibre radius. K is a
'propagation' parameter which we will discuss later (see the section *Velocity
of radial spread of excitation*). Using these parameters, eqn (5.4) becomes

$$\frac{\partial^2 V}{\partial R^2} + \frac{I}{R}\frac{\partial V}{\partial R} = v^2 V + \frac{\partial V}{\partial T}. \tag{6.18}$$

When the tubular membrane conductance is zero, the tubule space constant
becomes infinite and the term $v^2 V$ vanishes, leaving

$$\frac{\partial^2 V}{\partial R^2} + \frac{I}{R}\frac{\partial V}{\partial R} = \frac{\partial V}{\partial T}. \tag{6.19}$$

This equation is formally identical with that for another physical problem:
the inward flow of heat in a circular cylinder whose surface undergoes a step
change in temperature (Carslaw and Jaeger 1959, §7.6). As Eisenberg
and Johnson (1970) have recently emphasized, such analogies are very
useful since a large variety of heat flow problems have been analysed by Carslaw
and Jaeger. Eisenberg and Johnson also give a description of the way in
which the heat-flow solutions may be easily converted to solutions for electric
cables. In this case, the solution is given by

$$V(R, T) = V_a\left\{1 - 2\sum_{n=1}^{\infty} \exp(-\alpha_n^2 T)\frac{J_0(\alpha_n R)}{\alpha_n J_1(\alpha_n)}\right\}, \tag{6.20}$$

where $\alpha_1 \ldots \alpha_n$ are the positive roots of $J_0(\alpha_n) = 0$. This solution is given by
Falk (1968, eqn 13) but we have followed the terminology of Adrian,
Chandler, and Hodgkin (1969) since they also proceed to solve the more
general differential equation (eqn (6.18)).

The usefulness of eqn (6.20) depends, of course, on how reasonable it is to
assume $\bar{G}_W = 0$. The validity of neglecting $\bar{G}_W$ depends on its magnitude
compared to the luminal resistivity $\bar{R}_L$ (R_x in Falk's terminology), i.e. on the
magnitude of the tubule space constant which is determined by these parameters. As noted above (p. 123) the value of λ_T is still uncertain since the
experimental values estimated in Adrian, Costantin, and Peachey's experiments are smaller than the value expected theoretically. This may mean either

that $\bar{G}_W$ is larger than expected or that $\bar{R}_L$ is larger than that of the extra-cellular solution. The effects of both of these factors on current spread during brief surface potential changes will be considered later (see Fig. 6.9). At this stage, it will be assumed that $\bar{G}_W$ is not sufficiently small to be neglected in all circumstances. The effect of this conductance is to allow charge to leak through the tubule membranes. Provided that the membrane properties are uniform throughout the lattice, charge leakage will proceed at a uniform rate determined by the time constant $\tau_W = \bar{C}_W/\bar{G}_W$. As already shown in Chapter 5 (p. 90) the time course of the total charge is independent of the geometry of the system, so that for any particular initial distribution of charge the *decay* when no further charge is applied may be described by an exponential factor $\exp(-t/\tau_W)$.

Unfortunately, we cannot use this simple property to incorporate the effect of charge decay directly into eqn (6.20) since this equation is the solution for a *maintained* voltage change and the charge does not then decay to zero as $T \to \infty$. The desired result, however, may be obtained indirectly by making use of the property that the response to an infinitely brief application of charge (the delta function) may be obtained by differentiating the response to a maintained change (see Chapter 13 and eqn (3.49), Chapter 3). The method, therefore, is to first obtain the delta-function response. The effect of charge decay may then be incorporated by multiplying this solution by $\exp(-t/\tau_W)$. Finally, the response of a leaky lattice model to a maintained step change in voltage may be obtained by integrating the modified delta function response. Hence

$$V(R, T) = \int_0^t \frac{\mathrm{d}}{\mathrm{d}t}(V(r, t'))\exp(-t'/\tau_W)\,\mathrm{d}t', \qquad (6.21)$$

which may be integrated by parts to give

$$V(R, T) = (I/\tau_W) \int_0^t V(r, t')\exp(-t'/\tau_W)\,\mathrm{d}t' + V(r, t)\exp(-t/\tau_W). \quad (6.22)$$

The last expression is prescribed by Danckwert's method (Crank 1956) to deal with the additional term $\nu^2 V$ in eqn (6.18). Its validity may be checked by direct substitution of eqn (6.22) into eqn (6.18). The integration procedure corresponds to the Laplace-transform method for obtaining the step response of the two-dimensional model described in Chapter 5. In other words, Danckwert's method may also be expressed in terms of Laplace transforms. Thus, if $P(t)$ is the delta-function response for the case of no charge leakage, the step response of the leaky system corresponds to the inverse transform of

$$\bar{V}(s) = \frac{I}{s}\bar{P}(s+1). \qquad (6.23)$$

Adrian, Chandler, and Hodgkin (1969) have performed the integration in eqn (6.22) to give the general solution for the response to a step voltage change applied to the fibre surface (eqn 16 in their paper):

$$V(R, T) = V_a \left\{ 1 - 2 \sum_{n=1}^{\infty} \frac{\nu^2 + \alpha_n^2 \exp\left[-(\nu^2 + \alpha_n^2)T\right]}{\nu^2 + \alpha_n^2} \frac{J_0(\alpha_n R)}{\alpha_n J_1(\alpha_n)} \right\}, \quad (6.24)$$

which reduces to eqn (6.20) when $\nu = 0$. Fig. 6.7 shows computed solutions to this equation obtained by Adrian, Chandler, and Hodgkin for various times following the application of the step voltage change at the surface. The ordinate is the potential plotted as a fraction of the surface potential. The abscissa is the fractional radius R. The numbers on each curve give the time in milliseconds after the application of the surface voltage step. It can be seen that, in this case, more than 0·5 ms is required to bring the potential at the centre to half the surface potential, and 2–3 ms would be required to approximate closely to steady state conditions. Fig. 6.8 shows a plot of the time taken to reach half the surface potential at various distances from the surface. Since the steady state potential is nearly constant throughout the T

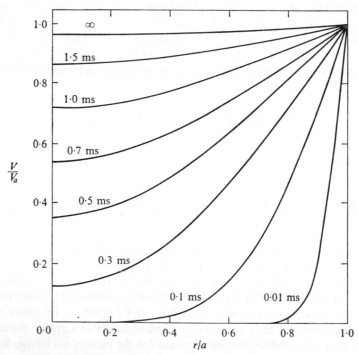

FIG. 6.7. Potential distribution in tubule at different times after sudden displacement of potential at surface of fibre (eqn (6.24)). The numbers on each curve are time in milliseconds after the step for a fibre with $a^2/k = 3\cdot2$ ms (see text). *Abscissa*: distance plotted as fraction of radius, surface of fibre at right, centre at left. *Ordinate*: potential as fraction of surface potential change. (From Adrian, Chandler, and Hodgkin 1969, Fig. 12.)

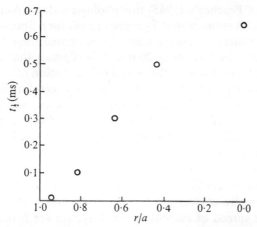

FIG. 6.8. Propagation of wave of depolarization in tubules. The time taken to reach half the surface potential change is plotted as a function of the distance (fraction of radius). As the wave approaches the centre, the 'velocity' increases. This effect is due to convergence of current at centre of T system and is not found in infinite cable (cf. Fig. 3.9).

system, this plot may be compared to that of the propagation of the normalized response in a one-dimensional cable (Fig. 3.9). The major difference is that the velocity of the wave increases as it approaches the centre of the fibre, whereas in the infinite one-dimensional cable the velocity approaches a constant value $2\lambda/\tau_m$. This difference is attributable to two factors which distinguish the T-system geometry from that of a one-dimensional infinite cable: (1) the area of membrane to be polarized decreases as the centre is approached and (2) the condition at the centre is formally similar to that of an open-circuit termination.

Velocity of radial spread of excitation. Although the velocity of the wave is not a constant, the factors determining the velocity may be determined in a way analogous to the treatment given in Chapter 3 (p. 34). The time parameter T in eqns (6.18)–(6.24) scales with real time t according to the factor K/a^2. Variations in K or a will change the time scale and so change the velocity of propagation. We may define the reciprocal of the scaling factor as a tubular redistribution time T_1 since this will determine the speed with which charge redistributes within the T system:

$$T_1 = a^2/K. \tag{6.25}$$

Since $K = 1/R_L C_W$

$$T_1 = a^2 R_L C_W.† \tag{6.26}$$

The corresponding time factor determining charge redistribution in a one-dimensional cable is (cf. eqns (9.8) and (9.9)):

$$T = aR_i C_m.† \tag{6.27}$$

† The reason for a^2 in eqn (6.26) is that C_W is defined in units of μF cm^{-3}, whereas C_m is in μF cm^{-2}.

On the basis of Peachey's (1965) morphological data, Adrian, Chandler, and Hodgkin (1969) estimate that T_1 is about 3 ms for a fibre of radius 40 μm. This value is considerably briefer than the estimated time constant for the tubule membrane, which is about 20 ms. This suggests that there should be very little charge leakage during the initial redistribution of charge within the T system. The radial propagation of the wave may therefore be regarded as a fairly high-frequency signal and, for many purposes, the charge leakage might be neglected. Adrian, Costantin, and Peachey's computations for brief surface potential changes also show that neglecting charge leakage has little effect (see Fig. 6.9).

The 'propagation' parameter K has units of cm² s⁻¹ which are the same as those for a diffusion coefficient. In fact, the comparison with a diffusion coefficient is formally correct since, as already noted above (p. 124), the solutions for the spread of current in the T system are formally analogous with those for the flow of heat in a cylinder. The same equations would be required for the diffusion of an ion within the fibre from the surface, which, as we noted at the beginning of this chapter, was the basis of some early theories of excitation contraction coupling. These theories were unsuccessful since the onset of contraction was found to be too rapid to allow time for such diffusion to occur. It is therefore of great interest to compare the values of K for the T system based on morphological data and reasonable assumptions for the electrical constants, with the diffusion coefficient of an ion like calcium. Adrian, Chandler, and Hodgkin (1969) obtained a value for K of 5×10^{-3} cm² s⁻¹ which, as they point out, is about 700 times greater than the diffusion coefficient of calcium ions in water. This factor may be regarded as the amount by which the T-system conduction speeds up the inward spread of excitation over and above that which could be achieved by free ion diffusion from the surface of the fibre (cf. Hill 1949).

Response to brief surface voltage change and to surface action potential. In a linear system, the response to one form of excitation, such as a voltage step or delta function, may be used together with superposition (or, in general, the convolution integral—see Chapter 13) to obtain the response to any form of excitation. In the case of the T system, the response of greatest interest is the response to a surface action potential. Also, for comparison with voltage-clamp experiments (see p. 119) it is of interest to know the response to brief step voltage changes. Adrian, Costantin, and Peachey (1969) have computed the responses for both cases, which are illustrated in Fig. 6.9. Fig. 6.9(a) shows the responses to a brief step change and Fig. 6.9(b) shows the responses to an action potential. In each case the solid curve is the response at the centre of the T system assuming that the tubular space constant is 60 μm (in a 100 μm fibre) and that the tubular lumen has a conductivity equal to that of the external solution. The dotted curve shows the response which is obtained when the tubular membrane conductance is neglected (i.e. the space

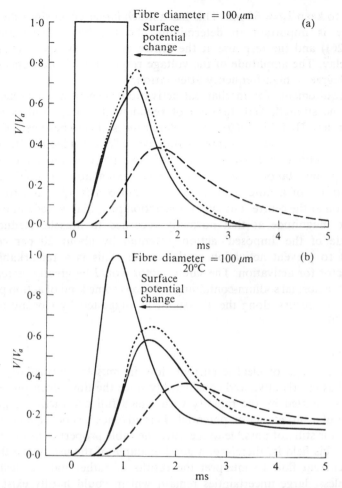

F I G. 6.9. (a) Response of centre of T system to short step change in potential at fibre surface. (b) Response at centre of T system to potential change at surface which has same shape and magnitude as the normal action potential at 20°C. In both cases, the continuous curve is calculated for a space constant of 60 μm and a tubular conductivity equal to that of the extracellular fluid. The dotted curve is response when the tubular conductance is neglected (space constant infinite). The interrupted curve shows response when tubular conductivity is reduced by about half to correspond to that of sarcoplasm, while adjusting the tubular membrane conductance to keep $\lambda_r = 60$ μm. (From Adrian, Costantin, and Peachey 1969, Fig. 2.)

constant becomes infinite, and there is no charge leakage). It can be seen that the responses differ only slightly from those assuming a space constant of 60 μm. For responses as brief as one or two milliseconds therefore, charge leakage in the T system does not have a large effect on the response. The interrupted curves show the responses obtained when the lumen conductivity is reduced to equal that of the sarcoplasm (cf. Schneider 1970), while G_W was adjusted to keep $\lambda_T = 60$ μm. This has a much larger effect since the tubular resistivity is important in determining the tubular redistribution time (eqn (6.26)) and the response at the centre of the T system occurs with a larger delay. The amplitude of the voltage response is also diminished by a greater degree of high-frequency attenuation.

The requirements for mechanical activation have been characterized by varying the strength and duration of rectangular voltage pulses (Adrian, Chandler, and Hodgkin 1969). From these results, it is to be expected that the electrical responses at the centre would be only just adequate to produce mechanical activation. This corresponds with Adrian, Costantin, and Peachey's (1969) experimental observation that imposing an action potential waveform at the surface of a fibre treated with tetrodotoxin is adequate to activate contraction at the centre, but only if the action potential waveform used is of the same magnitude as that normally recorded in muscle. Reducing the magnitude of the 'imposed' action potential by about 20 per cent was sufficient to prevent activation at the centre. This is a remarkably small safety factor for activation. The safety factor would be greatly increased, of course, if in normal sodium-containing solutions some kind of action potential propagation occurs along the T system, as suggested by Costantin's work (see p. 120).

Conclusion

Our knowledge of electric current flow in muscle, particularly skeletal muscle, has clearly advanced considerably since the time when physiologists were very puzzled by the rapidity with which full contraction follows the initiation of a surface action potential. This advance has occurred despite the fact that it is still not possible to measure the cable properties of the T system directly. This field is, therefore, a good example of the use of the theory of electric current flow to interpret the results of rather indirect techniques. Nevertheless, large uncertainties remain which would hardly exist were it possible to record directly from the T system and sarcoplasmic reticulum. Firstly, although Costantin's work suggests that the T-tubule membrane is excitable and may conduct action potentials, we do not yet know the extent to which its properties differ from those of the surface membrane. Second, we are only beginning to understand how depolarization of the transverse tubules might control the release of calcium from the sarcoplasmic reticulum (see Schneider and Chandler 1973). The electrical properties of the sarcoplasmic reticulum membrane itself are also relatively unknown (cf. Costantin and Podolsky 1967). For a recent review of these problems the reader is referred to Costantin (1975).

7. Mathematical models of the nerve cell

IN the preceding three chapters some of the features of current spread in a cell (or a functional syncytium), whose electrical geometry is more complicated than that of an infinite cable, have already been discussed. There the actual geometry to be considered was, at least, regular. Nerve cells are very diverse in appearance, but one common feature is the irregularity of their structure. Usually the nerve cell body gives rise to several dendrites whose orientation, length, and branching pattern may vary widely (see Ramon-Moliner 1962). Arising from the dendrites there are often dendritic spines (Gray 1959; Valverde, 1967; Scheibel and Scheibel 1968), which may show a systematic change in their form along the extent of the dendrite (Jones and Powell 1969; Peters and Kaiserman-Abramof 1970). In many nerve cells the majority of synaptic connections are to the extremities of these dendritic spines (see Scheibel and Scheibel 1968).

One of the reasons for taking a particular interest in the spread of current within a nerve cell is that it is in these cells that the summation of electrical activity takes place. It is this summation which is believed to underlie the characteristic behaviour of the nervous system. Any differences in the distribution of synaptic knobs (from different sources), and any consequent differences in the interaction of these synaptic effects, is obviously of possible importance in the behaviour of the cell. The difficulties in the way of establishing the exact distribution of one type of synaptic input, and exploring theoretically the functional significance of such a distribution, are formidable. The execution of this analysis may be divided into two major parts. The first problem is to determine the effective electrical geometry of the nerve cell (perhaps by a combination of histological and electrophysiological measurements) and then to determine the sites at which the various synaptic inputs are located. Usually this is also a necessary step to a secure analysis of the mechanism by which the synaptic input produces its effect. The second part is the subsequent modelling of the effects of activity, and hence interaction, when different patterns of synaptic input are delivered to the (model) nerve cell.

Only moderate progress has been made so far in developing the mathematical techniques which might suggest ways of making an experimental determination of the electrical geometry of a particular nerve cell. One reason for this is that the nerve cell structure only easily allows recording (and the passing of current intracellularly) from one site—the cell body or soma. Such a limitation on the experimental measurements leads to a severe restriction on the complexity of the model whose parameters are to be determined. Consequently, the only nerve cell model which has been used to make such determinations—the Rall model of the motoneurone (Rall 1959a, 1960, 1962a,

1969*a*)—was developed by making several simplifying assumptions. The four most important ones are as follows.

1. The membrane is assumed to be a simple, linear, R–C circuit with uniformity in the values of R_m and C_m.

2. Equipotentiality of the *entire external surface* of the membrane is assumed. This, of course, is equivalent to neglecting r_0—as in the preceding chapters.

3. The potential is assumed constant over the *internal surface of the cell body*—so that the soma can be represented as a lumped R–C element.

4. It is assumed that each dendritic tree can be represented as a cylinder of either finite or infinite extent, and each of these cylinders can in turn be lumped together so that the entire dendritic tree is represented by a single 'equivalent' cylinder (of either finite or infinite extent).

The latter two assumptions are novel to this chapter and we will discuss their justification, in turn, by considering the spread of current in a spherical cell and in a branching structure.

Part I: Spread of current in a spherical cell

In most cells it can be expected that the resistance to current flow across the membrane is much greater than the resistance to current flow within the cell, and it is therefore natural to assume that the displacement in membrane potential produced by current passed at any point within the cell will be fairly uniform. This is a comparable assumption to that made in earlier chapters where the membrane potential produced in a cylindrical cell by a point source was treated as a one-dimensional problem, i.e. there were no potential differences existing in the other two dimensions. The intuitive justification for this assumption was that the space constant was much larger than the diameter of the cell. The extent to which this assumption is justified for the cylindrical cell has been explored fairly carefully in several recent publications (see, e.g. Adrian, Costantin, and Peachey 1969; Rall 1969*b*; Eisenberg and Johnson 1970; Pickard 1971*a*). The formal mathematical solutions also exist for a spherical cell (Rall 1953; Hellerstein 1968; Eisenberg and Johnson 1970; Pickard 1971*b*) and a careful quantitative analysis has been given by Eisenberg and Engel (1970). They have calculated the distribution of the membrane potential for a point source of current applied just under the cell membrane. This is, of course, the circumstance in which there is most likely to be nonuniformity of the membrane potential—with the largest displacement of potential at the site of current injection.

One conclusion of their study is that, in the steady state (in response to a step of current), the effect attributable to three-dimensional spread for a typical cell is very small outside a limited area of the cell surface—the proportion of the total surface area whose membrane potential exceeds that expected from the simple equation

$$V_{\mathrm{m}} = \frac{IR_{\mathrm{m}}}{4\pi a^2} \qquad (a = \text{cell radius}) \tag{7.1}$$

is less than 0.2 per cent, for the usual values of R_{m}, R_{i}, and a (this value of surface area corresponds to having an angular separation between current-injection site and recording electrode of less than $5°$). Furthermore, the additional membrane potential produced in this way is established very quickly in response to a step of current—the time constant for a typical cell ($R_{\mathrm{m}} = 2000\ \Omega\ \mathrm{cm}^2$, $R_{\mathrm{i}} = 200\ \Omega\ \mathrm{cm}$, and radius $50\ \mu\mathrm{m}$) being about 2000 times briefer than the membrane time constant. Rall (1969b) had already reported a comparable figure for the cylindrical cell; the result is of importance in justifying the use of the slope of the transient response to a current step in determining the parameters of the Rall model (see later).

These results suggest that the assumption of isopotentiality for a spherical cell is largely justified. There are, however, at least three circumstances in which the three-dimensional spread of current may have to be considered. The first is in the use of a double-barrelled microelectrode (or a single electrode with a bridge) for current-passing and recording. If the electrode is just under the membrane surface, calculation of R_{m} using eqn (7.1) would lead to too large a value. This, and related problems, are discussed more fully in Eisenberg and Engel (1970), Engel, Barcilon, and Eisenberg (1972), and Peskoff and Eisenberg (1973).

Another example is in the recording of a synaptic potential generated by a knob on the cell body. In the unusual circumstance of the recording electrode penetrating the cell membrane close to the active knob it is possible that an extra effect would be prominent. Thus, if a synaptic potential were generated by current of brief duration in a spherical cell, the shape of the synaptic potential would be just as predicted for current injection into a simple R-C circuit except during the time of current injection, when an extra potential nearly identical in time course to the synaptic current would be observed. This effect is illustrated schematically in Fig. 7.1. The special effect illustrated in Fig. 7.1 will not be observed if the recording electrode is centrally placed An electrode at the centre of the sphere would record no three-dimensional effect since it is 'sampling' equally from all parts of the cell membrane.

A third circumstance in which the three-dimensional spread might be important is if the total amount of current injected by synaptic knobs is nearly sufficient to reach the voltage threshold for firing the cell. Such current injected at many different points over the cell surface might not reach threshold, whereas it is possible that the same current injected by a group of synaptic knobs located close together in a small region of a spherical cell might secure firing (this, of course, would be a deviation from constant charge threshold conditions—see Chapter 9). This is a restatement of a suggestion

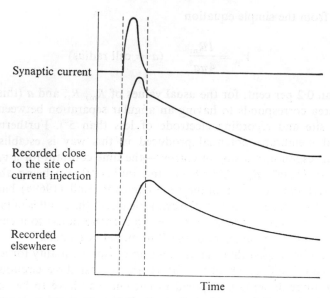

Synaptic current

Recorded close
to the site of
current injection

Recorded
elsewhere

Time

FIG. 7.1. Schematic illustration of the form of non-isopotentiality to be expected in a spherical cell. The three curves are, from above, the synaptic current, the membrane potential close to the site of injection, and the membrane potential remote from the synapse.

made many years ago by Lorente de No (1938) and which has already been discussed by Rall (1953, 1955).

Many nerve cell bodies diverge in appearance from a sphere, not only because they give origin to the dendritic trunks but also because they may be more ellipsoidal in shape. This may lead to quite significant potential drops between the two ends of the source. Rall (1959a,b) gives a calculation for the unfavourable case of a soma, whose major and minor diameters are 90 μm and 40 μm, with an electrode at one end and with most of the dendrites arising at the other end of the soma. As an example, if 20 times more steady current flows across the soma and into these dendrites as flows across the soma membrane then the difference in potential between the two ends of the soma could be as much as 2 per cent of the average displacement (assuming $R_m = 4000\ \Omega\ cm^2$ and $R_i = 50\ \Omega\ cm$).

Although this last calculation further indicates that in certain circumstances deviation from soma isopotentiality may become significant, it is still a fairly small effect; for the rest of this discussion such errors will be neglected.

Part II: Spread of current in a branching structure

Although dendritic trees show great diversity of configuration they invariably have branching and also commonly taper. Rall (1959a,b,1962a) has developed the mathematical methods appropriate to a description of the flow of current in such structures, and in this section we will follow his presentation closely. We first consider the steady state distribution of potential in a branching structure without tapering (i.e. each branch is a cylinder), and then show how the transient solution may be obtained for certain examples. Finally, a more general presentation which treats both tapering and branching (Rall 1962a) will be given.

Steady state in a structure with cylindrical elements

Formulation of the equations. Let us consider a single cylindrical element into which current is injected at the left-hand end and some proportion of that current leaves at the right-hand end (see Fig. 7.2). The general solution for the steady state voltage distribution is

$$V = A_1 e^{-x/\lambda} + A_2 e^{x/\lambda}$$
$$= A_1 e^{-X} + A_2 e^{X}. \tag{7.2}$$

Rall (1959a) has shown that it is convenient to use a different form of this solution, whose derivation from eqn (7.2) is given below.
Let

$$A_1 = \tfrac{1}{2}(C_1 + C_2)e^{l/\lambda}; \qquad A_2 = \tfrac{1}{2}(C_1 - C_2)e^{-l/\lambda},$$

then eqn (7.2) becomes

$$V = \tfrac{1}{2}(C_1 + C_2)e^{L-X} + \tfrac{1}{2}(C_1 - C_2)e^{X-L},$$

where $L = l/\lambda$,

$$= C_1\left(\frac{e^{L-X} + e^{X-L}}{2}\right) + C_2\left(\frac{e^{L-X} - e^{X-L}}{2}\right)$$

$$= C_1 \cosh(L-X) + C_2 \sinh(L-X).$$

Let V_1 be the voltage at $X = L$.

$$V_1 = C_1 \;(\text{since } \cosh(0) = 1,\; \sinh(0) = 0).$$

Therefore the general solution becomes

$$V/V_1 = \cosh(L-X) + C_2/V_1 \sinh(L-X)$$
$$= \cosh(L-X) + B \sinh(L-X). \tag{7.3}$$

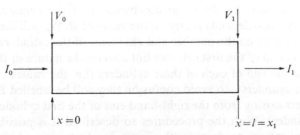

Fig. 7.2. A single cylindrical element labelled to indicate the significance of some of the symbols used in the text.

This is Rall's (1959a) eqn 2. As will be shown shortly the value of B is of crucial importance in describing the behaviour of the cable. In the special case of $X = 0$,

$$V_0/V_1 = \cosh L + B \sinh L$$

therefore

$$V = V_0 \frac{\cosh(L-X) + B \sinh(L-X)}{\cosh L + B \sinh L}. \tag{7.4}$$

The value of B depends on the character of the branches arising from the right-hand end of the cylinder. An appreciation of its significance may be gained by considering three examples.

(1) $B = 1$:

Eqn (7.4) simplifies to

$$V = V_0 e^{-X}:$$

In other words the distribution of voltage in this finite cylindrical element is identical to that which would be found if the cylinder was infinitely extended.

(2) $B = 0$

$$V = V_0 \frac{\cosh(L-X)}{\cosh L}.$$

This is identical to eqn (4.10); the distribution of voltage is that of a finite cable with an open-circuit termination at $X = L$.

(3) $B = \infty$:

Eqn (7.4) becomes

$$V = V_0 \frac{\sinh(L-X)}{\sinh L}$$

This is the same form as eqn (4.13) (when $s = 0$ the steady state solution is obtained—see Chapter 13), and gives the voltage distribution in a cable with a short-circuit termination.

Having explored the distribution of voltage along a cylindrical element and shown that it depends on the boundary condition at the end from which current leaves (i.e. on B) and the voltage at the site of current entry (i.e. V_0), we can in turn explore these two factors. The latter V_0 can be analysed by considering the input conductance of the cylinder. It will be shown that this also depends partly on the value of B. We will then show how B depends not only on the number and character of the cylinders attached to the right-hand end of the first cylinder but also on the nature of the boundary conditions at the end of each of these cylinders (i.e. the values of the 'B' for each of these cylinders—to avoid confusion they will be labelled B_{21} to B_{2n} for the n cylinders arising from the right-hand end of the first cylinder and B_1 for the first cylinder). With the procedures so described it is possible (by their iterative application) to describe the input conductance and voltage distribution when current is injected at any point in such a branching structure.

1. *Input conductance.* From Chapter 3, eqn (3.2), the axial current in a cylinder is given by

$$i_a = -\frac{1}{r_a}\left(\frac{dV}{dx}\right).$$

In cylinder 1 (of Fig. 7.2), from eqn (7.3),

$$V = V_1\left\{\cosh\left(\frac{l-x}{\lambda_0}\right) + B_1 \sinh\left(\frac{l-x}{\lambda_0}\right)\right\}$$

therefore

$$i_{a0} = \frac{V_1}{\lambda_0 r_{a0}}\left\{\sinh\left(\frac{l-x}{\lambda_0}\right) + B_1 \cosh\left(\frac{l-x}{\lambda_0}\right)\right\}, \tag{7.5}$$

where

$$\frac{1}{\lambda_0 r_{a0}} = \frac{\pi}{2}\sqrt{\left(\frac{d_0^3}{R_m R_i}\right)} = G_\infty$$

(G_∞ is the input conductance of an infinite cylinder with the same value of R_m, R_i, and diameter). The input current to cylinder I, I_0, is obtained from this equation by setting $x = 0$:

$$I_0 = V_1 G_\infty(\sinh L + B_1 \cosh L). \tag{7.6}$$

The input conductance G_0 of the cylinder is equal to I_0/V_0.

$$G_0 = \frac{V_1 G_\infty(\sinh L + B_1 \cosh L)}{V_1(\cosh L + B_1 \sinh L)}$$

$$= G_\infty \frac{B_1 + \tanh L}{1 + B_1 \tanh L}$$

$$= G_\infty B_0, \tag{7.7}$$

where

$$B_0 = (B_1 + \tanh L)/(1 + B_1 \tanh L). \tag{7.8}$$

Note that if L is small $B_0 \approx B_1$ because $(\tanh L) \to 0$ as $L \to 0$. On the other hand, if L is large $B_0 \approx 1$ regardless of the value of B_1, because $(\tanh L) \to 1$ as $L \to \infty$. For intermediate values of L, B_0 will be between the value of B_1 and 1.

Thus the input conductance of cylinder 1 has been expressed in terms of three factors:

(a) the input conductance of an infinite extension of that cylinder;

(b) the electrotonic length of the cylinder;

(c) the factor B_1, which is itself dependent on the relative amount of axial current leaving cylinder 1.

The dependence of B_1 on the subsequent branches will now be described.

2. *Dependence of B_1 on subsequent branches.* At the end of cylinder 1 the axial current flowing out enters n daughter cylinders, which will be labelled cylinders 11 to 1n (see Fig. 7.3). This current can be obtained from eqn (7.5) by setting $x = l$.

$$I_1 = V_1 G_\infty B_1. \tag{7.9}$$

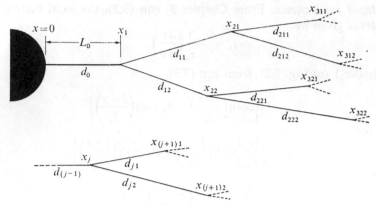

FIG. 7.3. This figure shows the convention for labelling dendrite branches. (From Rall 1959a.)

We now need to consider each of the branches 11 to $1n$ in order to determine how this current divides between them—it will do so, of course, in proportion to their input conductances. But for each of these branches we can formulate exactly the same equations as for cylinder 1. In other words we only need to change the subscripts to eqns (7.3), (7.4), (7.5), (7.6), (7.7), (7.8), and (7.9). Thus, for branch 11, having diameter d_{11}, length l_{11}, and characteristic length λ_{11}, the relation between current and voltage can be expressed as

$$I_{11} = \frac{V_{11}}{\lambda_{11}r_{a11}} \frac{B_{21} + \tanh(l_{11}/\lambda_{11})}{1+B_{21}\tanh(l_{11}/\lambda_{11})}. \tag{7.10}$$

Since

$$\frac{1}{\lambda_{11}r_{a11}} = \frac{1}{\lambda_0 r_{a0}} \frac{\lambda_0 r_{a0}}{\lambda_{11}r_{a11}} = G_\infty\left\{\left(\frac{2}{\pi}\sqrt{\frac{R_{m0}R_{i0}}{d_0^3}}\right)\bigg/\left(\frac{2}{\pi}\sqrt{\frac{R_{m11}R_{i11}}{d_{11}^3}}\right)\right\} \tag{7.11}$$

$$= G_\infty(d_{11}/d_0)^{\frac{3}{2}} \tag{7.12}$$

(providing $R_{m0} R_{i0} = R_{m11} R_{i11}$).

$$I_{11} = V_{11}G_\infty\left(\frac{d_{11}}{d_0}\right)^{\frac{3}{2}}B_{11},$$

where

$$B_{11} = \frac{B_{21} + \tanh(L_{11})}{1+B_{21}\tanh(L_{11})}.$$

Note that whereas the treatment until now only requires R_m to be uniform within each cylindrical branch, the simplification to eqn (7.12) (and its generalized application) requires that the $R_m R_i$ product be the same in all branches. Similar equations will hold for the other branches, so that we can now collect the expression for the current entering each daughter branch

$$\sum_n I_{1n} = V_1 G_\infty \sum_n B_{1n} \left(\frac{d_{1n}}{d_0}\right)^{\frac{3}{2}} = I_1. \tag{7.13}$$

As

$$I_1 = V_1 B_1 G_\infty$$

this leads to

$$B_1 = \sum_n B_{1n} \left(\frac{d_{1n}}{d_0}\right)^{\frac{3}{2}}. \tag{7.14}$$

Since each B_{1n} value will depend on the relevant B_{2m} values (by the same reasoning as that given above) it can be seen that B_1 not only depends on the nature of the primary branches, but in turn on the subsequent branches. This process can be repeated until the final branches are reached. The generalized formulae are

$$B_j = \sum_k B_{jk} \left(\frac{d_{jk}}{d_{(j-1)}}\right)^{\frac{3}{2}}, \tag{7.15}$$

where

$$B_{jk} = \frac{B_{(j+1)k} + \tanh(l_{jk}/\lambda_{jk})}{1 + B_{(j+1)k} \tanh(l_{jk}/\lambda_{jk})}, \tag{7.16}$$

$$\lambda_{jk} = \sqrt{\left(\frac{R_m d_{jk}}{4R_i}\right)}, \tag{7.17}$$

and k refers to the number of direct offspring.

There remains the problem of assigning a value to B_j at the termination of the final branches. We have already seen, in the discussion of the distribution of voltage along cylinder 1, that the value of B affects the distribution and, in particular, when $B = 0$ the voltage distribution is that of a cable with an open-circuit termination. We can now see that this is because assigning a value to B_1 of zero is the same as making the axial current at the termination zero ($I_1 = V_1 B_1 G_\infty$), which is the characteristic of an open-circuit termination. But the usual form of actual termination is for a cylinder to be 'sealed' by membrane which presumably has the same properties as membrane elsewhere. The current across this terminal disk would then be given by (see Rall 1959b)

$$I = V \pi d^2 / 4R_m \tag{7.18}$$

The axial current at the termination is given by

$$i_a = \left(-\frac{dV}{dx}\right) \frac{\pi d^2}{4R_i} \tag{7.19}$$

Combining these two equations gives the boundary condition at the end of the cylinder

$$\frac{dV}{dx} = -\frac{R_i}{R_m} V \tag{7.20}$$

The solution therefore is

$$V = V_0 \frac{\cosh(L-X)+(\lambda R_i/R_m)\sinh(L-X)}{\cosh(L)+(\lambda R_i/R_m)\sinh(L)}.$$ (7.21)

The coefficient $\lambda R_i/R_m$ (which is the value of B in this case) is small for most physiological situations. For example, with a terminal branch of diameter 9 μm with $R_i = 100\ \Omega$ cm and $R_m = 2500\ \Omega$ cm^2,

$$\frac{\lambda R_i}{R_m} = \sqrt{\left(\frac{R_i d}{4R_m}\right)} = \sqrt{\left(\frac{100\times9\times10^{-4}}{2500\times4}\right)}$$

$$\simeq 3\times10^{-3},$$

so that the solution will differ negligibly from

$$V = V_0\frac{\cosh(L-X)}{\cosh L}.$$

This is the basis for treating a cylinder with a 'sealed end' of membrane as being identical to an open-circuit termination.

We are now in a position to discuss some applications of the above formulae to actual dendritic trees. It will be shown that some simplification can be made, particularly when considering the peripheral spread of current.

Peripheral spread of current in a dendritic tree

Example of the distribution of voltage with distance. A simple application of the above equations can be shown by considering the steady state distribution of voltage in part of a dendritic tree. The example taken from Rall (1959b) is shown in Fig. 7.4. The first cylinder divides into two cylinders which, for simplicity, are assumed to have the same diameter, length, and terminal boundary condition. The length of all three cylinders is further assumed to be $0\cdot5\lambda$ and B_{21} and B_{22} are assumed equal to $1\cdot0$. Let us explore the voltage distribution for different values of the ratio d_0 to d_1, that is,

(a) $\dfrac{d_1}{d_0} = \dfrac{1}{4}$, (b) $\dfrac{d_1}{d_0} = \dfrac{1}{2}$, (c) $\left(\dfrac{d_1}{d_0}\right) = \left(\dfrac{1}{2}\right)^{\frac{2}{3}}$, and (d) $\dfrac{d_1}{d_0} = 1$.

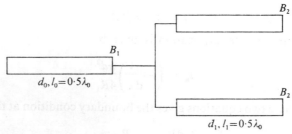

FIG. 7.4. The cylindrical element on the left gives rise to two equal-sized branches on the right.

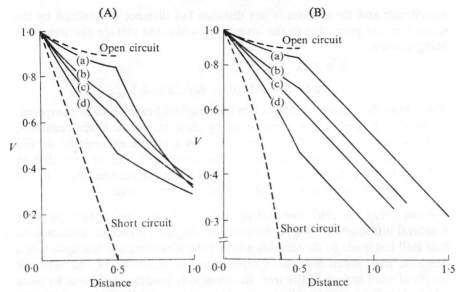

FIG. 7.5. Voltage distribution in some examples of the structure shown in Fig. 7.4. The continuous lines in the left-hand part of each figure are, from above down, for examples (a) to (d). Further description in the text.

Since we know the values of B_{21} and B_{22} we can easily calculate the values of B_1 for each of these examples from eqn (7.13):

(a) $B_1 = 2 \times (\tfrac{1}{4})^{\frac{3}{2}} = 0.125$;
(b) $B_1 = 2 \times (\tfrac{1}{2})^{\frac{3}{2}} = 0.63$;
(c) $B_1 = 2 \times (\tfrac{1}{2})^{\frac{2}{3}\cdot\frac{3}{2}} = 1$;
(d) $B_1 = 2 \times (1)^{\frac{3}{2}} = 2$.

We also need to know the value of λ_1 in relation to λ_0 so that the voltage distributions can be appropriately scaled.

$$\frac{\lambda_1}{\lambda_0} = \sqrt{\left(\frac{R_m d_1}{4R_i}\right)} \bigg/ \sqrt{\left(\frac{4R_i}{R_m d_0}\right)} = \left(\frac{d_1}{d_0}\right)^{\frac{1}{2}}$$

Therefore, for each example, we obtain a value for λ_1:

(a) $\lambda_1 = \lambda_0 \times (\tfrac{1}{4})^{\frac{1}{2}} = 0.5\lambda_0$;
(b) $\lambda_1 = \lambda_0 \times (\tfrac{1}{2})^{\frac{1}{2}} = 0.707\lambda_0$;
(c) $\lambda_1 = \lambda_0 \times (\tfrac{1}{2})^{\frac{1}{3}} = 0.79\lambda_0$;
(d) $\lambda_1 = \lambda_0 \times (1)^{\frac{1}{2}} = \lambda_0$.

In Fig. 7.5(A) the distribution of voltage with distance x is plotted. Note that it is a linear plot, so that the steady state decline in a simple unbranched infinite cable would be exponential, $V/V_0 = \exp(-x/\lambda)$. The dashed lines show the voltage distribution expected for an open-circuit ($B = 0$) and short-circuit ($B = \infty$) at the terminations. In Fig. 7.5(B) the ordinate is logarithmic and the abscissa is not distance but distance normalized by the space constant pertaining to the branch for which the voltage distribution is being plotted.

$$Z = x/\lambda_0 \quad \text{for} \quad x < 0\cdot5\lambda_0$$

$$= 0\cdot5 + (x - 0\cdot5\lambda_0)/\lambda_1 \quad \text{for} \quad x > 0\cdot5\lambda_0.$$

This allows the voltage distribution in the daughter branches to be compared, as the distribution is now normalized by their respective space constants. Since B_2 is set equal to 1 they now all show a simple exponential decline (straight lines) in the second branch. Furthermore, it can be seen that by defining the distance variable in this way the voltage decline when $B_2 = B_1 = 1$ (i.e., example (c)) remains a simple exponential throughout.

Reduction of a dendritic tree to an equivalent cylinder. The last example offers a natural introduction to describing one of the most powerful contributions that Rall has made to the simplification of the mathematical description of a dendritic tree. When distance is represented in terms of the characteristic length of each branch in the tree, the equations become equivalent to those which describe a simple cylinder providing that the following power law is obeyed

$$(d_{jk})^{\frac{3}{2}} = \sum_k (d_{(j+1)k})^{\frac{3}{2}} \tag{7.22}$$

It will be remembered (see eqn (7.11)) that this holds providing the membrane resistance is uniform. A more general form of this power law would be

$$\left(\frac{d_{jk}^3}{R_{mj} \cdot R_{ij}}\right)^{\frac{1}{2}} = \sum_k \left(\frac{(d_{(j+1)k})^3}{R_{m(j+1)k} R_{i(j+1)k}}\right)^{\frac{1}{2}}. \tag{7.23}$$

The above example was given for a branch and two branches where each value of B was 1. However, the power law allows the same reduction to be performed for dendritic trees which are finite in length, providing two other conditions are met: (1) that each path that can be traced from the parent trunk to the end of each terminal branch is of the same electrotonic length; (2) each termination is of the same form (e.g. open-circuit, so that each $B_{\text{term}} = 0$). We can most easily illustrate this by taking an example shown in Fig. 7.6.

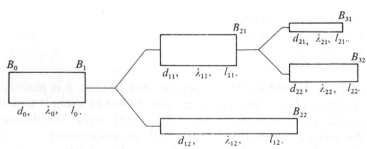

FIG. 7.6. The example treated in the text illustrating the significance of the symbols.

Let

$$L_{ij} = \frac{l_{ij}}{\lambda_{ij}}.$$

From eqn (7.15)

$$B_{21} = \frac{B_{31} + \tanh L_{21}}{1 + B_{31} \tanh L_{21}} \left(\frac{d_{21}}{d_{11}}\right)^{\frac{3}{2}} + \frac{B_{32} + \tanh L_{22}}{1 + B_{32} \tanh L_{22}} \left(\frac{d_{22}}{d_{11}}\right)^{\frac{3}{2}}.$$

If $B_{31} = B_{32}$ and $L_{21} = L_{22}$

$$B_{21} = \frac{B_{31} + \tanh L_{21}}{1 + B_{31} \tanh L_{21}} \left\{ \left(\frac{d_{21}}{d_{11}}\right)^{\frac{3}{2}} + \left(\frac{d_{22}}{d_{11}}\right)^{\frac{3}{2}} \right\}.$$

If the power law is obeyed, the term in braces is unity.

$$B_{21} = \frac{B_{31} + \tanh L_{21}}{1 + B_{31} \tanh L_{21}}.$$

Similarly

$$B_1 = \frac{B_{21} + \tanh L_{11}}{1 + B_{21} \tanh L_{11}} \left(\frac{d_{11}}{d_0}\right)^{\frac{3}{2}} + \frac{B_{22} + \tanh L_{12}}{1 + B_{22} \tanh L_{12}} \left(\frac{d_{12}}{d_0}\right)^{\frac{3}{2}}.$$

$$\frac{B_{21} + \tanh L_{11}}{1 + B_{21} \tanh L_{11}} = \frac{B_{31} + \tanh L_{21} + \tanh L_{11} + B_{31} \tanh L_{11} \tanh L_{21}}{1 + B_{31} \tanh L_{21} + B_{31} \tanh L_{11} + \tanh L_{21} \tanh L_{11}}$$

$$= \frac{B_{31}(1 + \tanh L_{11} \tanh L_{21}) + \tanh L_{21} + \tanh L_{11}}{(1 + \tanh L_{11} \tanh L_{21}) + B_{31}(\tanh L_{21} + \tanh L_{11})}$$

$$= \frac{\left\{ B_{31} + \dfrac{\tanh L_{21} + \tanh L_{11}}{1 + \tanh L_{11} \tanh L_{21}} \right\}}{\left\{ 1 + \dfrac{B_{31}(\tanh L_{21} + \tanh L_{11})}{1 + \tanh L_{11} \tanh L_{21}} \right\}}$$

$$= \frac{B_{31} + \tanh(L_{21} + L_{11})}{1 + B_{31} \tanh(L_{21} + L_{11})}.$$

The next simplification depends on the essential conditions that $B_{22} = B_{31}$, $L_{12} = L_{21} + L_{11}$, and $(d_{12})^{\frac{3}{2}} + (d_{11})^{\frac{3}{2}} = d^{\frac{3}{2}}$.

$$B_0 = (B_1 + \tanh L_0)/(1 + B_1 \tanh L_0)$$

can similarly be shown to become

$$\frac{B_{31} + \tanh(L_0 + L_{21} + L_{11})}{1 + B_{31} \tanh(L_0 + L_{21} + L_{11})}.$$

Thus it has been shown that the mathematical solution for the spread of current from left to right in the branching structure of Fig. 7.6 is identical to that for a simple extension of the first cylinder to a total electrotonic length of $L_0+L_{11}+L_{21}$ and with a termination condition described by the value of B_{31}. This illustrates the power of Rall's simplification. It brings a whole class of dendritic trees under the description of the simple formulae of one-dimensional cable theory. A corollary of this demonstration is that when a dendritic tree does obey the three requirements mentioned above, it will be impossible to determine anything about the pattern of branching by passing current and recording voltage centrally.

Calculation of the input conductance of a dendritic tree. It may be desirable to calculate the input conductance of a dendritic tree, whose geometry is described histologically, for given values of R_m and R_i. The procedure for doing this is to start at the terminal branches (as in the previous example) and to calculate back successively from B_{j+1} to B_j until B_0 is reached. This will involve assuming a value for the terminal B values and also, of course, means estimating the value of L_j for each branch. The input conductance is then defined by eqn (7.7). Rall (1959a) has given an example of such a calculation; though straightforward in principle it is rather laborious in practice.

A much simpler calculation can be performed if it can be established (from the histological data) that the dendritic branching obeys the $\frac{3}{2}$ power law mentioned above. It is then necessary to calculate the electrotonic distance of the various paths from the parent trunk to each termination. If these are all equal (or approximately equal) then the whole structure may be treated as an equivalent cylinder. The diameter of the cylinder is that of the parent trunk and the input conductance is given by the formula (assuming the B_{term} are all zero)

$$G_0 = \frac{\pi}{2}\sqrt{\left(\frac{d_0^3}{R_m R_i}\right)}\tanh L, \qquad (7.24)$$

where L is the electrotonic length of the equivalent cylinder.

Calculation of R_m from the input conductance of a nerve cell or dendritic tree. This is a problem closely related to the previous one (calculation of the input conductance) and is more likely to be required for experimental work since it is now possible to make electrophysiological measurements on particular nerve cells and then determine their structure by filling the cell with an opaque (or radio-opaque) material (Stretton and Kravitz 1968; Globus, Lux, and Schubert 1968; Barrett and Graubard 1970). Before considering the case of a nerve cell it will be useful to describe the procedure for a single dendritic tree.

If the Rall power law is not obeyed by the dendritic tree it is necessary to assume a value for R_m (and R_i) and then follow the method described in the previous section. This may then be compared with a measured value for the input conductance. On this basis the estimate of R_m can be modified and the procedure then repeated. After a few repetitions a fairly accurate match between measured and calculated input conductance should be obtained. The value of R_m with which to start the procedure can be obtained from eqn (7.8):

$$G_0 = B_0 G_0 = B_0 \frac{\pi}{2} \sqrt{\left(\frac{d_0^3}{R_m R_i}\right)} \tag{7.25}$$

therefore

$$R_m = \frac{B_0^2 \pi^2 d_0^3}{4 G_0^2 R_i} \approx \frac{\pi^2 d_0^3}{4 G_0^2 R_i} \tag{7.26}$$

The approximate formula given in eqn (7.26) is justified because B_0 can be assumed roughly equal to 1 when the power law is approximately obeyed and the electrotonic length of the dendrites is fairly long (see Rall 1959b).

When recording from nerve cells it is possible to measure their input conductance. This input conductance will be given by the sum of the soma conductance and of the input conductances of each of the dendritic trees.

$$G_N = G_{soma} + \sum_j G_{0j} = \frac{S}{R_m} + \sum_j B_{0j} \frac{\pi}{2} \sqrt{\left(\frac{d_{0j}^3}{R_m R_i}\right)}, \tag{7.27}$$

where G_N is the input conductance of the cell and S is the soma surface area.
This leads to a quadratic equation in $R_m^{\frac{1}{2}}$

$$G_N R_m - \left(\frac{\pi}{2 R_i^{\frac{1}{2}}}\right) \sum_j B_{0j} (d_{0j})^{\frac{3}{2}} R_m^{\frac{1}{2}} - S = 0 \tag{7.28}$$

or

$$G_N R_m - C D^{\frac{3}{2}} R_m^{\frac{1}{2}} - S = 0,$$

where

$$C = \pi / 2 R_i^{\frac{1}{2}} \quad \text{and} \quad D^{\frac{3}{2}} = \sum_j B_{0j} (d_{0j})^{\frac{3}{2}}.$$

that is,

$$R_m = \left[\frac{C D^{\frac{3}{2}} \pm \sqrt{\{(C D^{\frac{3}{2}})^2 + 4 G_N S\}}}{2 G_N} \right]^2. \tag{7.29}$$

For conventional physiological values the positive sign is appropriate, since otherwise $R_m^{\frac{1}{2}}$ would be negative. This may be shown by a numerical example, taken from Rall (1959a, b):

$$S = 1.25 \times 10^{-4} \text{ cm}^2, \quad \sum_j B_{0j} (d_{0j})^{\frac{3}{2}} = 2.5 \times 10^{-4} \text{cm}^{\frac{3}{2}},$$

$$C = 0.2 (\Omega \text{ cm})^{-\frac{1}{2}}$$

$$G_N = \frac{1}{1.2 \times 10^6} \text{ siemens}$$

(i.e. the input resistance $= 1.2$ MΩ). The two values of R_m calculated from eqn (7.28) are 3900 Ω cm² and 5.8 Ω cm². The second value implies that the soma conductance is greater than the total input conductance! Hence

Hence

$$R_m = \frac{C^2 D^3}{4 G_N^2} \left\{ 1 + \sqrt{\left(1 + \frac{4 S G_N}{C^2 D^3}\right)} \right\}^2 \approx \frac{C^2 D^3}{G_N^2} \left(1 + \frac{2 S G_N}{C^2 D^3}\right). \tag{7.30}$$

One practical problem about using this formula is the estimate of D; the same successive approximation described above may be required. Once again if the power law is obeyed and the dendrites are fairly long it is reasonable to assume each $B_{0j} = 1$ so that

$$D^{\frac{3}{2}} = \sum_j (d_{0j})^{\frac{3}{2}}.$$

In other circumstances the more laborious procedure described above (see p. 144) should be followed.

Central spread of current in a dendritic tree

When current is injected into a nerve cell by synaptic action, the flow of current in both central and peripheral directions must be considered. The general method for the description of this current is, of course, that described in the section *Formulation of the equations* on p. 135. However, it is possible to make a simplification of this procedure, not only for the case of the peripheral spread of current (section *Reduction of a dendritic tree to an equivalent cylinder*) but also for the equations describing the central spread (see also Rall and Rinzel 1973). The nature of the simplification for the central spread of current can be most easily described by reference to Fig. 7.7(a). Current I is injected in cylinder 11 at a distance y from its central junction with cylinders 1 and 12. Cylinder 1 terminates centrally in an impedance Z_0 which represents the soma and any other dendrites.

The simplification is that the central spread of current, along cylinder 1 *only* is described by exactly the same equations as for the model shown in Fig. 7.7(b), providing that the conditions for the reduction of the whole dendritic

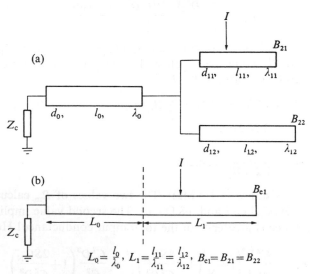

FIG. 7.7. Illustration of the two models discussed in the text.

tree to an equivalent cylinder (see *Reduction of a dendritic tree to an equivalent cylinder*) are fulfilled. In Fig. 7.7(b) an equal quantity of current (i.e. *I*) is injected into a cylinder (the 'equivalent' one, for *peripheral spread*, to the dendritic tree in Fig. 7.7(a) and terminating centrally in the same way, with the impedance Z_c) at an electrotonic distance X from the model soma which is the same electrotonic distance as in Fig. 7.7(a) (that is, $X = l_0/\lambda_0 + y/\lambda_{11}$). The proof of the identity of the equations (for the central length l_0) is similar in principle to the proof for peripheral spread outlined previously, and is omitted because it is lengthy but straightforward.

It is important to note that the equations describing axial current, membrane voltage, etc. for the regions beyond the central segment will be different in the two models. It will be obvious from inspection that the equations for the intermediary length between the site of current injection and the central length will be different. In the model of Fig. 7.7(a) the description of the voltage distribution for this intermediary length requires two equations, since it will decline centrally in cylinder 11 and peripherally in cylinder 12. In the model of Fig. 7.7(b) only one equation, describing a central decline in the voltage, is required. The similarities and differences in the voltage distribution for a particular case of these two models is shown in Fig. 7.8.

In order to obtain exactly the same voltage distribution as in Fig. 7.7(b) throughout the model of Fig. 7.7(a), it would be necessary to inject current not only into cylinder 11 but also into cylinder 12, in such a way that the voltage produced at the sites of injection (both an equal *electrotonic* distance from cylinder 1 and hence the soma) is the same; that is, the total current I would have to be subdivided and part of it injected into cylinder 11 at distance $(l_0/\lambda_0 + y/\lambda_{11})$ from the soma and the rest injected into cylinder 12 at distance $(l_0/\lambda_0 + z/\lambda_{12})$ (where $z = \lambda_{12} y/\lambda_{11}$), where the ratio of the two currents is affected by the input conductances (at the appropriate point) of the two cylinders.

Two conclusions follow; they only apply, of course, to the class of dendritic trees which can be represented as an equivalent cylinder. The first is that the voltage recorded at the central termination of a particular dendritic tree (i.e. at the soma) is only dependent on the magnitude of the current and the electrotonic distance at which it is injected. It will be independent of the proportion of dendritic branches at that electrotonic distance, into which the current is injected (providing, of course, that the membrane response remains linear). The second conclusion is a corollary: it is not possible to decide the number of sites at which current is injected into a dendritic tree (if all at the same electrotonic distance) unless recordings can be made at various points along the dendritic tree.

Transient solutions for a structure with cylindrical elements

All the equations so far given for current spread in a branching structure have described the steady state solutions when a constant current is injected at

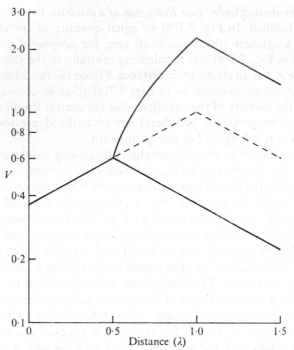

FIG. 7.8. This is an example of the voltage distribution in the two models shown in Fig. 7.7. The value of Z_c is taken to be equal to an infinite extension of the central cylinder. Similarly the value of B at the right-hand end is assumed equal to 1. The central branch is 0.5λ in length and the peripheral branches are 1.0λ in length. The continuous lines represent the voltage distribution for current injected into the smaller of two branches (input conductance one-third of the value for the 'equivalent' cylinder) at a distance of 0.5λ from the branch point. The dashed lines show the voltage distribution for the same current injected into the 'equivalent' cylinder model at the same electrotonic distance from the central end.

some point. In other words these equations are particular solutions of the general equation for steady state one-dimensional cable theory (see Chapter 3)

$$\mathrm{d}^2V/\mathrm{d}X^2 = V.$$

In order to determine the transient solutions we have to solve the partial differential equation
$$\partial^2V/\partial X^2 = V + \partial V/\partial T. \tag{3.13}$$

It has been shown already that use of the method of Laplace transformation converts eqn (3.13) into an ordinary differential equation

$$\mathrm{d}^2\bar{V}/\mathrm{d}X^2 = (s+1)\bar{V}$$
or
$$\mathrm{d}^2\bar{V}/\mathrm{d}y^2 = \bar{V},$$

where $y = \sqrt{(s+1)}\, X$.

It will be apparent, therefore, that the transient solutions in the Laplace-transform domain will bear a close relationship to the steady state solutions (providing the boundary conditions for the two cases are analogous). The remaining difficulty is that of converting these solutions back into the time domain. It is not always possible to do this easily; but those dendritic trees which satisfy the conditions for reduction to an equivalent cylinder for steady state current spread (see the section *Reduction of a dendritic tree to an equivalent cylinder*) can also be treated as equivalent cylinders in their transient response (providing the value of τ_m is the same throughout the tree). All the deductions which have been made in the section *Steady state in a structure with cylindrical elements* for the steady state response apply equally well to the transient response. The form of the proofs is exactly analogous to those illustrated earlier, but now made in the Laplace-transform domain instead of the time domain. In the next section an alternative proof (Rall 1962a) of this statement, for the case of the peripheral spread of current, will be quoted.

Structures with tapering as well as branching

Formulation of the equations. It is not surprising that the mathematical theory of tapered structures is much more complicated than that of cylinders; the extra difficulty created by this nonuniformity means that very few exact solutions have been obtained (see e.g. Gruner 1965; Ghausi and Kelly 1968; Rudjord and Rommetredt 1970). In this section we will consider structures which branch as well as taper; the primary aim being to determine what combinations of tapering and branching lead to differential equations which can be easily solved—in particular, of course, to find if there exist combinations of tapering and branching which will allow the dendritic tree to be treated as an equivalent cylinder. It would be a formidable task to determine the tapering condition for all possible forms of branching, but Rall (1962a) has given a detailed presentation of the case where all the branches are equal in diameter. Once again we will follow Rall's presentation very closely.

Rall assumes that both the radius r (common to all branches) and the number n of dendritic branches are functions of distance x from the soma. Each dendritic branch is assumed to be circular in cross-section, so that at any distance from the soma the total cross-sectional area of the dendritic branches is simply given by $\pi r^2 n$. The total axial resistance to current flow is therefore $R_i/\pi r^2 n$. Remembering the derivation of eqn (3.2), it will be apparent that

$$I_a = -\frac{\pi r^2 n}{R_i}\left(\frac{\partial V}{\partial x}\right), \qquad (7.31)$$

where I_a is the total axial current.

The membrane current per unit length i_m cannot be used in a nonuniform structure of this kind so that the membrane current density I_m is used. In a cylinder

$$i_m = -\partial i_a/\partial x \quad \text{and} \quad I_m 2\pi r = -\partial i_a/\partial x.$$

In a nonuniform structure the appropriate equation is

$$I_m(dA/dx) = -\partial I_a/\partial x. \qquad (7.32)$$

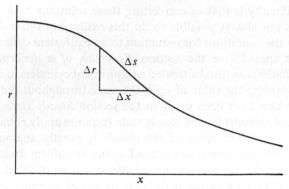

FIG. 7.9. Illustration of the symbols employed in order to calculate the membrane area in a tapered cylinder.

The increment of membrane area per unit distance can be readily calculated. Consider the tapering of a single branch, as illustrated in Fig. 7.9.

$$\Delta s^2 = \Delta x^2 + \Delta r^2,$$

therefore

$$\left(\frac{\Delta s}{\Delta x}\right)^2 = 1 + \left(\frac{\Delta r}{\Delta x}\right)^2.$$

In the limit, as $\Delta x \to 0$,

$$\left(\frac{\mathrm{d}s}{\mathrm{d}x}\right)^2 = 1 + \left(\frac{\mathrm{d}r}{\mathrm{d}x}\right)^2. \tag{7.33}$$

For a single branch, the increment in area $\mathrm{d}a$ is given by

$$\mathrm{d}a = 2\pi r\, \mathrm{d}s$$

therefore

$$\frac{\mathrm{d}a}{\mathrm{d}x} = 2\pi r \frac{\mathrm{d}s}{\mathrm{d}x} = 2\pi r \sqrt{\left\{1 + \left(\frac{\mathrm{d}r}{\mathrm{d}x}\right)^2\right\}},$$

therefore

$$\frac{\mathrm{d}A}{\mathrm{d}x} = 2\pi rn \frac{\mathrm{d}s}{\mathrm{d}x} = 2\pi rn \sqrt{\left\{1 + \left(\frac{\mathrm{d}r}{\mathrm{d}x}\right)^2\right\}}, \tag{7.34}$$

where $\mathrm{d}A$ is the sum of the area for all the branches.
From eqn (7.32) it follows that

$$I_\mathrm{m} = -\frac{\partial I_\mathrm{a}}{\partial x} \frac{1}{2\pi rn}\left(\frac{\mathrm{d}s}{\mathrm{d}x}\right)^{-1}. \tag{7.35}$$

Differentiation of eqn (7.31) with respect to x will give an expression for $\partial I_\mathrm{a}/\partial x$ in terms of V and x.

$$\frac{\partial I_\mathrm{a}}{\partial x} = -\frac{\pi}{R_\mathrm{i}}\left\{r^2 n \frac{\partial^2 V}{\partial x^2} + \frac{\partial V}{\partial x}\frac{\mathrm{d}(r^2 n)}{\mathrm{d}x}\right\}$$

$$= -\frac{\pi}{R}r^2 n\left\{\frac{\partial^2 V}{\partial x^2} + \frac{\partial V}{\partial x}\frac{\mathrm{d}(\ln(r^2 n))}{\mathrm{d}x}\right\} \tag{7.36}$$

since

$$\frac{d}{dx}\ln(f(x)) = \frac{1}{f(x)}\frac{d(f(x))}{dx}.$$

It was found useful in considering the steady state voltage distribution in cylindrical branches to introduce the variable Z, expressing the actual length normalized appropriately by the space constant pertaining to each branch. Before deciding what substitution might be appropriate here it is useful to consider some general properties of the relationship between Z and x.

$$\frac{\partial V}{\partial x} = \frac{\partial V}{\partial Z}\frac{\partial Z}{\partial x}, \tag{7.37}$$

by the chain rule.

$$\frac{\partial^2 V}{\partial x^2} = \frac{\partial Z}{\partial x}\frac{\partial(\partial V/\partial Z)}{\partial x} + \frac{\partial V}{\partial Z}\frac{\partial(\partial Z/\partial x)}{\partial x}$$

$$= \left(\frac{\partial Z}{\partial x}\right)^2\frac{\partial(\partial V/\partial Z)}{\partial Z} + \frac{\partial V}{\partial Z}\frac{\partial Z}{\partial x}\frac{\partial}{\partial x}\{\ln(\partial Z/\partial x)\}$$

$$= \left(\frac{\partial Z}{\partial x}\right)^2\left\{\frac{\partial^2 V}{\partial Z^2} + \frac{\partial V}{\partial Z}\left(\frac{\partial Z}{\partial x}\right)^{-1}\frac{\partial}{\partial x}\ln(\partial Z/\partial x)\right\}. \tag{7.38}$$

These relationships for $\partial V/\partial x$ and $\partial^2 V/\partial x^2$ (eqns (7.37) and (7.38)) can now be substituted in eqn (7.36).

$$\frac{\partial I_a}{\partial x} = -\frac{\pi r^2 n}{R_i}\left[\left(\frac{dZ}{dx}\right)^2\left\{\frac{\partial^2 V}{\partial Z^2} + \frac{\partial V}{\partial Z}\left(\frac{dZ}{dx}\right)^{-1}\frac{d}{dx}\ln\left(\frac{dZ}{dx}\right)\right\} + \frac{\partial V}{\partial Z}\frac{dZ}{dx}\frac{d}{dx}\ln(r^2 n)\right]$$

$$= -\frac{\pi r^2 n}{R_i}\left(\frac{dZ}{dx}\right)^2\left\{\frac{\partial^2 V}{\partial Z^2} + \frac{\partial V}{\partial Z}\left(\frac{dZ}{dx}\right)^{-1}\frac{d}{dx}\ln\left(r^2 n\frac{dZ}{dx}\right)\right\}. \tag{7.39}$$

The relationship between I_m and $\partial I_a/\partial x$ is given by eqn (7.35). Since, for a passive membrane

$$I_m = (V/R_m) + C_m(\partial V/\partial t),$$

we obtain the following differential equation

$$V + \tau_m\frac{\partial V}{\partial t} = \frac{rR_m}{2R_i}\cdot\left(\frac{ds}{dx}\right)^{-1}\left(\frac{dZ}{dx}\right)^2\left\{\frac{\partial^2 V}{\partial Z^2} + \frac{\partial V}{\partial Z}\left(\frac{dZ}{dx}\right)^{-1}\frac{d}{dx}\ln\left(r^2 n\frac{dZ}{dx}\right)\right\}. \tag{7.40}$$

Inspection of the right-hand side of the above equation indicates that a considerable simplification would be achieved if we made the relationship between Z and x obey the following equation

$$\frac{dZ}{dx} = \left(\frac{rR_m}{2R_i}\right)^{-\frac{1}{2}}\left(\frac{ds}{dx}\right)^{\frac{1}{2}}, \tag{7.41}$$

since this reduces eqn (7.40) to

$$V+\tau_{\mathrm{m}}\frac{\partial V}{\partial t} = \frac{\partial^2 V}{\partial Z^2} + \frac{\partial V}{\partial Z}\left(\frac{rR_{\mathrm{m}}}{2R_{\mathrm{i}}}\right)^{\frac{1}{2}}\left(\frac{\mathrm{d}s}{\mathrm{d}x}\right)^{-\frac{1}{2}}\cdot\frac{\mathrm{d}}{\mathrm{d}x}\ln\left\{r^{\frac{3}{2}}n\left(\frac{R_{\mathrm{m}}}{2R_{\mathrm{i}}}\right)^{\frac{1}{2}}\cdot\left(\frac{\mathrm{d}s}{\mathrm{d}x}\right)^{\frac{1}{2}}\right\},$$

i.e.

$$\frac{\partial^2 V}{\partial Z^2} + \frac{\partial V}{\partial Z}\left[\left(\frac{rR_{\mathrm{m}}}{2R_{\mathrm{i}}}\right)^{\frac{1}{2}}\left(\frac{\mathrm{d}s}{\mathrm{d}x}\right)^{-\frac{1}{2}}\frac{\mathrm{d}}{\mathrm{d}x}\ln\left\{r^{\frac{3}{2}}n\left(\frac{\mathrm{d}s}{\mathrm{d}x}\right)^{\frac{1}{2}}\right\}\right] = V + \frac{\partial V}{\partial T}, \tag{7.42}$$

since

$$\frac{\mathrm{d}}{\mathrm{d}x}\ln(af(x)) = \frac{\mathrm{d}}{\mathrm{d}x}\ln(f(x)).$$

We are now in a position to discuss various ways in which this equation may be reduced to simpler forms.

Condition for reduction to an equivalent cylinder. It will be remembered that the equation for a one-dimensional cylinder is

$$\partial^2 V/\partial X^2 = V+(\partial V/\partial T). \tag{3.13}$$

Eqn (7.42) will simplify to this equation when the coefficient of $\partial V/\partial Z$ is equal to zero; neither $\mathrm{d}s/\mathrm{d}x$ nor r can be zero for any realizable structure so that the condition is

$$\frac{\mathrm{d}}{\mathrm{d}x}\ln\left\{r^{\frac{3}{2}}n\left(\frac{\mathrm{d}s}{\mathrm{d}x}\right)^{\frac{1}{2}}\right\} = 0,$$

i.e. when

$$r^{\frac{3}{2}}n\left(\frac{\mathrm{d}s}{\mathrm{d}x}\right)^{\frac{1}{2}} = \text{constant } (C) \tag{7.43}$$

or

$$r^{\frac{3}{2}}n\left\{1+\left(\frac{\mathrm{d}r}{\mathrm{d}x}\right)^2\right\}^{\frac{1}{4}} = C.$$

Once a suitable choice of r and n as functions of x has been made it is necessary to determine the relationship between Z and x.

$$\frac{\mathrm{d}Z}{\mathrm{d}x} = \left(\frac{rR_{\mathrm{m}}}{2R_{\mathrm{i}}}\right)^{-\frac{1}{2}}\left(\frac{\mathrm{d}s}{\mathrm{d}x}\right)^{\frac{1}{2}} \tag{7.41}$$

$$= \left(\frac{R_{\mathrm{m}}}{2R_{\mathrm{i}}}\right)^{-\frac{1}{2}}r^{-2}n^{-1}C,$$

therefore

$$Z = C\left(\frac{R_{\mathrm{m}}}{2R_{\mathrm{i}}}\right)^{-\frac{1}{2}}\int_0^x \frac{\mathrm{d}y}{r^2 n}. \tag{7.44}$$

It can be shown that constant increments of Z, for this type of tapering and branching structure, correspond to constant increments of the membrane surface area. In other words

$$\mathrm{d}A/\mathrm{d}x \propto \mathrm{d}Z/\mathrm{d}x. \tag{7.45}$$

This follows because

$$\frac{dA}{dx} = 2\pi r n \frac{ds}{dx}$$

$$= 2\pi \left\{ r^{\frac{3}{2}} n \left(\frac{ds}{dx}\right)^{\frac{1}{2}} \right\} \left(\frac{1}{r} \cdot \frac{ds}{dx}\right)^{\frac{1}{2}}$$

$$= 2\pi C \left(\frac{1}{r} \frac{ds}{dx}\right)^{\frac{1}{2}}$$

$$\propto \frac{dZ}{dx}, \quad \text{since } \frac{dZ}{dx} \propto \left(\frac{1}{r} \cdot \frac{ds}{dx}\right)^{\frac{1}{2}}.$$

As Rall (1962a) pointed out this result is important for it shows that in any structure which can be mathematically described as an equivalent cylinder, equal steps of electrotonic length correspond to equal steps of dendritic surface area (and hence possible synaptic surface area).

In the special case of no tapering, $dr/dx = 0$ and therefore $ds/dx = 1$. The condition on r and n therefore becomes

$$nr^{\frac{3}{2}} = \text{constant}.$$

Since at $x = 0$ the left-hand side of the above equation $= r_0^{\frac{3}{2}}$, (if we are considering only one dendritic tree) the condition is

$$nr^{\frac{3}{2}} = r_0^{\frac{3}{2}}. \tag{7.46}$$

This is, of course, a special case of eqn (7.22).

We can also obtain the equation for Z for this special case:

$$\frac{dZ}{dx} = \left(-\frac{rR_m}{2R_i}\right)^{-\frac{1}{2}}$$

$$= \frac{1}{\lambda}, \tag{7.47}$$

where λ here represents the characteristic length of the appropriate cylinder element(s).

Hence

$$Z = \int_0^x \frac{1}{\lambda} \, dy. \tag{7.48}$$

An example. Rall (1962a) has given an example of a dendritic tree with cylindrical components, which behaves as an equivalent cylinder. It may help to illustrate the power (and limitation) of his analysis by giving examples of trees which have taper as well.

Let us consider three forms of tapering described by the following equations

$$r = r_0(1-ax), \tag{7.49}$$

$$r = r_0 \exp(-ax), \tag{7.50}$$

and

$$r = r_0(1+ax)^{-1} \tag{7.51}$$

where r_0 is the initial radius and a is the factor controlling the rate of taper with distance.

The forms of branching required for each of these tapering conditions, in order for the structures to behave as an equivalent cylinder, may be derived from eqn (7.43) and are given, respectively, by

$$n = n_0(1-ax)^{-\frac{3}{2}}, \tag{7.52}$$

$$n = n_0 e^{2ax}\left(\frac{a^2 r_0^2 + 1}{a^2 r_0^2 + e^{2ax}}\right)^{\frac{1}{4}}, \tag{7.53}$$

and

$$n = n_0(1+ax)^{\frac{5}{2}}\left\{\frac{a^2 r_0^2 + 1}{a^2 r_0^2 + (1+ax)^4}\right\}^{\frac{1}{4}}, \tag{7.54}$$

where n_0 is the number of branches at $x = 0$.

There remains the problem of determining the relationship between Z and x for these three examples. In the case of the tapering condition described by eqn (7.49) an exact solution can be provided

$$\frac{dZ}{dx} = \left(\frac{R_m}{2R_i}\right)^{-\frac{1}{2}}\left(\frac{1}{r}\frac{ds}{dx}\right)^{\frac{1}{2}} \tag{7.41}$$

$$= \left(\frac{R_m}{2R_i}\right)^{-\frac{1}{2}}\left\{\frac{(1+a^2 r_0^2)^{\frac{1}{2}}}{r_0(1-ax)}\right\}^{\frac{1}{2}},$$

$$Z = \left(\frac{R_m r_0}{2R_i}\right)^{-\frac{1}{2}}(1+a^2 r_0^2)^{\frac{1}{4}}\int_0^x (1-ay)^{-\frac{1}{2}}\,dy$$

$$= \frac{1}{\lambda_0}(1+a^2 r_0^2)^{\frac{1}{4}}\frac{2}{a}\{1-(1-ax)^{\frac{1}{2}}\}, \tag{7.55}$$

where $\lambda_0 (= (R_m r_0/2R_i)^{\frac{1}{2}})$ is the space constant of a cylinder with the same radius as the initial radius r_0 of the tree and the same values of membrane and axial resistivity.

For the other two cases approximate solutions may be obtained. Consider tapering of the form described by eqn (7.50)

$$\frac{dZ}{dx} = \left(\frac{R_m}{2R_i}\right)^{-\frac{1}{2}}\left\{\frac{e^{ax}}{r_0}(1+a^2 r_0^2 e^{-2ax})^{\frac{1}{2}}\right\}^{\frac{1}{2}},$$

$$\frac{dZ}{dx} = \frac{1}{\lambda_0}(e^{2ax}+a^2 r_0^2)^{\frac{1}{4}},$$

therefore

$$Z = \frac{1}{\lambda_0}\int_0^x (e^{2ay}+a^2 r_0^2)^{\frac{1}{4}}\,dy. \tag{7.56}$$

Provided that $ar_0 \ll 1$ this may be simplified to

$$Z \approx \frac{1}{\lambda_0} \int_0^\infty e^{\frac{1}{2}ay} \, dy$$

$$\approx \frac{2}{a\lambda_0}(e^{\frac{1}{2}ax}-1). \tag{7.57}$$

Similar reasoning leads to an approximate solution for the third example

$$Z \approx \frac{2}{3a\lambda_0}\{(1+ax)^{\frac{3}{2}}-1\}. \tag{7.58}$$

The three examples may be compared by considering a dendritic tree of total length 500 μm which has an initial radius of 11 μm and a radius at the end of each terminal branch of 1 μm. Fig. 7.10 shows the relationship, for each example, between the radius of the branches and distance, and also shows the number of branches needed to fulfil the condition for an equivalent cylinder. The curves display the expected property of increasing the number of branches inversely to the decrease in diameter of each branch. At the end of each model tree the diameter is the same (2 μm) and the number of branches

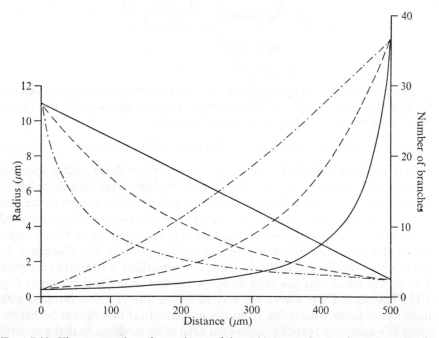

FIG. 7.10. Three examples of tapering and branching: continuous lines, eqn (7.49); dashed lines, eqn (7.50); dots and dashes, eqn (7.51). *Abscissa*: distance in micrometers from soma. *Ordinate*: left: radius in micrometres; right: number of branches required.

is closely similar. In fact, this is the same as that which would be expected if there were *no* tapering, but the branching of the cylindrical components followed the Rall power law (eqn (7.22)). This is so because the number of branches would be given by

$$n = n_0\left(\frac{r_0}{r}\right)^{\frac{3}{2}}. \tag{7.46}$$

In the case of a steady rate of tapering $ds/dx = $ constant and therefore eqn (7.43) becomes

$$nr^{\frac{3}{2}}K = C.$$

This also simplifies to eqn (7.46). When the rate of tapering is not constant it can still be shown that the same law holds approximately, providing the rate of tapering is small compared to the initial radius of the tree (that is) $ar_0 \ll 1$). For example, when $r = r_0e^{-ax}$

$$n = n_0e^{2ax}\left(\frac{1+a^2r_0^2}{e^{2ax}+a^2r_0^2}\right)^{\frac{1}{4}}$$

$$\approx n_0e^{\frac{3}{2}ax}$$

$$\approx n_0(r_0/r)^{\frac{3}{2}}.$$

Similarly, when $r = r_0(1+ax)^{-2}$

$$n = n_0(1+ax)^{\frac{5}{2}}\left\{\frac{1+a^2r_0^2}{(1+ax)^4+a^2r_0^2}\right\}^{\frac{1}{4}}$$

$$\approx n_0(1+ax)^{\frac{3}{2}}$$

$$\approx n_0(r_0/r)^{\frac{3}{2}}.$$

These illustrations are particular cases of the general conclusion which can be deduced from eqn (7.43). Whenever the rate of tapering with distance is small relative to the radius of the initial branch,

$$ds/dx \approx \text{constant,} \quad \text{and hence} \quad nr^{\frac{3}{2}} \approx \text{constant.}$$

This indicates that Rall's power law may be generalized (either approximately or exactly) to a much wider class of dendritic trees than those whose elements are all cylindrical. The above demonstration only applies to the case when all the branches are equal in diameter; furthermore, it has been assumed that n can take any value—whereas it can only strictly go up in binary steps. This latter limitation of the mathematical description is best illustrated by attempting to realize one of the dendritic trees described in the three examples. The way in which this has been done for the first example is shown in Fig. 7.11. The dashed line shows the way in which n is taken to increase with distance. A further deviation from the mathematical description is that the value of r near each branch point would need to be modified so that a smooth continuous tree results. The dotted line in Fig. 7.11 shows how r would have to change with x.

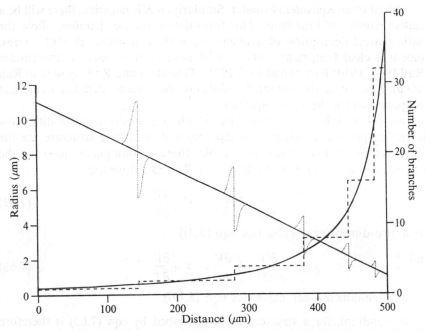

FIG. 7.11. The continuous lines are plotted as in Fig. 7.10. The dashed lines show how the number of branches would have to be changed in a physical realization of the mathematical results. Similarly the dotted line illustrates diagramatically how the radius of the branches would need to change in order that the dendritic tree should have a smooth contour.

Conditions for reduction to other solutions. On p. 152 the equation that Rall (1962*a*) had derived was given.

$$V+\frac{\partial V}{\partial T}=\frac{\partial^2 V}{\partial Z^2}+\frac{\partial V}{\partial Z}\left\{\left(\frac{rR_\mathrm{m}}{2R_\mathrm{i}}\right)^{\frac{1}{2}}\left(\frac{\mathrm{d}s}{\mathrm{d}x}\right)^{-\frac{1}{2}}\frac{\mathrm{d}}{\mathrm{d}x}\ln\left(r^{\frac{3}{2}}n\left(\frac{\mathrm{d}s}{\mathrm{d}x}\right)^{\frac{1}{2}}\right)\right\}. \tag{7.42}$$

Rall (1962*a*) pointed out that, if it was assumed that

$$r^{\frac{3}{2}}n\left(\frac{\mathrm{d}s}{\mathrm{d}x}\right)^{\frac{1}{2}} = Ce^{KZ}, \tag{7.59}$$

the differential equation simplifies to

$$V+\frac{\partial V}{\partial T} = \frac{\partial^2 V}{\partial Z^2}+K\frac{\partial V}{\partial Z}. \tag{7.60}$$

An intuitive understanding of the significance of this description can be obtained by considering the case of no tapering:

$$r^{\frac{3}{2}}n = r_0^{\frac{3}{2}}e^{KZ}. \tag{7.61}$$

Thus, if K is positive, the branching pattern will show flaring compared with the dendritic tree whose values of r have the same relationship to x, which can be reduced to an equivalent cylinder. Similarly, if K is negative, there will be a relative paucity of branching. This formulation should therefore allow the mathematical description of a much wider class of actual dendritic trees, since it is clear that many could not be reduced to an equivalent cylinder (Rall 1964, 1970; Barrett and Crill 1971). Goldstein and Rall (quoted in Rall 1969a) have recently obtained solutions and made detailed numerical computations for the above equation.

Two other differential equations which are obvious possibilities for describing particular examples of tapering and branching structure are the equations for the 'two-dimensional' cable (that is, a thin plane sheet) and the 'three-dimensional' cable (see Chapter 5). The equations are

$$\frac{\partial^2 V}{\partial Z^2} + \frac{1}{Z}\frac{\partial V}{\partial Z} = V + \frac{\partial V}{\partial T} \tag{7.62}$$

for a 'two-dimensional' cable (see eqn (5.4))

and

$$\frac{\partial^2 V}{\partial Z^2} + \frac{2}{Z}\frac{\partial V}{\partial Z} = V + \frac{\partial V}{\partial T} \tag{7.63}$$

for a 'three-dimensional' cable (see eqn (5.29)).

The condition for a structure to be described by eqn (7.62) is therefore

$$\frac{1}{Z}\frac{dZ}{dx} = \frac{d}{dx}\ln\left\{r^{\frac{3}{2}}n\left(\frac{ds}{dx}\right)^{\frac{1}{2}}\right\} \tag{7.64}$$

(see eqn (7.40)),

i.e.

$$\frac{d}{dx}\ln(Z) = \frac{d}{dx}\ln\left\{r^{\frac{3}{2}}n\left(\frac{ds}{dx}\right)^{\frac{1}{2}}\right\},$$

i.e.

$$Z \propto r^{\frac{3}{2}}n\left(\frac{ds}{dx}\right)^{\frac{1}{2}} \tag{7.65}$$

or

$$\int\left(\frac{1}{r}\frac{ds}{dx}\right)^{\frac{1}{2}}dx \propto r^{\frac{3}{2}}n\left(\frac{ds}{dx}\right)^{\frac{1}{2}} \tag{7.66}$$

Similarly for the structure to be described as a 'three-dimensional' cable,

$$Z \propto r^{\frac{3}{4}}n^{\frac{1}{2}}\left(\frac{ds}{dx}\right)^{\frac{1}{4}}. \tag{7.67}$$

It will be evident from inspection of eqns (7.65) and (7.67) that no realizable structure can obey these equations, because of the condition that Z must equal 0 when $x = 0$. Since $ds/dx (= \{1 + (dr/dx)^2\}^{\frac{1}{2}})$ will always be finite, this is equivalent to either n or r being zero at $x = 0$. Thus it has been shown that there is no realizable branching structure (which is described mathematically by the above formulation) that can be the equivalent of a 'two-dimensional' or 'three-dimensional' cable. This demonstration should not be surprising when it is remembered that the solution of these latter equations gives the steady state voltage (at $x = 0$) as infinite (see Chapter 5), whereas the above formulation makes the assumptions of 'one-dimensional' cable theory and hence does not lead to an infinite steady state voltage at $x = 0$.

We can explore how closely one kind of branching structure approaches eqns (7.62) and (7.63). Let us consider the case where there is no tapering (i.e. cylindrical elements) and the radii of all the branches are the same ($r = r_0$):

$$\frac{dZ}{dx} = \left(\frac{2R_i}{r_0 R_m}\right)^{\frac{1}{2}} = \frac{1}{\lambda_0},$$

that is,

$$Z = x/\lambda_0.$$

The coefficient of $\partial V/\partial Z$ in eqn (7.42) therefore becomes $\lambda d\,(\ln n)/dx$. In order for this to go exactly to expressions (7.62) and (7.63) we would require $n = ax$ for the two-dimensional case and $n = ax^2$ for the three-dimensional case, where a is a constant setting the rate of branching with distance. In both cases the assumption would lead to $n = 0$ at $x = 0$.

Let us explore the related branching conditions

$$n = n_0\,(ax+1) \tag{7.68}$$

and

$$n = n_0\,(ax+1)^2, \tag{7.69}$$

where n_0 is the initial number of branches, to see how closely these will approximate to the appropriate differential equation.

When $n = n_0\,(ax+1)$ and $r = r_0$ the differential equation which exactly describes this structure is

$$V+\frac{\partial V}{\partial T} = \frac{\partial^2 V}{\partial Z^2}+\frac{\partial V}{\partial Z}\left[\lambda_0\frac{d}{dx}\ln\left\{n_0(ax+1)r_0^{\frac{3}{2}}\right\}\right]$$

$$= \frac{\partial^2 V}{\partial Z^2}+\frac{\partial V}{\partial Z}\frac{\lambda_0 a}{ax+1}$$

$$= \frac{\partial^2 V}{\partial Z^2}+\frac{\partial V}{\partial Z}\frac{1}{Z+1/a\lambda_0}. \tag{7.70}$$

Thus, when $ax \gg 1$ (i.e. $n \gg n_0$), eqn (7.70) approximates to eqn (7.62).

Similarly it may be shown that, when the branching is described by $n = n_0\,(ax+1)^2$, the exact differential equation is

$$V+\frac{\partial V}{\partial T} = \frac{\partial^2 V}{\partial Z^2}+\frac{\partial V}{\partial Z}\frac{2}{Z+1/a\lambda_0}$$

so that, when $ax \gg 1$, this approximates to the 'three-dimensional' cable equation.

It is important to note that the above formulation does *not* treat the case of structures in which there is convergence as well as divergence in the possible paths of axial current. It may be possible, however, to show by a geometrical transformation that, in some cases, a structure with convergence behaves in the same way as one with divergence only. A simple example is the case where, for each point of convergence, the connections between the different branches are made at the same electrotonic distance from the point of current injection. Since the membrane voltage will be identical in the two branches, no current will flow across the connection and the structure will behave in the same way as a purely divergent one.

This is an example of the kind of justification required for the assumption made in Chapter 5 that eqns (7.62) and (7.63) may describe the electrical behaviour of certain classes of syncytia, except when recording very close to the site of current injection.

Part III: The Rall model of the motoneurone

Description and formal solutions

In the two preceding parts of this chapter the justification has been given for the two special assumptions which Rall made in his model of the moto-neurone, namely, isopotentiality of the soma and treatment of the whole dendritic tree as an 'equivalent cylinder'. Fig. 7.12 summarizes the trans-formations which these assumptions allow.

In the first transformation the soma is converted into an isopotential sphere (which can hence be represented by a single R–C element) and each dendrite into an equivalent cylinder. Notice that the diameter and length of each of the equivalent cylinders may be very different, the diameter will be equal to the trunk diameter of the dendrite which the cylinder represents. At this stage the model would still allow a description of the membrane voltage and current spread in individual dendrites, though not in their branches; in other words the effect of current injection into only one of the dendrites could be explored, not only at the soma but also in individual dendritic cylinders.

The second transformation assumes that all the equivalent cylinders representing each dendrite are of the same *electrotonic* length and have the same end-termination. They are consequently lumped into a single equivalent

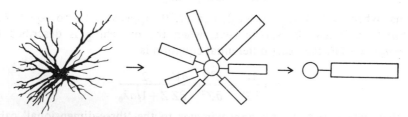

FIG. 7.12. Transformations made in the development of the Rall model. The nerve cell (left) is represented by a model with isopotential soma and each dendrite as an equivalent cylinder (middle). The next transformation is to assume that the individual cylinders, each representing a single dendrite, can all be lumped into a single equivalent cylinder.

cylinder of the same electrotonic length and with a diameter which obeys the Rall power law (see eqn (7.22)). In its reduced version the model formally only allows representation of current injection into the dendrites if the current is injected into all dendrites at the same electrotonic distance from the soma. However, this is less of a restriction than may appear. It is natural to focus attention on the voltage response at the model soma since this is usually the only part of a nerve cell from which intracellular records can be made. It can be shown that the voltage response at the *soma* of the reduced model is identical to that which would be obtained in the intermediate model if current of identical magnitude and time course were injected into only one of the equivalent cylinders of the intermediate model, provided, of course, that the requirements for making the second transformation hold (see Rall 1967; Jack and Redman 1971*b*).

Lumped soma semi-infinite cable model. The simplest possible form of the model is one in which each of the dendrites (and the axon) can be treated as cylinders of infinite extent. Rall made this assumption in his earliest description (Rall 1959*a*, 1960). The reduced model is then simply a semi-infinite cable terminated with a simple R–C element. Only one additional parameter is therefore required when compared to the description of a semi-infinite cable: this is a term giving a description of the electrical 'magnitude' of the terminating elements. Rall achieved this by defining a dendritic–soma conductance ratio ρ. Since the input conductance of a semi-infinite cable is given by $1/\sqrt{(r_m r_a)}$ (see eqn (3.26)),

$$\rho = 1/\sqrt{(r_m r_a)}G_s$$
$$= 1/r_a \lambda G_s, \tag{7.71}$$

where G_s represents the soma conductance. We can explore the dependence of ρ on the geometry of the cell in the following way:

$$\frac{1}{\sqrt{(r_m r_a)}} = \frac{\pi d^{\frac{3}{2}}}{2\sqrt{(R_i R_m)}}, \qquad \text{(see eqn (3.27))}$$

where d now symbolizes the diameter of the equivalent cylinder which represents all the dendrites. The soma conductance G_s may be written S/R_m, where S is the soma surface area. Hence eqn (7.71) becomes

$$\rho = \frac{\pi}{2} \frac{d^{\frac{3}{2}}}{S} \sqrt{\left(\frac{R_m}{R_i}\right)}. \tag{7.72}$$

It will be evident that the dendritic-to-soma conductance ratio is dependent on both a geometric $d^{\frac{3}{2}}/S$ and an electrical $(\sqrt{(R_m/R_i)})$ quantity, and therefore cannot be determined simply by histological measurement.

The mathematical formulation of the voltage response at the model soma is straightforward using the Laplace-transform technique, since it differs from the example of the semi-infinite cable (treated in Chapter 4) only in the nature of the terminal boundary condition. In this case the axial current at the end (the soma) is given by

$$i_a = V_8 G_8 + C_8(\partial V_8/\partial t), \tag{7.73}$$

where V_8 is the soma voltage and G_8 and C_8 are the (lumped) soma conductance and capacitance, respectively. Transforming into normalized time, eqn (7.70) becomes

$$i_a = V_8 G_8 + G_8(\partial V_8/\partial T)$$
$$= G_8\{V_8 + (\partial V_8/\partial T)\},$$

since

$$C_8/G_8 = \tau_m.$$

The equivalent boundary condition in the Laplace-transform domain is therefore

$$\bar{i}_a = \bar{V}_8 G_8 (1+s).$$

The voltage at the soma in response to a current $\bar{I}(s)$ injected at electrotonic distance X away from the soma can be shown to be

$$\bar{V}_8 = \frac{\bar{I}(s)}{G_8} \frac{\exp\{-X\sqrt{(s+1)}\}}{\sqrt{(s+1)}\{\rho+\sqrt{(s+1)}\}}. \tag{7.74}$$

The voltage response to a delta function can be obtained using the inverse transformation (Roberts and Kaufman, p. 248, pair 3.2.28),

$$\frac{\exp(-bs^{\frac{1}{2}})}{s^{\frac{1}{2}}(s^{\frac{1}{2}}+a)} \rightarrow \exp(ab+a^2 T)\mathrm{erfc}\left(\frac{b}{2\sqrt{T}}+a\sqrt{T}\right)$$

$$V_8 = \frac{Q_0}{2G_8\tau_m} \exp\{\rho X+(\rho^2-1)T\}\mathrm{erfc}\left(\rho\sqrt{T}+\frac{X}{2\sqrt{T}}\right). \tag{7.75}$$

The response to a step of current may also be obtained and is given by

$$V = \frac{r_a I\lambda}{2(r_a\lambda+G_8)}\left[\exp(-X)\mathrm{erfc}\left(\frac{X}{2\sqrt{T}}-\sqrt{T}\right)-\right.$$
$$-\left(\frac{\rho+1}{\rho-1}\right)\exp(X)\mathrm{erfc}\left(\frac{X}{2\sqrt{T}}+\sqrt{T}\right)+$$
$$\left.+\frac{2}{\rho-1}\exp\{\rho X+(\rho^2-1)T\}\mathrm{erfc}\left(\frac{X}{2\sqrt{T}}+\rho\sqrt{T}\right)\right] \tag{7.76}$$

(except when $\rho = 1$).

It will be noticed that eqn (7.76) has some resemblance to the step response of an infinite cable (eqn (3.24)). The response at the soma to a current step applied at the soma is

$$V = \frac{I}{G_8(\rho^2-1)}[\rho \, \mathrm{erf}(\sqrt{T})-1 + \exp\{(\rho^2-1)T\}\mathrm{erfc}(\rho\sqrt{T})]$$

($\rho \neq 1$). Inspection of this equation reveals that when $\rho = \infty$ (that is $G_8 = 0$) it can be simplified to the appropriate response for a simple cable,

$$V = r_a I\lambda \, \mathrm{erf}\sqrt{T} \qquad \text{(see eqn (3.28))}$$

since
$$\exp(\rho^2 T)\, \mathrm{erfc}(\rho\sqrt{T}) \to 0 \quad \text{as} \quad \rho \to \infty$$
and
$$1/G_s(\rho^2-1) \to 0 \quad \text{as} \quad G_s \to 0.$$

A ρ value of ∞ corresponds to the lumped soma becoming infinitesimally small; on the other hand, when $\rho = 0$ (i.e. no dendrites. $G_s r_a \lambda = \infty$), eqn (7.76) can be simplified to

$$V = \frac{I}{G_S}(1-e^{-T}).$$

Rall (1960) has already illustrated these properties of his model and his Fig. 1 is reproduced here (Fig. 7.13). For values of ρ lying between the two bounds of 0 and ∞, the time course lies between those given by an infinite cable and a simple R–C circuit. In a similar way the voltage response at the soma to a brief pulse of current applied at the soma will lie between the responses for the two limiting types of geometry. Fig. 7.14 shows the dependence of the decay time course on the value of ρ. The smaller the value of ρ, the slower the time course of decay. Intuitively, this simply means that after the charge is placed on the 'soma' capacitance, the rate at which charge redistributes along the dendritic equivalent cylinder is governed by its input conductance. This general effect of ρ on the time course of transients recorded at the soma holds also for the case when current is injected away from the soma. Fig. 7.15 shows how both the rise time and the half-width are increased by decreasing ρ, for any given value of distance.

Lumped soma short-cable model. Although representing the dendrites (and/or axon) as effectively infinite in length may give a satisfactory model of

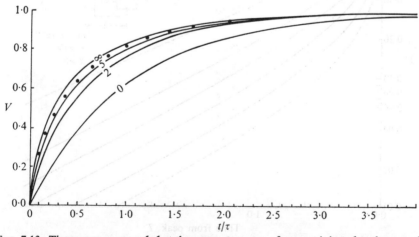

FIG. 7.13. The response, recorded at the soma, to a step of current injected at the soma in the infinite-cable model. The figures breaking the continuous curves indicate the appropriate value of ρ. The relation indicated by filled circles corresponds to a ρ value of 10. (From Rall 1960.)

certain nerve cells, it became evident to Rall (1962*a*,*b*,1964,1967) that motoneurone dendrites would be best represented as 'equivalent cylinders' whose electrotonic length is of the order of one or two space constants. This model therefore requires two additional parameters—one describing the electrotonic length of the cable and a second one for the end-boundary condition. In order to simplify the treatment it is convenient to take a single value for the end-boundary condition; Rall (1964, 1967, 1969*a*) assumed an open-circuit termination (i.e. $B_{term} = 0$). This can be an appropriate selection for a motoneurone model (Jack *et al.* 1971).

One consequence of a finite cable length is that the equation for ρ changes; the input conductance of a finite cable with an open-circuit termination is given by tanh $L/(r_a\lambda)$ (see eqn (4.11)),
therefore

$$\rho = \tanh L/r_a\lambda \, G_s. \tag{7.77}$$

As $L \to \infty$, tanh $L \to 1$, and this definition of ρ approaches that of eqn (7.71). It is sometimes convenient to use this latter definition, even when the cable is of finite length and it will be designated by the symbol ρ_∞, that is,

$$\rho_\infty = 1/r_a\lambda G_s. \tag{7.78}$$

Mathematical analysis of the short-cable model is quite complicated. The approach Rall (1962*a*,1969*a*) adopted is the classical method of

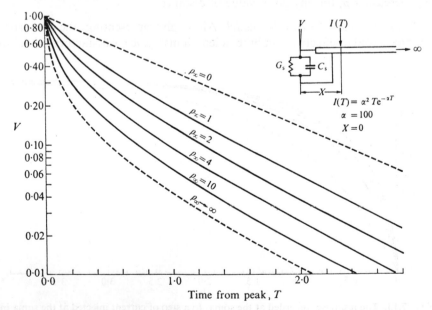

FIG. 7.14. The corresponding response in the infinite-cable model to a brief pulse of current delivered at the soma. The ordinate is V plotted logarithmically. (From Jack and Redman 1971*b*.)

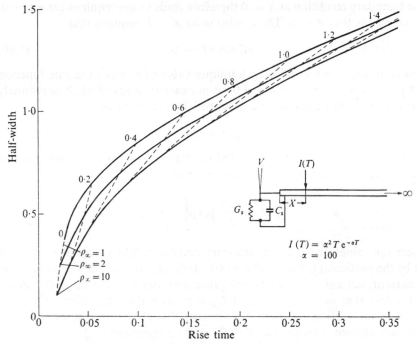

FIG. 7.15. The relationship between rise time and half-width for the infinite cable model when a brief pulse of current is injected at various distances from the soma. The three curves are for ρ values of 1, 2, and 10 (as labelled). The dashed lines join points on the three curves for a given electrotonic distance, indicated at the top of each dashed line. (From Jack and Redman 1971*b*.)

separating of variables and solving for the relevant boundary conditions. A detailed exposition is given in Rall (1969*a*), but it may be helpful to sketch the procedure for obtaining a solution. In order to make the working easy to follow, the case of a simple finite cable will be presented first.

A general solution to the cable equation (eqn (3.13)) is given by

$$V = (A \sin \alpha X + C \cos \alpha X)\exp\{-(1+\alpha^2)T\}, \tag{7.79}$$

where A, C, and α are arbitrary constants. We now need to introduce the boundary conditions so that these three constants are specified. If distance is defined so that the cable extends from $X = 0$ to $X = L$, the boundary condition for open-circuit termination means that

$$\partial V/\partial X = 0 \quad \text{at } X = 0 \quad \text{and at} \quad X = L.$$

From eqn (7.78) we obtain

$$\frac{\partial V}{\partial X} = (\alpha A \cos \alpha X - \alpha C \sin \alpha X)\exp\{-(1+\alpha^2)T\}.$$

The boundary condition at $X = 0$ therefore leads to the requirement $\alpha A = 0$, which means that $A = 0$. The condition at $X = L$ requires that

$$\alpha C \sin \alpha L = 0. \tag{7.80}$$

This equation is not satisfied by a unique value of α, since the sine function will periodically take the value of 0 for increasing values of αL. The infinitely many roots of this equation are described by the equation

$$\alpha_n = n\pi/L,$$

where n is a positive integer (including zero). This result means that the solution can be expressed as an infinite series

$$V = \sum_{n=0}^{\infty} C_n \cos\left(\frac{n\pi X}{L}\right) \exp\left\{-\left\{1+\left(\frac{n\pi}{L}\right)^2\right\}T\right\}, \tag{7.81}$$

where the values of C are still arbitrary constants. The values of C are then set by the particular initial condition selected, e.g. injection of a delta function of current, current step, etc. The elegance and usefulness of this solution lies in the fact that only the values of C_n vary with the exact form of current injection for a given location. This means that the voltage response to an arbitrary current waveform at $X = 0$, may be expressed as

$$V = C_0 \exp(-T) + C_1 \exp\left\{-\left(1+\frac{\pi^2}{L^2}\right)T\right\} + C_2 \exp\left\{-\left(1+\frac{4\pi^2}{L^2}\right)T\right\} + \cdots$$

$$= C_0 \exp\left(-\frac{t}{\tau_0}\right) + C_2 \exp\left(-\frac{t}{\tau_1}\right) + C_2 \exp\left(-\frac{t}{\tau_2}\right) + \cdots, \tag{7.82}$$

where $\tau_0 = \tau_m$, $\tau_n = \tau_m \Big/ \left(1+\frac{n^2\pi^2}{L^2}\right)$, and only the values of G_n are set by the initial conditions.

A similar partial solution may be obtained when one end of the cylinder is terminated in a 'soma', but now the values of the roots are more complicated; the boundary condition of the soma (eqn (7.71)) leads to an equation which differs from eqn (7.80). It is

$$\rho \sin \alpha L = -\alpha \cos \alpha L \tanh L$$

or

$$\beta \cot \beta = -\rho L/\tanh L, \tag{7.83}$$

where

$$\beta = \alpha L.$$

Some of the roots of this transcendental equation have been tabulated by Carslaw and Jaeger (1959, Table II, Appendix IV). These roots, when determined, set the value of the time constants in eqn (7.82),

$$\tau_n = \tau_m/\{1+(\alpha_n)^2\} \qquad (\alpha_0 = 0).$$

The way in which these results may be used to provide methods of experimentally determining the 'parameters' of a nerve cell will be presented in the next section. A complete solution by this method also requires the determination of the values of G_n. Rall (1962a, 1969a) has sketched the way this may be achieved but has not yet published the details.

An alternative way of obtaining a solution has recently been given by Jack and Redman (1971b). They use Laplace-transform methods but the complicated part of their treatment is the inverting of the transform back into the time domain†. This has been done successfully for the initial condition of a delta function of current applied to any point on the model. Using this result computations have been made of the voltage response at the soma to brief pulses of current at various distances, when L as well as ρ_∞ is varied. The general features of the effect of having a finite-length dendrite is exactly what one would expect by analogy with the properties of finite cables described in Chapter 4. The later decay of the voltage is a simple exponential (see also eqn (7.82)) while the earliest part of the voltage response is exactly comparable to that in a model with an infinite cable. The time at which transition occurs, for a given distance at which current is injected, depends on the electrotonic length of the dendrite—occurring earlier, the shorter the cable length, This is shown in Fig. 7.16 for current injection at the soma. Notice that these calculations have all been made for a constant value of ρ_∞ (not ρ) of 4. They correspond to ρ values of 3·95, 3·62, 3·05, 2·42, and 1·85, respectively, for the cable lengths of 2·5, 1·5, 1·0, 0·7, and 0·5.

Methods for determining the parameters τ_m, ρ, and L

It was mentioned at the beginning of this chapter that one aim of nerve cell modelling was to develop techniques which would allow the determination of the site of a synaptic action. It will be obvious from the descriptions given in Chapters 3 and 4, as well as those in this chapter, that the time course of the voltage transient can be very informative; but it is first necessary to determine the relevant nerve-cell parameters, namely, τ_m, ρ, and L, because each of these can have a significant effect on the voltage time course, irrespective of the distance at which the current is injected. The necessity to do this was the main spirit of Rall's (1960) paper in which he described transient solutions for the lumped soma infinite-cable model. In that paper he gave methods for determining τ_m and ρ. In a later paper Rall (1969a) provided suggestions for measuring τ_m, ρ, and L. Jack and Redman (1971b) have followed this lead, and their suggestions do not differ significantly from some of Rall's methods. All three parameters in principle can be measured by applying a brief pulse of current to the soma and studying the time course of the resulting voltage. An equivalent technique is to apply a step of current to the soma and make the calculations from the time course of the slope of the voltage response (i.e.

† See also Norman (1972).

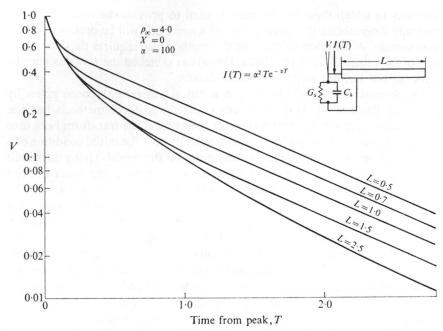

F IG. 7.16. A brief pulse of current is injected, at the soma, into the short-cable model. The decay time course of the voltage response is illustrated for various electrotonic lengths of the cable (value indicated on the figure). (From Jack and Redman 1971*b*.)

dV/dt), as originally suggested by Rall (1960). A brief pulse is, in some respects, more satisfactory as it is less likely to activate any time-dependent nonlinearity in the membrane response (see Nelson and Lux 1970).

Measurement of τ_{m}. The first step in the analysis is to determine the membrane time constant. There are two ways in which this may be done, the accuracy (or practicability) of each method depending on the cable length. If the dendritic length is long (e.g. $> 2\lambda$) a large part of the transient response will be identical to that of the lumped soma infinite-cable model (see Fig. 7.16). In 1960 Rall suggested that the time constant could be most readily obtained by plotting log $(V\sqrt{t})$ against t. The principle underlying this can be seen by considering the voltage response to a delta function at the soma of an infinite-cable model

$$V = \frac{Q_0}{2G_{\mathrm{s}}\tau_{\mathrm{m}}} \exp\{\rho X + (\rho^2 - 1)T\}\mathrm{erfc}\left(\rho\sqrt{T} + \frac{X}{2\sqrt{T}}\right)$$

$$= \frac{Q_0\rho_\infty}{2c_{\mathrm{m}}\lambda} \exp\{\rho X + (\rho^2 - 1)T\}\mathrm{erfc}\left(\rho\sqrt{T} + \frac{X}{2\sqrt{T}}\right).$$

By using the asymptotic expansion

$$\mathrm{erfc}(x) = \frac{1}{\sqrt{\pi}} . \exp(-x^2)\left(\frac{1}{x} + ...\right),$$

$$\underset{x \to \infty}{}$$

(13.85)

it can be seen that

$$V_\rho \sqrt{T \to \infty} = \frac{Q_0}{2c_m\lambda\sqrt{\pi}} \frac{\exp(-T)}{\sqrt{T}} \propto \frac{\exp(-t/\tau_m)}{\sqrt{t}}.$$

Hence if $\ln(V\sqrt{t})$ is plotted against t the resultant line should eventually decay with a slope of $1/\tau_m$; the larger the value of ρ, the earlier in time this occurs. Both Rall (1960) and Eccles (1961) have illustrated already the way in which the value of ρ affects this plot, as does Fig. 7.17.

This technique for measuring τ_m has been used by Lux and Pollen 1966; Burke 1968; Nelson and Lux 1970), although it may give an inaccurate value if the dendritic length is less than 2λ. The reason for this is that the voltage response may deviate away from the infinite-cable response at a time that is too early for the slope of the time constant to be reached (because ρ is relatively small or the pulse of current applied is not particularly brief). These effects are illustrated in Fig. 7.18 for two different values of cable length, where the end-termination is an open-circuit. The dashed line gives the slope of the membrane time constant; if the cable length is $2 \cdot 5\lambda$, this slope is reached even when $\rho_\infty = 2 \cdot 0$ ($\rho = 1 \cdot 97$). On the other hand, when $L = 1 \cdot 0$

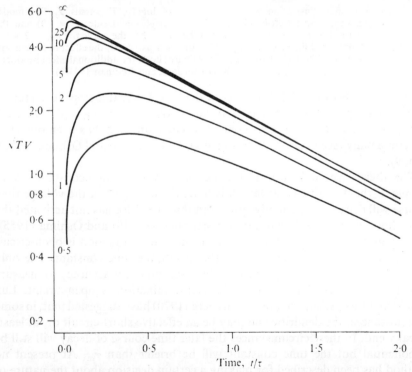

FIG. 7.17. An illustration of the behaviour of $\log(V\sqrt{T})$ versus T for the infinite-cable model in response to a delta-function current input. The curves from bottom to top are for ρ values of $0 \cdot 5$, 1, 2, 5, 10, 25, and ∞ respectively.

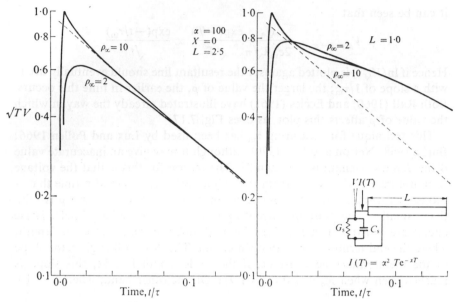

FIG. 7.18. A further illustration of the behaviour of $\log(V\sqrt{T})$ versus T. The model assumed has a short cable (left-hand side, $L = 2.5$; right-hand side, $L = 1.0$) and the current is a brief pulse ($I \propto Te^{-\alpha T}$, $\alpha = 100$). When $L = 2.5$, the curve for $\rho\infty = 2$ is not very different from that illustrated in Fig. 7.17; but when $\rho\infty = 10$, the effect of the current time course is obvious. The right-hand side of the figure shows the additional effect produced by a reduction in cable length. (From Jack and Redman 1971b.)

(right-hand side of Fig. 7.18), neither of the curves take this slope, even for an intermediary phase. Notice that, with larger values of ρ (e.g. $\rho\infty = 10$), the curve shows a distinct upwards concavity—such an effect has been observed experimentally (see Eccles (1961), Fig. 4 D, E, and F; Ito and Oshima (1965), Fig. 9).

The alternative method of measuring the membrane time constant is very simple. If the dendritic cable length is finite the later part of the voltage time course will decay exponentially (provided that the pulse has not activated the 'overshoot–undershoot' or other nonlinearities—see Ito and Oshima (1965), Nelson and Lux (1970)). When the dendritic end-termination is open-circuit this decay gives a direct measure of the membrane time constant. The only difficulty in this method (apart from obtaining sufficient accuracy to measure the slope) is in establishing that the end-termination is open-circuit. Lux (1967) and Lux, Schubert, and Kreutzberg (1970) have suggested that, in some circumstances, the dendritic end may be an effective short-circuit or, at least, a leaky end. In these circumstances the later time course of decay will still be exponential but the time constant will be briefer than τ_m. At present no method has been described for making a certain decision about the nature of the dendritic end-termination, although a technique which will decide in favourable circumstances has been given by Jack and Redman (1971a, b). It is essentially a compounding of the two methods for estimating τ_m given above. If the final exponential decay of the voltage with time yields a time constant τ_0 which is identical to the time constant of the intermediary phase of the log $(V\sqrt{t})$ plot, then the termination is (effectively) open-circuit and $\tau_0 = \tau_m$. A more detailed discussion of this method will be found in Jack and Redman (1971b).

Measurement of ρ. Study of the early part of the voltage-decay time course to a brief pulse of current (or step) can be used to estimate ρ, as reference to Fig. 7.14 will show. For a given time course of current application, the smaller the value of ρ, the slower the time course of decay. This method, suggested by Rall (1960), was made more explicit by Lux and Pollen (1966), who used a technique of successive approximation so that a new value of ρ is found which gives the best fit for the two points on the time course of the voltage response. A very similar technique is proposed by Jack and Redman (1971b) using the normalized half decay time. The general difficulty in obtaining an accurate result by this method is that it involves working with the early part of the time course which is more affected by technical artefacts.

A second method suggested by Rall in his original paper (Rall 1960) involves the application of sinusoidal current at various frequencies to the soma and measuring the phase shift in the voltage response. Nelson and Lux (1970) have obtained values for ρ in motoneurones by this technique. Rall has published results only for the theoretical relation between phase shift and frequency for the infinite-cable model. However, it has been pointed out that a finite length has less effect on the steady a.c. input admittance than on the d.c. input conductance of the cylinder (Rall 1969a; Jack and Redman 1971b). Lux (1967) has illustrated the phase-shift-frequency plots for different values of ρ and L, but he has assumed that the dendrites terminate in a short-circuit.

A third method has recently been described by Rall (1969a). It involves the successive determination of τ_m, L, and ρ. The solution given by Rall (1969a) for the voltage response to current injection (see eqn (7.82)) involved the sum of a series of exponential terms in which one decayed at a rate proportional to the membrane time constant and the rest with time constants successively briefer than τ_m. All these other time constants were dependent on both ρ and L, as well as τ_m. Analysis of the voltage response to current injection will therefore not provide a unique estimate of ρ and L from the comparison of τ_1 to τ_0 (i.e. τ_m) (see Fig. 1, Rall 1969a); but a unique estimate of L can be made by analysing the time course of the current that needs to be applied in order to obtain a voltage step (i.e. voltage clamp) at the soma. After the initial transient the time course of this circuit will be independent of ρ and simply dependent on L (and, of course, the end-termination).† Rall (1969a)

---

† It is perhaps worth noting here that if the dendrites are infinite in extent the time course of the applied current for a voltage step can be used to give a measure of τ_m. Rall (1960) has shown that, after the initial surge, the time derivative of the applied current is proportional to $-t^{-\frac{3}{2}} \cdot \exp(-t/\tau_m)$ Therefore a plot of $\log(t^{\frac{3}{2}} \, dI/dt)$ versus t would yield τ_m.

has shown that the exponential term with the slowest time constant is briefer than τ_m, being dependent on τ_m and L. Since τ_m has already been determined by the method of current injection, L may now be determined; this value of L can now be included in the consideration of the other time constants (τ_1, etc.) obtained from current injection, and hence ρ is specified.

Estimation of L. In the preceding paragraph the procedure for estimating L from the time course of a voltage-clamp current has been described. For technical reasons it is rather easier to observe the voltage response to current injection. As pointed out above the time constants obtained by this latter method depend on both ρ and L (and τ_m). If ρ has already been estimated by either of the first two methods described, it is then possible to make a unique estimate of L. In any case, as pointed out by Burke and ten Bruggencate (1971; see also Fig. 1 of Rall (1969a)), even if ρ has not been determined the estimate of L is unlikely to be in error by greater than 20 per cent.

A slightly different way of obtaining L has been suggested by Jack and Redman (1971b). Fig. 7.16 illustrates the fact that the time at which the voltage decay becomes exponential depends on the cable length; uniform distribution of charge (and hence, exponential decay) occurs earlier if the dendritic length is shorter. By making a measurement of the time at which passive decay is reached it is possible to obtain L. This method, to be accurate, also requires a previous estimate of the value of ρ.

Modifications and extensions to the Rall model

Detailed comparisons between the predictions of the Rall model and the actual behaviour of motoneurones have only begun to be made (Burke and ten Bruggencate 1971; Jack and Redman 1971b), and it is still too early to judge whether the model will be able to successfully simulate a variety of experimental results.† There are, however, some fairly obvious modifications and extensions which it would be useful to achieve.

In describing the transformations performed in obtaining the reduced Rall model, it was pointed out that an intermediate model, with isopotential soma and each individual dendrite represented by an equivalent cylinder (i.e. not lumped into a *single* equivalent cylinder), would allow a description of the voltage response in a cylinder representing a single dendrite. The lengths and end-terminations of individual dendrites could also be varied. This latter modification is likely to be particularly important in modelling cortical pyramidal cells for it seems likely, from their structure, that the basal dendrites and the apical dendrites will be very different in their electrotonic length.

These modifications are unlikely to present special mathematical difficulty, and already some progress has been reported. Rall (1969a) has shown how the

---

† Recently Iansek and Redman (1973) have reported that the model does not provide a fit to experimental transients generated by a brief pulse of current applied to the moto-neurone. Similar results have been obtained by Iles and Jack (in preparation).

time constants of a solution (of the form of eqn (7.82)) may be obtained. Jack and Redman (1971b) have performed some preliminary calculations on a model in which one cylinder is finite (representing the dendrites) and another infinite in extent (representing the axon). These models have additional parameters (ratio of each dendritic input conductance to the soma conductance and length and end-termination of each dendrite); an obvious task for any systematic study will therefore be to determine whether there are any practicable methods for uniquely estimating them.

Despite these hints of further progress there are other developments of the Rall model which seem less likely to yield either complete or partial analytic solutions. Models in which non-cylindrical dendrites are to be coupled to a soma, or in which the membrane time constant is not uniform, present difficulties owing to the complicated nature of the boundary conditions. Modelling a synaptic action by a conductance change rather than a current injection (see Chapter 3) leads to a differential equation which does not have constant coefficients and hence is unlikely to yield an analytical solution. In the face of these difficulties there may be no alternative to full numerical computations of the particular model selected, just as in the case of action-potential mechanisms (see Chapters 9, 10, and 11).

One way of performing these calculations deserves special mention because it has yielded a large amount of information. Rall (1964. 1967) has achieved considerable simplification in the numerical calculations by making the assumption that small segments of the nerve cell can be lumped into iso-potential units (which he calls 'compartments'). Each compartment is joined to adjacent ones by a series resistance; if each compartment is of equal size, and no branching occurs, this is equivalent to a cylinder in the limiting case of each compartment becoming infinitesimally small. It is possible in this way to take compartments of unequal size and an unlimited variety of branching patterns. Rall (1964, 1967) has given a careful presentation of the way in which conductance changes may be represented and a general description of the method of solution. The results of the computations Rall has made (Rall 1964, 1967) provide an extremely useful guide to such complicated problems as the interaction between synaptic potentials, the detectability of conductance changes, etc. Brief mention of some of these results will be made later in this chapter.

Part IV: Applications to junctional mechanisms

Determination of the location of synaptic action

In this section a review will be given of the electrophysiological techniques which are available for the determination of synaptic location, when recording intracellularly from the nerve cell soma. These methods do not have the direct elegance of an exclusively histological approach (see, e.g. Globus and Schiebel 1967; Conradi 1969; Rogers 1972) or of one combined with extracellular recording (see Andersen, Eccles, and Løyning 1963; Andersen, Blackstad and Lømo 1966); but they do have the advantage that the answer provided is expressed as an *electrotonic* length rather than a physical length or a particular location on the cell surface. The former measure, unlike the latter, allows quantitative judgements about the magnitude of the effect that current injected by the synapses would produce at the soma. The two types of methods are, of course, not mutually exclusive but, rather, complementary since intracellular techniques do not at present give information about the particular dendrite on which a synapse is placed.

The assumption common to all the methods to be described is that the synapses activated have a single location. There is, at the moment no electrophysiological method which allows the allocation of a set of distance estimates. This means that, unless spontaneous miniature junction potentials are being studied, the experimental design should lead to only one synapse, or a few closely adjacent synapses, being activated. There may be histological evidence that a population of functionally similar fibres have their endings close together; if this is not so, it is of extreme importance to know the mode of termination of individual fibres since the best that is likely to be achieved experimentally is the stimulation of a single nerve fibre supplying a cell. In the absence of any relevant histological information, it seems reasonable to assume a single location; the way this assumption may be checked and the consequences of it being false will be discussed later.

It is also important to all the methods that the synaptic potential should spread to the recording site by passive means, i.e. the membrane properties should be linear over the range of potential variation.

The other assumptions are particular to the method employed and will be described in the appropriate sections. Although there are a variety of ways in which a guide to location may be obtained, only two methods have so far been developed in such a way that an exact numerical estimate of the distance may be made. Both depend on the use of the Rall model, and they will be described first.

Analysis of the time course of a synaptic potential. The most developed method is to study the time course of a synaptic potential and from simple measurements, such as the rise time and half-width, infer the electrotonic distance at which the potential was generated. The crucial additional assumptions of this method are that the synaptic current has a monophasic time course and that it is of brief duration; if the current is monophasic, but cannot be assumed brief, it is still possible to set an upper limit (but not a lower limit) to the distance value.

The principle of this method has already been given in Chapter 3 and earlier in this chapter. It rests on the simple fact that any transient voltage change generated in one part of a cable structure has its time course slowed as it 'propagates' passively to other parts of the cell: the further the distance of propagation, the greater the relative slowing. The ideal procedure is to first determine the parameters of the nerve cell (i.e. τ_m, L, and ρ); for a particular value of ρ and L the normalized rise time and half-width will jointly provide a good estimate of the electrotonic distance of the synapse and a rough guide to the duration (within the 'brief' range) of the synaptic current. It has been pointed out already in Chapter 3 that the half-width is mainly dependent on the distance value and relatively insensitive to the exact duration of a brief current, whereas the rise time is affected by both; the same features hold true for the nerve cell model. In order to make these estimates a large number of curves would be required showing the dependence of both the normalized shape indices on ρ, L, X (distance), and current duration. For this reason Jack and Redman (1971b) decided to discard the dependence on current duration and thus aim to simply measure X (from the half-width). Fig. 7.19 provides four sets of curves which allow this to be done. The relationship between half-width and distance for intermediary values of L and/or ρ_∞ may be obtained by interpolation.

These curves have all been calculated for a current time course of general form $T \exp(-\alpha T)$; with $\alpha = 100$. If the current is slower the value of X

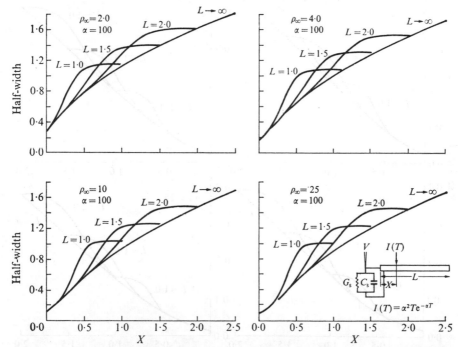

FIG. 7.19. The value of half-width for current injection at different distances along the dendritic cable. Each graph depicts a set of curves for one value of $\rho\infty$. The individual curves on each part of the figure are for different cable lengths, as indicated. The current time course is a brief pulse ($I \propto T \exp(-\alpha T)$, $\alpha = 100$). (From Jack and Redman 1971b.)

derived from these curves will be an overestimate, particularly at nearer distances (cf. Fig. 3.18).

It will be noticed that in all the examples of finite cables the half-width is practically the same when current is injected at any point within half a space constant of the end of the cable. In other words this method will be at its least accurate when attempting to judge location in the terminal part of the dendritic tree.

Another disadvantage of simply relying on the half-width is that it is not possible to check on the assumption that the synaptic potential is generated at a single location. A rough check on this may be obtained by using the rise time. Rall (1967) has pointed out that composite synaptic potentials usually have too brief a rise time for their half-width (see also Jack and Redman 1971*b*), so that it is a simple check to estimate X from the half-width and then establish, from Fig. 7.20, whether or not the rise time is too brief for this value of X. If it is, this means that one or other of the assumptions of the modelling is incorrect. Other evidence may help one to decide whether it is the

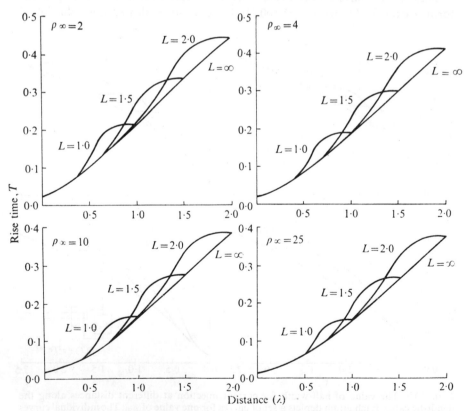

FIG. 7.20. The figure takes exactly the same form as Fig. 7.19, but shows the equivalent results for rise time rather than half-width. (S. J. Redman, unpublished calculations.)

assumption of a single location, of unequal dendrite lengths (see later), of membrane linearity, or of the form of synaptic current which is at fault, but in any case it would clearly be inappropriate to make a distance estimation in these circumstances. (If the explanation is the composite nature of the synaptic potentials, and the synaptic knobs are reasonably close to each other, then a distance estimate, if made, will tend to be larger than an average of the actual values. Jack and Redman (1971*b*) quote an example of synapses at 0·3, 0·4, and 0·5λ each injecting the same amount of current (of same time course) into the cell. Using the half-width a distance estimate of 0·45λ is obtained if $L = 1\cdot5$ and $> 0\cdot5\lambda$ if $L = 1\cdot0$ ($\rho_\infty = 10$ in both cases).)

If the rise time proves to be inappropriately large for the estimated distance value, the most likely explanations are either that the time course of the synaptic current is too prolonged (and/or of different form from $T\exp(-\alpha T)$ for the above method to work (in which case it may still be possible to make a distance estimate by the second method, to be described next) or else a membrane nonlinearity (such as the 'undershoot' phenomenon) has been secondarily activated by the synaptic potential. Of course, another possibility is that there is an accompanying synaptic action of opposite effect, which is either of longer latency or slower time course.

The method of distance estimation, as described above, is dependent for its accuracy on exact measures of the three nerve-cell parameters, τ_m, L, and ρ_∞. Experimentally these three parameters present increasing degrees of difficulty in their estimation and it is therefore worth considering the effect on the distance estimate if one or other of them is either inaccurately measured or assumed. Fortunately, it turns out that the most important parameter to measure is τ_m, followed by L and then ρ_∞. We can illustrate this point by the following example. Let us assume that a synaptic potential is generated in a cell by a brief pulse of current (of the form $T\exp(-\alpha T)$ with $\alpha = 100$) and the actual values of the nerve cell parameters are $\tau_m = 5$ (ms), $L = 1\cdot5$, and $\rho_\infty = 4\cdot0$. Let us consider in turn what distance value we would derive from the half-width if each of the parameters (taken one at a time) were thought to be as much as double, or as little as half, the actual value. Fig. 7.21 shows the range of possible distance estimates that could be made. The heavy line in the middle of the graph shows the relationship between half-width and X if all three parameters were estimated correctly. The cross-hatched area on either side of this line shows the range of X values which could be estimated from a particular value of half-width if ρ_∞ were thought to have a value somewhere between 2·0 and 8·0 (with L and τ_m having their correct values). Note that underestimating ρ_∞ leads to an underestimate of X for all values of half-width (and vice versa). Even so, the inaccuracy is, at most, of the order of 0·1λ. The stippled region on either side of the line shows the range of X values if L were thought to have a value lying somewhere between 0·75 and 3·0 (with τ_m and ρ_∞ having their correct values). The degree of inaccuracy depends on the distance away from the soma, but is generally less than 0·2λ. The thin

7

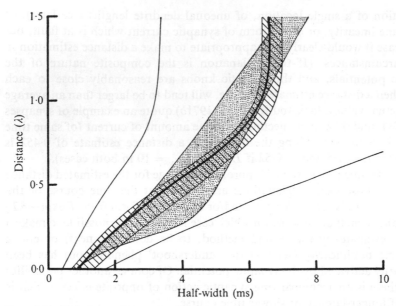

Distance (λ)

Half-width (ms)

Fig. 7.21. This figure shows the discrepancy between the true and estimated values of distance which can arise if one of the nerve cell parameters is incorrect. Further description in text.

lines on the graph show the range of X values when τ_m is assumed to lie between 2·5 ms and 10 ms (with L and ρ_∞ having their correct values). The range of error is now very great, increasing in absolute value with increasing distance from the cell soma. It may therefore be concluded that a distance estimate made without knowing the value of τ_m fairly accurately could be very misleading and that it is preferable to know L fairly exactly as well; but reasonable tolerance exists in the use of an inaccurate (or guessed) value of ρ_∞.

Another possible source of error arises if the individual dendrites of a nerve cell have different electrotonic lengths. Rall (1969a) has pointed out that the (single) value of L which is estimated will be approximately an average value of the different lengths. The synaptic potential being studied may be generated on a dendrite whose length is rather different from the average value. It is possible to give a qualitative indication of the sort of error in the distance value that would result from such a circumstance. The magnitude of the error will depend on the degree to which the half-width of the transient would be affected (in a cable of the approximate average electrotonic length) by voltage 'reflections' from the end of the cable (see Chapter 4). If the half-width is so affected then changing the cable length will have a substantial effect, since distance is judged from the half-width value. When the synaptic potential is generated on a dendritic cable whose electrotonic length is shorter than average, the half-width will be relatively large for that distance value. Consequently an overestimate of the distance value will be made. Similarly an underestimate will be made when the synapses are located on a dendritic cable longer than average. These errors will not be large unless the actual location of the synapses is within half a space constant of the end of the particular dendritic cable.

Measurement of the reversal potential. If a synaptic potential is produced by a conductance change to one or more ions, it is in principle possible, by displacement of the membrane potential, to reverse the direction of the junctional current and hence the polarity of the transient voltage displacement. The membrane potential at the junctional site, when the junctional current is zero, is called the reversal potential E_{Rev}. A steady voltage displacement at the soma (assumed to be the site where recording and polarizing electrodes are placed) suffers a decline in magnitude with distance along the dendrites. In order to displace the membrane potential at a synaptic site on the dendrites to the value of the reversal potential a larger displacement has to be imposed at the soma. The degree of discrepancy between the potential at the soma at which the reversal of polarity of the synaptic potential occurs and the true value of the reversal potential gives a theoretical method of determining the electrotonic distance which separates electrode and synapse (Calvin 1969*a*).

Let V_0 be the membrane potential displacement, recorded at the soma, at which reversal of the synaptic potential takes place. If the synapse is located at an electrotonic distance X away from the soma, the magnitude of the polarizing voltage at the synaptic site may be calculated.

$V_X = V_0 \exp(-X)$ for an infinite cable,

$V_X = V_0 \dfrac{\cosh(L-X)}{\cosh L}$ for an open-circuited short cable of electrotonic length L.

In both cases the value of the membrane potential has reduced the synaptic current to zero, that is, the driving voltage ($= E_m - E_{Rev}$) for the synaptic current is zero. If the normal driving voltage, when $E_m = E_r$, is expressed as V_{Dr} ($= E_r - E_{Rev}$) it may be concluded:

$$V_{Dr} = V_0 \exp(-X) \text{ for an infinite cable,} \tag{7.84}$$

$$V_{Dr} = V_0 \frac{\cosh(L-X)}{\cosh L} \text{ for a finite cable.} \tag{7.85}$$

The dependence of the ratio V_0/V_{Dr} on X is illustrated in Fig. 7.22. Note that, in an analogous way to the variation of half-width with distance (see Fig. 7.19), the ratio V_0/V_{Dr} is less sensitive to changes in X as the location approaches the end of the cable.

Thus if V_0 is measured and V_{Dr} assumed (perhaps by using the value for a somatically placed synapse of functionally similar action) it is possible to make a distance estimate. Although Fig. 7.22 could be used for this purpose,

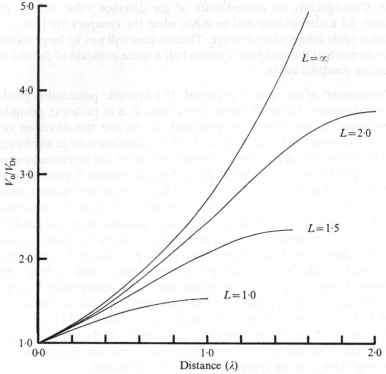

Fig. 7.22. The relative magnitude of the voltage at the site of current injection required to just cause reversal of synaptic current is plotted against the electrotonic distance between current-passing electrode and the synapse. The values for short-cable models ($L = 1\cdot0$, $1\cdot5$, and $2\cdot0$), as well as the infinite-cable model, are shown.

there is no need to do so since the distance may be calculated directly by manipulating eqns (7.84) and (7.85):

$$X = \ln(V_0/V_{Dr}) \text{ for an infinite cable,} \tag{7.84a}$$

$$X = L - \mathrm{sech}^{-1}\{(V_0/V_{Dr})\mathrm{sech}\, L\} \text{ for a finite cable.} \tag{7.85a}$$

Notice that this method only requires the direct measurement of one of the motoneurone parameters L and is independent of the time course of the conductance change. Measuring L (or obtaining evidence that the dendritic cable is sufficiently long to be treated as infinite) does, however, require the previous measurement of τ_m. However, the method does avoid the difficulties of measuring ρ (unless the method used for determining L requires ρ), although at the expense of needing an accurate value for V_{Dr}. The relative importance of having an exact value for V_{Dr} and L can be illustrated by an example similar to the one given in the previous section. In this example the nerve cell dendrites are assumed to have an electrotonic length of $1\cdot5\lambda$. As before the effect on the distance estimate of misjudging the dendritic length or the value of V_{Dr} is explored. Fig. 7.23 shows the result when the dendritic length is thought to be as little as one half, or as much as twice, the true value. It can be seen that substantial errors in the estimated value of X result. Nevertheless this error is relatively small compared to the effect of comparable deviations from the true value of V_{Dr}. Even an error of as little as 10 per cent in the value of V_{Dr} can lead to quite large errors in the value of X.

Just as in the case of a voltage time-course analysis, this method suffers from another possible error if the individual dendrites differ appreciably from the average electrotonic length. In this case, if the synapses are on a dendrite which is shorter than the average electrotonic length, the value of the polarizing voltage displacement (with an open-circuit termination of the dendrite) at the synaptic site will be larger than predicted. Hence the estimate of the distance will place the synapses nearer than they actually are; since this is an error in the opposite direction to that from a voltage time course analysis a very good check on the accuracy of the estimate can be made by comparing the results obtained by the two methods.

So far, discussion of this method has concentrated on the determination of the value of membrane potential at which the synaptic current reverses in

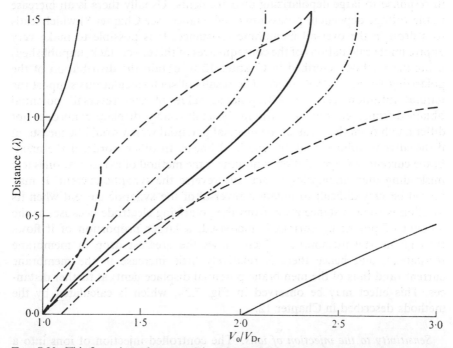

FIG. 7.23. This figure is analogous to Fig. 7.21. It shows the errors in the distance estimate that would arise if there were an error by a factor of 2 in the assumed value of the cable length (dashed lines) or of V_{Dr} (continuous lines). Even an error of 10 per cent in the value of V_{Dr} produces a substantial difference in the distance estimate (dashes and dots).

direction. It may not be possible to actually reverse the polarity of the synaptic potential. One reason is that the required voltage displacement cannot be produced by maximum current that can be passed through the microelectrode. In this event it is conventional to obtain a series of measurements of the relative size of the synaptic potential for membrane potential displacements much smaller than the actual value of the reversal potential; from these points a straight line may be extrapolated so that the value of the membrane potential at which the reversal would occur may be deduced.

This procedure relies on the assumption that at the synaptic site there is a linear relationship between the size of the synaptic current and the driving voltage. Takeuchi and Takeuchi (1959) have shown that this does hold for the end-plate current at the frog neuromuscular junction; it does not hold, however, for the inhibitory synaptic current produced in mammalian spinal motoneurones by activation of the appropriate group Ia fibres (Coombs, Eccles, and Fatt 1955*b*; Araki and Terzuolo 1962).

Another possible difficulty in seeking actually to reverse the synaptic potential, particularly for excitatory postsynaptic potentials which have their reversal potential well away from the resting potential, is that it is rare for excitable cells to have a steady state current–voltage relation which is linear in response to large depolarizing displacements. Usually there is an increase in the voltage-dependent potassium conductance (see Chapter 8) which leads to a drop in the over-all membrane resistance. It is possible to make very approximate calculations of the consequences of this effect (Jack, unpublished) using the method described in Chapter 12 to obtain the distribution of the polarizing voltage with distance. The results of such calculations support the natural intuition that the extrapolated value of the 'reversal' potential obtained from measurements with fairly small voltage displacements does not differ much from the value of the reversal potential which would be measured if the current–voltage relation remained linear. In other words, if the membrane current–voltage relation is nonlinear the method of extrapolation is less misleading than attempting to actually reverse the synaptic current. It may indeed be very difficult to obtain a reversal of the synaptic current when its location is some distance away from the polarizing electrode because, as the amount of polarizing current is increased, a larger proportion of it flows through nearby membrane (which shows the greatest drop in membrane resistance), and hence there is relatively little increase in the membrane current (and thus of the membrane potential displacement) at further distances. This effect may be observed in Fig. 7.24, which is calculated by the methods described in Chapter 12.

Sensitivity to the injection of ions. The controlled injection of ions into a nerve cell, the motoneurone, was first performed by Coombs, Eccles, and Fatt (1955*a*). They found that although the group Ia excitatory postsynaptic potential was relatively insensitive to ion injection (probably because of

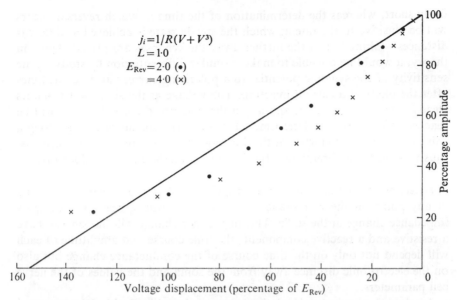

$$j_i = 1/R(V + V^3)$$
$$L = 1·0$$
$$E_{Rev} = 2·0 \; (\bullet)$$
$$= 4·0 \; (\times)$$

Percentage amplitude

Voltage displacement (percentage of E_{Rev})

FIG. 7.24. Calculations showing the effect of outward rectification on the measurement of the reversal potential. The current–voltage relation assumed is $j_i = (1/R)(V + V^3)$ (cf. Chapter 12). The synapse is assumed to be at the end of a sealed-end dendrite, represented by an equivalent cylinder one space constant in length (space constant defined in terms of the slope conductance when $V = 0$). The continuous line represents the results expected if the membrane remained linear, i.e. $j_i = V/R$. The abscissa represents the voltage at the site of current injection, normalized by the value of the reversal potential. The ordinate plots the amplitude of the synaptic effect, also recorded at the site of current injection.

quite rapid readjustments in the ionic concentrations of injected potassium or sodium) the group Ia inhibitory postsynaptic potential was very sensitive to chloride ions; an increase in the internal concentration of the ion lead to the inhibitory effect reversing from a hyperpolarization to a depolarization. On the assumption that all inhibitory actions on the motoneurone have the same ionic mechanism, several workers have used chloride injection as a means of deciding whether other inhibitory synapses are located near to or far from the soma (e.g. Llinas and Terzuolo 1965; Kellerth 1968). In a recent report Burke, Fedina and Lundberg (1971) have shown that, although the reversal potentials for group Ia and recurrent inhibition are similar, a separation in their behaviour can be observed by injecting chloride ions and observing the time at which the two inhibitory potentials reverse their polarity. The authors interpret this result as indicating that the recurrent inhibitory synapses are located further from the soma than the group Ia inhibitory synapses. If this explanation is correct then the differentiation between the locations of the two groups of inhibitory synapses by this method, but not by the reversal potential method, probably lies in the fact that the steady distribution with distance (of ion concentration) may be relatively flat, particularly if the dendrites are fairly short, whereas the determination of the time at which reversal occurs will be sensitive to the rate at which the steady state is achieved at different distances—taking longer the further away the synapses are placed. Thus, in theory, it would be possible to make a similar differentiation by studying the sensitivity of the synaptic potential to a polarizing current at different times after the onset of a current injection. The voltage at the soma will reach its steady value most quickly, and along the dendrite the steady value will be reached at progressively later times. However, the spread of charge along a cable is relatively rapid whereas the diffusion of ions is a much slower process, thus making the differentiation by the present method easier technically.

Detection of impedance or conductance change. A synaptic change occurring at any point on the membrane of a nerve cell will appear as a complex impedance change at the soma. This impedance change will, in general, have a resistive and a reactive component; the time course and amplitude of each will depend not only on the time course of the conductance change but also on the electrotonic distance away from the soma and the values of the nerve cell parameters.

An elegant experimental technique was developed by Smith, Wuerker, and Frank (1967) in which an a.c. impedance bridge utilizing a phase-sensitive detector was used to detect impedance changes in motoneurones accompanying postsynaptic potentials. The use of the phase-sensitive detector allowed measurements to be performed at a fairly low carrier frequency (100 Hz) and thus obtain reasonable sensitivity (higher carrier frequencies would be better suited to the detection of changes in the membrane capacitance). By combining this with an averaging technique they were able to obtain reasonable definition in the time course of conductance changes. Their experimental results showed that it was possible to detect impedance changes accompanying some forms of inhibition without difficulty, but group Ia excitatory post-synaptic potentials often produced no detectable impedance change. They concluded that this was compatible with the interpretation that some group Ia synapses were located on the dendrites; they were aware that an alternative explanation might be that this excitatory postsynaptic potential was not produced by an ionic conductance change but by electrical coupling between the group Ia fibre and the motoneurone.

Provided some other evidence can be adduced in favour of an ionic conductance change, this method could be used therefore to make a distance estimate. By making an assumption about the time course of the conductance change, and knowing the values of the nerve cell parameters, the time course of the resistive and reactive components of the impedance change could be analysed to obtain a distance value. As the authors point out, however, the signal-to-noise performance of the bridge limits the detectability of impedance changes so that small amplitude conductance changes at large electrotonic distances may be beyond the capabilities of the instrument.

Rall (1967) has examined the ease with which a transient synaptic conductance may be detected at different locations by searching for distortions in the voltage response at the soma to a current step. The emphasis of his discussion was on the problem of whether a transient conductance could or could not be detected, depending on the distance away from the soma. The results of his calculations showed that, within present experimental limits, it was probable that conductance changes in the distal electrotonic half of the dendritic tree could not be detected. It is unlikely, therefore, that this method can be readily developed to make other than qualitative allocations of distance to synaptic locations.

Effect of location on the magnitude of a synaptic potential

Once the location of a synaptic potential has been determined (or stipulated in a model) it is of some interest firstly to calculate the magnitude of the current producing the effect and secondly to calculate the voltage of the synaptic potential both at its site and in other parts of the nerve cell. The first calculation is of interest because it allows a judgement to be made about the relative effectiveness of synapses located on different parts of the nerve cell when there is a passive spread of the potential into the nerve cell soma (where it may algebraically summate with other potentials to produce, or inhibit, firing of the cell). The second calculation is a preliminary step to deciding whether there is likely to be a simple proportional relationship between the magnitude and time course of the synaptic conductance and the synaptic current, to the question of whether a nonlinear response of the dendritic membrane is likely to follow synaptic activation, and finally to the problem of the interaction between different synaptic potentials.

In this section attention will be devoted mainly to the first type of calculation because it is a more straightforward problem. Calculations of the second kind involve the use of a more complicated nerve cell model (at the least, the intermediate model of the nerve cell—see p. 160) to be realistic, and hence will not be explored in detail.

Voltage at the soma in response to current injection at different locations. It has already been pointed out that, provided the various assumptions of the Rall model hold, there is no difference in the voltage response at the soma between current injection at one point in the 'equivalent' cylinder of the reduced Rall model (formally identical to current injection, at the same electrotonic distance, into each of the dendritic branches of a more realistic model which represents each dendritic branch by a cylinder) and all of the current being injected at a single place in one branch in this more realistic model. Because of this fact, first proved by Rall (see Jack and Redman 1971*b*), all the calculations relevant to this section can be performed with the reduced Rall model.

1. *Steady state attenuation with distance.* Let us first consider the problem of a steady current *I* injected at different points in the nerve cell model. The

amplitude of the voltage at the soma can be calculated from the following
formulae,

$$V = \frac{I}{G_S} e^{-X} \left(\frac{1}{\rho + 1} \right) \quad \text{for the infinite-cable model,} \tag{7.86}$$

$$V = \frac{I}{G_S} e^{-X} \frac{(e^{2X-2L} + 1)}{\{1 + \rho_\infty + (1 - \rho_\infty) e^{-2L}\}} \quad \text{for a finite-cable model.} \tag{7.87}$$

Fig. 7.25 illustrates the relative magnitude of this voltage occurring in
these two models when the same magnitude of current is injected at various
distances. It will be noticed from eqns (7.86) and (7.87), that the magnitude of
ρ has no effect on the *relative* attenuation with distance, but simply affects the
absolute amplitude of the voltage.

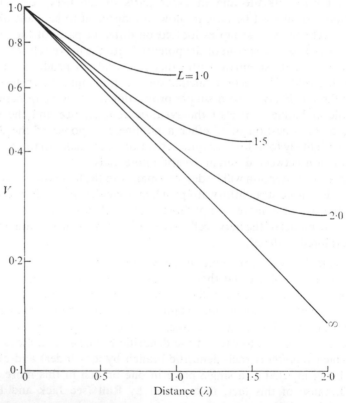

FIG. 7.25. A plot of the magnitude of the soma voltage in response to a steady current
injected at different distances from the soma. The ordinate expresses the voltage relative to
that produced when the current is injected at the soma. The abscissa is the distance, in space
constants, at which current is injected. Curves for dendrite lengths of 1·0, 1·5, and 2·0λ, as
well as the infinite cable, are shown for an open-circuited short cable (cf. Fig. 4.6).

2. *Area of a voltage transient.* Before considering the case of the attenuation of the peak amplitude of a transient it may be useful to consider the attenuation of the area of the transient. The mathematical treatment of this case is much simpler because it is possible to show that the attenuation of area is independent of the time course of the synaptic current. It is also of some relevance physiologically because, when a large number of small excitatory synaptic potentials are summating at the soma, the total area of each transient will be of more importance in reaching the voltage threshold for cell discharge than the individual peak amplitudes (Rall 1959*a*). It is possible to calculate the area of a voltage transient in the following way. Let $V(x, t)$ describe the voltage time course of the transient. Then the area A is given by

$$A = \int_0^\infty V(x, t)\, \mathrm{d}t.$$

But $V(x, t)$ is described, using the convolution theorem, (see eqn (13.10)) by the convolution of the delta-function response (denoted by $g(x, t)$) and the time course of the synaptic current (denoted by $i(t)$). In other words,

$$V(x, t) = \int_0^t g(x, z)i(t-z)\, \mathrm{d}z,$$

therefore

$$A = \int_0^\infty \int_0^t g(x, z)i(t-z)\, \mathrm{d}z\, \mathrm{d}t. \tag{7.88}$$

It can be shown that the right-hand side of eqn (7.88) transforms to

$$\int_0^\infty g(x, t)\, \mathrm{d}t \int_0^\infty i(t)\, \mathrm{d}t$$

(see Carslaw and Jaeger 1948, §33 p. 80 for the proof of a more general form of this theorem). But $\int_0^\infty i(t)\, \mathrm{d}t$ represents the total charge injected as synaptic current, and $\int_0^\infty g(x, t)\, \mathrm{d}t$ represents the area of the voltage transient produced by a delta function of current. It can be shown, in turn, that the area of a voltage transient generated by a delta function suffers the same attenuation with distance as the steady state potential. This follows because of the theorem in Laplace-transform theory (see eqn (13.13)),

$$\int_0^\infty g(t)\, \mathrm{d}t = \lim_{s \to 0} \bar{g}(s). \tag{7.89}$$

It is found that when s is put equal to 0 in the delta-function response (described in the Laplace-transform domain) the result is the same (except for the term I) as the steady state response. (see eqn (13.16d).) Thus it has been demonstrated that

$$A = \frac{Q}{G_S}e^{-X}\frac{1}{(\rho+1)} \quad \text{for the infinite-cable model,} \tag{7.90}$$

$$A = \frac{Q}{G_S}e^{-X}\frac{(e^{2X-2L}+1)}{\{1+\rho_\infty+(1-\rho_\infty)e^{-2L}\}} \quad \text{for the finite-cable model.} \tag{7.91}$$

A similar result has already been reported in Chapter 3 for the simple infinite cable.

The curves describing the relative attenuation of the area of a voltage transient are thus identical to those for the steady state response. It can be seen by inspection of Fig. 7.25 that the relative attenuation of the area of a synaptic potential in nerve cells such as the motoneurone (where L is usually less than 2·0—see Jack, Miller, Porter, and Redman (1971)) is not great—usually being less than a factor of $\frac{1}{4}$. This result is quite important because it has sometimes been suggested that synapses located on the dendritic terminals could produce no significant effect at the soma.

One advantage of having eqns (7.90) and (7.91) is that they can be manipulated so that an experimental determination of the amount of charge placed on the cell by a synaptic action may be calculated.

$$Q = AG_S(\rho+1)e^X = Ae^X/R_{in} \quad \text{for the infinite-cable model,} \tag{7.92}$$

where R_{in} is the input resistance of the cell measured at the soma.

$$Q = AG_S(\rho_\infty+1)e^X\left(1+\frac{1-\rho_\infty}{1+\rho_\infty}e^{-2L}\right)\Big/(1+e^{2X-2L}) \tag{7.93}$$

$$= \frac{A}{R_{in}}\frac{e^X(1+e^{-2L})}{(1+e^{2X-2L})} \tag{7.94}$$

for the finite-cable model. Thus in a nerve cell in which it can be shown that the dendritic cable is effectivtly infinite, it is a straightforward matter to calculate the charge by estimating the location of the synapse and measuring the cell's input resistance and the area of the synaptic potential.

Obtaining a charge measurement will be particularly useful in cases where different synaptic potentials are being compared, but the exact time course of the synaptic current may vary—for example, in comparing the number of quanta of synaptic action at different locations, when the timing of release of the individual quanta constituting a single synaptic effect may be variable (cf. Katz and Miledi 1965).

3. *Attenuation of the peak voltage with distance.* If favourable circumstances have allowed not only the allocation of a distance value to a synaptic effect, but also allowed an estimate of the time course of the synaptic current (e.g. a

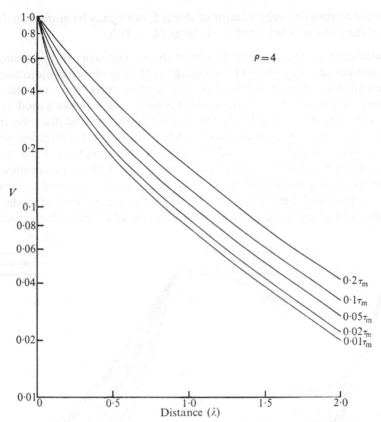

FIG. 7.26. Rectangular pulses of various durations are injected at various distances into an infinite-cable nerve cell model. The ordinate expresses the peak magnitude of the voltage produced at the soma, relative to the magnitude produced when the current is injected at the soma. The abscissa is electrotonic distance. The curves are, from above down, for pulse durations of 0·2, 0·1, 0·05, 0·02, and 0·01τ_m. (Unpublished computations of S. J. Redman.)

value to α, if the current is approximately described by $T\exp(-\alpha T)$ it is possible to calculate the relative reduction in the peak amplitude of the synaptic potential, when recorded at the soma, compared with the case of the same current injected at the soma. Just as in the case of the simple infinite cable (see Chapter 3) the relative attenuation will be greater, for a given distance, the briefer the current. Fig. 7.26 illustrates this in a simple way. The model assumed is an infinite-cable model with $\rho = 4$, and the current takes the form of rectangular pulses of duration $\Delta T = 0·01$, 0·02, 0·05, 0·1, and 0·2.

A more realistic illustration of the same effect is provided by Fig. 7.27. The effects of different dendritic lengths, three different values of α ($\alpha = 12$, 30, and 100) and of three different values of ρ_∞ ($\rho_\infty = 4$, 10, and 25) are illustrated. It can be seen that, with relatively prolonged current pulses, the degree of attenuation is by a factor of about 5, but it may be more than double that if the current is brief and ρ_∞ is large ($L = 1\cdot5$).

Calculation of the voltage at the site of current injection. In this section we will consider the magnitude of the voltage at the site of current injection and leave until later the calculation of its size in other parts of the cell. In order to make the mathematical work reasonably simple we will take a model of the nerve cell in which current is injected, at a single electrotonic distance, in one dendrite (if this dendrite is branched at the point of current injection we will assume that it is injected into all branches of that dendrite). Then the appropriate electrical model is that illustrated in Fig. 7.28. In this example, all the dendrites (represented as 'equivalent' cylinders) are assumed to have the same electrotonic length. The justifications for this general kind of model have already been given in the section on current spread in branching structures.

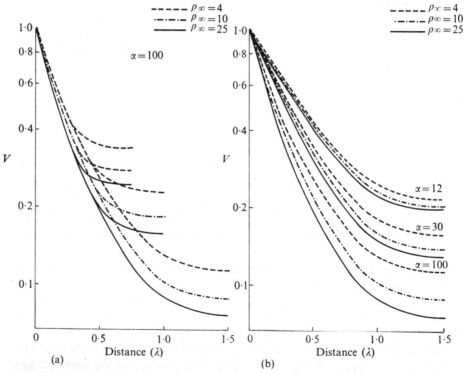

FIG. 7.27. (a) shows the effects of both ρ_∞ and L on the peak magnitude of the soma voltage, and (b) illustrates the effect of three different durations of current pulse. The ordinate and abscissa are as for Fig. 7.26. The current is of the form $T\exp(-\alpha T)$. In (a) α is always 100. The dashed lines are for $\rho_\infty = 4$ at the three dendritic lengths ($L = 0\cdot75$, $1\cdot0$, and $1\cdot5$); dashes and dots, $\rho_\infty = 10$ and continuous lines, $\rho_\infty = 25$. The same convention for values of ρ_∞ holds for (b); the dendrite length is $1\cdot5\lambda$ and the three sets of curves are, from above down, for $\alpha = 12$, 30, and 100 respectively. (Unpublished computations of S. J. Redman.)

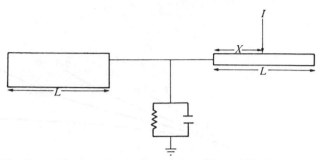

FIG. 7.28. The form of intermediate Rall model used for calculation of the voltage at the site of current injection, when current is injected into only a proportion of the dendritic tree.

Let us denote the input resistance of the cylinder on the right-hand side of Fig. 7.28 by $r_{a1}\lambda_1 \coth L$ and on the left-hand side by $r_{a2}\lambda_2 \coth L$.

It can be shown, using techniques similar to those already described, that the steady state voltage at the site of current injection is given by

$$V = \frac{r_{a1}I\lambda_1}{2}\frac{(1+e^{2X-2L})(1+Ke^{-2X})}{(1-Ke^{-2L})}, \tag{7.95}$$

where

$$K = \frac{r_{a2}\lambda_2 - r_{a1}\lambda_1 \tanh L - r_{a1}\lambda_1 r_{a2}\lambda_2 G_S}{r_{a2}\lambda_2 + r_{a1}\lambda_1 \tanh L + r_{a1}\lambda_1 r_{a2}\lambda_2 G_S}$$

$$= \frac{\rho_\infty \gamma - \rho_\infty(1-\gamma)\tanh L - 1}{\rho_\infty \gamma + \rho_\infty(1-\gamma)\tanh L + 1},$$

where

$$\gamma = \frac{r_a\lambda}{r_{a1}\lambda_1} \quad \text{and} \quad r_a\lambda = 1 \Big/ \left(\frac{1}{r_{a1}\lambda_1}+\frac{1}{r_{a2}\lambda_2}\right)$$

Fig. 7.29 illustrates some calculations that were made using this equation. The value of the soma conductance G_s and of the over-all size of the dendritic tree (expressed by $1/r_a\lambda = 1/r_{a1}\lambda_1 + 1/r_{a2}\lambda_2$) were kept constant. The voltages are all expressed relative to the size of the voltage obtained when the current is injected at the soma. The effects of varying the ratios of $r_{a1}\lambda_1$ to $r_{a2}\lambda_2$ and the dendritic length can be readily observed. Notice that the voltage generated does not vary greatly, for any particular set of parameters, when current is injected into the middle region of the dendritic tree. (This effect is much more marked for brief pulses of current—See Redman, (1973).) In other words, over this middle region the input resistance of the dendritic cable is roughly constant. Indeed, as would be expected, it can be shown that if X and L are both large the input resistance is approximately given by $\frac{1}{2}r_{a1}\lambda_1$—the same as for an infinite cable.

Two other points are worth noting in Fig. 7.29. The first is that the input resistance at the very end of the cable is approximately double its value in the middle—this result is to be expected for dendritic cables of reasonable length because, as compared with a cable of infinite extent, the current all spreads

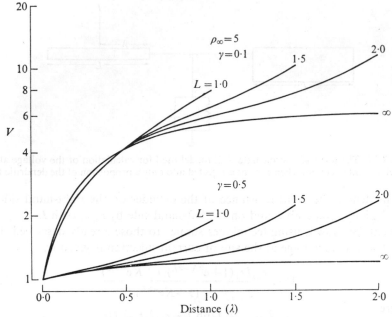

FIG. 7.29. The magnitude of voltage generated at the site of steady current injection, expressed relative to the magnitude produced at the soma by the same current injected there. For all the curves the value of ρ_∞ is 5. Dendritic lengths of $1\cdot0\lambda$, $1\cdot5\lambda$, $2\cdot0\lambda$, and of infinite length are illustrated. The lower set of curves are for $\gamma = 0\cdot5$ (i.e. $r_{a1}\lambda_1 = r_{a2}\lambda_2$) and the upper set of curves for $\gamma = 0\cdot1$ (i.e. $r_{a1}\lambda_1 = 9r_{a2}\lambda_2$). These correspond to current injection into one of two equal sized dendrites and one out of ten equal-sized dendrites respectively.

in one direction instead of dividing equally and only one half spreading towards the soma. The second point is the steep reduction in the value of the input resistance as the soma is approached, when γ is small. This result is not surprising since the soma and the other dendritic cylinder offer a very low resistance path for current flow.

It is also possible to obtain comparable solutions for the transient response (Redman 1973). In this way it becomes possible to compare the peak amplitude of a synaptic potential at its site of generation with its peak amplitude at the soma. The general features of these results are similar to those obtained for the steady state voltage although, of course, the difference in magnitude between the two voltages is relatively greater.

Interaction between synaptic potentials

Although it is now evident that there is great diversity in the effects that synaptic transmitters may have on a postsynaptic cell (Eccles 1964; Hauswirth, Noble, and Tsien 1968; Weight and Votava 1970; Siggins, Oliver, Hoffer, and Bloom 1971; Engberg and Marshall 1971; Weight and Padjen 1973), discussion here will be restricted to the case where the synaptic action is produced by an increase in the ionic conductance of a small patch of membrane. Two forms of interaction need to be considered: the degree of summation between two similar synaptic actions and the interaction between excitation and post-synaptic inhibition. The problems associated with the synapse terminating on a dendritic spine and of the reciprocal synapses will be briefly considered later. In this section we will assume that the postsynaptic cell behaves passively with an action potential arising only if the somatic potential displacement exceeds a certain value, the complications introduced if action potentials, or a more limited nonlinear response, arise in dendritic branches will be discussed later.

As mentioned earlier it has not yet been possible to obtain analytic solutions for a conductance change in finite geometries and so it is necessary to rely on numerical computations for an exact analysis. A few examples of such calculations have been published (Rall 1964, 1967; MacGregor 1968) and they provide an extremely useful insight into particular cases. A slightly more general, but qualitative, account will be attempted here in order that intuitions about a variety of interactions may be obtained.

Addition of functionally similar synaptic actions. In Chapter 3 an account was given of the nonlinearity in addition of the voltage generated by conductance changes of different size at one site. The form of the conductance change assumed was a rectangular pulse. A similar degree of nonlinearity of addition will occur for closely adjacent synapses, simultaneously activated, when they are located in the middle region of a dendritic cylinder (cf. Fig. 7.28) provided that both the cell soma and the dendritic termination are electrically remote (see comments in the previous section on the input resistance of a single dendrite). If, on the other hand, the synapses are close to the dendritic termination the nonlinearity will be greater because the degree of nonlinearity depends on two main factors: the ratio of the sum of the synaptic conductances G_j to the input conductance of the cable G_{in} and the duration of the conductance change (see Chapter 3). At the end of the cable the input conductance of the dendritic cable is half the value in the middle region (if the termination is open-circuit). It the synapses are located near the soma the input conductance of the dendrite at that point will be much greater, and so there will be a more linear addition. Rall (1967) has made calculations of the magnitude of the conductance change required at different locations in his compartmental model in order to produce a potential of the same peak amplitude at the soma. The degree of nonlinearity he observed is in accordance with the various features mentioned above. Kuno and Miyahara (1969) have interpreted the discrepancy they observed between estimates of the quantal size of unit synaptic potentials made from a study of amplitude fluctuations and of the proportion of failures in a number of trials as due to such nonlinear summation, but they did not attempt to calculate the effect of the location of the synapses.

A more difficult problem of interpretation arises when the conductance changes produced by two (or more) synapses are not simultaneous—or even if they are simultaneous but at different locations. In discussing this question it is useful to distinguish two features in the summation of synaptic potentials. The first is the degree of nonlinearity in the relationship between the magnitude of the conductance change and the magnitude of the synaptic current. This will be given by the amount of reduction in the driving voltage when (and where) the second conductance change occurs. It will be obvious that the amount of this nonlinearity will be less than the case of simultaneous conductances at the same site, since the combination of 'propagation' time and attenuation with distance will lead to a much smaller reduction in driving voltage, if the conductances are simultaneous; usually there will be most nonlinearity when the second conductance change occurs at the time of peak voltage of the first (depending on the time course of the conductance in relation to the time course of the voltage).

The second feature is the addition of the two (or more) potentials which occur as a result of the two (or more) currents which are injected. If the two synapses have adjacent locations the time courses of the two potentials, considered in isolation, will be roughly similar for similar time courses of conductance change; but they will be separated in time so that the peak amplitude of the composite potential, in general, will be much less than if they occurred simultaneously. The total area of the composite waveform, however, will give a good indication of the degree of nonlinearity that occurred in the generation of the synaptic current. MacGregor (1968) has made some interesting calculations of the summations of the potentials generated by a single synapse when activated at different frequencies (assuming the magnitude of each conductance change is the same).

When the synapses are at different locations (and perhaps also activated at different times) there will be relatively little nonlinearity, for the reason mentioned above. The calculations of Rall (1964) and MacGregor (1968) illustrated this for the case of synapses located on different dendrites. As pointed out by Rall *et al.* (1967), the most important consideration is the percentage change in the driving voltage that is produced by one synaptic input at the other location when a second synaptic conductance change occurs. This may be explored in a semi-quantitative way by considering the voltage distribution in a nerve cell model produced by the earlier synaptic conductance. If we calculate the steady state value of the voltage distribution this will give the maximum degree of reduction in the driving voltage at other places on the cell. Using the nerve cell model illustrated in Fig. 7.30 it can be shown that if V_0 is the voltage at the site of the synaptic current injection the voltage distribution in the 'active' dendrite is given by

$$V = V_0 \left(\frac{e^{-Y} + Ke^{Y-2X}}{1 + Ke^{-2X}} \right), \quad \text{for the distribution towards the soma,} \quad (7.96)$$

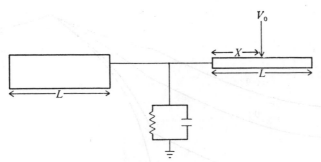

FIG. 7.30. The same model as in Fig. 7.28, which can be used to calculate the voltage distribution throughout the nerve cell model.

where

$$K = \frac{r_{a2}\lambda_2 - r_{a1}\lambda_1 r_{a2}\lambda_2 G_S - r_{a1}\lambda_1 \tanh L}{r_{a2}\lambda_2 + r_{a1}\lambda_1 r_{a2}\lambda_2 G_S + r_{a1}\lambda_1 \tanh L},$$

$$V = V_0 \left(\frac{e^{-Y} + e^{Y-2(L-X)}}{1 + e^{-2(L-X)}} \right) \quad \text{for the distribution towards the dendritic end,}$$

(7.97)

where Y indicates the distance from the synapse.
The soma voltage is given by

$$V_s = V_0 e^{-X} \left(\frac{1+K}{1+Ke^{-2X}} \right),$$

(7.98)

so that the distribution in the other dendrites is given by

$$V = V_0 e^{-X} \left(\frac{1+K}{1+Ke^{-2X}} \right) \left(\frac{e^{-Y} + e^{Y-2L}}{1 + e^{-2L}} \right),$$

(7.99)

where Y now expresses the distance from the soma. Fig. 7.31 gives a particular example for such a voltage distribution, when

$$X = 0.8, \quad L = 1.5, \quad \rho_\infty \left(= (1/r_{a1}\lambda_1 + 1/r_{a2}\lambda_2)/G_s \right) = 4,$$

and

$$\gamma = 0.1, \ 0.2, \ 0.5, \text{ and } 0.9.$$

(This would be equivalent to the nerve cell having ten dendrites, all of equal input resistance, with nine, eight, five, and one of them being represented by the cylinder on the left side of the figure.) The voltage in the peripheral part of the 'active' dendrite is high, but elsewhere its magnitude is relatively small, particularly when $\gamma < 0.5$. Thus a synaptic action on another dendrite would suffer only a small reduction in its driving voltage.

This example is for a steady state distribution of the voltage due to the first synaptic effect and would be misleading for a transient conductance change

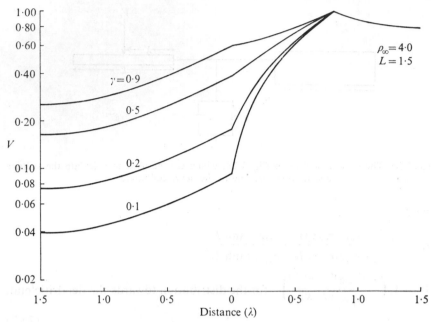

FIG. 7.31. The distribution of voltage at various points in a model like Fig. 7.30. In this case $\rho_\infty = 4$, $L = 1.5$, and current is injected in the right-hand dendrite 0.8λ from the model soma. The ordinate expresses voltage relative to its value at the site of current injection and the abscissa indicates electrotonic distance from the soma. The four curves to the left-hand side of the current injection site are, from above down, for $\gamma = 0.9$, 0.5, 0.2, and 0.1.

when the attenuation is even greater. The maximum degree of nonlinearity in this case would be observed when the second synaptic action occurred at the same time as the peak of the voltage of the first transient. Some insight into this problem of timing may be obtained by considering the question of when a transien potential has its peak value in different parts of a distributed structure. In Chapter 3 it was shown that in an infinite cable the time of peak amplitude of a transient (generated by a very brief pulse of current) propagated with a nearly constant conduction velocity equal to $2\lambda/\tau_m$. In other words, the peak takes one time constant to propagate two space constants. Although the propagation velocity of the peak of the transient is not given exactly by this formula for a nerve cell model, it can be used as a rough indication. Thus, in the above example, if a synapse was located on another dendrite at a distance of 0.7λ from the soma, the total electrotonic distance between the two synapses is 1.5λ. The time taken for the first synaptic potential to reach its peak value at the other site would then be roughly $0.75\tau_m$. The time for the first synaptic potential to reach its peak at the soma would be approximately $0.4\tau_m$. Thus, in an experimental situation, if one were seeking to detect some nonlinear addition between two synaptic potentials it may be quite inappropriate to adjust the timing of their action so that the peaks are simultaneous at the soma. In this example a much more favourable experimental test would be to time the second action so that its peak occurs about one-third of a time constant after the first. This is an extreme example but illustrates that attention must be paid to the timing of synaptic action if nonlinear effects are to be detected.

A related question of timing is that of achieving a maximal peak at the soma of a composite synaptic potential. Since the more distant components give rise to more delayed potentials at the soma, the optimal temporal pattern would be to first activate the most distant synapse and then, in succession, the nearer ones. Rall (1964, Fig. 7) has drawn attention to this. The result is a consequence of the way in which the individual components summate and is not dependent on the avoidance of any nonlinear interaction between the synaptic actions. Indeed, it will tend to maximise the degree of nonlinearity since the optimal temporal sequence of synaptic activation will involve timing each new synaptic conductance so that the reduction in driving voltage, at that location, produced by the earlier activation of other synapses will be maximal.

Interaction between excitation and post-synaptic inhibition. The characteristic feature of post synaptic inhibition is that the reversal potential for the ionic current flow is at or near the resting membrane potential. In their classical analysis of the group Ia inhibition of antagonist motoneurones Coombs, Eccles, and Fatt (1955d) drew attention to the fact that the time course of inhibitory action on a test monosynaptic reflex was not identical to the time course of the inhibitory post synaptic potential. They observed a much greater inhibition of the reflex during the time of the inhibitory conductance change. The subsequent time course of the inhibition followed the time course of the inhibitory post-synaptic potential. It is convenient, therefore, to divide the discussion of inhibitory action into the question of the summation of potentials and the effect of the inhibitory conductance. It is easiest to discuss the latter question by considering the simplest case, when the reversal potential for inhibition is the same value as the resting potential. The effect of inhibitory action is thus simply a change (lowering) of the value of R_m.

The only exact computations reported in the literature are those described by Rall (1962a, 1964). He considered two cases. In the first, the geometry of the cell was deliberately ignored and the interaction between excitation and inhibition in a uniform patch of membrane was calculated. This rather simple case lends itself to an exact solution and Rall (1964) has discussed it fully. Let us consider a slightly more complicated example where there is a steady, uniformly-distributed inhibitory action but excitation occurs at only one point on the cable. Then the only parameter that has changed, if we take an infinite cable as our model, is R_m (and hence both τ_m and λ). For this special case, then, the equations given in Chapter 3 still apply, but the change in τ and λ

lead to changes in the value of $X (= x/\lambda)$ for the location of synaptic action and in the duration of the current injected (when expressed in normalized time $T = t/\tau_m$). If the response to a brief pulse of current is being considered, it is worth remembering that the degree of attenuation that the potential suffers in the recording site is dependent on two opposing factors: the electrotonic distance and the duration of the synaptic current. The further the electrotonic distance, the greater the attenuation but the longer the duration (in normalized time), the less the attenuation. A reduction in R_m reduces both τ_m and λ and hence increases both relative electrotonic distance of the synapse (leading to greater relative attenuation) and the normalized duration of current (leading to relatively less attenuation). The results illustrated in Fig. 7.32 will therefore not be a surprise. The lower half of the figure plots the relative reduction in peak amplitude and area of a transient potential or the steady value in response to a maintained current when compared with the peak amplitude and area in the uninhibited state. The reduction in R_m assumed during inhibition is one half its normal value. Note that the peak amplitude is very little reduced, particularly for near distances, although the area is substantially reduced. For completeness, the effect of doubling R_m is also shown in Fig. 7.32. The same type of result is found. Similar calculations can be readily performed for more complicated structures such as a nerve cell model. The change in R_m will also affect the value of the parameters ρ_∞ and L. It would be expected that the general trend of the results would be similar to those illustrated; a reduction in R_m will lead to a reduction in ρ_∞ (and hence relatively less attenuation) and an increase in L (hence relatively more attenuation). The relative sizes of these two opposing effects will, of course, depend on the exact values selected for each of the parameters.

There remains the more difficult question of both the excitation and the inhibition being restricted to particular parts of a cable structure. Some points can be made on the basis of previous discussion. The first is that an inhibitory conductance change in one dendrite is very unlikely to affect an excitatory potential generated on a different dendrite. The reason for this is the same as for the difficulty in detecting a conductance change any distance out in a dendrite, even with an electrode placed in the soma (see the section *Determination of the location of synaptic action*). In such circumstances the only way an inhibitory action could have an effect would be by the hyper-polarization it produced; and there would be nearly linear summation of the two potentials (cf. Rall *et al.* 1967, Fig. 7). Secondly, an inhibitory synapse located on the somatic side of an excitatory synapse will be more effective in reducing the excitatory potential recorded at the soma than if it were located an equivalent distance from the excitatory synapse, but on the non-somatic side. This follows from considering the proportion of the total current injected which spreads along the dendrite to the soma, instead of 'leaking' across the membrane resistance. If the inhibitory conductance is on the somatic side it will cause a leak of a substantial proportion of the current, spreading directly

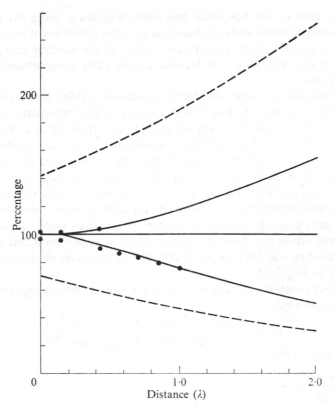

FIG. 7.32. This figure illustrates the effect of halving or doubling R_m on the magnitude of potentials recorded at one point in an infinite cable (*not* a nerve cell model) when current is injected at various distances away from the recording site. The ordinate expresses the voltage relative to that obtained without the change in R_m. The abscissa indicates the distance at which current injection takes place, expressed in terms of λ (original value of R_m). In the lower half of the figure the lower (dashed) line shows the reduction in area of a transient potential (or of the steady state amplitude in response to maintained current injection) when R_m is halved. The continuous line in the lower half of the figure shows the effect on the peak amplitude of the delta-function response, and the filled circles the effect on the peak amplitude of the response to a rectangular pulse of current of duration $0.05\tau_m$.

The upper half of the figure shows the comparable results for a doubling of R_m.

towards the soma. If it is on the non-somatic side it will allow a similar proportion of the peripherally spreading current to cross the membrane. But in a finite model a small proportion of the axial current which reaches the dendritic termination is 'reflected' back, and a small part of it reaches the soma; it is this current which will be reduced by a peripherally placed inhibition. Thus if the excitatory potential is a transient it is unlikely that the peak amplitude at the soma will be reduced, although the later time course of decay will be. Rall has illustrated exactly this result (Rall 1964, Fig. 8).

There remains the problem of whether a maximal inhibition of a dendritically located excitatory synapse is achieved by a synapse close by, or on, the soma. Rall (1964) has made one such calculation using his compartmental model. It was found that there was a greater reduction of the excitatory potential when inhibitory action was located in the somatic compartment than when the inhibition was located in the same compartments as the excitatory action.

This result may be surprising since it is natural to think that an inhibitory conductance is most effective if it is located in the immediate vicinity of the excitatory synapse. In order to explore this result it is convenient to subdivide the effect of the inhibitory conductance into two parts: (1) the reduction in the input impedance at the site of the synaptic excitation, and (2) the effect on the attenuation of the excitatory voltage in spreading to the soma from its place of generation. Rall's (1964) result is for transients of excitation and inhibition, but it is easier to calculate these effects in a model which assumes steady values of excitation and inhibition. If we model synaptic excitation by a constant current injection at one point and inhibition by a conductance at another point then the appropriate electrical circuit is illustrated in Fig. 7.33.

The input resistance of this model, at the site of current injection, when $G_i = 0$ is given by

$$R_{in} = \frac{1+e^{-2(L-X)}}{2\rho_\infty G_S} \left\{ \frac{\rho_\infty+1+(\rho_\infty-1)e^{-2X}}{\rho_\infty+1-(\rho_\infty-1)e^{-2L}} \right\} \tag{7.100}$$

and, when G_i is finite,

$$R_{in} = \frac{1+e^{-2(L-X)}}{2\rho_\infty G_S} \times$$

$$\times \frac{\rho_\infty+1+(\rho_\infty-1)e^{-2X}+\dfrac{G_i}{2G_S}(1+e^{-2Z}-e^{-2(X-Z)}-e^{-2X})+ \\ \qquad\qquad +\dfrac{r_a\lambda G_i}{2}(1-e^{-2Z}-e^{-2(X-Z)}+e^{-2X})}{\rho_\infty+1-(\rho_\infty-1)e^{-2L}+\dfrac{G_i}{2G_S}(1+e^{-2Z}+e^{-2(L-Z)}+e^{-2L})+ \\ \qquad\qquad +\dfrac{r_a\lambda G_i}{2}(1-e^{-2Z}+e^{-2(L-Z)}-e^{-2L})} \tag{7.101}$$

when $Z \leqslant X$ and

$$R_{in} = \frac{\{\rho_\infty+1+(\rho_\infty-1)e^{-2X}\}\times \\ \quad\times\left\{1+e^{-2(L-X)}+\dfrac{r_a\lambda G_i}{2}(1-e^{-2(L-X)}-e^{-2(Z-X)}+e^{-2(L-Z)})\right\}}{2\rho_\infty G_S\left\{\rho_\infty+1-(\rho_\infty-1)e^{-2L}+\dfrac{r_a\lambda G_i}{2}(1+e^{-2(L-Z)}-e^{-2L}- \\ \qquad -e^{-2Z})+\dfrac{G_i}{2G_S}(1+e^{-2(L-Z)}+e^{-2Z}+e^{-2L})\right\}} \tag{7.102}$$

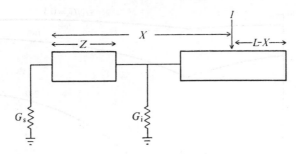

FIG. 7.33. The form of model assumed for studying the effectiveness of an inhibitory conductance on synaptic excitation (steady current injection).

when $Z > X$. The relative reduction in input resistance produced by G_i is therefore given by

$$(\rho_\infty+1-(\rho_\infty-1)e^{-2L})\Big\{\rho_\infty+1+(\rho_\infty-1)e^{-2X}+$$

$$+\frac{G_i}{2G_s}(1+e^{-2Z}-e^{-2(X-Z)}-e^{-2X})+\frac{r_a\lambda G_i}{2}(1-e^{-2Z}-e^{-2(X-Z)}+e^{-2X})\Big\}$$

$$\overline{(\rho_\infty+1+(\rho_\infty-1)e^{-2X})\Big\{\rho_\infty+1-(\rho_\infty-1)e^{-2L}+}$$

$$+\frac{G_i}{2G_s}(1+e^{-2Z}+e^{-2(L-Z)}+e^{-2L})+\frac{r_a\lambda G_i}{2}(1-e^{-2Z}+e^{-2(L-Z)}-e^{-2L})\Big\}$$

$$\tag{7.103}$$

when $Z \leqslant X$ and

$$(\rho_\infty+1-(\rho_\infty-1)e^{-2L})\times$$

$$\times\Big\{1+e^{-2(L-X)}+\frac{r_a\lambda G_i}{2}(1-e^{-2(L-X)}-e^{-2(Z-X)}+e^{-2(L-Z)})\Big\}$$

$$\overline{(1+e^{-2(L-X)})\Big\{\rho_\infty+1-(\rho_\infty-1)e^{-2L}+} \tag{7.104}$$

$$+\frac{r_a\lambda G_i}{2}(1+e^{-2(L-Z)}-e^{-2L}-e^{-2Z})+\frac{G_i}{2G_s}(1+e^{-2(L-Z)}+e^{-2Z}+e^{-2L})\Big\}$$

when $Z > X$

It can be shown that the maximal reduction of the input resistance is produced when $Z = X$. An example of the effect is shown in Fig. 7.34 for $G_i = 0{\cdot}1G_s$, $0{\cdot}5G_s$, and $2{\cdot}0G_s$. In each case $\rho_\infty = 5$, $L = 1{\cdot}5$, and $X = 0{\cdot}5$.

On the other hand, when considering the effect of G_i on the voltage attenuation, exactly the opposite result holds—maximal effect is achieved when the

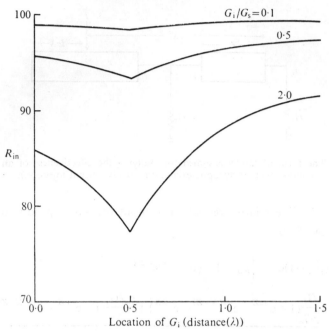

FIG. 7.34. Reduction in the input resistance of the model shown in Fig. 7.33 by three different magnitudes of inhibitory conductance (G_i/G_s = 0·1, 0·5, and 2·0). Input resistance is measured one-third of the way between the soma and the end of the dendritic cable ($X = 0·5$, $L = 1·5$). The value of ρ_∞ is 5. The ordinate plots the input resistance relative to its value without an inhibitory conductance and the abscissa the electrotonic distance at which the inhibitory conductance is located.

inhibitory conductance is located at the soma ($Z = 0$) and, of course, there is no voltage attenuation if $Z \geqslant X$. This can be seen from the following equations. When $G_i = 0$ the equations of V_0 (voltage at site of excitatory current injection) and V_s (voltage at soma) are

$$V_0 = \frac{I\{1+e^{-2(L-X)}\}}{2\rho_\infty G_s} \frac{\{\rho_\infty+1+(\rho_\infty-1)e^{-2X}\}}{\{\rho_\infty+1-(\rho_\infty-1)e^{-2L}\}}, \qquad (7.105)$$

$$V_s = \frac{Ie^{-X}}{G_s} \frac{\{1+e^{-2(L-X)}\}}{\{\rho_\infty+1-(\rho_\infty-1)e^{-2L}\}}. \qquad (7.87)$$

The voltage attenutation (V_s/V_0) is therefore given by

$$\frac{V_s}{V_0} = \frac{2\rho_\infty e^{-X}}{\rho_\infty+1+(\rho_\infty-1)e^{-2X}}. \qquad (7.106)$$

The same value for voltage attenuation is found when G_i is non-zero, but $Z \geqslant X$.

When $Z < X$ the equations are

$$V_0 = \frac{I\{1+e^{-2(L-X)}\}}{2\rho_\infty G_s} \times$$

$$\times \frac{[\rho_\infty+1+(\rho_\infty-1)e^{-2X}+(G_i/2G_s)\{1+e^{-2Z}-e^{-2(X-Z)}-e^{-2X}\}+}{[\rho_\infty+1-(\rho_\infty-1)e^{-2L}+(G_i/2G_s)\{1+e^{-2Z}+e^{-2(L-Z)}+e^{-2L}\}+}$$
$$\frac{+(r_a\lambda G_i/2)\{1-e^{-2Z}-e^{-2(X-Z)}+e^{-2X}\}]}{+(r_a\lambda G_i/2)\{1-e^{-2Z}+e^{-2(L-Z)}-e^{-2L}\}]}$$

$$(7.107)$$

and

$$V_s = \frac{Ie^{-X}}{G_s} \frac{\{1+e^{-2(L-X)}\}}{[\rho_\infty+1-(\rho_\infty-1)e^{-2L}+(G_i/2\rho_\infty G_s)\times}$$
$$\times\{(\rho_\infty+1)(1+e^{-2(L-Z)})+(\rho_\infty-1)(e^{-2Z}+e^{-2L})\}]} \qquad (7.108)$$

and the voltage attenuation is therefore given by

$$\frac{V_s}{V_0} = \frac{2\rho_\infty e^{-X}}{\rho_\infty+1+(\rho_\infty-1)e^{-2X}+(G_i/2\rho_\infty G_s)\times} \qquad (7.109)$$
$$\times\{(\rho_\infty+1)(1-e^{-2(X-Z)})+(\rho_\infty-1)(e^{-2Z}-e^{-2X})\}$$

The attenuation factor produced by G_i can be calculated therefore from eqns (7.106) and (7.109),

$$\frac{(V_s/V_0)_{G_i\text{finite}}}{(V_s/V_0)_{G_i=0}} = \frac{\rho_\infty+1+(\rho_\infty-1)e^{-2X}}{\rho_\infty+1+(\rho_\infty-1)e^{-2X}+(G_i/2\rho_\infty G_s)\times} \qquad (7.110)$$
$$\times\{(\rho_\infty+1)(1-e^{-2(X-Z)})+(\rho_\infty-1)(e^{-2Z}-e^{-2X}\}$$

Maximal voltage attenuation produced by G_i will occur when the denominator of eqn (7.110) is maximum. This occurs, for any value of G_i when $(\rho_\infty-1)\exp(-2Z)-(\rho_\infty+1)\exp\{-2(X-Z)\}$ is maximum; i.e. $Z = 0$ (since $0 < Z < X$). (The only exception to this conclusion is if $\rho_\infty < 1$, in which case maximal voltage attenuation is achieved when

$$Z = \frac{X}{2} - \frac{1}{4}\ln\left(\frac{1+\rho_\infty}{1-\rho_\infty}\right).$$

In order to satisfy this, $\rho_\infty < \tanh X$.) An example of the amount of relative attenuation produced by G_i, depending on its placement, is given by Fig. 7.35.

There are therefore two conflicting factors in considering the optimal site for inhibitory synapses in order to reduce the soma voltage maximally. For a given excitatory current, maximal reduction in V_0 is obtained by placing G_i at the location of excitatory current injection. On the other hand, placing G_i at the soma gives a maximal reduction in the ratio V_s/V_0 (with the exception noted above). The relative quantitative importance of these two effects can be

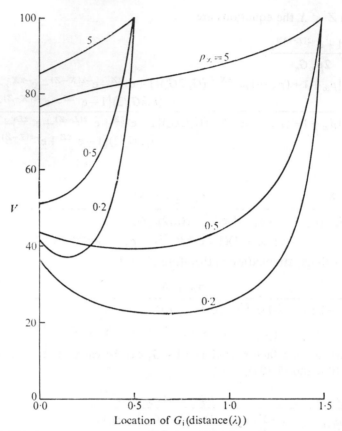

F IG . 7.35. This figure illustrates the additional reduction in voltage, recorded at the soma, as a result of a change in the degree of attenuation when spreading from the site of current injection to the soma. The model assumed is that shown in Fig. 7.33, with $L = 1\cdot5$. Examples of current injection at $X = 0\cdot5$ and $X = 1\cdot5$ are given. The two upper curves are for $\rho_\infty = 5$ and they demonstrate that the voltage attenuation factor is maximal when the inhibitory conductance is placed at the soma. The other curves are for $\rho_\infty = 0\cdot5$ and $\rho_\infty = 0\cdot2$ and show that for such low values of ρ_∞, voltage attenuations may not be maximal when the inhibitory conductance is at the soma.

estimated by combining the two factors in a single equation, i.e. by studying the soma voltages produced by a single value of current, with and without G_i (i.e. eqns (7.87) and (7.108))

$$\frac{V_{s(G_i \text{ finite})}}{V_{s(G_i=0)}} = \frac{\rho_\infty+1-(\rho_\infty-1)e^{-2L}}{\rho_\infty+1-(\rho_\infty-1)e^{-2L}+(G_i/2\rho_\infty G_s)\times} \qquad (7.111)$$
$$\times\{(\rho_\infty+1)(1+e^{-2(L-Z)})+(\rho_\infty-1)(e^{-2Z}+e^{-2L})\}$$

$(Z \leqslant X)$.

Deductions can be made from this equation about the optimal placing of G_i for maximal reduction in the soma voltage. The degree of attenuation is maximal when $(\rho_\infty+1)\exp\{-2(L-Z)\}+(\rho_\infty-1)\exp(-2Z)$ is maximal. If $\rho_\infty < 1$ it is easy to show that inhibition will always be most effective when $Z = X$. This therefore sets one fundamental constraint on the design of nerve cells, namely that, in order for a somatically placed inhibition to be more effective than one placed in the vicinity of the excitatory synapse, it is required that $\rho_\infty > 1$. It will be shown later that this constraint on ρ_∞ is a necessary but not a sufficient condition for a soma inhibition to be most effective.

From eqn (7.111) we can also compare the effectiveness of a soma inhibition with one located at the same site as the excitatory action. The two equations (when $Z = 0$ and $Z = X$) are

$$V_{s(G_1 \text{ at } Z=0)} = \frac{Ie^{-X}}{G_s} \frac{(1+e^{-2(L-X)})}{\{\rho_\infty+1-(\rho_\infty-1)e^{-2L}+(G_i/G_s)(e^{-2L}+1)\}} \tag{7.112}$$

and

$$V_{s(G_1 \text{ at } Z=X)} = \frac{Ie^{-X}}{G_s} \frac{(1+e^{-2(L-X)})}{[\rho_\infty+1-(\rho_\infty-1)e^{-2L}+(G_i/2\rho_\infty G_s)\times}$$
$$\times\{(\rho_\infty+1)(1+e^{-2(L-X)})+(\rho_\infty-1)(e^{-2X}+e^{-2L})\}] \tag{7.113}$$

so that inhibition at the soma is more effective only if

$$2\rho_\infty(e^{-2L}+1) > (\rho_\infty+1)(1+e^{-2(L-X)})+(\rho_\infty-1)(e^{-2X}+e^{-2L}),$$

i.e. if $e^{4X}-e^{2X}\{1+(\rho_\infty-1)/(\rho_\infty+1)e^{2L}\}+(\rho_\infty-1)/(\rho_\infty+1)e^{2L} < 0.$

Provided that $\rho_\infty > 1$ (see above) this leads to a condition on X:

$$X < L-\tfrac{1}{2}\ln(\rho_\infty+1)/(\rho_\infty-1). \tag{7.114}$$

It is also possible to show that, for a given value of X,

$$\left(X > \tfrac{1}{2}L-\tfrac{1}{4}\ln\left(\frac{\rho_\infty+1}{\rho_\infty-1}\right)\right)$$

inhibition is least effective when

$$Z = \tfrac{1}{2}L-\tfrac{1}{4}\ln(\rho_\infty+1)/(\rho_\infty-1) \qquad 0 < Z < X \tag{7.115}$$

which, since $Z \not< 0$ will only hold when

$$2L > \ln(\rho_\infty+1)/(\rho_\infty-1),$$

i.e.

$$\rho_\infty > \coth L \quad \text{or} \quad \rho > 1. \tag{7.116}$$

This is because if $\rho_\infty < \coth L$ inhibition is most effective when $Z = X$ rather than when it is somatically placed. For example, if $L = 1{\cdot}5$, ρ_∞ must be greater than 1·1 in order for soma inhibition to be more effective and further-more it will only be so provided X is less than a certain value. If $\rho_\infty = 4, 10$, and 25, respectively, X has to be less than $(L-0·25)$, $(L-0·09)$, and $(L-0·04)$ respectively. Thus it can be seen that all synaptic excitations except in the most peripheral part of the dendritic tree are more reduced by a somatic inhibition than one placed elsewhere on the nerve cell. Thus if the location of inhibitory synapses is designed in order to secure most effective operation it would be expected that they would be in two sites only: on, or very near, the soma of the nerve cell and also on the distal tips of the dendrites. The only

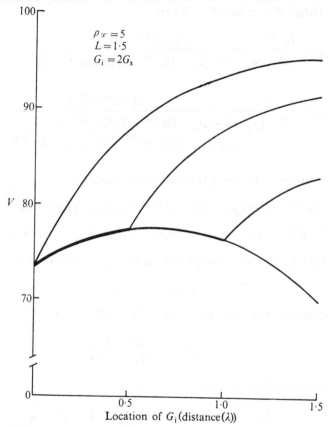

FIG. 7.36. The total effect of an inhibitory conductance on the soma voltage is shown in this figure. The ordinate is the soma voltage, relative to that without an inhibitory conductance, in response to current injection. (Abscissa expresses electrotonic distance of the inhibitory conductance from the soma.) Four different sites of current injection are illustrated: $X = 0, 0·5, 1·0$, and 1·5. The model parameters are $\rho_\infty = 5$, $L = 1·5$, and $G_i = 2G_s$. The lowermost curve shows the effect of inhibition when $Z < X$, so that the complete curve is for $X = 1·5$. The lines rising up from this curve show the attenuation for $Z > X$ (i.e. when inhibition is further out on the dendrite than the current injection point). These three lines, therefore, describe, from right to left, the effect of inhibition when $X = 1·0, 0·5$, and 0 respectively.

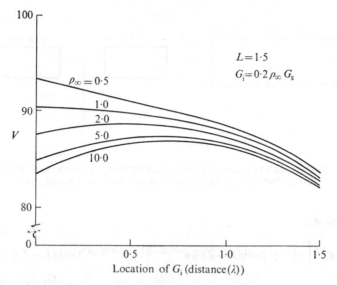

FIG. 7.37. This figure illustrates the total effect of an inhibitory conductance when it is located between the site of current injection and the soma. The model assumed is shown in Fig. 7.33, and the ordinate and abscissa are the same as in Fig. 7.36. The model parameters are $L = 1.5$ and $G_i = 0.2\, \rho_\infty G_s$. Each curve shows the effect of different values of ρ_∞: 0.5, 1, 2, 5, and 10 from above down. When $\rho_\infty = 2$, 5, and 10 the curve shows a minimum effectiveness, as described by eqn (7.117). No such minimum occurs when $\rho_\infty = 0.5$ or 1, and the conductance is most effective when placed at the site of current injection.

exception to this rule that would be expected on the above reasoning is if the nerve cell body were very large in relation to the dendritic trunk diameters (or, more exactly $\rho \ll 1$), in which case the inhibitory synapses might be expected to be in close proximity to the excitatory synapses. Examples of these effects are given in Figs. 7.36 and 7.37.

Several major qualifications must be made to this conclusion. The model which leads to these calculations assumes that, if the nerve cell has more than one dendrite, the excitatory and inhibitory synapses are located similarly on all the dendrites. Since this may be an unwarranted assumption it is necessary to check with a more complicated model, of the kind illustrated in Fig. 7.38.

The solutions for the soma voltage in this case are as follows

$$V_s = \frac{I\gamma}{G_s} \frac{e^{-X}\{1+e^{-2(L-X)}\}}{(1+\rho_\infty\gamma+\rho_\infty(1-\gamma)\tanh L)\left[\gamma(1-Ke^{-2L})+\dfrac{G_i}{2\rho_\infty G_s}\times\right.}$$

$$\left. \times\{1+e^{-2(L-Z)}+Ke^{2Z}+Ke^{-2L}\}\right]$$

$$(7.117)$$

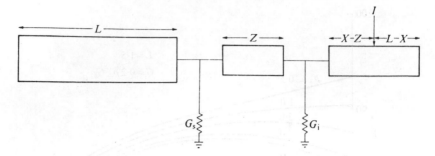

FIG. 7.38. Model used to explore the interaction between excitation and inhibition when they are both placed on the same fraction of the dendritic tree.

when $Z < X$;

$$V_s = \frac{I}{G_s} \frac{e^{-X}\left[\gamma\{1 + \tanh(L-Z)+e^{-2(Z-X)}-e^{-2(Z-X)}\tanh(L-Z)\}+\dfrac{G_i}{\rho_\infty G_s}(1-e^{-2(Z-X)})\right]}{(1+\rho_\infty\gamma+\rho_\infty(1-\gamma)\tanh L)\left[\gamma\{1 + \tanh(L-Z)+Ke^{-2Z}-Ke^{-2Z}\tanh(L-Z)\}+\dfrac{G_i}{\rho_\infty G_s}(1+Ke^{-2Z})\right]}$$

(7.118)

when $Z > X$, where K and γ are the same as in eqn (7.95).

Although inhibition will have its strongest effect when it is placed either at the soma or on the right-hand dendrite ($Z < X$, i.e. eqn (7.117)) we also need to consider, for completeness, the effect of G_i when it is located on the left-hand dendrite. In this case the model is shown in Fig. 7.39.

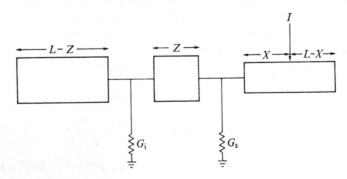

FIG. 7.39. Model in which excitation is located on one fraction of the dendritic tree and inhibition exclusively on the other fraction.

The equation for V_s in this circumstance is

$$V_s = \frac{I}{G_s} \frac{e^{-X}\{1+e^{-2(L-X)}\}(1-qe^{-2Z})}{[1-qe^{-2Z}+\rho_\infty\{1+(1-2\gamma)qe^{-2Z}\}](1+oe^{-2Z})}, \qquad (7.119)$$

where

$$q = \frac{(G_i/2\rho_\infty G_s)\{1+e^{-2(L-Z)}\}-(1-\gamma)e^{-2(L-Z)}}{(G_i/2\rho_\infty G_s)\{1+e^{-2(L-Z)}\}+(1-\gamma)}$$

and

$$o = \frac{1-qe^{-2Z}+\rho_\infty(1-2\gamma+qe^{-2Z})}{1-qe^{-2Z}+\rho_\infty(1+(1-2\gamma)qe^{-2Z})}$$

Finally we should consider the case when G_i is located on only one dendrite, but a different one from that which the excitatory synapses are located (Fig. 7.40).

The equation is

$$V_s = \frac{\rho_\infty I e^{-X}(1+e^{-2(L-X)})(1-n)}{2\gamma G_s(1+ne^{-2L})}, \qquad (7.120)$$

where

$$n = \frac{1+\rho_\infty(1-\beta-\gamma)\tanh L+\rho_\infty\beta\left(\dfrac{1+me^{-2Z}}{1-me^{-2Z}}\right)-\rho_\infty\gamma}{1+\rho_\infty(1-\beta-\gamma)\tanh L+\rho_\infty\beta\left(\dfrac{1+me^{-2Z}}{1-me^{-2Z}}\right)+\rho_\infty\gamma}$$

and

$$m = \frac{(G_i/2\rho_\infty G_s)\{1+e^{-2(L-Z)}\}-\beta e^{-2(L-Z)}}{(G_i/2\rho_\infty G_s)\{1+e^{-2(L-Z)}\}+\beta}.$$

$$\gamma = \frac{r_a\lambda}{r_{a_1}\lambda_1} \quad \text{and} \quad \beta = \frac{r_a\lambda}{r_{a_2}\lambda_2}.$$

In order to determine the optimal site for inhibition we need only consider eqn (7.117), since the previous analysis still applies, that is, showing that the two effects of an inhibitory conductance, namely, reduction in input resistance and increased voltage attenuation in spreading to the soma, are achieved best by locations at, or between, $Z = X$ and $Z = 0$ respectively. V_s is minimal in eqn (7.117) when $K\exp(-2Z)+\exp\{-2(L-Z)\}$ is maximal. The

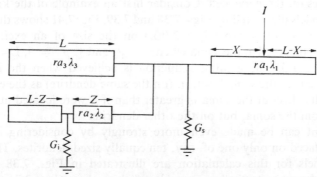

FIG. 7.40. Model in which both excitation and inhibition go to only a small fraction of the dendritic tree, but not the same fraction.

8

first differential of this expression with respect to Z is $2\exp\{-2(L-Z)\}-2K\exp(-2Z)$. If K is negative the slope of the expression is always positive and therefore there will be most attenuation when $Z = X$. K is negative if

$$\gamma < \frac{\rho_\infty \tanh L+1}{\rho_\infty(\tanh L+1)} \qquad \left(\text{or } \rho_\infty < \frac{1}{\gamma-(1-\gamma)\tanh L}\right) \qquad (7.121)$$

For experimentally observed values of L and ρ_∞ ($L > 1{\cdot}0$, $\rho_\infty > 4$), this means that $\alpha < 0{\cdot}5$–$0{\cdot}6$. But if there are more than two dendrites, of roughly equal trunk diameter, this condition is met; so that the soma placing is shown to be generally not optimal.

If $\gamma > \{\rho_\infty\tanh L+1\}/\{\rho_\infty(\tanh L+1)\}$ then the further conditions to be met before a somatically placed inhibition is more effective (than one at the same site as the excitation) are:

$$\rho_\infty > \frac{1}{2\gamma-1} \quad \text{and} \quad L > \tfrac{1}{2}\ln\!\left(\frac{\rho_\infty(2\gamma-1)+1}{\rho_\infty(2\gamma-1)-1}\right), \qquad (7.122)$$

$$X < L-\tfrac{1}{2}\ln\!\left(\frac{\rho_\infty\gamma+\rho_\infty(1-\gamma)\tanh L+1}{\rho_\infty\gamma-\rho_\infty(1-\gamma)\tanh L-1}\right). \qquad (7.123)$$

For example, if $\gamma = 0{\cdot}6$, then ρ_∞ must be greater than 5; if $\gamma = 0{\cdot}6$ and $\rho_\infty = 10$ then L must be greater than $0{\cdot}55$; if $\gamma = 0{\cdot}6$, $\rho_\infty = 10$ and $L = 2{\cdot}0$ then the excitatory synapses must be no more than $1{\cdot}19\lambda$ from the soma.

Thus the previous conclusions are strikingly modified when considering the case of a nerve cell with more than one dendrite and in which synaptic actions are localized on only one dendrite. Nevertheless, it may still be a sensible general design to have inhibitory synapses located on or near the soma if it is not possible, with respect to the factors controlling the formation of synaptic connections, to determine on which of the dendrites a particular excitatory input will be located. This is most easily illustrated by considering examples of the effectiveness of inhibition when it is placed at various sites on the nerve cell. Consider first an example of the kind which uses the models illustrated by Figs. 7.38 and 7.39. Fig. 7.41 shows the effect of an inhibitory conductance ($G_i = 2{\cdot}0G_s$) on the size of an excitatory depolarization recorded at the soma when $\gamma = \tfrac{1}{10}$ and $\tfrac{1}{2}$, $\rho_\infty = 5$, $L = 1{\cdot}5$, and $X = 0{\cdot}5$ and $1{\cdot}5$. The greatest reduction is achieved when the inhibitory conductance is at the same location (on the same dendrite) as the excitation. However, its effect at the soma is greater than if it were located at a similar distance from the soma, but on the other dendrite.

The point can be made even more strongly by considering inhibitory synapses placed on only one of, say, ten equally sized dendrites. The appropriate models for this calculation are illustrated in Figs. 7.38 and 7.40. Fig. 7.42 shows a particular example of this when $G_i = 2G_s$, $\rho_\infty = 10$, $L = 1{\cdot}0$, and $X = 0$, $0{\cdot}25$, $0{\cdot}75$, $0{\cdot}5$, and $1{\cdot}0$. If the inhibitory synapses are

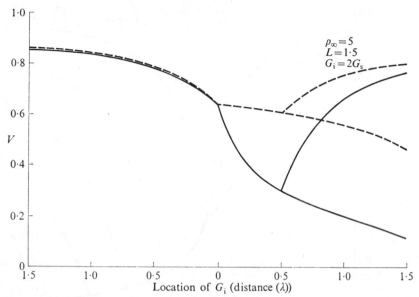

$$\rho_\infty = 5$$
$$L = 1\cdot5$$
$$G_i = 2G_s$$

V

Location of G_i (distance (λ))

FIG. 7.41. The effect of inhibition when it is located either on the same dendrite as the injected current, or on the rest of the dendritic tree (that is, the models of Figs. 7.38 and 7.39) The ordinate and abscissa have the same significance as before (that is as in Fig. 7.36). The right-hand half of the figure is similar to Fig. 7.36; the model parameters chosen are $\rho_\infty = 5$, $L = 1\cdot5$, and $G_i = 2G_s$. The dashed lines show the effect of inhibition (with current injection at $X = 0\cdot5$ and $X = 1\cdot5$) when current injection and the inhibitory conductance are restricted to half the total dendritic tree ($\gamma = 0\cdot5$). Notice that the curve for $X = 1\cdot5$ now does not show a minimum of effectiveness in the centre of the cable and for current injection at the end of the dendrite a local inhibitory conductance is substantially more effective than one at the soma. This effect is much more pronounced when the inter- action is restricted to one tenth of the dendritic tree ($\gamma = 0\cdot1$, continuous lines). The left- hand half of the figure shows the effect of inhibition when it is located in that part of the dendritic tree which does not receive the current injection (i.e., Fig. 7.39). There is a negligible (approximately 1 per cent) difference in effectiveness depending on the value of γ.

on a dendrite ($0\cdot5\lambda$ from the soma) other than that on which the excitatory synapses end, its effect is very much smaller than when placed on the soma.

A further factor which must be considered is whether a constant current injection is a suitable model for the excitatory synaptic action. One feature that the 'equivalent' cylinder model of a single dendrite tends to disguise is that, with extensive peripheral branching, if the excitatory synapses end on only one of the daughter branches, the ratio of the excitatory synaptic conductance to the input conductance of the dendrite at that point may be- come very high. In this case a more appropriate model for excitatory action may be a voltage clamp (see Chapter 3). The effect of the inhibitory conduct- ance in reducing the input resistance will then be much less important, and the only factor affecting the reduction in the soma voltage will be the relative attenuation in spreading to the soma. It has been shown already for a single-cylinder nerve cell model that voltage attenuation is maximal when

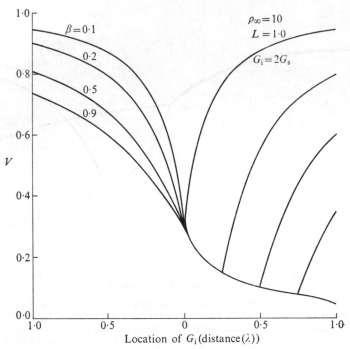

FIG. 7.42. The effect of a conductance when it is located either on the same dendrite as the injected current or on a variable proportion of the rest of the dendritic tree (i.e. the models of Figs. 7.38, 7.39, and 7.40). The ordinate and abscissa have the same significance as in Fig. 7.36. The model parameters are $\rho_\infty = 10$, $L = 1\cdot0$, and $G_1 = 2G_s$. The dendrite on the right, in which current injection takes place ($X = 0$, $0\cdot25$, $0\cdot5$, $0\cdot75$, and $1\cdot0$), is one-tenth of the total dendritic tree. The curves on the left-hand side of the figure show the effect of differing locations of the conductance when it is on the remainder of the dendritic tree (lowermost curve) or on a variable proportion of it (from above down $\beta = 0\cdot1$, $0\cdot2$, and $0\cdot5$ respectively—the lowermost curve is, of course, for $\beta = 0\cdot9$).

the inhibitory conductance is on the soma provided $\rho_\infty > 1$. In the case of a multi-cylinder model the condition can be shown to be

$$\rho_\infty > 1/\{\gamma - (1-\gamma)\tanh L\}.$$

Since it is likely that this is not so (see above) it is important to know what the optimal placing of the inhibitory synapses is in this circumstance (that is, when $\gamma < (\rho_\infty \tanh (L)+1)/(\rho_\infty(\tanh (L)+1))$). This can be shown to be

$$Z = \left(\frac{X}{2}\right) - \tfrac{1}{4}\ln\left\{\frac{1+\rho_\infty(1-\gamma)\tanh L + \rho_\infty\gamma}{1+\rho_\infty(1-\gamma)\tanh L - \rho_\infty\gamma}\right\}, \qquad (7.124)$$

so that a placing of the inhibitory synapses, on the same dendrite, approximately halfway between the soma and the excitatory synapses is optimal.

We may conclude therefore that the most favourable places to locate inhibitory synapses, in order for them to be most effective, are either on the soma and proximal part of the dendrites or, for the most distally located excitatory synapses, near to the extreme tips of the dendrites. Unfortunately, it is still too early in the experimental analysis of the location of synapses to know whether their distribution on real nerve cells will match these predictions. These suggestions are, moreover, based on the assumption that there is no question of selective interaction between excitatory and inhibitory synapses of distinct origins.

Two additional points might be made about these conclusions. In the description of inhibitory synaptic action a rather artificial distinction between hyperpolarization and conductance change was made, and the discussion focused exclusively on the effects of the conductance change. If the inhibition is accompanied by a true hyperpolarization (that is, the reversal potential for inhibition is more internally negative than the resting potential) then this effect will be maximal when the synapses are on the soma, since the hyper-polarization will suffer attenuation in spreading to the soma from any other point. Furthermore, the soma is usually the site of lowest input resistance in the cell, so that there will be least nonlinearity in the relationship between the inhibitory conductance size and the size of the inhibitory synaptic current when the synapses are on the soma; this may be an important effect since the difference between the membrane potential and the inhibitory reversal po-tential is often small and hence the nonlinearity is likely to be considerable. These two considerations therefore reinforce the general conclusion that the soma is a very favourable place on which to locate inhibitory synapses, if it is required that the inhibition be effective against all excitatory actions delivered to the cell. By the same token, the above analysis shows that if the inhibition is to be selective in its action then the best location is adjacent to the appropriate excitatory synapses—and such selectivity is best achieved when the excitatory synapses are near the tips of the dendrites.

Action potential mechanisms in dendrites

In discussing the magnitude of voltage produced at the place of current injection it was pointed out that the input resistance varies considerably at different sites on the nerve cell model (see Fig. 7.29). The input resistance is highest at the end of a dendrite, approximately half that value in the middle region of the dendrite, and then drops steeply in the proximal part of the dendrite to its lowest value at the soma. This variation in the magnitude of the input resistance would be very much greater in a more realistic model of the nerve cell, in which individual dendrites were not 'collapsed' into 'equivalent' cylinders, since the input resistance of a particular branch will always be much higher than that of the cylinder which represents the total number of daughter branches at that electrotonic distance away from the soma. The difference in input resistance between the tip of a dendritic branch and the soma might then be of the order of 100-fold or more. If excitatory synapses, wherever their location, cause approximately the same amount of depolarizing current to enter the cell, there will be equivalently large variations in the magnitude of the voltage generated at different sites. It is therefore natural to anticipate that it would be easier for an excitatory synapse to exceed the voltage threshold for the Hodgkin–Huxley sodium conductance, the greater their distance from the soma. This prediction assumes, however, that the voltage threshold for net inward ionic current is constant at different sites.

Although it is extremely difficult, for technical reasons, to obtain information about the voltage threshold at different sites on the nerve cell, it is now fairly clear that there are differences. The first evidence (Fuortes, Frank, and Becker 1957; Coombs, Curtis, and Eccles 1957) came from a careful study of the form of the action potential recorded at the soma when it was generated by various means. The conclusion was that there was a region of the nerve cell (probably the initial segment of unmyelinated axon) which had a much lower voltage threshold (about 10 mV) than other regions, i.e. the soma and perhaps proximal dendrites (about 30 mV). This result, first obtained for the motoneurone, was confirmed in the same cell by the voltage-clamp studies of Araki and Terzuolo (1962). There is still no direct experimental evidence concerning the voltage threshold of the dendritic membrane.

The consequence of this difference in voltage threshold is that an excitatory synaptic potential generated in the dendrites may not exceed the threshold locally but, after passive propagation to the soma, the potential might still be of sufficient magnitude to exceed the threshold for the initial segment region (that is, provided its attenuation in peak amplitude is by less than a factor of 3). Even if a synaptic potential generated on one dendrite did not exceed the voltage threshold of the initial segment, it could algebraically summate with synaptic potentials generated on other dendrites, which have also propagated passively to the soma, and hence cause an action potential to be produced by the nerve cell. This 'global hypothesis' (Eccles 1957) for action-potential generation is assumed to be the normal mechanism for many nerve cells, and is the justification for devoting so much attention to the passive propagation of, and interaction between, synaptic potentials given in this chapter.

It could be questioned whether the 'global hypothesis' is an exhaustive account of the possible ways in which action potentials may be generated. If the synaptic potential generated on the dendrite did exceed 30 mV this may be greater than the voltage threshold for net inward ionic current. Would this lead to an action potential in the dendrites? The first uncertainty is that the value for the voltage threshold is not known, and may progressively increase with distance from the soma. Let us first consider the question of whether an action potential would be generated if the voltage threshold for net inward current was exceeded. As discussed in Chapter 9, the conditions for action potential generation when current is injected at a single point in a distributed structure is *not* that voltage at which the net ionic current becomes inward. The minimal condition is that the net ionic current generated by the whole structure should become inward. We can translate this in qualitative terms by saying that action potential generation depends on three factors: (1) the extent to which the injected current exceeds the value of current threshold for the generation of inward current at the same point—excitation being favoured the greater the threshold is exceeded; (2) the exact form and time-dependence of the current voltage relation; (3) the degree of 'loading' of the region generating inward current by other regions carrying outward current—the larger the membrane area carrying substantial amounts of outward current, the higher the effective voltage threshold for excitation (see Chapters 9, 10, and 12 for a discussion of these factors). It is difficult to give any more than a rough indication of how these factors would affect the threshold for action potential initiation in a dendritic tree. As pointed out above, the voltage threshold for inward current may vary systematically along the length of the dendrite, so that the gain from having a higher input resistance is largely cancelled by a higher voltage threshold. Similarly, the degree of passive 'loading' will depend on the exact pattern of branching in the dendritic tree; in particular, the degree of 'loading' will depend on the proximity of the branch which is the parent of the one in which synaptic action occurs. One general rule is that in the proximal part of the dendrite (before any branching occurs) the passive loading will become less the further away the site is from the soma. Hence, other things being equal, it should be easier to evoke an action potential in the most distal part of the unbranched proximal dendrite (the input resistance will also be higher here).

Once an action potential has been initiated there is still the problem of how easily it can propagate, both towards the soma and towards the dendritic tips. Rall (1964) has pointed out that if his power law is obeyed the safety factor for propagation is the same as for a cylinder (e.g. an axon) with an identical current–voltage relation. This conclusion will apply to the propagation towards the dendritic tips, although it would have to be modified if the membrane properties were not uniform. Propagation to the soma may not be successful because, if the action potential is only generated in a single daughter branch of the dendritic tree, it will meet regions of lower safety factor at each central branch-point. With uniform membrane properties, therefore, conduction in the peripheral direction would be favoured over conduction in the central (somatic) direction. Of course, if the membrane properties are not uniform and there is an increasingly lower voltage threshold for net inward current towards the soma, central spread of the action potential could be favoured over peripheral spread. Even so, it is quite likely that the action potential would fail as it conducts along the proximal dendrite towards the soma, as the inward current generated in the dendrite may not be sufficient to raise the voltage in the most proximal part of the dendrite above the threshold, owing to the relative short-circuiting effect of the soma and other dendrites. All that would be seen when recording from the soma is a voltage waveform much smaller in amplitude and rather slower in time course than an action potential. If failure occurred at a considerable electrotonic distance from the soma the time course of the waveform might be indistinguishable from that of a synaptic potential.

Finally there is the possibility of responses which are intermediary between the non-decremental conduction of action potentials and the fairly sharply decrementing propagation of passive potentials. Lorente de No and Condouris (1959) first suggested that dendrites might support decrementing waves which were aided by nonlinear mechanisms. This possibility is discussed further in Chapter 10.

These considerations suggest some experimental methods by which an active mechanism in the dendrites might be detected. Two basic phenomena have been described. Firstly, the threshold for inward current may be exceeded but no action potential initiated. In this case the synaptic excitatory current has added to it further depolarizing current (inward sodium movement) and, perhaps, subsequently some outward potassium current. Two sites were suggested where this could easily occur (even if the ratio of the maximal sodium conductance to the resting conductance were similar to that in an axon): in the proximal dendrite near the soma or further out in the dendritic tree, where the synaptic action takes place just distal to a branch point. In the former case the additional current may be evident from the time course of the synaptic potential since, instead of a smooth rise and fall, there may be a variable hump of depolarization near the peak of the synaptic potential. A steady hyperpolarizing current injected into the soma should, by displacing the membrane potential away from the voltage threshold for inward current, eliminate this irregular component. When the synaptic potential is generated further out on the dendrites the voltage waveform may be quite smooth, and it would require a much larger hyperpolarizing current to eliminate the extra current. A reduction in the peak amplitude of distal excitatory synaptic potential by a hyperpolarizing current is, however, very strong evidence for such a mechanism, unless the cell's current–voltage relation shows the presence of an in-going rectifier (see Chapter 8). Another form of evidence would be provided by showing that two excitatory synaptic potentials, when activated together, produced a larger depolarization than the sum of the two when separately evoked. One obvious difficulty about such an experiment is the requirement that both sets of synapses would need to be distally located on the same dendrite of the cell.

The second possible phenomenon described is that an action potential is initiated at one point in the dendritic tree but tends to fail at some distance before it reaches the soma. If failure occurs quite close to the soma, the voltage waveform may be unmistakably action-potential-like in shape. This amounts to the time course of the voltage not being explained by the mono-phasic injection of current—in other words, unlike most synaptic potentials (cf. Wachtel and Kandel 1971), a diphasic injection would be required in order to produce the observed voltage time course. Further evidence for such a phenomenon would come from the observation that the potential tended to be 'all-or-nothing' or to show refractory behaviour on repeated trials. Hyperpolarizing currents need not completely abolish the potential, unless they are very strong, because they might simply cause the action potential to fail slightly further away from the soma.

If the action potential fails at a considerable electrotonic distance from the soma it may be difficult from an analysis of its voltage time course to decide whether it has been generated by a purely passive, synaptic, mechanism or has an additional contribution from the Hodgkin–Huxley conductances. Once again, it may only be by the use of strong hyperpolarizing currents that a definite decision can be made.

Although these comments have been made without reference to any pertinent experimental data many of the phenomena described above have been observed. There is, in other words, fairly clear evidence that inward sodium current (with or without a propagated action potential) can be activated in some nerve-cell dendrites. Purpura (1967) has given a clear review of this evidence and interested readers are referred to his paper and to subsequent discussions by Diamond and Yasargil (1969), Kuno and Llinas (1970a,b) and Nicholson and Llinas (1971). Apart from studies with intracellular microelectrodes there is a considerable amount of experimental evidence derived from the use of extracellular recording, although a recent controversy has highlighted the interpretive difficulties inherent in this technique (Llinas, Nicholson, Freeman, and Hillman 1968; Calvin and Hellerstein 1968; Llinas et al. 1969; Zucker 1969; Calvin 1969b; Hellerstein 1969; see also Rall and Shepherd 1968; Llinas, Nicholson, and Precht 1969; Nicholson and Llinas 1971).

The functional role of action potentials or nonlinear responses in dendrites is still not clear. It is obvious that they could act as an amplifying device, particularly suitable for increasing the effect of synapses when they are a substantial electrotonic distance away from the soma. Indeed, Rall (1970) has predicted that they may only normally be found in nerve cells which have a dendrite (or dendrites) greater than 2 or 3 space constants in length; as indicated earlier it is only then that the passive attenuation of synaptic potentials becomes very severe. It has also been suggested that a combination of spikes, either in different parts of the same dendrite or in two dendrites, might secure discharge of the cell when either dendritic spike alone would not. This would offer complex possibilities for the use of nerve cells in pattern detection, etc.

A quite different way in which action potential or nonlinear mechanisms in dendrites may be very important comes from recent observations that dendro-dendritic synapses are present in many parts of the nervous system (Rall, Shepherd, Reese, and Brightman 1966; Ralston 1968, 1971; Lund 1969; Famiglietti 1970; Wong 1970; Harding 1971; Sloper 1971). Although the relationship between prejunctional potential and transmitter release of course is not known for these synapses, it is likely that it takes the same exponential form as at peripheral junctions (Katz and Miledi 1967; Kusano 1970). As Ralston (1971) points out, it is quite possible that potentials generated by purely passive means could attain sufficient size to cause the release of some transmitter; but the amount released, and hence the effectiveness of the synapse, will be greatly increased by any boosting of the passively generated potential. Apart from the local generation of an active response, the occurrence of dendrites as presynaptic structures also emphasizes the importance of the possible dendritic invasion of an action potential generated in the initial segment by 'conventional' means. In most nerve cells it is extremely difficult to determine the extent of antidromic invasion of dendrites by an action potential (see Purpura 1967) because the only method usually available is that of extracellular recording.

Types of synaptic action

Throughout this chapter it has been assumed that the way in which synaptic action is produced on the soma-dendritic membrane is by a change in conductance to one or more ions. Although this is the commonest mechanism so far described for the vertebrate nervous system, the description tends to conceal some of the diversity. There is, for instance evidence that one form of synaptic excitation on sympathetic ganglion cells takes place by means of a reduction in the value of the resting potassium conductance (Weight and Votava 1970). In addition, there are transient synaptic effects which result from a decrease in the magnitude of tonic inhibitory or excitatory conductance effects (e.g. Wilson and Burgess 1962; Llinas 1964). None of these actions present any special difficulty of understanding, however, since they can be analysed using the principles already described (see also Brown, Maller, and Murray 1971).

There are two forms of synaptic action which do deserve brief discussion because of the nature of the geometry of the ending: synapses terminating on dendritic spines and reciprocal synapses.

Synapses on dendritic spines. In most parts of the mammalian nervous system the commonest way a synapse is formed between two nerve cells is by the presynaptic component making contact with an evagination of the dendritic (or somatic) surface of the postsynaptic cell. The postsynaptic evagination is usually called a dendritic spine. Although these structures show considerable variation in size and shape, their general form is that of a tenuous stalk with an ellipsoidal terminal enlargement. The presynaptic component makes synaptic contact with the ellipsoid. (Fig. 7.43 is a schematic diagram, taken from Jones and Powell (1969). which illustrates some of the forms of dendritic spine synapse (see also Scheibel and Scheibel 1968)).

Two predictions follow directly from the structure of this synaptic connection. First, the input resistance of the dendritic spine head will be much

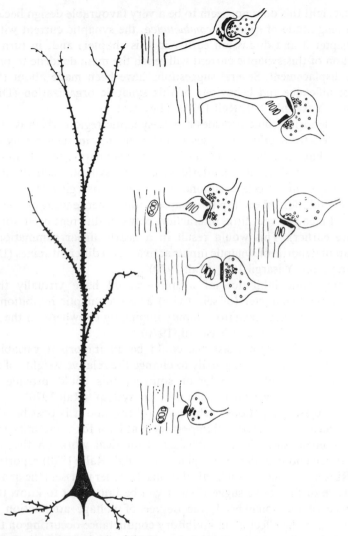

F I G. 7.43. On the left-hand side of the figure is shown the structure of a typical pyramidal cell of the somatic sensory cortex of the cat. The incidence of spines on various parts of the dendritic tree can be seen. On the right-hand side are shown the variety of shapes of dendritic spines. The placing of the various types of spine indicates where they are commonly seen. (From Jones and Powell 1969.)

higher than for the dendrite at the point at which the spine stalk becomes continuous with the main dendritic trunk. For a given amount of synaptic current this will lead to the development of a large voltage displacement in the spine head. Secondly, there will be a reduction in the amount of synaptic current, injected into the spine head, which reaches the dendritic trunk: the thinner and longer the stalk, the greater the reduction.

At first sight this does not seem to be a very favourable design because, for a given magnitude of synaptic *conductance*, the synaptic current will be less (see Chapter 3 and discussion earlier in this chapter) and, in turn, only a proportion of this synaptic current will reach the main dendrite to produce a voltage displacement. Several suggestions have been made about the functional significance and behaviour of this synaptic organization (Diamond, Gray, and Yasargil 1970; Rall 1970). They are:

1. The dendritic spine provides a post-synaptic region which is effectively isolated from other synapses in the neurone, in such a way that the immediate and the long-term effects of pre-synaptic activity in the spine occur with little or no interference from synaptic activity generated elsewhere in the cell (Diamond, Gray, and Yasargil 1970).

2. The 'isolation' of each spine synapses from others ending on the neurone would lead to very little interaction between different excitatory inputs (see earlier). This would result in a nearly linear summation in the parent dendrite of synaptic inputs activated on different spines (Diamond Gray, and Yasargil 1970; Rall 1970).

3. Postsynaptic inhibition of a spine would have virtually the same functional character (of selectivity) as a presynaptic inhibition since it would not affect excitatory inputs impinging elsewhere on the neurone (Diamond, Gray, and Yasargil, 1970).

4. '... the spine stem resistance could be an important variable which might be used physiologically to change the relative weights of synaptic inputs from different afferent sources; this could provide a basic mechanism for learning in the nervous system' (Rall 1970).

None of these suggestions exclude the others, and it is possible that they might all have some functional importance, at least for some spine synapses. At the moment there is no published theoretical work on the predicted electrical behaviour of dendritic spines, although Rall (1970) reports that he and J. Rinzel have started such calculations. In order to assess the quantitative significance of the above suggestions it will be important to know the input impedance of the spine head, the degree of voltage attenuation in both directions, and the effect of an inhibitory conductance occurring on the spine head, on the input impedance of the parent dendrite. These calculations can be made quite readily for steady state conditions, and the relevant equations are listed below. In all of them it has been assumed that the spine head can be treated as isopotential, the spine stalk as a simple short cable, and the parent dendrite as an infinite cable.

The input resistance of the spine head is given by

$$R_{\text{in}} = \frac{r_{a_1}\lambda_1\{r_{a_1}\lambda_1 + \frac{1}{2}r_{a_2}\lambda_2 - (r_{a_1}\lambda_1 - \frac{1}{2}r_{a_2}\lambda_2)e^{-2L}\}}{(r_{a_1}\lambda_1 + \frac{1}{2}r_{a_2}\lambda_2)(1 + r_{a_1}\lambda_1 G_h) + (r_{a_1}\lambda_1 - \frac{1}{2}r_{a_2}\lambda_2)(1 - r_{a_1}\lambda_1 G_h)e^{-2L}}, \quad (7.125)$$

where $r_{a_1}\lambda_1$ describes the spine stalk cable, $r_{a_2}\lambda_2$ describes the parent cable, G_h is the total conductance of the spine head membrane, and L is the electrotonic length of the spine stalk cable. The voltage attenuation from spine head to parent dendrite is given by:

$$V_{den}/V_{sp\,h} = e^{-L} \frac{2}{(2r_{a_1}\lambda_1/r_{a_2}\lambda_2)+1-\{(2r_{a_1}\lambda_1/r_{a_2}\lambda_2)-1\}e^{-2L}}. \quad (7.126)$$

The input resistance of the parent dendrite, at the point from which the spine takes origin, is given by

$$R_{in} = \frac{r_{a_1}\lambda_1\{(r_{a_1}\lambda_1 G_h+1)-(r_{a_1}\lambda_1 G_h-1)e^{-2L}\}}{(1+r_{a_1}\lambda_1 G_h)\{1+(2r_{a_1}\lambda_1/r_{a_2}\lambda_2)\}-(1-r_{a_1}\lambda_1 G_h)\{1-(2r_{a_1}\lambda_1/r_{a_2}\lambda_2)\}e^{-2L}}, \quad (7.127)$$

and the voltage attenuation from parent dendrite to spine head is given by

$$\frac{V_{sp\,h}}{V_{den}} = e^{-L} \frac{2}{r_{a_1}\lambda_1 G_h+1-(r_{a_1}\lambda_1 G_h-1)e^{-2L}}. \quad (7.128)$$

Notice that eqns (7.126) and (7.128) imply that the voltage attenuations in the two directions are not the same unless $G_h = 2/r_{a_2}\lambda_2$, which is very unlikely. If we take, as an example, a parent dendrite diameter of 5 μm, the spine stalk diameter of 0·2 μm and length of 2 μm, the surface area of the spine head as 5 μm², and assume the membrane and axial resistance are similar to other excitable cells (e.g. $R_m = 5000\ \Omega$ cm² and $R_i = 200\ \Omega$ cm), we find $G_h = 10^{-11}$ S (S = Ω^{-1}) and $2/r_{a_2}\lambda_2 = 3·5 \times 10^{-8}$S. The electrotonic length of the spine stalk is 0·018 λ. The voltage attenuation from spine head to parent dendrite would be 0·1827, whereas from parent dendrite to spine head it is only 0·9986. For this example, therefore, the synapses on the spine are not isolated from voltages in the parent dendrite; electrical isolation would require a very long thin spine stalk, or very different values of the electrical parameters R_m and R_i (cf. suggestions 1 and 2 above).

With the same assumed values eqns (7.125) and (7.127) can be used to calculate the input resistance of the spine head (1·555 × 10⁸ Ω) and of the parent dendrite (2·846 × 10⁷ Ω). If an excitatory synapse on the spine head, with reversal potential 75 mV away from the resting potential, produced a steady conductance of 10⁻⁹ S the synaptic current would be 64·9 × 10⁻¹² A, the voltage in the spine head would be 10·09 mV, and the voltage in the parent dendrite 1·84 mV. The same conductance occurring directly on the parent dendrite would produce a current of 72·9 × 10⁻¹² A and a local voltage of 2·08 mV. It may be noticed that the main cause of the smaller voltage displacement in the parent dendrite when the synapse is on the spine head is the reduction in synaptic current (owing to the much greater local input resistance); in this example only about 2 per cent of this synaptic current does not reach the parent dendrite.

The degree of selectivity achieved by having a postsynaptic inhibitory synapse on a spine head can be estimated by assuming that the inhibitory action is a steady conductance change of, say, 10^{-8} S. If this were located on the parent dendrite it would change the local input resistance from 2.846×10^7 Ω to 2.216×10^7 Ω. An adjacent excitatory synapse (with characteristics as above) would then produce a local voltage of 1.63 mV—a reduction to 78.3 per cent of the value without inhibition. The same excitatory conductance acting adjacent to the inhibitory conductance on the spine head would produce only 4.30 mV, instead of 10.09 mV (0.79 mV and 1.84 mV in the parent dendrite, respectively)—a reduction to 42.6 per cent. The inhibitory conductance on the spine head would produce a change from 2.846×10^7 Ω to 2.53×10^7 Ω in the input resistance of the parent dendrite, so that excitatory synapses ending directly on the parent dendrite or on other adjacent spines would have their depolarizing action reduced by as much as 10 per cent. This example illustrates that locating both excitatory and inhibitory synapses together on a spine head can enhance the selectivity of action of postsynaptic inhibition, without making it absolute (cf. suggestion 3).

Finally, we can calculate the consequences of a change in the electrical length of the spine stalk. Let us assume that the electrical length doubles. The input resistance of the exemplar spine discussed above would change from 1.555×10^8 Ω to 2.822×10^8 Ω and the voltage attenuation from 18.27 per cent to 10.05 per cent. An excitatory synapse (10^{-9} S) on the spine head would now produce a synaptic current of 58.49×10^{-12} A, a local voltage of 16.51 mV and a voltage displacement in the parent dendrite of 1.66 mV. Thus the net effect would be a reduction in the voltage in the parent dendrite by about 10 per cent.

One possible complication that has not been discussed is action-potential generation in the spine head. It will be evident from the above calculations that a small excitatory conductance can produce a large voltage displacement in the spine head. If the spine head membrane is equipped with Hodgkin–Huxley conductances, an action potential could be generated. Diamond and Yasargil (1969) have provided strong circumstantial evidence than an action potential does occur in a dendritic spine of motoneurones in the fish spinal cord. This could then act as an amplifying device, tending to offset the reduction in synaptic current if it simply spread by passive means to the parent dendrite. If this mechanism does occur generally it would mean that there may be an optimal value for spine stalk electrotonic length. An increase in spine stalk electrotonic length above a certain value (for a given value of spine head conductance) might lead to a sufficient increase in the input resistance locally so that a synaptic excitation just generates sufficient voltage displacement to reach threshold for the action potential to be produced (or vice versa). Using the example above, the voltage threshold in the spine head might be 15 mV. Initially the excitatory synaptic potential is 10.09 mV locally, and it would propagate passively to the parent dendrite to produce a 1.84 mV displacement. After a doubling in electrotonic length of the stalk the synaptic conductance would lead to a potential of 16.51 mV in the spine head, which in turn would lead to a, say, 100 mV action potential. Even without further regenerative activity in the spine stalk this could produce a potential displacement in the parent dendrite of up to 10 mV (depending on the degree of additional attenuation due to the action potential not being a steady voltage).

This final possiblility emphasizes that it will require very subtle experimental techniques to estimate the relevant electrical characteristics of the spine; but only then will the various possibilities discussed above be put on a proper empirical basis.

Reciprocal synapses. Reciprocal synapses have some features of geometry in common with the dendritic spine synapse. The usual pattern is that a dendrite spine of one cell is not only postsynaptic to the dendrite of another cell, but also presynaptic to it. The typical electronmicroscopic appearance of this synaptic complex suggests that the dendritic spine (usually called a gemmule, to distinguish it from the more common dendritic spine which is not presynaptic) receives excitatory action from the other dendrite whereas it has an inhibitory action on the other dendrite.

This interesting and unexpected organization was first analysed in a remarkable investigation which combined electrophysiological and ultrastructural data with a theoretical analysis (Rall, Shepherd, Reese, and Brightman 1966; Rall and Shepherd 1968). Their analysis of extracellular fields in the olfactory bulb gave strong evidence for an excitatory action of the mitral-cell dendrite on the granule cell. When this result was combined with electronmicroscopic data and other electrophysiological studies the simplest interpretation was that the electrophysiological results could be explained by an observed reciprocal synapse whose action was as just described. The ultrastructural appearance of the reciprocal synapse was in accord with the postulated mechanism, since the synapse from mitral-cell to granule-cell gemmule displayed asymmetrical thickening of the synaptic membranes and spherical vesicles, whereas the synapse from the gemmule to the mitral-cell dendrite had symmetrical thickening and flattened vesicles (cf. Uchizono 1968; Gray 1969; Price and Powell 1970*a,b*).

Theoretical analysis of the mode of action of the excitatory synapse terminating on the gemmule is identical to that already given for a synaptic ending on a dendritic spine. In the case of the reverse arrangement the novel question is how transmitter is caused to be released, since its postsynaptic action will presumably be the same as for other dendritic inhibitions. Reference has already been made to the exponential relationship between the amount of transmitter released and the prejunctional depolarization. It is evident from the complexity of the synaptic geometry that there is more than one way in which this might be achieved. If an action potential (or passive depolarization), generated elsewhere on the cell, propagated along the gemmule's parent dendrite branch there would be a spread of depolarization into the gemmule which might, or might not, cause it to generate an action potential. The amount of transmitter released would then depend on whether such an event occurred. Alternatively, an excitatory synapse ending on the gemmule might produce sufficient depolarization to either secure an action potential in the gemmule head or release a small amount of transmitter in its own right. In this respect it is natural, in those cases where the reciprocal synapse does occur on the gemmule, to view the function of the gemmule as providing a sufficiently high input resistance to allow the return release of transmitter from the gemmule. Furthermore, the other synapses on, or near, the gemmule could serve to determine whether an action potential was generated in the gemmule head—thus providing a powerful modifying action on the effectiveness of this synapse. The same synapses, ending near a reciprocal synapse, might also determine whether the excitatory action from the other dendrite to the gemmule head was localized in its effect to that single reciprocal synapse or spread in the gemmule-bearing cell to activate additional gemmule synapses, which may be making contact with other cells. If the latter possibility occurred it would be comparable in effect to more conventional neuronal circuits for recurrent inhibition.

Electrical synapses. A completely different form of mechanism is found at some synapses. Displacement of the membrane potential of the postjunctional cell is the result of the direct injection of current from the presynaptic axon or dendrite (see Eccles 1964; Bennett 1966, 1968). This effect is achieved by means of a special region of contact between the two cells (e.g. a 'tight junction'): sometimes the resistance of this junctional region is rectifying so that current spread in one direction is favoured (Furshpan and Potter 1959).

The problem of the spread of current across electrical synapses has been explored quantitatively by Bennett (1966). The similarities and differences between electrical and chemical synapses are reviewed by Rall, Burke, Smith, Nelson, and Frank (1967) and Bennett (1968).

8. Nonlinear properties of excitable membranes

IN previous chapters we have assumed that the cell membrane resistance is constant, that is, that there is a linear relation between membrane voltage and membrane current. In most cases, however, this is an approximation and, in excitable cells, the range of potentials over which the membrane may be assumed to be linear is very restricted. In the following chapters we shall derive and discuss equations which are applicable to nonlinear cables. In order to prepare the way for doing this, the present chapter will be concerned with an account of the most commonly occurring nonlinearities. We shall briefly review the origins and nature of these nonlinearities and indicate how they may be described, at least approximately, by relatively simple models.

Membrane conductance

Linear membrane

A linear membrane is one which shows a linear relation between membrane ionic current I_i and the transmembrane potential:

$$I_i = G_m V, \tag{8.1}$$

V is the potential expressed as a deviation from the quiescent (resting) potential and G_m is a constant, the *membrane conductance*. This is the reciprocal of the membrane resistance R_m used in previous chapters. The derivative dI_i/dV is called the *membrane slope conductance*. In the case of a linear membrane, (8.1) may be differentiated to give

$$dI_i/dV = G_m. \tag{8.2}$$

In other words, for a linear membrane, the membrane slope conductance dI_i/dV turns out to be equal to G_m. G_m is often called the *membrane chord conductance* to distinguish it from dI_i/dV. The distinction between slope and chord conductance becomes more clear when one considers the more general case, the nonlinear membrane, where the chord and slope conductance are not identical.

Nonlinear membrane: I_i a function of V only

In this section and the one which follows it, we shall introduce various terms and definitions which are commonly used in describing nonlinear properties. If I_i is a function of V only then we may write

$$I_i = F(V), \tag{8.3}$$

where F is an arbitrary nonlinear function. We may then define the chord conductance G_m as

$$G_m = I_i/V. \tag{8.4}$$

However, the nonlinearity assumed in eqn (8.3) means that G_m will not be a constant. From eqns (8.3) and (8.4),

$$G_m = F(V)/V \tag{8.5}$$

eqn (8.3) may be differentiated to give the membrane slope conductance

$$dI_i/dV = F'(V). \tag{8.6}$$

Clearly, G_m and dI_i/dV are not identical. This is illustrated in Fig. 8.1 which shows a typical excitable membrane current–voltage relation (cf. Fig. 8.11).

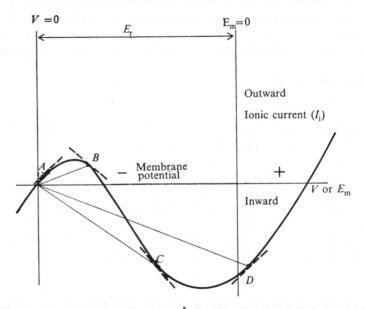

FIG. 8.1. Chord and slope conductances. The diagram shows a current–voltage relation of the kind frequently shown by excitable cells (cf. Fig. 8.11). The curve intersects the voltage axis at three points. The most negative intersection is the resting potential at which the system is quiescent and V is defined to be zero. Chord conductances are given by the ratio I_i/V of the coordinates of the curve. Slope conductances are given by the slopes of tangents to the curve (interrupted lines). For the points labelled we have

	chord conductance	slope conductance
A	positive	positive
B	positive	negative
C	negative	negative
D	negative	positive

Note that the definition of chord conductance is partly arbitrary. Different chords would be drawn if the positive intersection were defined as $V = 0$. However, excitable cells rarely show more than transient stability at positive potentials so that the use of chord conductances defined with a positive value of quiescent potential is very limited.

Apart from points, such as A, near the resting potential E_r, the chord and slope conductances are very different. At B the chord conductance is positive while the slope conductance is negative. At C both conductances are negative. At D the chord conductance remains negative but the slope conductance is positive. It will now be evident why the terms *chord* and *slope* are used: G_m is given by a chord from E_r to the point concerned, whereas dI_i/dV is the slope of the current voltage relation at that point. In general, the chord conductance is the important parameter when the absolute value of the ionic current is required, whereas the slope conductance determines the response to small variations in voltage or current about any point. It is also worth noting that there is a certain ambiguity about the definition of chord conductance for a current–voltage relation, like that in Fig. 8.1, which crosses the voltage axis at several points. Thus, if the quiescent potential were defined as the right-hand intersection (which is also a stable point—see p. 251) points C and D would become points of positive chord conductance, whereas points A and B would become points of negative chord conductance. This ambiguity arises from the fact that current–voltage relations of this kind must be formed by the addition of more than one ionic current–voltage relation (cf. Fig. 8.11) and it will be shown (see p. 229) that the ambiguity disappears when current–voltage relations for individual ionic species are considered.

Nonlinear membrane: I_i a function of V and t

In many cases, I_i is not only a function of V, it is also a function of t, that is,

$$I_i = f(V, t), \tag{8.7}$$

where f is an arbitrary nonlinear function. In this case we must distinguish between two kinds of slope and chord conductances.

(1). *Steady state (low-frequency) conductances.* These are obtained by allowing the membrane voltage to be held long enough at each value for a steady state current flow to be achieved. Alternatively, an applied oscillating voltage or current may have a frequency which is low enough for the system to approximate to steady state conditions at all times. We then have

$$I_\infty = f(V, t = \infty), \tag{8.8}$$

which is a form similar to eqn (8.3) since I_∞ is not a function of time. We may then define the *steady state chord conductance* by

$$G_{m_\infty} = I_\infty/V = f(V, \infty)/V \tag{8.9}$$

and the *steady state slope conductance* by

$$dI_\infty/dV = f'(V, \infty). \tag{8.10}$$

(2). *Instantaneous (high-frequency) conductances.* These are obtained by changing the voltage so rapidly that the time-dependent processes do not have time to occur. Then at any particular time t^* we have

$$I^* = f(V, t^*).$$ (8.11)

Once again, this equation is of the same form as eqn (8.3) since t^* is a constant and I^* is a function of V only. In this case, the *instantaneous chord conductance* is defined by

$$G_m^* = I^*/V = f(V, t^*)/V$$ (8.12)

and the *instantaneous slope conductance* by

$$dI^*/dV = f'(V, t^*).$$ (8.13)

It is important to note that, although there may be only one pair of values of steady state conductances at each potential (neglecting the ambiguity of definition referred to earlier), there can be an infinite number of pairs of values of instantaneous conductances since, at any time, the system may be in any one of an infinite number of states between one steady state condition and another. In systems of the complexity of excitable cells, it would be exceedingly cumbersome to attempt to characterize the system by sets of values of conductances under a sufficiently wide variety of conditions, and it becomes necessary to obtain more general descriptions. One of the aims of the electrophysiological analysis of membrane nonlinearities carried out over the last two decades has been to obtain relatively simple general accounts of the time-dependent behaviour of the membrane current, so that it is not necessary to actually measure a very large number of instantaneous conductances in order to fully characterize the system. Having obtained functions which adequately describe the time dependence of the membrane current (see below) we may then calculate any particular conductances which may be required.

Ionic conductances

All the above variables may also be defined in the case of individual ionic currents I_{Na}, I_K, I_{Ca}, etc. However, two restrictions must then be introduced. First, most of the voltage-dependent ionic currents which have been investigated so far are passive flows of current down electrochemical potential gradients. In such cases, the current flow must be zero when the electrochemical potential gradient is zero. Thus, for potassium (K) ions we may write

$$I_K = G_K(E - E_K),$$ (8.14)

where E is the transmembrane potential (inside potential minus outside potential) and E_K is the potassium equilibrium potential defined by the Nernst equation

$$E_K = \frac{RT}{F} \ln \frac{[K]_0}{[K]_i}.$$ (8.15)

Hence, in the nonlinear case, we have

$$I_K = f(E, E_K).$$ (8.16)

As before, f is a nonlinear function. However, its form is no longer completely arbitrary since, as already noted, f must be zero when $(E-E_K)$ is zero. Moreover, the sign of f must also be determined by the sign of $(E-E_K)$ since current always flows in the same direction as the electrochemical potential gradient. Thus, f must be positive for positive values of gradient and negative for negative values of the gradient. Hence the current–voltage relation must cross the voltage axis at only one point E_K and the *potassium chord conductance* may never be negative, although the *potassium slope conductance* may be negative (see p. 238).

Of course, these restrictions need not apply to current flow produced directly by active transport processes. However, so far very little is known concerning the voltage-dependence of electrogenic pumping activity, and such activity is not thought to be essential for the characteristic nonlinear properties of excitable membranes to be displayed.

Membrane rectification

A full discussion of membrane rectification is beyond the scope of this book, and we shall restrict the present section to the major forms of rectification which are observed in excitable membranes. More advanced accounts are given in Goldman (1943), Cole and Curtis (1941), Hodgkin and Huxley (1952), Grundfest (1966), Cole (1965, 1968), Conti and Eisenman (1965), Adrian (1969), Woodbury (1971), and Haydon and Hladky (1972).

Terminology. It is now customary to characterize membrane rectification according to the direction in which the membrane most easily passes current. Thus, a membrane which passes outward current more easily than inward current is said to show *outward-going rectification*, whereas a membrane which passes inward current more easily is said to show *inward-going rectification*. Fig. 8.2 shows the symbols frequently used in electrical equivalent circuits to represent these forms of rectification. As we will show later (p. 249) it is possible for a membrane to show both forms of rectification at different ranges of voltages or to show one form of rectification for instantaneous current–voltage relations and another form for steady state current–voltage relations.

Models of ion permeation

In order to illustrate possible mechanisms of nonlinearity we will first consider a very simple model of the permeation process which may then be developed to produce more complex models of the kind which are used in electrophysiology. The membrane conductance is dependent on the ease with which ions permeate the cell membrane.

Single energy-barrier model. Suppose that the membrane provides an energy profile to ion movement across it of the type shown in Fig. 8.3(a). Provided that the energy ΔG_0 of the largest barrier far exceeds the energy of the

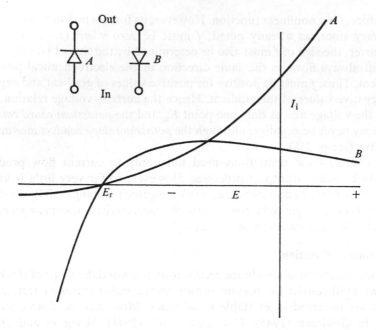

FIG. 8.2. Symbols and typical current–voltage relations for rectification properties shown by cell membranes. *A* passes outward current more easily than inward current and is called an outward-going rectifier. *B* passes inward current most easily and is called an inward-going rectifier. Inward-going rectifiers sometimes show a negative slope conductance at some potentials, as shown (see also Fig. 8.6).

other barriers, the membrane will provide only one rate-limiting step for ion movements. The movement of ions from one solution to the other will therefore obey first-order kinetics (see Chapter 2, p. 10) such that

$$M_{12} = k_1(a)_1 \tag{8.17}$$

and

$$M_{21} = k_2(a)_2, \tag{8.18}$$

where M_{12} and M_{21} are the unidirectional fluxes of ions, k_1 and k_2 are rate coefficients, and $(a)_1$ and $(a)_2$ are the activities (or concentrations, if the solutions are sufficiently dilute) of the ions in their solutions. In the absence of a transmembrane potential k_1 and k_2 may be related to the free energy of activation in crossing the membrane, so that

$$k_1 = A \exp(-\Delta G_0/RT) \tag{8.19}$$

and

$$k_2 = A \exp(-\Delta G_0/RT), \tag{8.20}$$

where A is a constant.

As already noted in Chapter 2 (p. 23), the rate of permeation of ions across the membrane is extremely low compared to the rate of diffusion across a layer of salt solution of similar thickness. Therefore ΔG_0 must be very large.

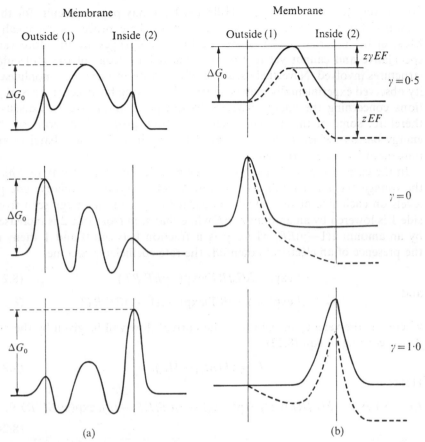

Fɪɢ. 8.3. (a) Energy diagrams for permeation mechanism involving two surface energy-barriers and one internal energy-barrier. The surface reactions may correspond to ion-carrier complex formation or to adsorption of ions to a fixed membrane site. The internal barrier may correspond to the energy required to transfer ion-carrier complex across membrane. Three diagrams are shown representing situations in which one or other of the energy barriers is rate-limiting. This energy barrier sets the value of ΔG_0. (b) Simplification of three-barrier model in which only the rate-limiting barrier is represented. The effect of an electric field is represented by the interrupted curves. When the internal energy-barrier is rate-limiting, both efflux and influx will be field-dependent ($\gamma = 0.5$). When a surface reaction is rate-limiting, the field will influence the flux largely in one direction ($\gamma = 0$ or 1).

However, the temperature dependence of ion permeation is usually very low (the Q_{10} of membrane conductance is usually about 1·3), so that most of the free energy of activation must correspond to a large negative entropy of activation. This would be expected if the sites at which ions may cross the membrane are extremely sparse (see p. 23), so that only those ions colliding with the membrane at certain points and, perhaps, in certain directions may cross. This point is of some importance since it implies that the processes which underlie nonlinearities in cell membranes may control the current flow at very few sites indeed (cf. Hille 1970). It may prove difficult, for this reason if for no others, to isolate or identify these processes chemically. Most of the theories of membrane-conductance changes are therefore very speculative and cannot yet rely on a detailed knowledge of the molecular structures involved. Nevertheless, some of the forms of membrane nonlinearity observed experimentally may be derived by making fairly general assumptions concerning the energy barriers to ion permeation. We will discuss, therefore, some of the models which have been proposed in terms of the energy barriers which they assume and the way in which these barriers are influenced by the electric field.

In the case of the single energy-barrier model, the simplest way in which the voltage may influence the rates k_1 and k_2 is by varying the relative energy levels on each side of the barrier. Thus, if the activation energy seen from side 1 is lowered by an amount $z\gamma EF$ while that seen from side 2 is increased by an amount $z(1-\gamma)EF$, where γ is a fraction between 0 and 1 then, in the presence of an electrical potential, the rate coefficients become

$$k_1 = A \exp(-\Delta G_0/RT)\exp(z\,\gamma EF/RT) \tag{8.21}$$

and

$$k_2 = A \exp(-\Delta G_0/RT)\exp\{-z(1-\gamma)EF/RT\}, \tag{8.22}$$

where z is the valency of the ions. The current flow will be given by the net flux according to eqn (8.23),

$$I = zF(M_{21}-M_{12}). \tag{8.23}$$

Hence

$$I = zFA \exp(-\Delta G_0/RT)[(a)_2 \exp\{-z(1-\gamma)EF/RT\}-(a)_1 \exp(z\gamma EF/RT)].$$
$$\tag{8.24}$$

It can be seen that, in this simple case, the current–voltage relation is always nonlinear. However, the degree of nonlinearity and its nature are very dependent on the value of γ. Over the usual physiological range of potentials (-100 mV to $+50$ mV), the degree of nonlinearity displayed is fairly small when $\gamma = 0.5$. This is shown in Fig. 8.4, in which it can be seen that the direction of rectification is determined by γ. γ may be regarded as a measure of the degree of asymmetry of the system. Thus, if the largest energy barrier to ion permeation occurs about halfway across the electrical field then γ will be about 0.5. However, if the barrier (or the rate-limiting barrier in the case of a multi-barrier model—see below) is very near one side of the electric field, γ will tend towards 0 or 1, and a fairly marked degree of rectification may then occur. This may arise if the rate-limiting process is one which occurs at or close to one of the membrane surfaces (see also the section *Carrier models*, p. 238).

Multiple barrier models. It is, of course, unrealistic to suppose that ions will encounter only one energy-barrier in moving across a membrane about 100 Å thick. The model described in the previous section would be expected to

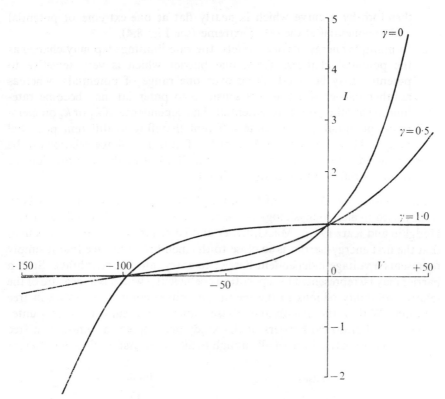

FIG. 8.4. Current–voltage relations given by single barrier model (eqn (8.24)) for $a_1/a_2 = 50$ and $\gamma = 0$, 0·5, and 1·0.

apply only when one of the energy barriers is very much greater than any of the others. When the energy barriers are more nearly equal in height, the equations governing the rate of ion permeation become more complex. Woodbury (1969, 1971) has developed a model which postulates the existence of four energy barriers and which may show various forms of rectification depending on the relative heights of the barriers. In particular, he has determined the energy barriers which must be assumed in order to give linear current–voltage relations of the kind observed for the instantaneous conductances in squid nerve membrane. Clearly, there is a large number of possible ways in which four energy-barriers might be arranged and an even larger number of physical mechanisms to which these models might correspond. However, it is possible to make some generalizations about models of this kind.

1. The maximum degree of rectification is obtained when the highest energy-barrier is assumed to be so close to one or other side of the membrane that the model approximates to the single energy-barrier model when $\gamma \to 0$ or 1. The maximum degree of rectification is given

therefore by a curve which is nearly flat at one extreme of potential and exponential at the other extreme (see Fig. 8.4).

2. In multiple energy-barrier models, the rate-limiting step may change as the potential changes. Thus, one barrier which is very sensitive to potential may be rate-limiting over one range of potentials, whereas another barrier which is less sensitive to potential may become rate-limiting at other ranges of potential. The dependence of k_1 or k_2 on membrane potential may be quite different therefore at different potential ranges. This allows a whole variety of current voltage relations to be produced within the limits described in (1). For further information the reader is referred to Woodbury (1971).

Constant-field model. One of the multiple barrier models which has been widely used in electrophysiology is the constant-field model (Goldman 1943; Hodgkin and Katz 1949*a*; MacGillivray and Hare 1969). This model assumes that the first energy-barrier is so close to the membrane surface that it simply represents a voltage-independent barrier to entry into the membrane. Such a barrier may be represented by a partition coefficient β which is a measure of the relative solubility of ions in the membrane phase compared to that in free solution. Within the membrane phase, ions are assumed to encounter diffusion and electrical barriers of the kind encountered in diffusion in free solution. These barriers are small enough to allow the assumption that there is

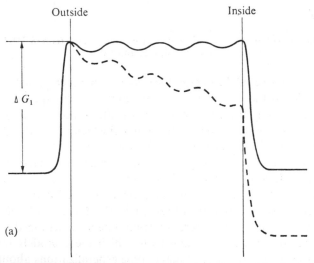

FIG. 8.5. (a) Energy profile assumed in constant-field theory. The largest barrier ΔG_1 occurs at the edge of the membrane and is included in the partition coefficient β (see text). The smaller energy-barriers are those for electrodiffusion across the membrane and are assumed to be sufficiently small and numerous to be included in a mobility term u. In the presence of an electric field, the energy within the membrane falls linearly but the value of ΔG_1 remains unchanged, that is, the electric field is assumed to influence the electrodiffusion process but not to affect penetration into the membrane.

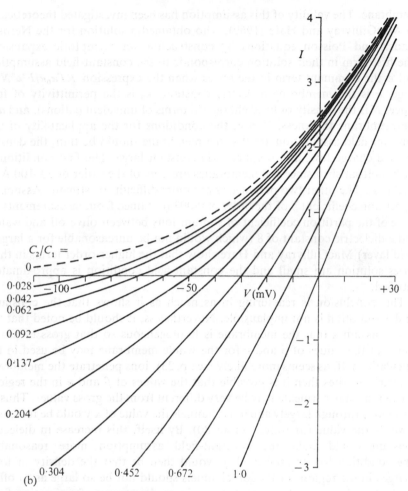

(b)

FIG. 8.5 (b) Current–voltage relations given by eqn (8.27) for various values of C_2/C_1
Note that rectification occurs only when the ion concentrations are unequal, and that the
relations all tend towards a common asymptote at very positive potentials.

an approximately smooth variation of energy across the membrane. The
movement of ions within the membrane phase is assumed therefore to obey
the Nernst Planck equation for electrodiffusive flow

$$I = zuRT \, dC/dx + z^2 FuC \, dE/dx, \qquad (8.25)$$

where I is the current flow from inside (side 2) to outside (side 1), dC/dx is the
gradient of concentration within the membrane phase measured in the
direction 1 to 2 (i.e. x is distance across membrane from side 1), dE/dx is the
electric field within the membrane, and u is the ionic mobility within the
membrane.

The constant-field model assumes that dE/dx is constant within the membrane. The validity of this assumption has been investigated theoretically by MacGillivray and Hare (1969), who obtained a solution for the Nernst–Planck and Poisson equations by constructing an asymptotic expansion. The first term in their solution corresponds to the constant-field assumption and is the dominant term in the series when the expression $RT\kappa_0\kappa/F^2a^2N$ is large. κ is the membrane dielectric constant, κ_0 is the permittivity of free space, N is the density of fixed charge (in terms of univalent cations), and a is the membrane thickness. Hence, the conditions for the applicability of the constant-field assumption are that the membrane should be thin, the density of fixed charge low, and the dielectric constant large. The first condition is fairly well satisfied since cell membranes are only of the order of 50–100 Å in thickness. The other two parameters are more difficult to estimate. Assuming a partition coefficient β of the order of 0·003 (obtained from measurements by Hare of the partition coefficient of sodium ions between olive oil and water) and a dielectric constant of 8 (which would not be unreasonable for a largely lipid layer) MacGillivray and Hare show that the higher-order terms in their series solution are small and the constant-field condition is approximately satisfied.

This conclusion is reassuring in as much as it shows that the constant-field assumption is not implausible. Nevertheless, it should be noted that the theory assumes that the membrane is homogeneous so that gross measurements of the values of β and κ for the whole membrane may be used in the calculations. If, as seems more likely (see p. 23), ions penetrate the membrane at restricted sites then it is possible that the values of β and κ in the regions where ions may permeate may be very different from the gross values. Thus, if ions move through largely aqueous channels the value of κ would be increased towards the value for water (about 80). By itself, this increase in dielectric constant would make the constant-field assumption more reasonable. The condition for its applicability would then be that the density of fixed charges in the regions of the ion channels should not be so large as to offset the large value of κ. At present, there is no direct way of estimating fixed charge-density in ionic channels (see Passow 1969), so that the question of whether the constant-field assumption is correct for these channels must remain an open one.

When dE/dx is constant, eqn (8.25) may be simplified, since we may then equate dE/dx to E/a, where now E is redefined as the total transmembrane potential. Hence

$$I = zuRT\, dC/dx + z^2FuCE/a. \tag{8.26}$$

Now, the concentrations just within the boundaries of the membrane, $C_{x=0}$ and $C_{x=a}$ are related to the extracellular C_2 and intracellular C_1 concentrations by the partition coefficient β. β may be related to the difference in free energy between the solution phase and the membrane phase, i.e. the first energy-jump ΔG_1 in Fig. 8.5. Assuming that the net flow of ions is small enough for the concentrations immediately inside and immediately outside the membrane to be at equilibrium,

$$\beta = \exp(-\Delta G_1/RT),$$

$$C_{x=0} = \beta C_1,$$

and $\qquad\qquad C_{x=a} = \beta C_2.$

We may then integrate eqn (8.26) with these boundary conditions to give an equation for the transmembrane ionic current in terms of the values of C_1, C_2, and E (Goldman 1943; Hodgkin and Katz 1949a):

$$I = zPF\xi\frac{C_2 \exp(\xi)-C_1}{\exp(\xi)-1}, \tag{8.27}$$

where $P = \beta uRT/aF$ and $\xi = EF/RT$.

Eqn (8.27) is plotted for various values of C_i and C_0 in Fig. 8.5. It can be seen that the degree of rectification depends on the ratio C_i/C_0. When this ratio equals 1 (i.e. equal ion concentrations) eqn (8.27) simplifies to

$$I = zCPF\xi, \tag{8.28}$$

which is linear. At ion concentrations corresponding to a negative equilibrium potential, the rectification predicted is in the outward-going direction. The physical reason for this rectification is simple: when current flows from the high concentration towards the low concentration, ions initially enter the membrane more quickly than they leave it. The mean ionic concentration within the membrane therefore rises towards a limit which is determined by βC_i. Since the ionic mobility within the membrane is assumed to be constant, the membrane conductance rises. Conversely, when current flows in the opposite direction the mean membrane concentration falls towards the lower limit given by βC_0 so that the conductance becomes smaller.

The constant-field equations have been extensively used in electrophysiology. They provide a satisfactory description of the chloride current in skeletal muscle (Hodgkin and Horowicz 1959; Hutter and Noble 1960), although only at neutral pH (Hutter and Warner 1969). The instantaneous current voltage relations for sodium and potassium ions in myelinated nerve are also well described by the constant-field equations (Frankenhaeuser 1960, 1962). In an experimental study of a 'porous' lipid bilayer membrane, Andreoli and Watkins (1973) found that the chloride conductance displayed rectification similar to that predicted by the constant field theory. However, in many other cases the constant-field equations are not very useful and more complex models are then required.

Fixed charge models. As already noted, the constant-field theory predicts a linear current–voltage relation at equal ion concentrations. In many cases, this is found not to be the case. Moreover, in some cases (e.g. the instantaneous current–voltage relations in squid nerve) the current–voltage relation may be linear at ion concentrations which are unequal.

One of the simplest ways of allowing for these deviations is to assume the presence of fixed charges at the membrane surface (Frankenhaeuser 1960; Chandler, Hodgkin, and Meves 1965; Adrian 1969). There will then be a potential hump or well at the membrane surface which will lower or raise the ion concentrations at the surface compared to that in free solution. If we assume that the flow of ions through the membrane still obeys the constant-field theory then the current-voltage relation will be given by eqn (8.27) with one or both of the concentration terms modified to allow for the fixed-charge layer. This gives eqn (8.29) (Adrian 1969, eqn 7.2):

$$I = PF(\xi + \xi^a)\frac{C_2 \exp(\xi) - C_1}{\exp\{(\xi + \xi^a)\} - 1},$$ (8.29)

where $\xi^a = E^a F/RT$. E^a is the surface potential in region 2 due to the fixed charge layer at $x = a$. Adrian (1969) has plotted this equation (see his Fig. 1) for various values of surface charge. The rectification may vary from outward-going to inward-going, but is not very marked unless extremely high densities of surface charge are assumed. Moreover, as in all the models discussed so far, the current–voltage relations always have a positive slope conductance.

In addition to the possibility that charge may occur at the membrane surface, charge may also occur inside the membrane phase. Various models have been developed for this situation both for fixed charge and moveable charges (Teorell 1935, 1951, 1953; Meyer and Sievers 1936; Schlogl 1954; Conti and Eisenman 1965). The theory of these models is beyond the scope of this book and the reader should consult the original papers or the reviews by Cole (1965), Eisenman and Conti (1965), and Haydon and Hladky (1972).

Carrier models. The degree of rectification which is predicted by models using ion or fixed charge concentration asymmetries is rather limited. In particular, such models do not account for negative slope conductances which have been observed in the current–voltage relations for potassium ions in some muscle membranes (Adrian and Freygang 1962; Noble and Tsien 1968). Adrian (1969) has reviewed this problem and has developed models which assume that ions may cross the membrane in combination with a charged carrier molecule. We will briefly discuss one of Adrian's models in order to illustrate how nonlinear properties may be produced by carrier systems. The model is illustrated in Fig. 8.6.

Potassium ions are assumed to cross the membrane in combination with a divalent negative carrier, one potassium ion being transported by each carrier molecule. The rate-limiting step is assumed to be the formation of the ion-carrier complex. The transfer across the membrane is then assumed to be sufficiently fast for the ratio of concentrations of ion-carrier complex at

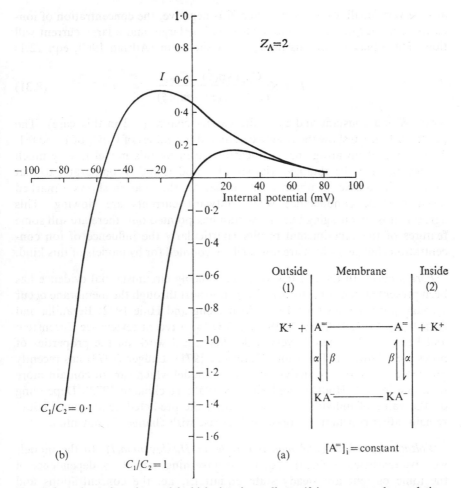

FIG. 8.6. A charged-carrier model which gives inwardly-rectifying current–voltage relations similar to those observed experimentally in muscle membranes. (a) Diagrammatic representation of kinetics of model (see text). (b) Current–voltage relations for $[K]_o/[K]_i = 1$ and $0 \cdot 1$ (Adrian 1969).

each side of the membrane to obey the Boltzmann distribution. In this particular model it is also assumed that the concentration of carrier molecules A on the inside surface is buffered so that $[A]_2^=$ is constant. The rate of formation of ion-carrier complex $[AK]_2^-$ at the inside surface of the membrane will be determined therefore by $[K]_2$ and the rate coefficient α. Moreover, for a given value of $[AK]_2^-$, the value of $[AK]_1^-$ will be given by the Boltzmann relation

$$[AK]_1^- = [AK]_2^- \exp(-\xi). \tag{8.30}$$

Hence, when E is positive, the concentration of ion-carrier complex on the outside surface will be very small and the outward flow of potassium ions will also be very small. Conversely, when E is negative, the concentration of ion-carrier complex on the outside surface will be large, and a large current will flow. The equation for this model is of the form (Adrian 1969, eqn 12.1)

$$I = K\frac{C_{in} \exp(\xi) - C_{out}}{\exp(-z_A\xi) + \exp(\xi)},$$ (8.31)

where K is a constant and z_A is the carrier charge (-2 in this case). The curves in Fig. 8.6 show the results calculated for values of C_1/C_2 of 1 and 0·1. The degree of inward-going rectification given by this model is very much greater than may be obtained with simpler models and is comparable to that observed in muscle membranes. In particular, the relation shows a marked negative slope conductance when outward currents are flowing. This agreement is encouraging but, as Adrian has pointed out, there are still some features of the experimental results (particularly the influence of ion concentration changes) which are not easily accounted for by models of this kind.

Pore models. In the last few years convincing circumstantial evidence has been presented that some forms of ion transport through the membrane occur through pores (Armstrong 1971; Armstrong and Hille 1972; Bezanilla and Armstrong 1972; Hille 1971, 1972, 1973; for a recent review see Ehrenstein and Lecar 1972). There is very little theoretical work on the properties of pores (see, however, Barry and Diamond 1971). Laüger (1973) has recently presented a rate theory analysis of a pore model which cannot contain more than one ion (cf. Hodgkin and Keynes 1955; Heckmann 1972). Depending on the value of individual rate constants, the predicted conductance could remain either constant, decrease or increase with changes in potential.

Voltage-dependent 'gate' models (Hodgkin–Huxley model). In the models we have discussed so far, the equations governing the voltage dependence of the ionic current are steady state equations, i.e. the concentrations and potentials at any point in the membrane are assumed to be independent of time. We have not treated the transients which occur as the system changes from one steady state to another. Moreover, unless the membrane mobilities are assumed to be several orders of magnitude smaller than in free solution, the time taken for such transients to occur should be very small indeed compared to times of physiological interest ($> 10\ \mu s$). Thus, if the ion mobilities in the membrane phase are assumed to be similar to those in free solution (about $10^{-3}\ cm^2s^{-1}\ V^{-1}$) a membrane 100 Å thick would require only about 0·01 μs to change from one steady state to another (Cole 1965). In view of the low temperature dependence of membrane conductances, it seems unlikely that the ion mobilities within the membrane are sufficiently small to increase this estimate to times of physiological importance. The equations which we have discussed so far are more likely to be used in describing the instantaneous behaviour of the ionic currents (see p. 228).

However, it is very characteristic of excitable cells that the ionic currents are not only instantaneous functions of the membrane potential but are also relatively slow functions of time. Thus, in nerve cells several milliseconds are required for the ionic current to reach a steady state following step changes in potential and, in muscle membranes, much longer times are required. Moreover, although the absolute magnitudes of the ionic conductances are not very temperature dependent, the time dependence is usually very highly temperature-dependent (Hodgkin and Huxley 1952; Frankenhaeuser and Moore 1963a, b; Moore 1958; Tsien and Noble 1969). This suggests that the processes responsible for the time-dependent behaviour of the ionic current are separable from the processes responsible for the instantaneous behaviour, that is, that there are additional time- and voltage-dependent processes which act as 'membrane gates' controlling the flow of ionic current. These processes were first studied quantitatively when the voltage-clamp technique was introduced (Cole 1949; Marmont 1949; Hodgkin, Huxley, and Katz 1952; Hodgkin and Huxley 1952a,b,c,d). Since then, a variety of models have been developed which may account for the results (Goldman 1964; FitzHugh 1965; Hoyt 1963, 1968; Mullins 1968; Tille 1965) but, as yet, there is very little additional experimental evidence to distinguish critically between them (but see Goldman and Schauf 1972, 1973). The model originally proposed by Hodgkin and Huxley is a relatively simple one, and with minor modifications we shall use it in this section to give an account of the forms of nonlinearity which are generated.

We suppose that, in addition to encountering energy-barriers of the kinds already discussed, ions also encounter essentially steric barriers which are very sensitive to the membrane potential. For example, whether ions may enter a particular channel in the membrane may be determined by the position or shape of a membrane structure which moves, rotates, or undergoes any other kind of conformational change in response to variations in the electric field. The structure itself may be charged, in which case the electric field may influence its position directly, or it may undergo changes which are consequent upon other voltage-sensitive membrane reactions (e.g. electron or proton transfer reactions). We call the state which allow conduction of ions the a-state and the state which blocks ion flow the b-state. It is also assumed that there is a first-order reaction between these states

$$a \underset{\alpha}{\overset{\beta}{\rightleftharpoons}} b$$

If the fraction of structures in the a-state is y the fraction in the b-state will be $(1-y)$ (a negligible fraction is supposed to be in any transition state) so that

$$dy/dt = \alpha(1-y) - \beta y. \tag{8.32}$$

If we assume that there is only one rate-limiting energy-barrier between the a- and b-states we may treat the kinetics in the same way as the kinetics of

9

ion movement across a single energy-barrier (see *Single energy-barrier model*, p. 229). However, unlike ion permeation, the energy barrier in this case must be attributed mainly to a large heat energy (enthalpy) of activation since the Q_{10} of the process is large. In the case of a single energy-barrier, the voltage dependence of the rate coefficients will be given by

$$\alpha = \alpha_0 \exp(\gamma z \Delta \xi) \qquad (8.33)$$

and

$$\beta = \beta_0 \exp\{-z(1-\gamma)\Delta \xi\}, \qquad (8.34)$$

where α_0 and β_0 are the values of the rate coefficients at the voltage $V_{\frac{1}{2}}$ at which $\alpha = \beta$ (that is, $\alpha_0 = \beta_0$) and $\Delta \xi = \Delta V F / RT$, where $\Delta V = V - V_{\frac{1}{2}}$ (see Fig. 8.7). In general, $V_{\frac{1}{2}}$ is not equal to zero and is usually negative. Moreover, $V_{\frac{1}{2}}$ varies greatly. It is not the same for all voltage-dependent permeability reactions (even for reactions in the same membrane), and it is also a function of agents (e.g. divalent ions) which might influence the membrane surface charge. It is possible, therefore, that $V_{\frac{1}{2}}$ is a measure of the surface charge in the vicinity of the ionic channels, but this cannot be regarded as certain since it is also possible that the energies of the *a*- and *b*-state may be different for chemical reasons so that an energy difference exists even when the electric field is zero.

The factor γ is the fraction of the membrane potential which influences the transition from the *b*-state to the activated state at the peak of the energy barrier. γ is therefore a measure of the symmetry of the system. z is the 'valency' of the reaction, i.e. the amount of charge which moves through the electric field during the $b \rightarrow a$ or $a \rightarrow b$ transitions. As shown in Fig. 8.7 this model, in its simplest form, gives exponential functions for the rate coefficients whose steepness is determined by the factors γ and z. It must be emphasized that the treatment we have given here is oversimplified for the purposes of illustration. In fact, although some of the rate coefficients of conductance changes measured experimentally are roughly exponential functions of membrane potentials, many are not. As in the case of the ion permeation process itself (see previous sections) we may suppose that the more complex functions arise when more than one energy-barrier is significant in the $a \rightarrow b$ transition (see Tsien and Noble 1969).

The steady state variation in the fraction of 'open' structures, i.e. y_∞ may be obtained by setting $dy/dt = 0$ in eqn (8.32):

$$y_\infty = \alpha/(\alpha+\beta). \qquad (8.35)$$

Hence, from eqns (8.33), (8.34), and (8.35) we obtain

$$y_\infty = 1/\{1+\exp(-z\Delta \xi)\}. \qquad (8.36)$$

Eqn (8.36) is a sigmoid function of V (see Fig. 8.7), whose steepness is determined by z. In practice, estimates of z (which vary between 2 and 6 for different voltage-dependent reactions) are obtained from experimental measurements of the steepness of the y_∞–V relation.

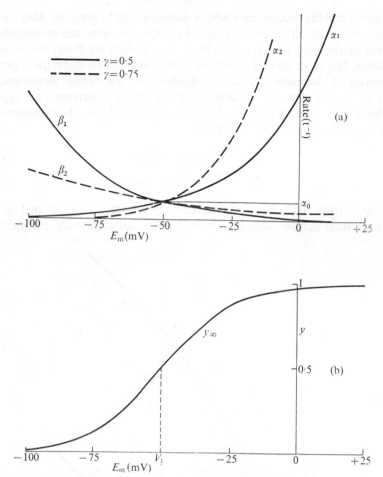

FIG. 8.7. Simplified kinetics of voltage sensitive 'gate' model (Hodgkin–Huxley model). (a) Voltage dependence of rate coefficients assuming one energy-barrier to be rate-limiting in $a \leftrightarrow b$ transition (eqns (8.33) and (8.34)). It is assumed that $z = 2$ and $\gamma = 0.5$ or 0.75. $V_{\frac{1}{2}}$ is assumed to be -50 mV. (b) Voltage dependence of steady state degree of activation, $y_\infty = \alpha/(\alpha+\beta)$. Note that y_∞ does not depend on γ.

The transient changes occurring following step changes in potential may be obtained by integrating eqn (8.32) to give

$$y = y_0 - ([y_0 - y_\infty][1 - \exp\{-(\alpha+\beta)t\}]), \qquad (8.37)$$

which means that y will change exponentially from its initial value y_0 to its steady state value y_∞.

In order to illustrate the general form of the current voltage relations generated by mechanisms of this kind, we will consider the simple case of a membrane with sodium and potassium channels whose instantaneous current–voltage relations are linear. The potassium equilibrium potential will be negative and the sodium equilibrium potential will be positive. Also we will assume that each 'channel' is 'gated' by only one structure undergoing a reaction of the kind we have described. As we shall see later, this is over-simplified. The experimental results are more complicated and require further assumptions to be made (see section *Applications to excitable membranes*).

Let y_K be the 'gating' variable for the potassium current and y_{Na} the variable for the sodium current. Also let $\bar{G}_{Na}$ and $\bar{G}_K$ be the conductances when y_{Na} and y_K are equal to 1. Then

$$I_K = \bar{G}_K\, y_K\, (E - E_K) \tag{8.38}$$

and

$$I_{Na} = \bar{G}_{Na}\, y_{Na}\, (E - E_{Na}). \tag{8.39}$$

These relations have been plotted for steady state conditions in Fig. 8.8. The potassium current–voltage relation shows outward-going rectification.

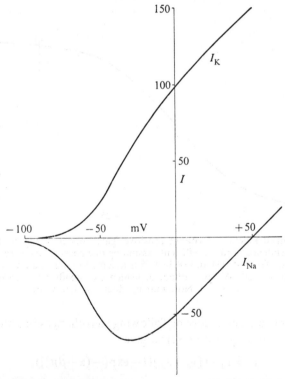

FIG. 8.8. Ionic current–voltage relations given by eqns (8.38) and (8.39). It is assumed that $\bar{G}_K = \bar{G}_{Na}$; $E_K = -100\ \text{mV}$, $E_{Na} = +50\ \text{mV}$, and the y variables are steady state variables given by Fig. 8.7. $I_K(V)$ shows outward-going rectification. $I_{Na}(V)$ is a more complicated function and cannot be simply characterized in terms of rectification.†

† This diagram is purely illustrative and is not based on any observed current–voltage relations. However, the general form of the curves is similar to that of relations obtained experimentally in excitable cells.

The sodium relation is also very nonlinear, but in this case it cannot be described simply in terms of rectification.

Descriptions of excitable membranes

A variety of excitable membranes have been analysed using voltage-clamp techniques and a large amount of information now exists on the kinetics of conductance changes. In this section we will briefly summarize the results in terms of the various forms of nonlinearity described in previous sections.

Squid nerve. The first quantitative analysis was done on the squid nerve membrane (Hodgkin and Huxley 1952*d*). In retrospect, it can now be seen that this was a particularly fortunate choice. The instantaneous current–voltage relations for sodium and potassium ions at physiological concentrations are linear so that the instantaneous behaviour at any particular time may be characterized by one variable—the chord conductance. However, the time-dependent behaviour is more complex than in the simplified model described in the previous section.

First, the ionic current changes which follow step changes in membrane potential are not simple exponentials. The response to a positive voltage step is sigmoid. It may be fitted by using a power function of an exponential. Hodgkin and Huxley interpreted this to mean that more than one voltage-dependent 'gating' reaction occurs at each conducting site and that all the 'gates' must be in the *a*-state in order that ions may cross the membrane. Thus, in the case of potassium ions they used the equation

$$I_K = \bar{G}_K n^4 (E - E_K), \tag{8.40}$$

where $\bar{G}_K$ is the maximum instantaneous conductance, E_K is the potassium equilibrium potential, and n is a variable obeying equations of the same form as eqns (8.32)–(8.37). This implies that four charge-transfer reactions must occur at each site for it to become permeable to potassium ions.

The sodium conductance in squid also requires a power function. In this case Hodgkin and Huxley used

$$I_{Na} = \bar{G}_{Na} m^3 h (E - E_{Na}), \tag{8.41}$$

where $\bar{G}_{Na}$ is the maximum instantaneous sodium conductance, E_{Na} is the sodium equilibrium potential, and m and h are variables obeying equations of the same form as eqns (8.32)–(8.37). m is a variable which, like n, increases on depolarization. However, h differs from n and m in that its voltage dependence is in the *opposite* direction, that is, depolarization reduces h. A possible physical interpretation of this equation is that each conducting site is 'gated' by three m gates and one h gate, all of which must be in the *a*-state for the site to conduct sodium ions. The m gates are much faster than the h gates so that the result of depolarization is an increase in sodium conductance corresponding to the activation of m followed by an inactivation of the sodium conductance corresponding to the decline in h. The processes determining the sodium and potassium conductances in squid nerve are summarized in Figs. 8.9 and 8.10.

The second major difference from the simple model described in the previous section is that, although some of the rate coefficients are simple exponential functions of membrane potential (β_n, β_m, and α_h could all be fitted by exponentials), others are not. α_n and α_m become linear at extremes of potential and are better fitted by equations of the constant-field form (see p. 234). β_h saturates at strong depolarizations. These deviations from the simple exponential form suggest that more than one energy barrier is rate-limiting in the $a \leftrightarrow b$ transition.

Myelinated nerve. The instantaneous current voltage–relations in this case are nonlinear and are fitted quite well by the constant-field equations (see p. 234). The time-dependent behaviour is similar to that in squid, except that lower powers (2 in the case of n and m) are required to fit the sigmoid current changes in response to positive voltage-steps. Most of the rate coefficients are well fitted by equations of the constant-field form. A third, non-specific current component (labelled p) is also present. Frankenhaeuser and Huxley (1964) have summarized the voltage-clamp data in the form of a set of differential equations and have reproduced the action potential and other excitation phenomena in amphibian nerve.

Lobster nerve. The ionic currents in this case are similar to those in squid membrane (Julian, Moore, and Goldman 1962) except that the delayed potassium current is much smaller than the initial sodium current, and its magnitude is much more variable than in squid nerve. This variation is associated with a corresponding variation in the speed of repolarization of the action potential.

Skeletal muscle. The potassium current underlying repolarization in skeletal muscle is similar to that in nerve in having a sigmoid time course of activation (Adrian, Chandler, and Hodgkin 1966; Ildefonse and Rougier 1972). However, it is not a pure potassium current and has a reversal potential at about 20 mV more positive than the resting potential. This point corresponds to the beginning of the after potential in skeletal muscle, which may therefore be attributed to the decay of delayed rectification. Adrian, Chandler, and Hodgkin (1970) and Adrian and Peachey (1973) have summarized the membrane currents as a set of differential equations and have given solutions for the skeletal muscle action potential.

Cardiac muscle. The sodium current in cardiac muscle appears to behave in a similar way to that in nerve (Weidmann 1955; Dudel, Peper, Rüdel, and Trautwein 1966), although the effects of cooling are strikingly different from what might be expected (Dudel and Rüdel 1969). In some cases there is also a second, slower inward current (Reuter and Beeler 1969; Rougier, Vassort,

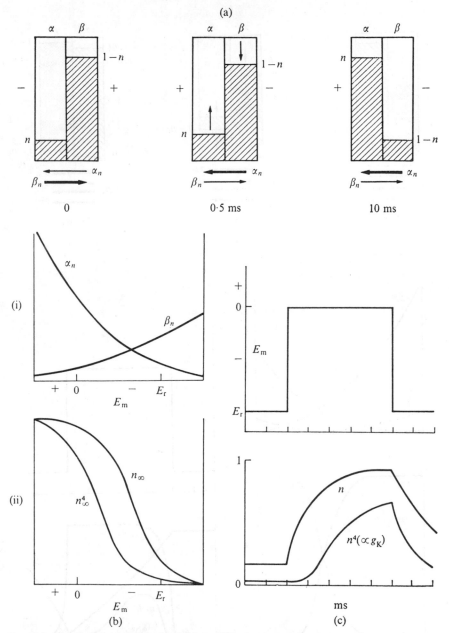

FIG. 8.9. Processes assumed by Hodgkin and Huxley to describe the potassium conductance in squid nerve. (a) Response of n reaction (n is a variable obeying equations similar to those for y described in text) to sudden depolarization. Initially, fraction n of structures in the a-state is small since α_n is small and β_n is large. Depolarization increases α_n and decreases β_n so that n rises exponentially to a larger value. Relative magnitudes of rate coefficients are indicated by thickness of arrows. (b) (i) Variation of rate coefficients with E_m. (ii) Variation of n_∞ and $(n_\infty)^4$ with E_m. (c) Response of n and G_K to sudden depolarization (based on Noble 1966)†.

† This diagram is illustrative only; the curves are not accurate solutions of the Hodgkin–Huxley equations.

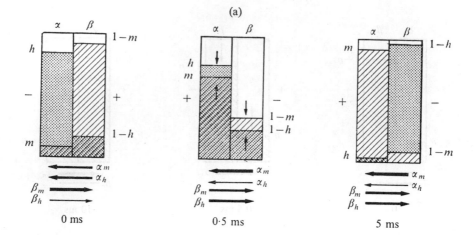

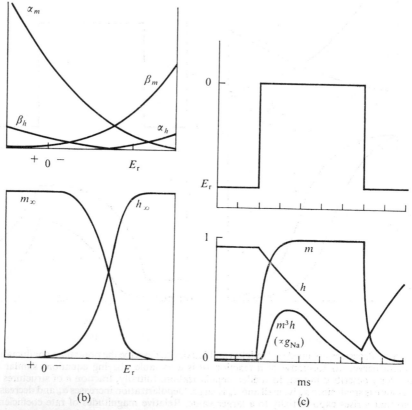

FIG. 8.10. Processes assumed to determine sodium conductance in squid nerve. (a) Responses of m (activation) and h (inactivation) reactions to sudden depolarization. Initially m is small and h is large. On depolarization α_m becomes very large so that m rapidly increases; β_h also increases but, since it is smaller than α_n, h does not fall greatly before most of the increase in m has occurred. (b) Variations of rate coefficients and steady state values of m and h with E_m. (c) Response of m, h, and G_{Na} to sudden depolarization (based on Noble 1966).

Garnier, Gargouïl, and Coraboeuf 1969) which is at least partly conducted by calcium ions.

The most striking differences between nerve and cardiac muscle lie in the behaviour of the outward current. Most of the quantitative kinetic analysis has been done on the Purkinje fibre system. This shows no less than four separate components of outward current which are time-dependent and which may be analysed in terms of variables obeying equations of the same form as eqns (8.32)–(8.37). Two of the components may normally be neglected since they either activate far too slowly to be involved in normal electrical activity or are largely inactivated at the frequencies at which cardiac action potentials normally occur. The remaining two components of outward current are important in the termination of the action potential plateau and in the generation of pacemaker activity. Their time constants are of the order of hundreds of milliseconds (or even seconds), compared to the millisecond time constants of nerve. Moreover, the time courses are simple exponentials so that there is no need for power functions. Otherwise, the kinetics of the 'gating' mechanisms controlling these currents are very similar to those of currents in other excitable membranes. However, the instantaneous current–voltage relations are very nonlinear. The pacemaker potassium current shows inward-going rectification with a marked negative slope conductance at strong depolarizations. The plateau current (which is not a pure potassium current) shows inward-going rectification but does not exhibit a detectable negative slope conductance (Noble and Tsien 1968, 1969a).

The current–voltage relations in cardiac muscle are, therefore, considerably more complex than in nerve. In particular, cardiac membranes may show inward-going rectification in response to sudden changes in membrane potential but outward-going rectification when time has been allowed for delayed rectification to occur. The total steady state current–voltage relations are usually N-shaped and often show a negative slope conductance at one range of potentials. McAllister, Noble, and Tsien (1975) have reconstructed the normal electrical activity of Purkinje fibres using the experimental information at present available. Beeler and Reuter (1975) have given equations for ventricular muscle.

Simplified current–voltage relations

Although the mechanisms underlying the time-dependent changes in conductance in excitable cells obey equations of similar form, it is clear that there are large and important quantitative differences. In particular, there is a large degree of variation in the rates of the 'gating' reactions. The sodium activation reaction m occurs with a time constant of the order of 100 μs, whereas the potassium current underlying the Purkinje fibre pacemaker potential changes with a time constant of several seconds. This large degree of variation raises the possibility that, for many purposes, one or more of the reactions may be assumed to occur infinitely quickly or infinitely slowly.

Thus, for many purposes, in cardiac membranes the sodium reactions may be assumed to occur instantaneously; this greatly simplifies the theory (Noble 1962*a*; Noble and Tsien 1969*b*, 1972). Despite the quantitative variations, the current–voltage relations obtained by making simplifying assumptions of this kind have certain features in common which may be used to give a semi-quantitative account of excitable membranes. We will illustrate these features using one example.

We assume that the *m* reaction in nerve membranes occurs sufficiently quickly compared to the *n* and *h* reactions for it to be regarded as occurring instantaneously. *m* will then have its steady state value at each potential, and the sodium current–voltage relation before inactivation occurs will have the form already shown in Fig. 8.8. Since this relation is not strictly speaking instantaneous, we will refer to it as a momentary current–voltage relation. The instantaneous potassium current–voltage relation is assumed to be given by a straight line, as in squid. The total ionic current will be given by the sum of the individual ionic currents

$$I_i = I_{Na} + I_K. \tag{8.42}$$

For the sake of simplicity, we have neglected other smaller components (e.g. leak currents, anion currents).

The momentary current–voltage relation obtained in this way has the form shown in Fig. 8.11. The relation is N-shaped and crosses the voltage axis at three points, listed below.

   1. The left-hand intersection occurs where the current–voltage relation has a positive slope. Small deflections in either direction from this point

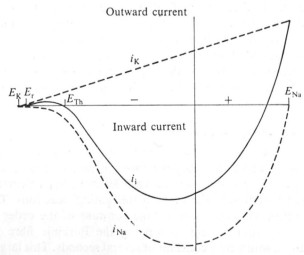

FIG. 8.11. Diagram illustrating form of simplified (momentary) current–voltage relation obtained by allowing fast sodium activation reaction (*m*) to be in a steady state while slower reactions (*h* and *n*) are held constant.

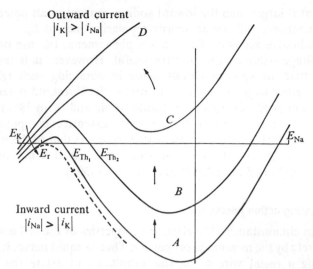

Fig. 8.12. Diagram illustrating change in momentary current–voltage relations with time on depolarization.

produce currents which restore the membrane potential to its original value. Thus, a small depolarization produces an outward ionic current which tends to repolarize the membrane, whereas hyperpolarizing deflections produce inward ionic current which depolarizes the membrane. This point is therefore stable and its forms the resting potential E_r.

2. The middle intersection occurs at a point at which the slope conductance is negative. A small depolarizing deflection will produce inward current which will further depolarize the membrane, whereas a small hyperpolarization will produce outward ionic current which will further hyperpolarize the membrane. This point is clearly unstable and forms the critical point beyond which the membrane changes its potential regeneratively. It is therefore the voltage threshold for initiating an action potential. The current threshold is given by the minimum current required to reach this potential which is therefore the peak outward current between E_r and the voltage threshold E_{Th}.

3. The right-hand intersection occurs at a positive slope conductance point. In principle, therefore, this point is stable, and a membrane obeying a current–voltage relation of this form would show two stable states. However, in practice, the stability of the third point is transient since, once the depolarization occurs, the potassium activation n and sodium inactivation h reactions will occur, and the current–voltage relation will shift spontaneously in the outward direction, as shown in Fig. 8.12. The third intersection therefore will become more negative. Eventually, this point will disappear completely (when the outward potassium

current is larger than the inward sodium current at all potentials), and the membrane will spontaneously repolarize back to E_r.

As a qualitative account of excitation phenomena, the use of simplified current–voltage relations can be very useful. However, it is important to remember that the approximations made in obtaining such relations are sometimes rather large. Thus in squid nerve, the threshold potential is not very accurately predicted by this method (Noble and Stein 1966). For more accurate work it is necessary to use more extensive computations. The simplest case, which does not involve cable theory, is that of the non-propagating action potential, which is sometimes also called a *membrane action potential* (Hodgkin and Huxley 1952d; Huxley 1959a).

Non-propagating action potentials

In certain circumstances, the electrical properties of the cell may be determined entirely by the membrane properties. Thus in squid nerve, it is possible, by inserting a metal wire along the axoplasm, to excite the membrane uniformly so that cable complications are eliminated. These complications may also be eliminated, to a certain extent, in cases where the cell is short enough for the membrane potential to be effectively uniform during the action potential. In these cases, the relation between ionic current flow across the membrane and membrane potential is very simple (see Fig. 8.13). All the ionic current is used to charge the local cell membrane capacity and none flows as local circuit (cable) current flow. The equation for this case is

$$I_m = C_m \, dV/dt + I_i, \tag{2.1}$$

where I_m is the stimulus current. Once the action potential has been initiated by passing sufficient depolarizing current to exceed threshold, I_m may be set to zero and we then have

$$dV/dt = -I_i/C_m. \tag{8.43}$$

I_i will be given by appropriate combinations of equations of the kind discussed in this chapter. Eqn (8.43) may then be integrated to compute the voltage–time course. Such solutions have been obtained numerically for a variety of conditions using ionic current equations obtained for squid nerve (Hodgkin and Huxley 1952; Huxley 1959; FitzHugh 1960; Cooley and Dodge 1966), frog myelinated nerve (Frankenhaeuser and Huxley 1964), frog skeletal muscle (Adrian, Chandler, and Hodgkin 1970; Adrian and Peachey 1973), and mammalian cardiac muscle (Noble 1962a, 1966; McAllister, Noble, and Tsien 1975). A review of this work is beyond the scope of this book and readers interested in the applications to problems which do not involve cable theory should consult the references cited, and the books by Hodgkin (1964), Cole (1968), and Adelman (1971).

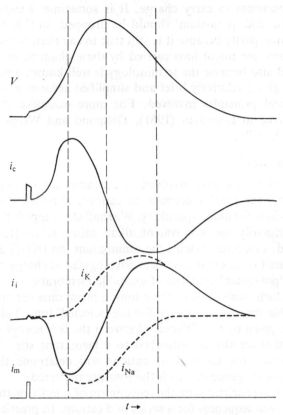

V

i_c

i_i

i_K

i_m i_{Na}

$t \rightarrow$

FIG. 8.13. Variations in voltage and currents during a 'membrane' (uniform) action potential initiated by a brief square pulse current. The top diagram shows the potential change as a function of time. The second diagram shows the capacity current, given by by $C_m dV/dt$. The initial, nearly square, wave is attributable to the applied current (see bottom diagram), most of which flows as capacity current when brief pulses are used. The third diagram shows the total ionic current (continuous curve) and its potassium and sodium components (interrupted curves). The bottom diagram shows the total membrane current $(i_c + i_i)$, which is zero apart from times at which the stimulus is applied. The three vertical interrupted lines indicate the times at which dV/dt is maximal, zero, and minimal. This diagram may be compared with the corresponding one for a propagated action potential shown in Fig. 10.2.

Ionic selectivity

As noted previously in this chapter (p. 227), some of the important forms of nonlinearity displayed by excitable cells, in particular current–voltage relations with three zero current points, arise from the summation of ionic currents flowing through membrane channels of differing selectivity. The mechanism of ionic selectivity is therefore of some importance in the generation of these forms of nonlinearity. In this section we shall show that the sodium and potassium channels of excitable membranes allow ions other than sodium or potassium to carry charge. It is sometimes suggested that the names 'sodium' and 'potassium' should be dropped. In this book we shall retain the names, partly because it is still true to say that, in normal physiological solutions, the major ions carried by these channels are sodium and potassium and also because the terminology is well known and understood. Also we shall give a relatively brief and simplified account of some possible physico-chemical principles involved. For more extensive treatments the reader is referred to Eisenman (1961), Diamond and Wright (1969), and Williams (1952, 1970).

Selectivity sequences

Since most of the currents involved in excitation are carried by cations, we shall restrict the present account to cations, although the same basic principles also hold for anion specificity. We shall also simplify the account by considering primarily the monovalent alkali cation series (Li^+, Na^+, K^+, Rb^+, Cs^+) and, to a lesser extent, the ammonium ion (NH_4^+) and its derivatives. These are the ions that have been investigated as charge carriers in the 'sodium' and 'potassium' channels of excitable membranes.

In order to distinguish between these ions, a membrane site must provide a more favourable free-energy change for the 'selected' ions. This free-energy change will be given by the difference between the free energy of the ion in solution and that for the ion adsorbed to a membrane site. The ratios of free-energy change for the various cations will determine the specificity sequence, i.e. the sequence in which the ions are preferred.

If there were no restrictions on the mechanism of selectivity, there might be up to 5! (120) such sequences for a series of 5 cations. In practice, only 11 of these sequences normally occur. Moreover, these 11 sequences may be arranged in an order (see Fig. 8.14) such that only one pair of ions is inverted between each sequence. The first sequence is in the order of decreasing ion crystal sizes. The last sequence is the reverse order. The intermediate sequences are obtained by first interchanging the order of the larger ions, Cs^+, Rb^+, and K^+. Later sequences are obtained by the smaller ions, first Na^+ and then Li^+, crossing from low selectivity to high selectivity. This arrangement of sequences corresponds to an underlying physical order, as is shown for example, by Eisenman's (1965) work on selectivity in glass membranes, where it is possible to shift from one sequence to the next in the order shown in Fig. 8.14 simply by varying the pH. Thus, protonation of negative binding sites, which would weaken their negativity by shielding the sites, leads to low-numbered sequences, whereas alkalinization shifts the selectivity toward high-numbered sequences.

This strongly suggests that one of the important factors determining selectivity is the field strength of the binding site. Strong negative sites (weak acid groups) preferably bind small cations, while weak sites (strong acid groups) prefer to bind large cations. This general rule is also found in the

I	Cs > Rb > K > Na > Li	(Lyotropic series)
II	$\boxed{\text{Rb} > \text{Cs}}$ > K > Na > Li	
III	Rb > $\boxed{\text{K} > \text{Cs}}$ > Na > Li	
IV	$\boxed{\text{K} > \text{Rb}}$ > Cs > Na > Li	
V	K > Rb > $\boxed{\text{Na} > \text{Cs}}$ >· Li	
VI	K > $\boxed{\text{Na} > \text{Rb}}$ > Cs > Li	
VII	$\boxed{\text{Na} > \text{K}}$ > Rb > Cs > Li	
VIII	Na > K > Rb > $\boxed{\text{Li} > \text{Cs}}$	
IX	Na > K > $\boxed{\text{Li} > \text{Rb}}$ > Cs	
X	Na > $\boxed{\text{Li} > \text{K}}$ > Rb > Cs	
XI	$\boxed{\text{Li} > \text{Na}}$ > K > Rb > Cs	(series of atomic radii)

FIG. 8.14. The selectivity sequences commonly found for the alkali metal cations. The sequences are numbered following Eisenman's notation. Each sequence is similar to the one above it apart from one inversion (indicated by box in each case).†

stability constants for the salts of strong and weak acids (see e.g. Williams (1970) p. 341).

Free energy of binding. To quantify the hypothesis that the strength of the binding site determines the selectivity sequence, we require equations both for the energy of hydration ΔG_H of the cations in solution and for the energy of binding ΔG_s to negative sites of differing strengths. The difference between these energies, $\Delta G = \Delta G_s - \Delta G_H$, will give the free-energy change on binding to the site from aqueous solution.

Fig. 8.15 shows the way in which hydration energy depends on the cation size. Water molecules are most strongly bound to small cations whose field strength is very large, since the oxygen of the water molecules can approach closer to the centre of positive charge than it can in the case of large cations. There are various ways of calculating the energy of hydration, involving differing degrees of theoretical sophistication. The simplest uses the Born equation for the energy of Coulombic attraction between ligands of opposite charge (Phillips and Williams 1965, p. 161):

$$-\Delta G = A/(r^+ + r^-), \qquad (8.44)$$

† (1) There is one inversion (indicated by box) between each sequence. (2) As series moves from I to XI (i.e. increasing strength of site binding) large ions invert first. (3) There are very few examples known of 'abnormal' sequences (i.e. other 109 permutations of 5 cations). The biological examples nearly all involve an abnormal position for Li; that is, Li is sometimes treated as a larger ion.

	Li	Na	K	Rb	Cs	NH₄
Pauling radii (Å)	0·60	0·95	1·33	1·48	1·69	1·45
†ΔH^0_{hyd} (kJ mol⁻¹)	−500	−391	−307	−281	−248	
‡ΔH^0_{hyd} (calculated) (kJ mol⁻¹)	−483	−391	−323	−303	−277	

† Assumes hydration energy for $H^+ = -1079$ kJ mol⁻¹.
‡ Using Born equation:

$$\Delta H^0_{hyd} = -7 \times 10^3 \frac{z^2}{r^+ + 0.085} \text{ kJ mol}^{-1}.$$

0·085 nm = radius of O in water molecule.

FIG. 8.15. The atomic sizes and hydration energies of the alkali metal cations and of ammonium ion.

where A is a constant determined by the charge on the ligands and r^+ and r^- are the radii of the cation and anion. In the case of water the appropriate radius is that of the oxygen atom, assuming that the centre of the negative charge of the water dipole is at the centre of the oxygen atom.

It is found experimentally that the variation of ΔG_H with cation size is fitted by the equation†

$$-\Delta G_H = \frac{7 \times 10^3}{r^+ + 0.085} \text{ kJ mol}^{-1}, \tag{8.45}$$

where 0·85 Å (0·085 nm) has been taken to be the required radius of the oxygen atom. The justification for this equation is discussed in Phillips and Williams (1965, pp. 160–3).

If we rewrite eqn (8.45) as

$$-\Delta G_H = \frac{7 \times 10^3}{r_{eff}} \text{ kJ mol}^{-1}, \tag{8.46}$$

where $r_{eff} = r^+ + 0.085$ we obtain a linear relation between $-\Delta G_H$ and $1/r_{eff}$ as shown in Fig. 8.16(a).

Similarly, the energy of adsorption to a negative site will be given by

$$\Delta G_s = -nB/(r^+ + r^-), \tag{8.47}$$

where B is a constant determined by the charge on the site and n is the number of sites to which the ion may bind simultaneously (this is known as the co-ordination mumber and will be discussed further below).

---

† Eisenman's treatment of ΔG differs from the simple account given here. In particular, Eisenman uses different equations for the variation in force with distance near a membrane site and near a water molecule. His equations are based on the fact that hydration involves an ion interacting with a dipole rather than a single charge and that the force then falls off more rapidly with distance. In terms of predicting the selectivity sequence, this assumption produces the same result as that of assuming the water molecule to be represented as a very small negative site.

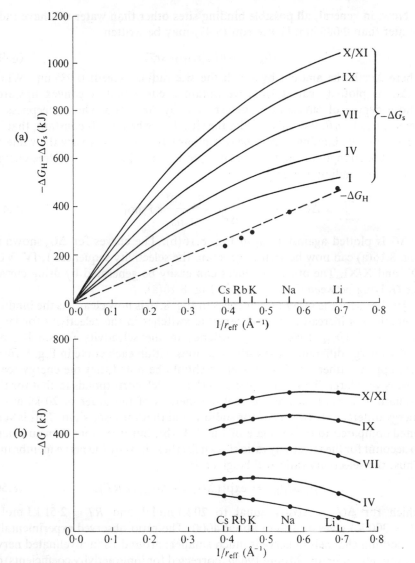

FIG. 8.16. Diagrams illustrating how the selectivity sequences shown in Fig. 8.14 are generated by a simple coulombic model of ion binding (cf. Williams 1970). (a) The points show the experimental values for hydration energies of the alkali metal cations. The interrupted line is given by eqn (8.46). The curves illustrate binding energy as a function of $1/r_{eff}$ for sites of various field strengths. (b) The differences between the points and each curve shown in (a) are plotted as points in (b). The strongest sites can now be seen to display high-numbered sequences while weak sites display low-numbered sequences.

Now, in general, all possible binding sites other than water will have radii greater than 0·085 nm. Hence eqn (8.47) may be written

$$\Delta G_s = -nB/(r_{\text{eff}} + \Delta r^-), \tag{8.48}$$

where Δr^- is the amount by which the site radius exceeds 0·085 nm. When $-\Delta G_s$ is plotted against $1/r_{\text{eff}}$ we obtain a curve that is convex upward, whose degree of curvature is determined by Δr^-, and whose steepness is determined by nB. We will consider the situation when Δr^- is constant, that is, we assume the binding sites to be of similar size and simply vary the value of B, i.e. the strength of the binding site (see Fig. 8.16(a)). Now the free-energy change on binding will be given by

$$\Delta G = -\frac{nB}{r_{\text{eff}} + \Delta r^-} + \frac{7 \times 10^3}{r_{\text{eff}}} \text{ kJ mol}^{-1}, \tag{8.49}$$

$-\Delta G$ is plotted against $1/r_{\text{eff}}$ in Fig. 8.16(b). The curves for ΔG_s shown in Fig. 8.16(a) can now be seen to generate the selectivity sequences I, IV, VII, IX, and X/XI. The other sequences can easily be generated by using curves for G_s lying between those shown in Fig. 8.16(a).

The basis for the order of sequences in Fig. 8.16 is now clear. As the binding strength B is increased we first generate switches in the selectivity for large ions (small $1/r_{\text{eff}}$) followed by switches in the selectivity for small ions.

The energy differences for various cations within each curve in Fig. 8.16(b) may appear rather small. However, it should be noted that the energy scale used is very large. Thus, in sequence X/XI (which corresponds to that for the 'sodium' channel in nerve—see below) there is of the order of 20 kJ mol⁻¹ energy difference between ΔG for sodium and that for potassium. This is very small compared to the ordinate of Fig. 8.16(b), but it is more than sufficient to account for the selectivity ratio shown for these ions by the nerve membrane. Thus, the selectivity ratio will be given by

$$\beta_{\text{Na}^+}/\beta_{\text{K}^+} = \exp\{(\Delta G_{\text{K}^+} - \Delta G_{\text{Na}^+})/RT\}, \tag{8.50}$$

which for $\Delta G_{\text{K}^+} - \Delta G_{\text{Na}^+}$ equal to 20 kJ mol⁻¹ and $RT = 2·51$ kJ mol⁻¹ ($T = 300$ K) gives a ratio of about 4000. The ratio observed experimentally for sodium channels in nerve is much smaller: around 18 in myelinated nerve (Frankenhaeuser and Moore 1963b; corrected for ionic activity coefficients) or 12 in squid axon (Chandler and Meves 1965). Although there is clearly a large *quantitative* discrepancy in the degree of selectivity, it is worth noting that the *sequence* of permeabilities conforms with the selectivity sequence for a relatively high field strength site. Chandler and Meves found that the alkali cations fall in the permeability sequence Li:Na:K:Rb:Cs = 1·1:1:1/12:1/40: 1/60, in agreement with selectivity sequence X/XI. Although it is quite likely that interaction with a high field strength site plays a role in generating selectivity in the sodium channel, a variety of arguments suggest that other factors must also be involved.

One important factor is the size of the cation. This is clear from Hille's (1971) study of the degree of permeability of the sodium channel to various organic cations. Many cations appear to be excluded on the basis of size alone. Hille has combined ideas about steric and field strength affinity mechanisms in an imaginative model of the ion-selective region of the sodium channel. This hypothetical structure, and various physical explanations for selectivity have been recently reviewed by Hille (1974).

Turning to the 'potassium channel', it has been known for some time that in nerve Rb^+ ions can carry current about as effectively as K^+ (Pickard, Lettvin, Moore, Takata, Bernstein, and Pooler 1964; Müller-Mohussen and Balk 1966), and that ammonium ions are somewhat permeable (Lüttgau 1961; Binstock and Lecar, 1969). However, in some contrast to the sodium channel, the potassium channel is very poorly permeable to almost all other cations examined thus far. In the most extensive study to date, Hille (1973) found only four 'permeant' cations, falling in the sequence Tl:K:Rb:ammonium, with relative permeabilities of 2·3:1:0·91:0·13. Cs and Na are each very poorly permeant. These results are consistent with a selectivity sequence IV, corresponding to interaction with an intermediate field strength site. In addition to this mechanism, it is likely that in the case of the potassium channel (as for the sodium channel) the size of the cation may be critical. A number of lines of evidence (Bezanilla and Armstrong 1972; Hille 1973) suggest that the 'selectivity filter' of the potassium channel may be a narrow pore (diameter 0·3 nm).

Comparing the selectivity behaviour of the ionic channels leads to the expectation that the potassium channel binding site(s) should be stronger acids than the sodium channel binding site(s). Titration experiments in nerve (Hille 1968; Drouin and Thé 1969; Gilbert and Ehrenstein 1970; Woodhull 1973; Hille 1973) have shown this to be the case: the apparent pKs are 4·4 (K) and 5·2 (Na). Although this qualitative agreement is encouraging, the difference in pKs is rather small in view of the large increase in binding energy required to shift from sequence IV to sequence X (see (see Fig. 8.16). Once again, this suggests that affinity for the titratable binding site may be only one of a number of factors governing the over-all selectivity (see Hille 1974). Finally, it is most important to remember that the channel sequences are obtained either by measuring current carried (as done by Chandler and Meves) or by measuring reversal potentials (as done by Hille). Neither of these measurements leads to results that are directly comparable to those obtained from binding energies since the binding energy is only one of the factors that may determine the current flow. Thus, the current carried also depends on the mobility of the ions in the channel. In general, the larger the binding energy, the lower will be the mobility since it will be more difficult for ions to move from site to site. Hence the ratio of currents carried will be lower than the selectivity ratio for binding. Indeed, it is possible that the conductance may decrease as the binding increases so that conductance sequences opposite to the binding sequence may be obtained (Woodbury and Miles 1973). For reasons of this kind Armstrong (see Appendix to Bezanilla and Armstrong 1972) has suggested that models very different from those based on equilibrium binding constants are more relevant to ion transport across pores. The only case for which the conductance and binding sequences are identical is that of an isosteric carrier for which the mobility is independent of the ion carried (Szabo, Eisenman and Ciani, 1969). As we have already mentioned, it is more likely that the sodium and potassium channels in nerve membranes carry current through pores rather than via carriers.

Cyclic ligands: effect of co-ordination number

One way in which the strength of binding may be varied without changing the strength of individual binding sites is to vary the number n of sites to which the cation may bind. This may be done by arranging several sites in a ring whose size determines whether a cation may co-ordinate with all the sites (when the cation is well fitted by the ring), or with few sites (when the cation is too small to co-ordinate fully with all sites simultaneously), or with none at all (when the cation is too large to enter the ring). A fairly large number of cyclic ligands are now available that will bind cations, the best known being valinomycin, gramicidin, enniatin, the actin antibiotics (non-, mon-, din-, and trin-actin), and the cyclic polyethers synthesized by Pedersen (1970). Most of these ligands provide about 6 oxygen atoms to which cation binding may occur and they show greater selectivity towards K^+ than Na^+. Thus, the $K^+:Na^+$ selectivity ratio for valinomycin is 400:1, and for nonactin is 100:1. In order to obtain a sodium selective ligand, a smaller ring with fewer oxygen sites is required (e.g. the cyclic polyether II synthesized by Pedersen).

In view of the fact that the difference in pK between the sodium and potassium sites in nerve is too small to allow the difference in strength of the individual ligands to be the sole factor determining selectivity, it seems likely that steric factors of the kind involved in the cyclic ligands are also involved. Of course, the membrane molecules concerned need not be cyclic. It is also possible that a multiple-site ligand similar to the cyclic ligands may occur naturally as a result of several molecules with negative binding sites being brought together in an appropriate way. This idea is attractive since it allows the conductance of the channels to be determined by the positions of the molecules involved and, if these were field-dependent, some of the nonlinear phenomena described earlier in this chapter would occur (see voltage-dependent 'gate' models—Hodgkin–Huxley theory).

9. Nonlinear cable theory: excitation

ONE of the most important functional properties of excitable cells is that they conduct all-or-nothing non-decremental signals known as action potentials. In Chapter 8 it was shown that this property depends on certain kinds of non-linearity displayed by excitable cell membranes. In order to give a mathematical account of the conduction of action potentials, it is necessary to combine analyses of membrane nonlinearities of the kind discussed in Chapter 8 with the description of spatial spread of currents presented in earlier chapters. The result may be called nonlinear cable theory since it concerns the analysis of cables with nonlinear membranes. In this chapter and in Chapter 10 we will give explanatory and relatively brief accounts of the initiation and conduction of action potentials. In Chapter 11 we will give a more extensive analysis of repetitive activity. Chapter 12 will be concerned with a more advanced analytical approach to initiation and conduction and in that chapter we will discuss the current–voltage relations of nonlinear cables in some detail. However, in order to discuss the threshold conditions for excitation it will be necessary to give a preliminary account of current–voltage relations in cables.

The conditions for excitation

In this section we will discuss the conditions for excitation of a cable by current applied at one point $x = 0$. In Chapter 8 it was shown that the condition for excitation of a uniformly polarized membrane is that the voltage should exceed a critical value at which the net ionic current becomes inward. The corresponding condition in a cable is that the voltage should exceed a critical value at which the net ionic current generated by the cable *as a whole* becomes inward. We will show later that this is, in fact, only a minimal condition. However, as a first approximation to determining the conditions for exciting a cable by current applied at one point it is useful to obtain an expression for the *input current–voltage relation*. This is the relation between the applied current at $x = 0$, I, and the voltage at $x = 0$.

The Cole theorem

It was shown in Chapter 3 that in a linear cable the input current–voltage relation must also be linear with a slope and chord conductance equal to $2/\sqrt{(r_m r_a)}$. If the membrane current–voltage relation is nonlinear, the cable current–voltage relation must be nonlinear. However, the cable properties have a large influence on the shape of the input current–voltage relation (see Chapter 12) and on the value and nature of the excitation threshold. As a result of decrement of voltage along the cable, each area of membrane will be at a different point on the membrane current–voltage relation; the total current from the source will be equal to the sum of the currents flowing through all the areas of membrane.

Initially we will assume that the membrane obeys a time-independent current–voltage relation (cf. the simplified current–voltage relations discussed in Chapter 8) and that this relation $i_i(V)$ is identical for all points on the cable. Consider one point ($x = 0$) at which the voltage and axial current are initially V_0 and $(i_a)_0$, respectively. The variation of voltage and current with distance must satisfy the equations (see Chapter 3)

$$\mathrm{d}i_a/\mathrm{d}x = -i_i(V) \tag{3.3a}$$

and

$$\mathrm{d}V/\mathrm{d}x = -r_a i_a, \tag{3.2a}$$

where ordinary derivatives are used since we are considering the steady state when there are no variations with time.

Now let the voltage and current at $x = 0$ be increased by increments δV and δi_a. Provided that the increments are sufficiently small to neglect higher-order terms, we have

$$\delta i_a = -i_i(V)(\delta x)_1 \tag{3.3b}$$

and

$$\delta V/i_a = -r_a(\delta x)_2, \tag{3.2b}$$

where $(\delta x)_1$ is the distance at which the current is now equal to $(i_a)_0$ and $(\delta x)_2$ is the distance at which the voltage is now equal to V_0. We now show that in the case of the infinite cable $(\delta x)_1 = (\delta x)_2$. This may be done simply by noting that at locations positive to $(\delta x)_1$, the cable in the direction of increasing values of x is in the same state as the cable starting at $x = 0$ before the application of the current increment. This must be the case since the membrane characteristic is independent of x, and because the cable is infinite so that there are no effects due to terminations to be taken into account. Hence the voltage at $(\delta x)_1$ must equal V_0 and $(\delta x)_1$ must equal $(\delta x)_2$. Therefore we may divide eqn (3.3b) by eqn (3.2b) to obtain

$$i_a \delta i_a/\delta V = i_i/r_a$$

or, letting $\delta V \to 0$,

$$i_a \mathrm{d}i_a/\mathrm{d}V = i_i/r_a. \tag{9.1}$$

It is important to note that this result does not apply to terminated cables. Thus, if the cable is terminated at $x = l$, then at $(\delta x)_1$ the cable in the direction of increasing values of x is no longer equivalent to the cable starting at $x = 0$ before the application of the current increment. In fact, it is shorter by the distance $(\delta x)_1$ so that the termination effects (i.e. those due to the 'reflection' terms—see Chapter 4) will no longer be identical. Hence V at $(\delta x)_1$ after the current increment will not equal V_0, so that $(\delta x)_1 \neq (\delta x)_2$. The appropriate equation relating i_a and i_i may not then be obtained directly from eqns (3.2b) and (3.3b). The derivation of equations for terminated cables will be given in Chapter 12 (p. 386).

Eqn (9.1) was first obtained and used in electrophysiology by Cole and Curtis (1941), and is sometimes called the Cole theorem. It has usually been used to obtain the relation $i_i(V)$ from experimental measurements of V and i_a at one point (near the point of current application where $i_a = \frac{1}{2}I$). Provided that the measurements are not made during a regenerative excitation (i.e. during an action potential), the applicability of eqn (9.1) does not depend on the degree of nonlinearity involved. This may appear surprising since the magnitudes of δx in eqns (3.2b) and (3.3b) will vary with the degree of non-linearity in the $i_i(V)$ relation. Thus, when i_i decreases on depolarization (i.e. the current–voltage relation may exhibit inward-going rectification or the voltage may be approaching an excitation threshold), the voltage will decrement more slowly so that the distance $(\delta x)_2$ required for a given decrement δV will increase. The important point to note is that, in an infinite cable, the distance required for the current decrement $(\delta x)_1$ will increase by exactly the same amount. Hence δx may always be eliminated as in the derivation of eqn (9.1).

Voltage threshold

For the present purpose, we require the inverse relation for obtaining $I\ (= 2i_a$ at $x = 0)$ from a theoretical $i_i(V)$ relation. This may be obtained (cf. Noble and Stein 1966) by integrating eqn (9.1) to give

$$\int_{V_r}^{V} i_a \frac{di_a}{dV}\, dV = \frac{1}{r_a} \int_{V_r}^{V} i_i\, dV, \qquad (9.2)$$

where V_r is the resting potential (cf. Fig. 8.11, where the resting potential is E_r). This equation may be rewritten

$$i_a^2 - i_0^2 = \frac{2}{r_a} \int_{V_r}^{V} i_i\, dV, \qquad (9.3)$$

where i_0 is the value of i_a when $V = V_r$. In our case, $i_0 = 0$ (i.e. no current is required to keep the cable at the resting potential) so that

$$I = 2\left(\frac{2}{r_a} \int_{V_r}^{V} i_i\, dV\right)^{\frac{1}{2}}. \qquad (9.4)$$

In Chapter 12 we shall consider analytical solutions of this equation for a variety of nonlinear $i_i(V)$ relations. The present discussion will be limited to some general properties relevant to the threshold conditions. We may, in fact, draw some immediate conclusions from eqn (9.4).

1. Since I depends on an integral of the $i_i(V)$ relation, the cable input current–voltage relation will generally be less nonlinear than the $i_i(V)$

relation itself. In fact, if the nonlinearities in the membrane current–voltage relation are not very marked, the cable current–voltage relation may appear almost linear.

2. Since I becomes zero when the integral term is zero, the voltage threshold for cable excitation, expressed in terms of the voltage at the point of current injection, must be larger than the threshold for uniform membrane polarization as shown in Fig. 9.1. Moreover, provided that the $i_i(V)$ relation is independent of time and the system is allowed to approach a steady state before excitation occurs, the voltage threshold V_{Th} will be given by

$$\int_{V_r}^{V_{\mathrm{Th}}} i_i \, \mathrm{d}V = 0. \tag{9.5}$$

If there is no voltage V_{Th} which satisfies this equation, excitation cannot occur. It should be noted that this means that excitation may fail to occur even when the membrane current–voltage relation has a negative chord conductance region. The important criterion is the magnitude of the integral of this region relative to that of the positive conductance region immediately positive to the resting potential. It is possible, therefore, for a membrane to be in a state such that it may be excited by current applied uniformly but cannot be excited nonuniformly. These differences between uniform and nonuniform excitation become even more marked in multi-dimensional systems (see p. 385).

Eqn (9.5) may be more readily understood in terms of the corresponding curves for voltage and current as functions of distance from the current source. Analytical solutions for these curves will be derived in Chapter 12,

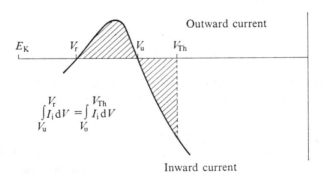

FIG. 9.1. Relation of voltage threshold for cable excitation (V_{Th}) to membrane current–voltage relation. V_r is the resting potential and V_u is the voltage threshold for excitation by uniform polarization. V_{Th} is given by the point at which the integral $\int_{V_r}^{r} i_i \, \mathrm{d}V$ becomes zero. Note that this is the *minimum* value for the voltage threshold. When brief currents are applied, the voltage at the point of excitation may initially exceed V_{Th}, as shown in Fig. 9.3. (From Noble 1966.)

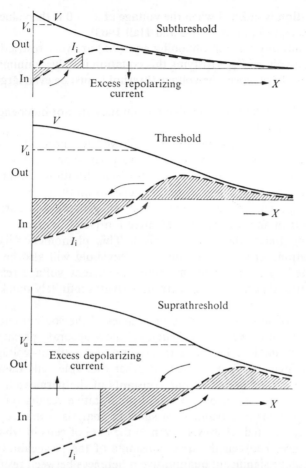

FIG. 9.2. Spatial variation in membrane current for subthreshold, threshold, and supra-threshold depolarizations. When the depolarization is subthreshold, the amount of outward (repolarizing) current generated by areas of membrane polarized below V_u exceeds the amount of inward current generated near the polarizing electrode. Threshold is reached when the amounts of current are equal. A suprathreshold depolarization produces excess inward (depolarizing) current. (Noble 1966.)

but it is helpful at this stage to give a semiquantitative analysis of the kind illustrated in Fig. 9.2. This shows the voltage and current curves correspond-ing to subthreshold, threshold, and suprathreshold currents, applied at $x = 0$. It can be seen that the condition corresponding to eqn (9.5) is that the integral of current with respect to distance should be zero, so that no further current is required from the stimulating electrode, i.e.

$$\int_0^\infty i_i \, dx = 0. \tag{9.6}$$

This equation is satisfied when the voltage at $x = 0$ is the value V_{Th} which also satisfies eqn (9.5) (see Noble and Hall 1963).

The assumptions made in obtaining eqn (9.5) are very restrictive, and the existence of a voltage V_{Th} satisfying this equation is only a minimal condition for excitation. In practice, the voltage threshold must be even larger than V_{Th} since:

(a) the $i_i(V)$ relations in excitable membranes are not independent of time (cf. Chapter 8) and

(b) the assumption that the cable is allowed to approach a steady state before excitation occurs will not be valid for short-duration stimuli and will only be approximately valid for long-duration stimuli.

The importance of the first restriction depends on the speed with which the current–voltage curves change with time. In general, the current–voltage relations shift in an outward current direction (see Chapter 8, Fig. 8.12) so that, at long times, the threshold rises. This phenomenon is known as accommodation. At very short times the threshold will also be larger than that predicted by a simplified (momentary) current–voltage relation, since the assumption that the sodium current activates infinitely quickly will then be invalid.

In the case of squid nerve, the time courses of the sodium and potassium activation reactions overlap too much to allow accurate predictions of the excitation threshold to be made from simplified current–voltage relations (Noble and Stein 1966). However, in other excitable cells there are much larger differences between the time constants of the permeability reactions, and it is then possible to obtain useful quantitative results (cf. Noble and Hall 1963). The major advantage of such an analysis is that relatively little computing is required. However, even in the case of cells in which there are large differences between the time constants of the permeability reactions, there may still be significant quantitative differences between results obtained using an approximation of the kind involved in using simplified current–voltage relations and results obtained by solving the full cable equations (e.g. Hall and Noble 1963). These differences are attributable to the second restriction mentioned above.

This restriction becomes severe at short stimulus durations when the cable deviates to a great extent from steady state conditions. The charge applied to the cable will be concentrated near the stimulating electrode so that application of the same quantity of charge will raise the voltage further at $x = 0$. At the termination of a very brief stimulus the charge will rapidly redistribute itself by flowing away from the current source, so that the potential will fall rapidly before excitation occurs (cf. Fig. 9.3). If we assume (as will be shown in the section *Charge threshold*) that the threshold condition is determined by the amount of charge applied to the cable, the voltage threshold, measured as the voltage at the stimulating electrode at the termination of a threshold stimulus, will rise steeply as the stimulus duration is decreased.

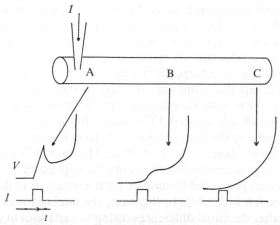

FIG. 9.3. Diagram illustrating charge redistribution which occurs following brief current pulses. Near the electrode (A) the voltage rapidly rises to a value larger than the threshold for long duration currents. It then rapidly falls as charge flows along the cable away from the current electrode. The potential may then change fairly slowly before excitation occurs. At a certain distance (B) away from the current source the voltage will show a small hump, while at a long distance away (C) it will simply rise smoothly. At this distance virtually no charge flows during the current pulse and the depolarization occurs as a consequence of cable propagation of the excitation. The voltage waveform then approaches the shape obtained for a uniformly propagating action potential (see Fig. 10.2).

Charge threshold

Hodgkin and Rushton (1946) suggested that the threshold condition for excitation of a cable is that a constant amount of charge should be applied rather than that the voltage near the stimulus should reach a particular value. The proof (cf. Noble and Stein 1966) makes use of the superposition principle (see Chapter 13). We may represent a rectangular current pulse of duration δt as the sum of two step currents of infinite duration but opposite in sign, displaced with respect to each other by the time δt. If superposition holds the response to the current pulse follows directly from the step current response. As already shown in Chapter 3, as δt becomes small we obtain

$$V = Q_0 \, \partial V_1 / \partial t, \tag{3.49}$$

where $Q_0 \, (= I\delta t)$ is the amount of charge applied and V_1 is the response to a long-lasting current step of unit intensity. For short-duration pulses, therefore, the voltage at all points will depend only on the applied charge and not on the strength or duration of current provided that these are always related by $Q_0 = I\delta t$, i.e. provided that the intensity of current is reduced as the duration is increased.

In order that the superposition principle holds, the voltage response should scale linearly with the applied current in the cable equation

$$\frac{1}{r_a} \frac{\partial^2 V}{\partial x^2} = c_m \frac{\partial V}{\partial t} + i_i + I(t, x = 0), \tag{9.7}$$

which is the equation previously derived (eqn (3.11)), with the addition of the current term $I(t, x = 0)$ corresponding to the applied current at $x = 0$. The membrane current term i_i is generally nonlinear. However, since the membrane chord conductance is bounded i_i must become negligibly small since the rapid changes in charge distribution on the cable at short times will ensure that the differential terms are very large at small values of δt. At early times, therefore, the voltage at each point should depend only on Q_0 and not on the way in which it is applied. Of course, after a certain period of time (determined largely by the time constants of the m reaction—see Chapter 8) the nonlinearities will begin to have an effect. However, it is possible that by this time the charge redistribution following the application of a fixed value Q_0 will have already proceeded far enough for the responses to different values of I to converge (cf. Fig. 3.14). In this case, the nonlinearities will begin to exert an effect after the initial differences owing to variations in the pattern of charge application have disappeared.

It is clear from eqn (3.49) that this must be the case for very short pulses. Moreover, provided that charge redistributes sufficiently quickly after the termination of the pulse, we may expect the responses to converge before nonlinear effects become apparent even when the current durations are too long for eqn (3.49) to apply accurately. The time factor for charge redistribution may be obtained by setting the current terms in eqn (9.7) to zero and rearranging to give

$$\frac{\partial^2 V}{\partial x^2} = \frac{\partial V}{\partial(t/r_a c_m)} \tag{9.8}$$

from which it is clear that the product $r_a c_m$ is important. This is expected since charge redistributes on the capacitance c_m by flowing through the resistance r_a. However, the product $r_a c_m$ is not in the form of a time constant since its units (see the section *Notation and definitions*, p. xiii, and Chapter 3, p. 42) are inappropriate (time/distance²). In its reciprocal form it has the same dimensions as a diffusion coefficient (distance²/time). We can convert this into a time factor in various ways. For example, we can normalize distance by the fibre radius a (in an analogous way to that done in Chapter 6, where radial distance r was normalized in terms of the fibre radius, i.e. $R = r/a$). Eqn (9.8) can be rewritten

$$\frac{\partial^2 V}{\partial(x/a)^2} = \frac{\partial V}{\partial(t/a^2 r_a c_m)} = 2 \frac{\partial V}{\partial(t/aR_i C_m)}. \tag{9.8a}$$

Therefore, the time factor for spread of charge in a one-dimensional cable, with distance normalized by the fibre radius, is $aR_i C_m$.

On the other hand, it was shown in Chapter 3 that the spatial spread of voltage scales as $\sqrt{a}$ so that the first term in eqn (9.8) will only be comparable for fibres of different sizes if the distance is normalized with respect to a parameter that varies as $\sqrt{a}$. The choice of normalizing parameter is important since, as we shall see, it determines the magnitude of the time factor for charge redistribution. This must be so since the time taken for charge to flow along the fibre must increase as the distance involved increases.

Thus, for a passive cable, it might be appropriate to choose the resting space constant λ. Then, by multiplying eqn (9.8) by λ^2 we obtain

$$\frac{\partial^2 V}{\partial X^2} = \frac{\partial V}{\partial (t/r_a c_m \lambda^2)} = \frac{\partial V}{\partial T} \tag{9.8b}$$

from which we obtain a time factor $r_a c_m \lambda^2$, which is the membrane time constant τ_m. In the form $r_a c_m \lambda^2$ this constant can also be seen to correspond to the 'time constant' for the charging or discharging of the capacitance per space constant $c_m \lambda$ through the axial resistance per space constant $r_a \lambda$. The charging of one space constant with a time constant τ_m may be compared with the velocity of 'propagation' of passive and 'active' potentials (see Chapter 3, p. 34 and Chapter 6, p. 117). It would be misleading to think of $r_a c_m \lambda^2$ as a simple time constant, however, because the redistribution process is not a simple exponential one. It does, however, indicate the order of magnitude of the time taken for the process to occur.

The length of fibre over which the relevant charge distribution process occurs in an excitable cell is not necessarily of the same order of magnitude as the resting space constant. A more relevant length is that over which the excitation process occurs at threshold, i.e. the liminal length for excitation x_{LL} (see Rushton 1937; Noble 1972; Chapter 12, p. 413), since it is the charge distribution time for this length that determines whether convergence towards a common spatial distribution following threshold currents of different durations occurs within the time taken for the excitatory sodium conductance to activate. We may obtain the time factor in this case by normalizing the distance with respect to x_{LL}. (Note that this parameter scales as $\sqrt{a}$, as required, since it is proportional to the threshold 'space constant' λ_B—see Chapter 12, eqn (12.36), noting that $x_{LL} = \lambda X_{LL}$). Eqn (9.8) then becomes

$$\frac{\partial^2 V}{\partial (x/x_{LL})^2} = \frac{\partial V}{\partial (t/r_a c_m x_{LL}^2)}, \tag{9.9}$$

and the relevant time factor for charge redistribution is of the order of $r_a c_m (x_{LL})^2$.

Using values of r_a, c_m, and x_{LL} for squid nerve and for cardiac Purkinje fibres, Noble (1972) obtained time factors of the order of 100 μs, similar to the delay in the activation of the sodium conductance. Notice also that the time factor for charge redistribution is much smaller than the resting membrane time constant, which is usually measured in milliseconds.

The range of stimulus durations for which the charge threshold is constant will depend on a number of factors, including the cable constants and the membrane nonlinearities, which may vary greatly between excitable cells.

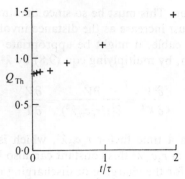

Q_{Th}

t/τ

FIG. 9.4. Dependence of charge threshold on duration of current pulse. At short durations, the charge threshold is nearly constant. When the current duration is increased beyond about $T = 0.5$ the charge threshold progressively increases. (Noble and Stein 1966.)

Although the analysis we have described so far is of use in understanding the nature of the excitation threshold, it will be necessary in particular cases to obtain numerical solutions to eqn (9.7). In general, this is done by replacing eqn (9.7) by a corresponding difference equation in δx and δt which is then solved numerically on a computer (Hall and Noble 1963; Cooley, Dodge, and Cohen 1965; Cooley and Dodge 1966; Noble and Stein 1966). Fig. 9.4 shows the computed charge threshold as a function of stimulus duration for the case of squid nerve using the original Hodgkin–Huxley equations for i_i. It can be seen that the charge threshold is virtually constant for about one-third of a time constant and then increases as the stimulus duration is prolonged.

The strength–duration curve

It was shown in Chapter 3 that the total charge on a cable increases exponentially in response to a step current applied at one point (eqn (3.39)). In Chapter 5 it was shown that this result also applied to more complex geometries. If we assume that the charge threshold is strictly constant, then the magnitude and duration of just-threshold currents must be related by the equation

$$Q_{\text{Th}} = \tau_m I \{1 - \exp(-t/\tau_m)\}, \tag{9.10}$$

where Q_{Th} is the charge threshold. We may rearrange this equation to give

$$I = \frac{Q_{\text{Th}}/\tau_m}{1 - \exp(-t/\tau_m)}. \tag{9.11}$$

Q_{Th}/τ_m is a constant and is equal to the amount of current required to maintain the charge Q_{Th} in the steady state, i.e. the just-threshold current at very long current durations. This current is called the rheobasic current I_{Rh}.

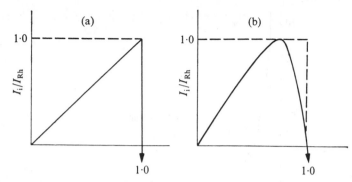

FIG. 9.5. (a) Current–voltage relation that must be assumed for eqn (9.12) to be obeyed accurately. The current increases linearly up to the voltage threshold (defined here as $V = 1$) and then suddenly become inward (cf. the switch-line model described in Fig. 10.3) (b) The current–voltage relations obeyed by excitable cells are linear only for small depolarizations. The peak current is reached before the voltage threshold is reached (cf. the experimental relations shown in Fig. 12.19). (From Noble and Stein 1966.)

Hence

$$I = \frac{I_{Rh}}{\{1 - \exp(-t/\tau_m)\}}, \tag{9.12}$$

which is the classical strength–duration equation (Lapicque 1907).

The assumptions made in deriving eqn (9.12) are very severe. In fact, the only system for which eqn (9.12) is strictly obeyed is one in which the membrane current–voltage relation is virtually linear up to threshold and in which the threshold is extremely sharp (see Fig. 9.5). Moreover, it must also be assumed that no accommodation occurs. In view of these restrictions it is unlikely that eqn (9.12) will be obeyed very accurately by excitable cells. Nevertheless it is very striking that an equation of the form of eqn (9.12) *is* obeyed, at least approximately, in virtually all cases. We shall consider now the reasons for this surprising result by discussing the effects of relaxing the assumptions made in deriving eqn (9.12).

Effects due to membrane nonlinearities. The effects due to membrane nonlinearities may be isolated from other effects by considering a uniformly polarized membrane. Noble and Stein (1966) have derived the strength–duration equation for a uniformly polarized membrane with a nonlinear current–voltage relation given by a fairly simple function (see Fig. 9.6). They found that this relation has almost the same *shape* as that for a linear membrane. However, the time constant in the strength–duration curve is no longer equal to the membrane time constant τ_m. They therefore defined a strength–duration time constant τ given by

$$\tau = \lim_{t \to 0} It/I_{Rh}. \tag{9.13}$$

In general, the relation between this time constant and that of the membrane is complex. It depends partly on the nature of the membrane nonlinearities

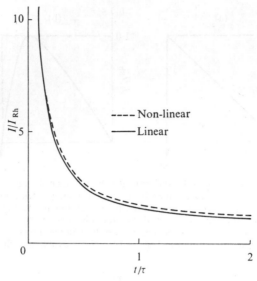

FIG. 9.6. Effect of nonlinearity on the shape of the strength–duration curve. The strength–duration curves are plotted on dimensionless coordinates in order to compare the shapes rather than the absolute values of time-constants or thresholds. I is expressed as a fraction of I_{Rh} and t is expressed as a fraction of the strength–duration time constant given by eqn (9.13). The nonlinear curve was computed using a sinusoidal relation for I as a function of V between $V = 0$ and $V = V_{Th}$. Note that the shape change in the strength–duration curve is quite small (Noble and Stein 1966).

and partly on the geometry of the system (see *Effects due to cable properties* and Chapter 12). In general, therefore, experimentally determined strength–duration curves do not give reliable information on the true membrane time constant.

Effects due to membrane accommodation. As already noted previously, the threshold in excitable cells rises at long times after the application of a current pulse as a consequence of the progressive shift of the momentary current–voltage relation in an outward current direction (Chapter 8, Fig. 8.12). That this phenomenon changes the form of the strength–duration curve has been known for a long time, and Hill (1936) and others have attempted to give a quantitative analysis of the effect. Hill assumed that the accommodation process is simply a rise in voltage threshold as the potential approaches threshold. Using a fairly simple functional relation between the time course of membrane potential and the rise in threshold, Hill developed a modification of the classical strength–duration equation which included two time constants. This model was very successful since all the known strength–duration curves could be fitted and the two time constants (one of which is related to the accommodation process) could be calculated. Moreover, the model was attractive since it allowed a fairly quantitative measure of the accommodation process to be obtained from the strength–duration curve.

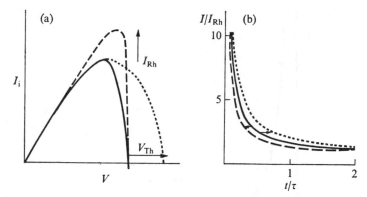

FIG. 9.7. Diagrams comparing the effects of increases in I_{Rh} and V_{Th} (a) on the shape of the strength–duration curve (b). An increase in I_{Rh} with little or no change in voltage threshold (that is, an increase in i_K in a cell with a sharp and virtually constant sodium threshold) steepens the strength–duration curve. An increase in voltage threshold (that is, a shift in sodium threshold with no change in i_K) makes the curve more shallow. If both effects are assumed to occur simultaneously (as in Hill's model of accommodation) large changes in threshold can occur without greatly influencing the shape of the strength–duration curve (Noble and Stein 1966).

Unfortunately, it is now clear that the success of this model is, to a great extent, fortuitous. In the first place, the nature of the change in shape of the strength–duration curve depends critically on whether the accommodation process is largely an increase in current threshold or largely an increase in voltage threshold (see Fig. 9.7). There is no simple relation between accommodation and the form of the strength–duration curve. Secondly, the deviation from the classical strength–duration curve predicted by the Hodgkin–Huxley equations for the case of a uniformly polarized membrane is actually outside the range of possible curves given by the Hill model (Cooley and Dodge 1966; Noble and Stein 1966). The predicted relation falls within the range of the Hill model only when nonuniform cable excitation is considered (see Fig. 9.8).

Effects due to cable properties. It has already been noted that, on theoretical grounds, the strength–duration time constant will not equal the membrane time constant in nonlinear systems. Experimental evidence has also indicated that there must be only a very tenuous relation between the two time constants, if any at all. Thus, the strength–duration time constant depends greatly on the way in which the current is applied to the cell (Davis 1923; Grundfest 1932; Fozzard and Schoenberg 1972), whereas the membrane time constant for small currents is a characteristic of the membrane and not of the stimulus parameters. This result suggests that the cable spread of current must also influence the shape of the strength–duration curve. This is indeed the case and the computed strength–duration time constant is shorter for cable excitation than for uniform excitation (Cooley and Dodge 1966; Noble and

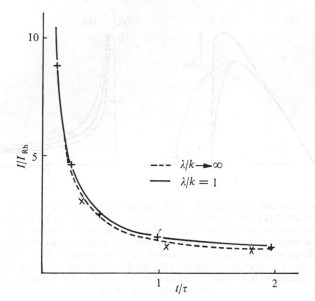

FIG. 9.8. Strength–duration curves for squid nerve polarized uniformly ($\times$) and at one point in a cable ($+$). The continuous and interrupted curves show the limits of the range of shapes given by Hill's model. Note that the cable results fall within the range of Hill's model. The uniform polarization results fall outside the range (the curve is too steep) which suggests that an increase in I_{Rh} (activation of i_K) predominates in the case of squid nerve (cf. Fig. 9.7) (Noble and Stein 1966).

It is clear from these curves that the effect of varying Hill's parameter λ/k from 1 (its lowest value—corresponding to maximum accommodation) to ∞ (no accommodation) is quite small when the strength–duration curves are plotted on dimensionless coordinates. This method of plotting the results conceals large changes in time constant and rheobasic current. It should also be emphasized that the differences between the two sets of computed points are also minimized. The absolute time constants are in fact different for the uniform and cable cases (see text). Differences of this kind are best represented in plots of the kind shown in Fig. 12.21.

Stein 1966; Fozzard and Schoenberg 1972). Although appropriate computations have not yet been done, there is no doubt that in a multi-dimensional system the strength–duration time constant for excitation at one point would be even briefer, since the voltage changes near the current electrode are then very fast indeed compared to the membrane time constant (see Chapter 5, Figs. 5.3 and 5.4).

Since the time course of total charge on the membrane is independent of the way in which current is applied (see Chapter 5), these results suggest that the assumption that the *total* charge on the membrane is the important parameter in determining the excitation threshold cannot be strictly correct. For a given quantity of total charge, some distributions may be more effective in producing excitation than others. That this should be the case is reasonable in the light of the nature of the membrane nonlinearities underlying excitation. In general, the quantity of excitatory sodium current and the speed with which it is activated both increase with increasing depolarization (Chapter 8, p. 243). Hence, for a given amount of charge applied to the system the most effective distribution of charge will be one in which a considerable fraction is concentrated in a small area, so that the membrane in this area is strongly depolarized. In this way the charge can activate a larger excitatory current than it would if it were more evenly distributed. The important parameter determining excitation is therefore not the total charge but rather the charge on an area close to the stimulating electrode (cf. Rushton 1937; and Chapter 12, pp. 417). Now, as shown in Chapters 3 and 5, the time taken to charge the local area of membrane is less as the system becomes increasingly multidimensional. Thus membrane potentials in the vicinity of an electrode polarizing a two-dimensional preparation (eqn (5.14)) rise more rapidly towards the steady state value than in a one-dimensional cable (eqn (3.24)) which, in turn, polarizes more rapidly near the electrode than does a uniformly polarized membrane (eqn (2.3)). The reason for this effect is that as the number of dimensions of current flow increases, the voltage levels off more rapidly to its steady state value since a larger fraction of current is drawn by distant regions of the preparation. In general, therefore, the strength–duration time constant will be briefer in the cable and two-dimensional cases than in the uniform case. This phenomenon will be discussed in more detail in Chapter 12 (Part IV).

It is worth noting that the time course of the *local* charge will not be strictly exponential, so that effects on the shape of the strength–duration curve due to redistribution of charge are to be expected in the cable and two-dimensional cases. In fact these effects are quite large. In the case of the squid nerve equations, they are sufficient to counteract some of the effects due to the large degree of accommodation in the Hodgkin–Huxley equations (Noble and Stein 1966).

It may now be seen that the strength–duration curves for a large variety of excitable cells are fairly well fitted by the classical strength–duration equation, and even more accurately by Hill's two time-constant modification, because of the interplay of a large number of factors. At present it does not seem possible to give a general treatment without relying on detailed numerical computations since the important parameters, (I_{Rh} and τ) are not simply related to single variables in the equations describing the membrane nonlinearities and the cable properties of the cell. Nevertheless, gross changes in the strength–duration parameters may be related in a semiquantitative way to the underlying cell properties. The dependence of τ on the geometry of the system has already been related above to the expected effects of charge redistribution. Another example which may be given a fairly simple interpretation is the large variation of τ with temperature (Cooley, Dodge, and Cohen 1965), which may be related to the temperature dependence of the time constants of the reactions determining the membrane permeability (Chapter 8, p. 241).

Influence of fibre size on excitation threshold

Threshold for current applied intracellularly

The total amount of current which must be applied in order to achieve excitation depends on the fibre size and on the way in which the current is applied. We will consider first the effect of fibre size on the threshold for current applied intracellularly at one point. This may be obtained by noting that increasing the fibre size a has two effects on the current distribution and density:

1. As already shown in Chapter 3 (p. 37), if the applied current is suitably scaled to give a particular voltage at $x = 0$, the pattern of voltage spread will be independent of fibre size provided that the longitudinal axis is scaled according to the factor $\sqrt{a}$. Thus a four-fold increase in a doubles the spatial spread of voltage. Hence a longer length of fibre is polarized and correspondingly more membrane current must flow.

2. The membrane area per unit length of fibre scales as the circumference, i.e. as a. Hence a four-fold increase in a reduces the current density by 75 per cent for a given value of i_a. In order to maintain the current density I_m, therefore, the membrane current per unit length i_m must be scaled as the factor a.

Hence the total applied current required to produce a given voltage pattern must scale as $a\sqrt{a} = a^{\frac{3}{2}}$. This result may also be obtained from eqn (9.4) which may be rewritten

$$I = 2\left(\frac{2\pi a^2}{R_i} \int_{V_r}^{V} 2\pi a I_i \, dV\right)^{\frac{1}{2}}$$

or

$$I = 4\pi a^{\frac{3}{2}}\left(\frac{1}{R_i} \int_{V_r}^{V} I_i \, dV\right)^{\frac{1}{2}}.$$

However, this proof is less general since, as noted in its derivation, eqn (9.4) applies to steady state polarization only.

In general, therefore, the current threshold will be much larger in large axons than in small axons. This result applies not only to applied currents but also to excitatory currents generated by synaptic or sensory mechansims.

Threshold for current applied extracellularly

If a given quantity of current is applied extracellularly to a muscle or nerve trunk containing fibres of different sizes, the current will not flow equally through all the fibres. The smaller fibres have a larger resistance to current flow, and so a smaller fraction of the total current will enter them. Provided that the stimulating electrodes are separated by a distance which is large enough for the current flow to be largely in the direction of the fibre axis then the main resistance to current flow will be provided by the axial resistance r_a.

This resistance is proportional to a^{-2} (Chapter 3 p. 27), so that the relative current flow will scale as a^2. Combining this result with that given above for the amount of intracellular current required to excite, we obtain the result that the relative thresholds for initiating action potentials by extracellular stimulation will vary as $a^{\frac{3}{2}} a^{-2}$. Hence the threshold will vary inversely as $\sqrt{a}$ so that, in contrast to the result obtained for intracellular stimulation, the larger fibres will have lower thresholds than the smaller fibres to applied external currents.

These results apply only if it is assumed that the specific membrane properties are independent of fibre size. As will be discussed later (p. 295) this assumption may not be valid in the case of very small axons when it is found that the ionic currents underlying excitation are less intense than in larger axons. This factor will further increase the difficulty of exciting small axons with extracellularly applied current. It should be noted also that the results have been obtained for non-myelinated fibres. For myelinated fibres, the same qualitative arguments apply but, in order to obtain quantitative relations, the geometry and electrical properties of the myelin sheath must also be taken into account, since the over-all fibre diameter includes a substantial thickness of myelin (see p. 296). The results obtained will depend partly on whether the internal or external fibre radius is used. At present, the quantitative aspects of myelinated nerve excitation and conduction are not sufficiently well known to derive simple relations comparable to those which may be obtained for non-myelinated fibres.

10. Nonlinear cable theory: conduction

ONCE the threshold for excitation has been exceeded, an all-or-nothing potential wave propagates from the point of excitation to other areas of the cell. The theory that this propagation is mediated by cable current flow from excited to resting regions was suggested at the turn of the century by Hermann. However, it was not until very much later (Hodgkin 1937) that direct experimental proof became available, and a quantitative theory of conduction based on experimental measurements of the membrane current and the cable constants did not emerge until 1952 (Hodgkin and Huxley 1952*d*). Full numerical solutions for excitation and conduction using eqn (9.7) appeared relatively recently (Cooley, Dodge, and Cohen 1965; Cooley and Dodge 1966; Noble and Stein 1966).

In this chapter we shall give an account of the conduction process. This will be done initially at a qualitative level. We shall then discuss the determinants of conduction velocity from a quantitative point of view.

The direction of the membrane current and its relation to potential changes

Before explaining the current flows during propagation of the action potential, it may be helpful to clarify the general relationship between current flows and potential changes. This relation is sometimes found confusing, since it appears possible for currents flowing in opposite directions across the membrane to change the membrane potential in the same direction. Thus, an *outward* stimulating current must be applied to excite the cell, whereas an *inward* sodium current then continues the depolarization. This fact is puzzling unless it is understood that there is an important difference between an applied current originating in an external circuit, or from other regions of the cell, and current which is generated by the local membrane e.m.f..

The potential is a measure of the net difference of charge between the inside and outside of the cell. This difference exists between fluids separated by the dielectric layer formed by the cell membrane. The thickness and dielectric constant of this layer determine the membrane capacitance (Chapter 2), which in turn determines how much charge separation must occur to generate a particular potential:

$$\Delta V = \Delta Q / C_{\mathrm{m}}. \tag{10.1}$$

The direction of potential change is determined by the change in Q with time, that is, the capacity current I_{c}:

$$\partial V / \partial t = I_{\mathrm{c}} / C_{\mathrm{m}}. \tag{10.2}$$

In order therefore to change the membrane potential in a positive direction a positive capacity current must flow. This involves the addition of positive charge to the inside of the cell and may be achieved in either of two ways.

1. Current may be applied to the cell from an external circuit via an electrode. In this case, in order to change the cell potential in a positive direction, it is necessary to apply a positive current to the intracellular fluid. Initially this current will simply add charge to the inside of the cell and so depolarize it. As the potential change occurs, current will also flow out across the cell membrane resistance (see Chapter 2, Fig. 2.4). This current is an outward ionic current. However, it is important to note that it is not this outward current which depolarizes the cell. The situation is rather the other way round: an outward ionic current flows because the cell is depolarized (we are assuming here that it has not yet been depolarized beyond the threshold for generating inward ionic current). This point may become clear by noting that the application of a positive current to the inside of the cell would depolarize it even if the membrane resistance were infinite so that no outward ionic current could flow. The membrane current would then be entirely capacitive and the change in the membrane potential would be simply proportional to the amount of charge applied, as in eqn (10.1).

An applied positive current may arise from sources other than an external circuit connected to the cell via electrodes. Current flow from another region of the cable which is already depolarized will have the same effect (see p. 280).

2. Charge may also be transferred across the membrane as current flow generated by the local membrane e.m.f.. Thus, when the sodium permeability of the membrane is increased, sodium ions flow inwards down their concentration gradient. Provided that this current is not collected (e.g. by an intracellular electrode forming part of a voltage-clamp circuit) it will add positive charge to the inside of the cell and so depolarize it. In this case, the ionic current flow inwards across the cell membrane is used to change the charge on the local membrane capacity. Since it does so in the same direction as an applied outward current, the effect corresponds to an opposite flow of capacity current to that of ionic current. The capacity current flow produced by the inward sodium current is therefore outward. It is important to note, however, that ions do not leave the cell as a consequence of this outward capacity current, any more than they do during the application of a positive current from an external source to a cell with an infinite membrane resistance. In both cases, the outward capacitive flow represents the accumulation of positive charge inside of the cell.

It may help to understand these points to consider a simple electrical model of an excitable cell as shown in Fig. 10.1. An external current generator is connected to a model consisting of a resistance and capacitance in parallel. In addition, to represent the sodium current system, a battery (the sodium potential) and resistance (the sodium resistance) are also placed in parallel with the capacitance. This element of the circuit is controlled by a voltage-sensitive switch (corresponding to the sodium channel 'gates' discussed in Chapter 8) which closes when the level of depolarization reaches threshold.

When a positive current is applied to the 'inside' of the model, positive

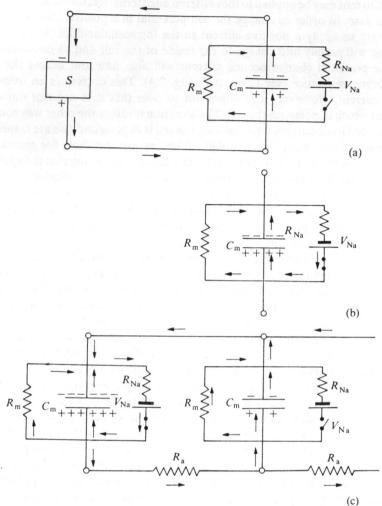

(a)

(b)

(c)

FIG. 10.1. Simple electrical circuits representing excitation and conduction. (a) An external current generator applies charge to the capacitance of the model circuit. When the potential across the circuit reaches a threshold value, the switch on the sodium circuit closes and the sodium battery passes current into the cell. (b) The external current generator is removed. The potential change on the capacitance continues as current flows through the sodium circuit and replaces the external current generator as the source of excitatory current. (c) The 'excited' circuit is connected to a 'passive' circuit and now acts like the external current generator in applying current to the passive circuit. Thus, the excitation may propagate from circuit to circuit. Since the sodium current in the 'excited' circuit is the only inward current flowing, R_{Na} must be low (i.e. g_{Na} high) in order to allow sufficient sodium current to flow to continue charging the local capacitance and to excite the passive circuit. Hence, there will be a minimal value of g_{Na} below which propagation cannot occur (see Fig. 10.11).

charge accumulates on the inside of the membrane capacitance. The potential changes in a positive direction and, as a consequence, outward current flows across the membrane resistance. When the threshold voltage is reached the voltage dependent switch closes and an inward current is generated across the sodium resistance. It is assumed that the external current then terminates, that is, the stimulating current is just threshold. The inward flow of current across the sodium circuit now takes over the function of the external circuit in applying positive charge to the inside of the membrane capacitance. Moreover, it is evident that, if this circuit were now connected to a similar circuit which has not yet been excited, the current flowing across the sodium circuit could also apply positive charge to the capacitance of the resting circuit. The resting circuit would then also become excited and so a wave of excitation could be passed on from circuit to circuit. In fact, such an arrangement of serially connected circuits can be used to represent the propagation process (see p. 286).

Current flows during propagated action potential

We may now consider the currents flowing during a propagated action potential. These are shown in Fig. 10.2. This figure is not an accurate solution of the equations for a propagated action potential. It has been drawn to illustrate the important features. The details will vary depending on which kind of excitable cell is being considered.

It is assumed that the action potential is conducting from right to left and that the site of initiation is far away to the right, so that the action potential is already conducting as a wave of constant shape and velocity. The first sign of arrival is an exponential rise in potential known as the 'foot' of the action potential. As indicated by the current diagram at the bottom of Fig. 10.2 this phase corresponds to the flow of outward current resulting from current flowing from the active regions of membrane further to the right. During this phase, the active regions are acting as an 'external circuit' applying depolarizing current to the resting regions ahead. The current flow is almost entirely capacitive, and the membrane conductance and ionic current are both very small. This phase of the action potential has already been analysed in Chapter 6 (p. 116).

After a certain period of time, which is determined by the time constants of the m reaction (see Chapter 8), the sodium conductance rapidly increases. The ionic current becomes large and inward. This inward flow of sodium current then further depolarizes the membrane.

As the sodium conductance inactivates and the potassium conductance increases, the ionic current becomes less inward. It should be noted, however, that repolarization begins before the ionic current becomes outward. The reason for this is that, in addition to adding charge to the local membrane capacity, the inward flow of ionic current is also supplying local circuit current flow to regions of membrane which are already repolarizing. It

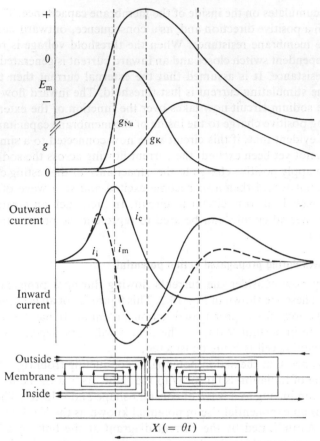

Fig. 10.2. Currents flowing during propagation of action potential. The top curve shows the membrane potential. Immediately below are shown the changes in g_{Na} and g_K. The middle diagram shows the changes in i_m and its components, i_i and i_c. The bottom diagram is a schematic representation of the local circuits during propagation. The wave is propagating from right to left and, as noted in the text, the abscissa may also be regarded as time since $x = \theta t$. The interrupted vertical lines connect the points corresponding to the maximum depolarization rate ($i_m = 0$, i_c at maximum), peak of action potential ($i_m =$ minimum, $i_c = 0$), and the maximum repolarization rate ($i_m = 0$, i_c at minimum). Further discussion in text. (From Noble 1966.)

continues to do so beyond the peak of the action potential, and for a period of time the local capacity and ionic currents flow in the same direction. Eventually the ionic current becomes outward, and the membrane potential returns towards its resting value.

It is worth noting that the current curves shown in this figure may all be obtained experimentally from measurements of the propagating action potential (cf. Cole and Curtis 1939). The capacity current is obtained by differentiating the action potential

$$i_o = C_m \, \partial V / \partial t. \tag{10.3}$$

Moreover, the total membrane current may be obtained from the second derivative since, when the wave conducts with a constant shape and velocity, the time and distance scales are proportional, the constant of proportionality being the conduction velocity (see the section *Equations for constant conduction velocity*, below). Hence eqn (3.5) becomes

$$i_m = (1/r_a\theta^2)\, \partial^2 V/\partial t^2. \tag{10.4}$$

The ionic current may then be computed as the difference between i_m and i_c. Of course, this analysis does not allow the ionic current to be separated into individual components, such as sodium and potassium currents.

It is now possible to compare the current flows during propagated and non-propagated ('membrane') action potentials. During a non-propagated action potential, i_m becomes zero as soon as the current stimulus is terminated. The ionic and capacitive currents are then always equal and opposite (see Chapter 8, p. 252, eqn (8.43)). This contrasts with the more complex time course of currents during the propagated action potential and, in particular, with the fact that during part of the repolarization phase the capacitive and ionic current flow in the same direction.

Equations for constant conduction velocity

In this section we shall consider the propagation of action potentials from a quantitative point of view and, in particular, we shall derive equations relating the conduction velocity to various cable and membrane parameters. In order to do this we shall consider the case of an action potential which conducts along a fibre at a constant conduction velocity θ. This will occur in a uniform cable when the response has propagated a sufficient distance away from the site of initiation for the boundary disturbances to become negligibly small. The shapes of the curves relating V to t, at constant x, and V to x, at constant t, are then identical. The partial derivatives of V with respect to x and t must then be simply related to each other. This relation may be obtained in the following way. Let

$$V = f(u),$$

where $u = x - \theta t$.

This relation expresses the fact that, when the wave conducts at a constant velocity θ, the voltage at a given point x_1 at time t_1 must be the same as the voltage at another point x_2 at time t_2, where $(t_2 - t_1)$ is the time taken for the wave to propagate from x_1 to x_2, i.e. $x_1 = x_2 - \theta(t_2 - t_1)$. Applying the chain rule and using the values $\partial u/\partial x = 1$ and $\partial u/\partial t = -\theta$, we may show that

$$\partial^2 V/\partial x^2 = \partial^2 V/\partial u^2, \quad \text{and} \quad \partial^2 V/\partial t^2 = \theta^2\, \partial^2 V/\partial u^2.$$

Hence

$$\partial^2 V/\partial x^2 = (1/\theta^2)\, \partial^2 V/\partial t^2. \tag{10.5}$$

Using eqn (10.5), eqn (3.12) becomes

$$\frac{a}{2R_i\theta^2}\frac{\mathrm{d}^2 V}{\mathrm{d}t^2} = C_m\frac{\mathrm{d}V}{\mathrm{d}t} + I_i. \tag{10.6}$$

This is an ordinary differential equation, and its solution requires very much less computing time than does eqn (9.7). The major disadvantage is that a value of θ is required in advance in order to solve the equation. The method used by Hodgkin and Huxley (1952; Huxley 1959) is to choose a trial value of θ, and solve the equations numerically. The trial value will almost certainly be wrong, and the solution for V as a function of t then becomes unbounded. It approaches $+\infty$ or $-\infty$ according to whether the guessed value is too high or too low. When the difference between the trial value and the correct value is large the solution becomes unbounded very quickly. As the trial value approaches the correct value, V remains bounded for an increasing period of time. By a process of successive approximation, therefore, the correct value of θ may be obtained to an arbitrary degree of accuracy.

Hodgkin and Huxley used this iterative method in their 1952 paper to carry out one of the most stringent tests then available of their equations for the membrane current of squid nerve. They obtained a predicted value of θ which was quite close to the experimental value. This agreement, together with the strong resemblance of the calculated action potential to experimental recordings, provides impressive support for the validity of their voltage-clamp analysis of the ionic currents (see Chapter 8).

Although the iterative procedure is the most general method for solving eqn (10.6), it suffers from an important drawback: the conduction velocity must be determined very accurately in order to obtain a finite solution for the complete action potential waveform. In practice, this degree of accuracy requires considerable computing time.

Semi-analytical methods for estimating conduction velocity

The difficulties of the full iterative procedure have stimulated efforts to determine θ by semi-analytical methods. The major aim is to obtain an equation which gives θ explicitly as a function of the fibre properties. Even if the estimate obtained were much less accurate than that given by the iterative method, it could still be very useful for physiological purposes. A more analytical approach might also clarify the interplay between the passive and active cell properties which generate the propagated spike.

One of the earliest examples of this approach was an ingenious analysis by Rushton (1937), who obtained analytical solutions for the conduction velocity in a cable whose membrane e.m.f. was assumed to change suddenly at the threshold voltage. No change in membrane resistance was assumed to occur. Following Cole and Curtis's (1939) observation that the transverse impedance of the squid axon falls forty-fold during the passage of an impulse, Offner, Weinberg, and Young (1940) modified this approach. The sharp change in e.m.f. at the voltage threshold was retained but the membrane conductance was also assumed to increase (see Fig. 10.3). The current–voltage relation of the circuit is discontinuous, but otherwise resembles the N-shaped curves which momentarily describe the properties of real excitable membranes

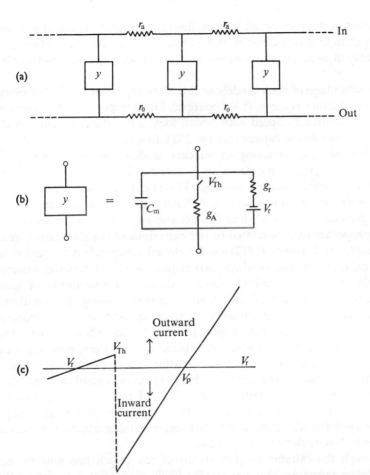

FIG. 10.3. The switch-line model of conduction. (a) general circuit in which the membrane is represented by elements y connected via internal r_a and external r_0 resistances. (b) Circuit for y used by Offner, Weinberg, and Young. Switch is assumed to close when the voltage reaches V_{Th}. (c) Representation of this circuit in terms of current–voltage diagram.

(see Chapter 8). The membrane patch circuits were incorporated into a distributed structure, which was capable of undergoing an all-or-nothing excitation. Offner, Weinberg, and Young obtained the following approximate formula for the conduction velocity,

$$\theta = \frac{K}{\sqrt{(1-K)}} \frac{1}{C_m} \left(\frac{g_A}{r_0+r_a} \right)^{\frac{1}{2}}, \qquad (10.7)$$

where

$$K = V_{Th}/(V_p - V_{Th}) \quad \text{and} \quad Kg_A \gg g_r.$$

This scheme is one example of a larger class of models which can reproduce non-decremental propagation. Following electrical engineering terminology, these models may be called 'switch lines', since their essential feature is a sudden change in conductance at a critical voltage threshold. More recently the theory of switch lines has been extended by Scott (1963, 1964) and Richer (1966).

One advantage of such models is that they represent the barest essentials of the conduction process. It is apparent, for example, that the propagation velocity is limited by speed with which local circulating current can charge the nearby membrane capacitance (p. 281). In switch-line models, the critical delay between the switching of successive elements must scale with the constant $(r_0+r_a)c_m$, since no delay is assumed in the closing of the switch once the threshold voltage is reached. This is not strictly the case in excitable cells, where the 'switching' delay may have important consequences (p. 432).

The periodicity of switch-line models also suggests how the specific membrane properties may be related to the behaviour of the distributed structure (cf. Chapter 6). Nasonov (1952) has developed a diagrammatic representation of the conduction process which makes full use of the periodic structure of excitable cells. The cylindrical fibre is divided into segments of identical length, which may each be described by a stimulus-response relation. The local circulating current produced by a given segment may be regarded as the stimulus for the next segment which, in turn, produces further local circulating current (response). Neglecting interactions between non-neighbouring patches, the condition of non-decrementing conduction is equivalent to a stimulus–response ratio of unity. FitzHugh (1969) has employed the Nasonov diagram to explain qualitatively the effects of temperature on conduction (see p. 289). He also points out how the stimulus–response ratio may be a useful way of describing decremental or incremental conduction, as well as the concept of 'safety factor' (see p. 424).

Although the extreme simplifications of the switch line may be helpful for didactic purposes, they are hardly likely to predict a realistic value of conduction velocity. Their most severe limitation is the assumption of a step change in membrane conductance, which generates a sudden change in the rate of depolarization of the propagating waveform.

Pickard (1966) has produced a more sophisticated treatment of nerve conduction which attempts to overcome this limitation. His model incorporates an explicit, continuous variation of sodium conductance, using an equation similar to that used in Chapter 3 (eqn (3.50)) to represent a transient flow of current. Pickard's equation is

$$g_{Na}(t) = (g_{Na})_r + g_0' t \exp(-\beta t), \tag{10.8}$$

where $(g_{Na})_r$ is the resting conductance, g_0' is the maximum rate of rise in conductance, and β is a positive constant. The rise in sodium conductance is assumed to begin when the voltage reaches a threshold V_{Th}. In order to use eqn (10.8), this time is defined as $t = 0$. The slope of the rising exponential foot is matched at V_{Th} to the slope of the subsequent depolarization generated by $g_{Na}(t)$. The boundary conditions at the threshold potential determine the conduction velocity. Pickard obtained an analytical solution for the membrane voltage as a function of time in the vicinity of $t = 0$. For the case of squid nerve, he was also able to derive an approximate formula (his equation 27) which gives θ explicitly as a function of g_0', C_m, r_a, r_0, and V_{Th}. Using values appropriate to squid nerve, the formula gives a reasonable estimate of θ although, as Pickard points out, the sensitivity of the estimate of θ to reasonable variations in the measured parameters makes it impossible to test the formula rigorously.

Although this agreement is encouraging, there are serious objections to the basic assumptions of this type of analysis. The most serious objection concerns the concept of 'threshold' and its applicability to a propagating action potential. Pickard uses a value of V_{Th} which is 25 mV positive to the resting potential. This might be an appropriate value of threshold for the initiation of an action potential in a point–polarized axon, but it is not at all realistic to use this value to define the voltage when the sodium conductance increases during a propagating spike. The model, like the switch-line models, requires that the sodium conductance should activate as soon as a particular value of voltage is reached. Until this occurs the system behaves as a linear cable in response to the current flow from excited regions and in this case, as shown previously (Chapter 6, p. 116), the voltage rises exponentially. As soon as the sodium conductance is activated, the system becomes nonlinear and the voltage no longer follows a simple exponential time course. In theory, therefore, the voltage at which the exponential time course ceases should give the value of V_{Th} required for Pickard's model. However, the value obtained experimentally in this way is very much more positive than any reasonable estimate of the voltage threshold for cable excitation. It usually lies near 0 mV, or at least 50 mV positive to the resting potential (see Rosenblueth *et al.* 1948; Hodgkin and Katz 1949*a*; see also Fig. 10.4). Pickard has also noted in a later paper (Pickard 1969) that the deviation from an exponential time course occurs at a much more depolarized potential than the one used in his formula for estimating the value of θ. Unfortunately, this imposes a serious limitation on the usefulness of the formula. Unlike a true threshold, V_{Th} in this approach is not simply related to the $i_i(V)$ relation (see Chapter 12, p. 425, and Fig. 12.23).

The large discrepancy between V_{Th} and the point at which the time course ceases to be exponential arises from the nature of the sodium permeability increase in real axons. Under normal conditions, the exponential depolarization is sufficiently rapid to make the time constant of m activation (see Chapter 8) a significant quantity. In contrast to Pickard's assumptions, the sodium conductance g_{Na} rises with a delayed sigmoid time course, and the degree of sigmoidness is increased by the cubic dependence of g_{Na} on m. Thus, the concept of a voltage 'threshold' is misleading in this context. As an operational definition (e.g. the voltage when an arbitrary degree of deviation

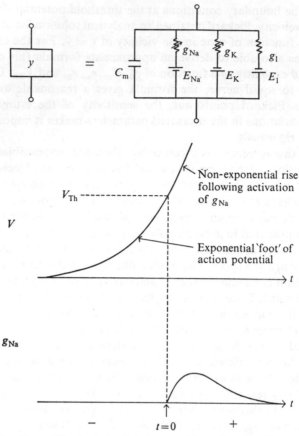

FIG. 10.4. Pickard's modification of switch-line model to allow for delay in activation of g_{Na}. As shown in bottom diagrams, the sodium channel is still assumed to be triggered when the voltage reaches a preset voltage V_{Th}, but the conductance does not switch on instantaneously.

from exponentiality occurs) it would still depend on a comparison of the activation time constant with the exponential time constant of the rising foot. The latter is, in turn, dependent on the conduction velocity itself (eqn (6.13)). There is little hope of avoiding the circularity in such a definition of 'threshold' since the propagation process itself is inherently cyclic. This criticism could be applied to any treatment which assumes an explicit time course for $g_{Na}(t)$ which is triggered by a 'threshold' voltage. Beyond the problem of defining a 'threshold' lies a potentially misleading division of the action potential upstroke into 'passive' and 'active' phases. It is tempting to suppose, for example that, in the 'active' phase, the maximal rate-of-rise is simply related to the maximal inward ionic current. This is certainly the case for a uniform membrane action potential. However, in the propagating action potential, $I_{i(max)}$ occurs some time after $(dV/dt)_{max}$ (see Fig. 10.2 and Fig. 12.24). Under varying experimental conditions (e.g. varying $[Na]_0$) the relative magnitudes of $I_{i(max)}$ and (dV/dt) will also depend on the extent of the lag between voltage changes and sodium conductance changes. One effect of the delayed rise in sodium conductance is to shift the peak inward ionic current toward the peak of the spike. This means that local circulating currents generated by the maximal ionic current will be more broadly distributed along the fibre, tending to flow away from the more immediate regions which have already been depolarized.

We shall return to a discussion of analytical approaches to the propagation process in Chapter 12 where we will discuss the use of models that represent the membrane current by simple polynomial functions of the membrane potential. These models avoid the assumption of sudden current changes inherent in switch-line models and they may be modified more readily to include time-dependence in order to reproduce the phenomena of activation and inactivation. Such modifications, however, also make it more difficult, though not impossible (p. 432), to obtain analytical solutions for propagation. In many cases that are of most interest from the experimental point of view, therefore, there is still no fully satisfactory substitute for the full iterative procedure using numerical computations.

Variation of conduction velocity with temperature

Huxley (1959) has used the iterative method to investigate the dependence of conduction velocity on temperature and on a number of other cable and membrane parameters. The effect of temperature may be computed relatively easily since, in squid nerve, all the time constants of the permeability reactions (m, n, and h—see Chapter 8) have virtually the same value of Q_{10}, namely 3. Huxley's result is shown in Fig. 10.5. θ is found to be an increasing function of temperature up to about 33 °C. Fairly close agreement between observed and predicted conduction velocity in the squid axon over the range 5–25 °C has been reported by Chapman (1967). Chapman also allowed for changes in R_i and the magnitude of the conductances with temperature, so that his curve has a steeper slope than that shown in Fig. 10.5 (filled circles). Above the value of 33 °C, there is no value of θ which satisfies the equations, signifying that an all-or-nothing response cannot be initiated. This corresponds to the condition of heat block observed experimentally by Hodgkin and Katz (1949).

This result arises from an important property of the ionic conductances which has already been described in Chapter 8. The absolute magnitudes of the ionic conductances are not very temperature-dependent. By contrast, the kinetic processes which 'gate' the conductances are highly temperature-dependent. Hence, when the temperature is increased, the rate of activation of the sodium and potassium conductances greatly increases, but the maximum conductances which may be activated do not change very much. The maximum rate at which the membrane capacity may be charged or discharged

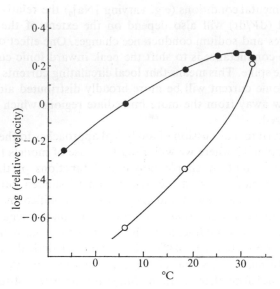

FIG. 10.5. Computed variation in conduction velocity with temperature. The filled circles are values for conduction velocity of the action potential expressed in terms of $\theta_T/\theta_{6.3}$, where $\theta_{6.3}$ is the conduction velocity (12·4 m s⁻¹) at 6·3 °C. (From Huxley 1959a.) The open circles show values obtained by Huxley (1959b) for the conduction velocity of the sub-threshold wave.

is determined by the value of the membrane capacity and of the ionic conductance. Once the latter has reached its peak, the rate of depolarization cannot be further increased. Hence, beyond a certain temperature, the rate of depolarization will not increase with temperature (Huxley 1959a, Fig. 19). However, the speed of onset of the repolarizing processes (activation of g_K and inactivation of g_{Na}) continues to increase so that the repolarization process occurs at an increasingly 'earlier' stage of the action potential as the temperature rises. The height of the action potential therefore falls, as shown in Fig. 10.6. Eventually, the repolarization process occurs too early for sufficient inward current to be generated for propagation and the conduction of an all-or-nothing response becomes impossible.

A surprising feature of Fig. 10.5 is that the curve relating θ to temperature curves back. Huxley (1959b) has shown that there are two values of θ at each temperature which satisfy eqn (10.6). The second value is much lower than the first, and the height of the response is also much smaller. This solution corresponds to the conduction of a subthreshold wave which would be highly unstable in a system which is capable of generating a large inward current during the full action potential. However, it may be possible for fibres which generate less excitatory current to conduct waves of this kind for some distance before they decay or 'blow up' into full action potentials (see the section *Subthreshold responses in nonlinear systems*).

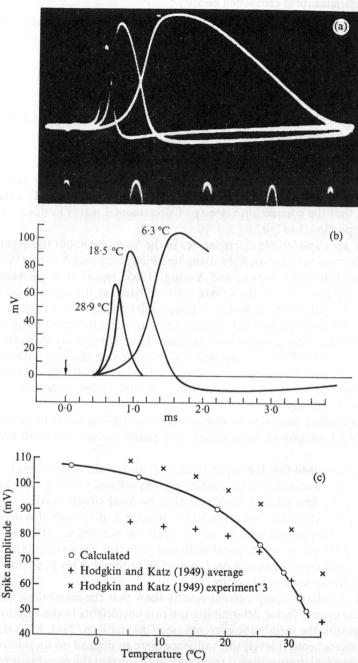

FIG. 10.6. (a) Recorded action potentials at 32·5 °C (*A*), 18·5 °C (*B*), and 5 °C (*C*). (From Hodgkin and Katz 1949*b*.) (b) Propagated action potentials computed at 28·9 °C, 18·5 °C, and 6·3 °C. (From Huxley 1959*a*.) (c) Relation between amplitude of propagated action potential and temperature. The curve shows computed values obtained by Huxley. The crosses show experimental values obtained by Hodgkin and Katz. (From Huxley 1959*a*.)

Other determinants of conduction velocity

We have already discussed the large effect which temperature has on conduction velocity. In this section we shall discuss the effects of cable and membrane current parameters. Eqn (10.6) may be rewritten as

$$\frac{1}{K}\frac{d^2V}{dt^2} = \frac{dV}{dt} + \frac{I_i}{C_m},$$ (10.9)

where $K = 2R_i\theta^2 C_m/a$. If C_m and the functional relationship between I_i and V and t remain unchanged, then

$$\theta \propto \sqrt{(a/R_i)} \propto 1/\sqrt{(ar_a)}.$$ (10.10)

If R_i is also assumed to be constant, we obtain the result that the conduction velocity is proportional to the square root of the fibre radius. In Chapter 3 we showed that the conduction velocity of decremental waves in linear cables is also proportional to $\sqrt{a}$.

There are considerable discrepancies in the literature about the dependence of conduction velocity on fibre diameter. Pumphrey and Young (1938) and Burrows, Campbell, Howe, and Young (1965) report that the velocity is roughly proportional to the square root of diameter for cephalopod axons, but Hodes (1953) has reported a linear relationship for the giant axons. Gasser (1950) concluded that mammalian unmyelinated nerve fibres showed a linear rather than a square root relation. In a recent study Pearson, Stein, and Malhotra (1970) have suggested that in cockroach and locust the relationship is an intermediate power of about 0·75 (i.e. $\theta \propto D^{0.75}$). Pearson *et al.* noted that the duration of the action potential varied systematically with conduction velocity.

It is therefore necessary to discuss the assumptions made in deriving the square root relation in more detail. The points to be considered are listed below.

1. It is assumed that the dependence of I_i on V is the only means by which propagation is achieved so that active regions influence resting regions of the fibre only by first altering their potential by local circuit current flow. If a more strictly chemical process were involved, e.g. if I_i were also dependent on some chemical released by active areas of membrane, other equations including the effects of chemical diffusion (or other forms of transport) and of this hypothetical chemical on I_i would be required. This possibility seems unlikely for at least two reasons.

(a) The voltage-clamp current results show that the membrane voltage is the crucial factor determining the rate coefficients in the reactions that govern the membrane currents (see Chapter 8, p. 241). Even if, at the microchemical level, this influence were to depend on an intermediate chemical 'transmitter' it remains true to say that the membrane currents may be completely specified once the history of the membrane potential is known.

(b) There is very strong experimental evidence that local circuit currents are the only effective means by which one area of a fibre can influence the excitability of other areas (Hodgkin 1937a,b; Tasaki 1939, 1953; Lorente de Nó 1947; Goldman 1964). Moreover, the accuracy with which the conduction velocity may be predicted when the Hodgkin–Huxley equations are combined with the cable equations (see p. 284) reinforces this experimental evidence.

It seems likely, therefore, that the basic theory is correct in predicting that the square-root dependence of conduction velocity on diameter should hold, all factors being equal apart from fibre size. Therefore we must consider the question whether any of the cable or current parameters may systematically, vary with fibre size.

2. It is assumed that R_i is independent of a. This seems to be a reasonable assumption in a given species, although there are large variations between species, e.g. between salt-water animals and others.

3. It is assumed that the extracellular resistance to local circuit current flow is negligible. This may be true in many situations in vitro, and perhaps also in vivo in some cases. However, it may not necessarily be true in all cases since the extracellular space around nerve axons is sometimes quite restricted. In such cases, θ should also depend on r_0 (Hodgkin 1939) as well as r_a (cf. eqn (10.10)):

$$\theta \propto 1/\sqrt{\{a(r_a+r_0)\}}. \tag{10.11}$$

Only if r_0 is always negligibly small or varies with a as r_a does (that is, as $1/a^2$) will the square-root relation still hold. The latter condition, however, would require a very unlikely situation: that the effective thickness of the extra-cellular fluid available for current flow should greatly decrease as a decreases. Although Gasser (1950) suggested that variations in r_0 might account for the results in vertebrate non-myelinated axons, it seems more likely that variations in r_0 would have an effect opposite to that of linearizing the relation between θ and a. Thus, if axons of various sizes share a common extracellular space, the resistance of the extracellular space is likely to become significant compared to r_a only in the case of the larger axons (where r_a is smallest). This would make θ proportional to an even lower power of a than 0·5 for the larger axons. In fact, the deviation from the square-root dependence has been observed for small diameter axons where the exponent becomes greater than 0·5 (Pearson, Stein, and Malhotra 1970).

4. It is assumed that C_m is constant and that the functional dependence of I_i on V and t does not vary systematically with a. Huxley (1959a) has analysed the influence of variations in C_m and I_i on conduction velocity and it will be convenient here to use his nomenclature and equations. We consider the effects of:

(a) increasing C_m by a factor γ. If we suppose, for the sake of convenience, that the 'standard' membrane capacity is 1 μF cm^{-2} (see Chapter 2) then we can equate the capacity numerically to γ and use γ in place of C_m in the equations;

(b) increasing the absolute values of all ionic currents by a factor η, i.e.

$$I_i = \eta f(V, t);$$

(c) accelerating all permeability changes by a factor ϕ, i.e.

$$I_i = \eta f(V, \phi t). \tag{10.12}$$

In the case of the Hodgkin–Huxley equations this involves multiplying all the rate coefficients (αs and βs) by ϕ. However, it should be emphasized that the analysis does not depend on assuming any particular theory of the membrane nonlinearities. Its validity depends only on the assumptions discussed in 1, 2, and 3 above, and other equations than the Hodgkin–Huxley equations could equally well be used to describe the membrane ionic current.

It is possible to introduce these three scaling factors directly into the differential equation for the uniformly propagating action potential. Expressing I_i as a function of τ and rewriting $t = \tau/\phi$, eqn (10.9) becomes

$$\frac{\phi^2}{K}\frac{d^2V}{d\tau^2} = \phi\frac{dV}{d\tau} + \frac{\eta}{\gamma}\frac{I_i(V, \tau)}{C_m}. \tag{10.13}$$

Dividing all terms by ϕ and letting $\beta = \eta/\gamma\phi$,

$$\frac{\phi}{K}\frac{d^2V}{d\tau^2} = \frac{dV}{d\tau} + \beta I_i(V, \tau). \tag{10.14}$$

For any value of β for which conduction is possible there will be a corresponding value (or, more correctly, two values—see p. 290) of K/ϕ which keeps the solution for V bounded. Hence

$$K/\phi = F(\beta). \tag{10.15}$$

The function F will depend on the functional relationship between I_i and V, and t. It must be determined separately for individual cases. However, once $F(\beta)$ has been evaluated, the conduction velocity θ follows directly from the definition of K.

$$\theta^2 = (a/2R_i\gamma)\phi F(\beta). \tag{10.16}$$

Huxley has obtained the function $F(\beta) = K/\phi$ numerically for the squid-axon equations (see Fig. 10.7). $F(\beta)$ increases monotonically as β increases, and follows a fairly simple curved relation when plotted on double logarithmic scales. Huxley's analysis enables the following additional predictions concerning the conduction velocity to be made:

1. Since γ appears in the denominator of eqn (10.16) and since β, and hence also $F(\beta)$, decreases as γ is increased, θ must always decrease as the membrane capacity is increased. The functional relationship between θ and γ is not obtainable analytically but by using his numerical values

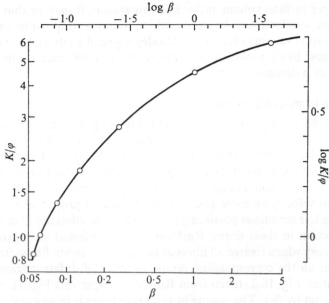

FIG. 10.7. Relation between $F(\beta) = K/\phi$ and $\beta = \eta/\gamma\phi$ plotted on double logarithmic scales. (From Huxley 1959a.)

for $F(\beta)$ Huxley has shown that the effect of changes in the membrane capacity will be greatest when the ionic currents are smallest, that is, when η is small.

2. Since ϕ is present both in the numerator of eqn (10.16) and in the denominator of the argument in $F(\beta)$, the effects of changing the rate coefficients of the ionic currents will not be as simple as those of changing the membrane capacity. In particular, if $F(\beta)$ is a monotonically increasing function of β, there must be an optimal value of ϕ at which θ is maximal. Thus, since temperature changes exert their effects largely through influencing ϕ (see p. 241) there must be a temperature at which θ is maximal.

3. An increase in the absolute value of the ionic currents (that is, an increase in η) must always increase θ since η is present only in the numerator of the argument of $F(\beta)$.

We may return now to the question of the relation between θ and a.[†] It is clear that if any of the factors η, ϕ, or γ vary systematically with a, the result will be a deviation from the square-root relation of eqn (10.10). Variations in γ are unlikely. However, it may be that ϕ and/or η decrease as a decreases, since there is evidence that the action potentials in small fibres are systematically slower than in large fibres (Paintal 1967; Pearson, Stein, and Malhotra 1970). A possible explanation of this systematic variation is that in small fibres the ionic concentration gradients may be less well maintained because

† See also Jack (1975), *Br. J. Anaesth.* (in press).

of the larger surface–volume ratio. For this reason, it may be that the ionic currents are smaller in smaller axons, which would correspond to a reduction in η. Stein and Pearson (1971), using Huxley's type of analysis, have suggested that the most likely explanation of the results in cockroach axons is that η decreases as a decreases.

Conduction in myelinated axons

In myelinated axons the action currents are not generated uniformly along the fibre (Tasaki 1953, 1959; Huxley and Stampfli 1949), and thus the analysis of the factors determining conduction velocity given in the previous sections of this chapter does not apply. At present there is no theory of myelinated nerve conduction that successfully treats the dependence of conduction velocity on cable and membrane current properties (see Pickard 1966). One less ambitious possibility is to explore the effects of fibre size alone on conduction in these fibres. Rushton (1951) presented an ingenious and elegant theory which treated all fibres as being in 'corresponding states', which means that all the corresponding points in fibres of different sizes must be equipotential. Fig. 10.8, taken from Rushton's paper (his Fig. 1), illustrates what is meant by this. The scaling of the nerve fibres is in units of internodal length and both P_1, P_2, and Q_1, Q_2 are pairs of corresponding points.

The main conditions necessary for the assumption of 'corresponding states' to be true are as follows.

1. In different fibres the specific properties, both of the myelin and of the nodal membranes, are identical.

2. The area of excitable membrane at the node A should be proportional to the square of the internal diameter d divided by the internodal length l. The reason for this requirement is that the same membrane potential at two

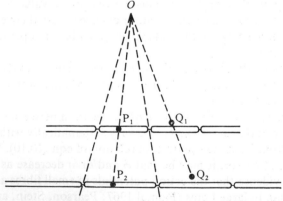

FIG. 10.8. Diagram illustrating Rushton's 'corresponding states' hypothesis. Two myelinated nerves with different internodal lengths are shown. Each line from the point O joins pairs of points (e.g. P_1 and P_2) which are assumed to be undergoing exactly the same membrane potential changes. (From Rushton 1951).

corresponding nodes will produce similar current densities, and hence the nodal current will be proportional to the nodal membrane areas. The membrane potential difference between adjacent nodes in the same fibre will be proportional to the axial resistance and hence will scale as l/d^2. In order to hold the membrane potential difference between adjacent nodes constant in different fibres the product of current and resistance must be constant. Hence

$$A \propto d^2/l. \tag{10.17}$$

3. The internodal length should be related to the internal (axon) and external (fibre) diameters by the following relation:

$$l \propto d\{\ln(D/d)\}^{\frac{1}{2}}, \tag{10.18}$$

where D = fibre diameter.

This relationship follows from the fact that, since the axial current will scale in proportion to d^2/l, the *change* in axial current with distance will have to scale as d^2/l^2. This places a requirement on the change in axial current per unit distance, which is equal to the total current flow (that is, the sum of ionic and capacitive currents) per unit distance across the myelin. Both the conductance and capacitance per unit length for a cylindrical shell are proportional to $1/\{2\ln(D/d)\}$ so that if nerves are in 'corresponding states' (i.e. both the voltage and rate of change of voltage at corresponding points are identical) the radial leak and capacitive currents will both be proportional to this factor. This leads to the requirement that

$$d^2/l^2 \propto 1/\{\ln(D/d)\}. \tag{10.19}$$

Rushton (1951) examined the histological data relevant to the third condition and showed that, although there appeared to be exceptions, eqn (10.18) held for fibres whose external diameter D was greater than $4\,\mu$m. No adequate experimental data was available to Rushton concerning the first two conditions.

The theory of 'corresponding states' assumes that space relations will be the same if scaled in units of internodal length so that the conduction velocity of different fibres should be proportional to their internodal length. A further consequence of the theory is that, if the ratio of the internal to external diameters $d/D\ (=g)$ is constant, the conduction velocity should also be directly proportional to the fibre diameter. This follows from the fact that eqn (10.18) can be rewritten

$$l/D \propto g \sqrt{(-\ln g)}, \tag{10.20}$$

so that if g is constant $D \propto l$.

The proportionality between conduction velocity and fibre diameter still may be roughly obeyed if g is not quite constant for different fibres. Rushton showed that the right-hand side of eqn (10.20) changes by less than 5 per cent as g is varied between 0·47 and 0·74. The maximum value occurred at $g = 0·6$. Rushton pointed out that this value is optimal for the spread of current down the internode, and this led him to suggest that it is the attainment of maximal conduction velocity for a given external diameter which has been the controlling factor in the evolution of myelinated nerve structure. Hodgkin (1964) noted that the observed value for g for large myelinated fibres was 0·7 or more. In a simple cable the constant determining the spread of charge with time is $1/r_a c_m$ (see p. 42), but in a myelinated nerve fibre allowance should be made for the capacitance of the node as well as that of the internode. Hodgkin calculated that, if a value of 0·4 was assumed for the ratio of nodal capacity to internodal capacity, then the most rapid spread of potential would be achieved if $g = 0·7$.

A more quantitative exploration of the sensitivity of the $\theta(D)$ relation to changes in g has been made by Goldman and Albus (1968). They have obtained digital computer solutions of the cable equations for myelinated nerve (see FitzHugh 1962), making only the first two of Rushton's three assumptions. The membrane currents generated at the node were calculated by using the equations of Frankenhaeuser and Huxley (1964). Fig. 10.9 shows the effect of changes in g, for constant D, on the conduction velocity. The curve has been drawn from points measured from Fig. 3 of Goldman and Albus (1968). The insights of Rushton and Hodgkin are confirmed by these results: θ changes very little for g values between 0·6 and 0·75, but falls off more steeply as the deviation from this optimal range increases.

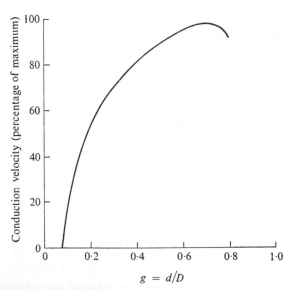

FIG. 10.9. The effect of changes in g (the ratio of internal to external diameter) on the conduction velocity of myelinated nerve. The curve is drawn from points obtained from Fig. 3 of Goldman and Albus (1968).

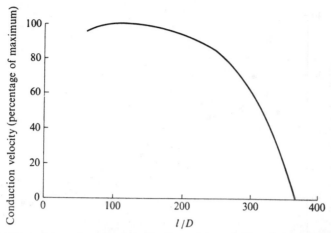

FIG. 10.10. Dependence of conduction velocity on internodal length (expressed in terms of fibre diameters).

In an earlier analysis Huxley and Stämpfli (1949) had suggested that a comparable result would hold for variations in the internodal length of a fibre of a given diameter. They gave analytical arguments which suggested that there would be an optimal internodal length for attainment of maximal conduction velocity, but that the relationship might show a broad plateau. Goldman and Albus (1968) have also performed the relevant computations to test this expectation. The conduction velocity does have a maximum at a particular node spacing, which roughly corresponds to normal spacing, and is nearly maximal over a broad range. Fig. 10.10 shows, for instance, that a doubling of the internodal length from its normal value ($l = 92D$) leads to a reduction in conduction velocity of only about 10 per cent. This theoretical result is in accord with the experimental observations of Sanders and Whitteridge (1946), who found that in regenerated nerves the internodal length was about half its normal value, but the fibres conducted at nearly normal conduction velocity.

The general conclusion that may be drawn from these considerations is that, unless Rushton was substantially incorrect in his three assumptions, his conclusion follows, namely, that myelinated nerves appear to be constructed so that their dimensions favour the most rapid conduction velocity for a given external diameter. The classical observations for both mammalian (Hursh 1939) and frog myelinated nerve fibres (Tasaki 1953) that there is a roughly linear relation between conduction velocity and fibre diameter is therefore no surprise. Although recent work on frog fibres has supported this relationship (Hutchinson, Koles, and Smith 1970), it has been questioned for both mammalian motor (e.g. Boyd 1965) and muscle afferent fibres (Coppin and Jack 1972). Both of the latter workers suggest that smaller fibres conduct less fast relative to their fibre diameter than larger fibres.

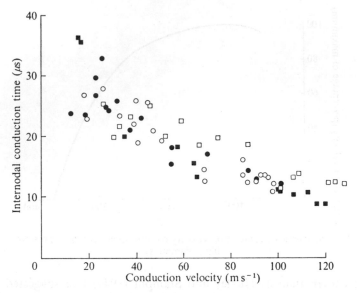

FIG. 10.11. Observed variation of internodal conduction time with conduction velocity
(Coppin and Jack, unpublished).

If these recent results are correct there is the problem of determining which
of Rushton's assumptions is incorrect. Coppin and Jack (1972) showed that
there was a nonlinear (logarithmic) relation between θ and l, where l had been
determined by methods analogous to those used by Lussier and Rushton
(1952). This means that the internodal conduction time is shorter in the larger
fibres, as illustrated in Fig. 10.11. There could be more than one explanation
for this more rapid spread of a threshold quantity of charge between each
node. The two most obvious possibilities are as follows:

1. The axial resistance per internode is not inversely proportional to D,
   being relatively higher in smaller fibres. This could be a result of a
   nonlinear relation between l and D (see Coppin and Jack 1972) and/or
   because g was smaller in small fibres. One difficulty in accurately
   estimating g is that electron microscopic observations on peripheral
   nerves, apparently free of fixation artefacts, reveal that some nerve
   fibres are not circular in cross-section (Berthold 1972, 1.A; Boyd,
   personal communication). The relevant way to estimate a g value would
   then be to measure the circumference and axon area. If smaller fibres
   deviated relatively further from a circular cross-section their axial
   resistance would be relatively greater than the inverse of the square of
   their circumference; such a trend would explain the above results.
2. A quite different explanation is that larger fibres generated relatively
   more inward current at each node, thus providing more charge to
   spread along the internode than expected on Rushton's dimensional

theory. For example, the nodal area might scale as expected but the maximal possible sodium conductance per unit area of nodal membrane might be greater in larger fibres (cf. p. 295).

At present, the experimental evidence does not allow us to decide between these and other possibilities. The possibility that there are variations in the ionic current characteristics of nodal membranes is supported by recent voltage-clamp studies. Bergman and Stämpfli (1966) have reported differences in the current–voltage relations of nodes in sensory and motor fibres. Frankenhaeuser and Vallbo (1965) have described differences in the kinetics of the voltage-clamp currents which would result in differences of action potential shape and which are probably related to experimentally observed differences in accommodation in different myelinated nerve fibres. But these results are not directly relevant to the question of ionic currents in relation to fibre size, since the differences have all been observed in large fibres and the kinds of differences described may not affect the relationship.

An additional observation which conflicts with the theory of 'corresponding states' is that the time course of the action potential varies systematically with conduction velocity; fibres with slower conduction velocity have action potentials with a slower rise time and longer duration (Paintal 1966; Coppin and Jack, unpublished observations). Rushton's theory predicts that the time relations should be the same in all fibres. This observation, however, does not help to distinguish between the two main possibilities mentioned above.

It also seems likely that Rushton's theory does not apply to myelinated fibres within the central nervous system. Waxman and Bennett (1972) have reviewed the evidence with particular reference to Rushton's suggestion that there is a critical diameter below which an unmyelinated fibre would conduct faster than a myelinated fibre of the same diameter.

Thus, although Rushton's elegant theory gives an approximate description of the effect of fibre size on the properties of medullated axons, there are important factors which were unknown to him and could not be taken into account. Unfortunately, it is not obvious how the theory may be amended to deal with this greater heterogeneity of fibre structure and properties, so that further theoretical work may be forced to rely on numerical computations of the kind performed by FitzHugh (1962) and Goldman and Albus (1968).

Subthreshold responses in nonlinear systems

The differences between decremental conduction in linear cables (Chapters 3, 4, and 5) and action-potential (non-decremental) conduction in nonlinear systems are very striking. The space constant for small currents (when nonlinear systems may be treated as linear) is usually very short so that the linear response becomes negligibly small at a relatively small distance from the point of current injection. The action potential, by contrast, may propagate an unlimited distance. Moreover, the sharpness of the threshold in most excitable cells ensures that the probability of occurrence of waves intermediate between full action potentials and rapidly decrementing subthreshold responses is exceedingly small. Thus, in the case of the squid-axon equations, the excitation current must be adjusted with a degree of accuracy which cannot be achieved experimentally in order to reveal the fact that the threshold is a pseudo-threshold and that intermediate responses are, at least mathematically, possible (FitzHugh 1955; FitzHugh and Antosiewicz 1959; Cooley, Dodge, and Cohen 1965). Also, the subthreshold propagated waves computed by Huxley (1959—see p. 290) are highly unstable, although Cooley and Dodge (1966) have shown theoretically that more stable waves may be expected to occur in narcotized axons in which all the ionic conductances are reduced (see below).

However, not all excitable cells have such large safety factors for conduction, that is, such large excesses of excitatory current over and above that which is minimally required to ensure non-decremental conduction. Some slow muscle fibres, for example, exhibit insufficient inward ionic current for the generation of propagated action potentials. Moreover, some parts of nerve cells (sensory endings, dendrites) may be relatively or completely inexcitable. Lorente de Nó and Condouris (1959) suggested that dendrites might be capable of at least some degree of nonlinear behaviour and Llinas and Nicholson (1971) have obtained experimental evidence for the occurrence of action potentials of some kind in Purkinje cell dendrites. However, the excitability of dendrites may be much lower than that of axons for two reasons. First, the large surface-volume ratio of dendrites might prevent them from maintaining ionic concentration gradients as large as in axons. Second, a large proportion of the dendrite membrane area is covered by synaptic endings. If these areas act as linear shunts they could greatly reduce the nonlinearity of the total current–voltage relation. Both of these effects would favour some sort of intermediate conduction.

Cooley, Dodge, and Cohen (1965) have computed responses in a cable obeying the Hodgkin–Huxley equations, but with the conductance scaling factor η reduced to values less than 1. Fig. 10.12 shows the computed variation in θ with η. Above $\eta = 0.26$ there are two values of θ, the upper one corresponding to the normal action-potential, the lower one corresponding to the unstable wave computed by Huxley (1959b). At $\eta = 0.26$ the values coincide, and there is then no difference between the 'action potential' and the 'unstable wave'. Below $\eta = 0.26$, only decremental waves are possible. However, there will be a range of values of η near 0.26 which will give rise to waves which propagate a considerable distance before decrementing significantly. Moreover, for values equal to and slightly greater than 0.26 the non-decremental waves are relatively small and can supply very little extra current over and above that required to maintain conduction (Fig. 10.13). It remains to be seen whether such waves serve any functional role in physiological systems.

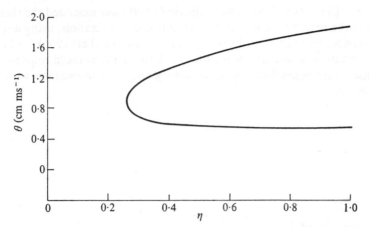

FIG. 10.12. Variation of computed value of conduction velocity for action potential (upper values) and unstable wave (lower values) when the ionic conductances are reduced by factor η. (Cooley, Dodge, and Cohen 1965.)

Oscillatory responses to subthreshold currents

When currents which are well below threshold are applied to nerve membranes, the membrane may behave in a linear manner even if the nonlinear sodium and potassium conductances are partially activated. Hodgkin and Huxley 1952d (Fig. 23) noted that the conductance changes in response to small depolarizations give rise to damped oscillatory responses similar to that of an inductive system, and that this behaviour underlies the inductive phenomena observed by Cole and Baker in their impedance studies (see

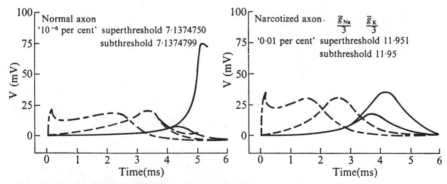

FIG. 10.13. Computed responses near threshold for normal nerve (left) and one in which the conductances have been reduced to 0·3 of normal values. The uneven–dash curves show responses at current electrode. Note the charge redistribution effect occurring when current pulse is terminated (cf. Fig. 9.3). The dashed lines show responses 1 cm away and the continuous curves show responses 2 cm away from current electrode. In normal axon, the subthreshold wave propagates at least 1 cm before decaying or generating a full action potential. However, the 'stimulus' must be adjusted to within 10^{-8} to produce this effect. In axon with reduced conductances less accuracy is required. (Cooley, Dodge, and Cohen 1965).

Chapter 2, Fig. 2.10). Sabah and Liebovic (1969) have extended the theory to deal with the response of a cable to nonuniform polarization, using a numerical inversion procedure (cf. Bellman, Kaliba, and Lockett 1966) for Laplace transforms to obtain solutions for the oscillatory subthreshold responses. We shall discuss the general theory of oscillatory responses in excitable cells in the next chapter.

11. Repetitive activity in excitable cells

THE transmission of information by the nervous system over distances greater than a few millimetres is dependent on the propagation of action potentials. As shown in Chapter 10, the propagation mechanism does not normally transmit signals of more than one amplitude, which means that the signal amplitude itself does not carry any information. Therefore the information must be coded in terms of the interval between successive action potentials or its reciprocal, the frequency. Many nerve cells are capable of generating trains of impulses at constant or varying frequencies and it is well known that this frequency may be modulated by sensory stimuli, synaptic potentials, or applied current. Physiological experiments frequently involve the use of steady applied current as a means of modulating the action potential frequency (e.g. Hodgkin 1948; Fuortes and Mantegazzini 1962; Tomita and Wright 1965; Kernell 1965a,b,c; Chapman 1966; Connor and Stevens 1971a,b,c) which is usually found to increase progressively over a substantial range of frequencies as the magnitude of the applied current is increased.

In addition to cells which respond repetitively to steady applied stimuli there are also cells which exhibit spontaneous rhythmic firing, such as the pacemaker cells of the heart and some invertebrate nerve cells. In the former case, the spontaneous frequency may be modulated by transmitters (in particular adrenaline and acetylcholine) and by temperature. It is also possible to modulate the frequency using applied current (see e.g. Trautwein and Kassebaum 1961; Brown and Noble 1969).

In this chapter we will discuss some of the models which have been advanced to account for repetitive activity in excitable cells. In doing so, we will only discuss mathematical models and will ignore the various electrical, chemical, and hydraulic models which have been proposed (see e.g. Harmon and Lewis 1966). For most of this chapter we will concentrate on models in which firing either occurs spontaneously or in a response to a steady current injection. Mathematical models which attempt to account for the firing of excitable cells when the current stimulus varies in intensity with time require more complicated treatment. They will be briefly discussed at the end of the chapter.

Although subthreshold oscillatory activity may be treated semi-analytically using the equations for the time dependence of the ionic current given by Hodgkin and Huxley (see Chapter 10, p. 303), it is more difficult to analyse oscillatory activity in response to larger currents near or beyond the threshold for initiating action potentials. If the Hodgkin–Huxley and cable equations are used, it is necessary to resort almost entirely to numerical computations (see the section *Numerical computations of repetitive activity*, p. 331). Such computations are tedious and do not easily allow more general conclusions to be drawn. These difficulties have prompted the development of alternative models which do not specify the detailed behaviour of the ionic current variables m, h, and n. In the first part of this chapter we shall discuss simplified models in order to illustrate the general conditions which must be satisfied for repetitive activity to occur. We shall then describe numerical computations which include voltage and time-dependent conductances and cable equations.

Simplified models of repetitive firing

The generation of an indefinite train of impulses requires a continuous expenditure of energy. In excitable cells this energy is derived from the ionic concentration gradients established by active ion transport. Provided that at least two ionic concentration gradients are established which produce current flows in opposite directions, the total membrane current–voltage relation may display a region of negative conductance (see Chapter 8, Fig. 8.11). In such regions, the net ionic current flows in a direction which is opposite to that which would be expected if the energy for the current flow derived entirely from the applied stimulus. The negative conductance may be regarded therefore as a manifestation of the cells' ability to provide the energy to generate its own electrical activity.

The regions of negative conductance not only allow activity to be generated; they also limit its magnitude. An excitable membrane exhibits negative conductance properties only over a certain range of potentials and, as will be shown later (p. 328), this is a sufficient condition for keeping the potential excursion within certain bounds. In this respect excitable membranes resemble a variety of mechanical and electrical oscillatory mechanisms, such as the valve oscillator. These physical systems have been extensively studied from a mathematical point of view (van der Pol 1926; van der Pol and van der Mark 1928; Minorsky 1947; Bonhoeffer 1948; Jones 1961). We will follow FitzHugh (1960, 1961, 1969) in using this theory to develop some properties of excitable membrane oscillations; in particular we will obtain a condition for the magnitude of the negative conductance in order that sustained oscillations may occur.

Before giving an account of oscillator theory it may be useful to describe even simpler models. The simplest possible one is to ignore all the voltage–time-dependent processes which underlie the action potential and assume that an action potential is generated whenever a threshold voltage displacement is exceeded.

Voltage–threshold model

In this model the membrane is represented by a simple R–C circuit. A steady depolarizing current is applied and the membrane potential is displaced until the threshold voltage is reached. Following an action potential the membrane potential is assumed to be reset to its resting level, and another cycle of membrane potential displacement then starts. These assumptions

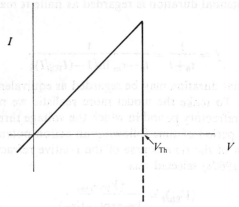

FIG. 11.1. Assumed current–voltage relation for the voltage-threshold model. The membrane behaviour remains linear until a sharp threshold is reached.

are equivalent to assuming that the time constants of the Hodgkin–Huxley conductance changes are all much briefer than the time constant of the membrane. It will be assumed also that the passive membrane resistance behaves linearly so that the form of the membrane current–voltage relation is that illustrated in Fig. 11.1.

The first case to consider is when the membrane is uniformly polarized by steady current. Such models have been discussed by Stein (1967a). In order to calculate the frequency of firing predicted by the model we need to calculate the interval between each action potential. The membrane potential displacement V is given by

$$V = IR_m\{1-\exp(-t/\tau_m)\}.$$

The interval is given by the value of t when $V = V_{Th}$, so that

$$t = -\tau_m \ln\left(1-\frac{V_{Th}}{IR_m}\right). \tag{11.1}$$

The smallest value of current at which an action potential can be generated may be defined as the rheobasic current I_{Rh} (when $t \to \infty$),

$$I_{Rh} = V_{Th}/R_m. \tag{11.2}$$

Substituting in eqn (11.1),

$$t = -\tau_m \ln\{1-(I_{Rh}/I)\}. \tag{11.3}$$

If the duration of the action potential is regarded as infinitesimally small, the frequency of firing of the cell is simply given by the reciprocal of the interspike interval

$$f = \frac{1}{-\tau_m \ln\{1-(I_{Rh}/I)\}}. \tag{11.4}$$

If the action potential duration is regarded as finite it may be represented by t_0.

Then

$$f = \frac{1}{t_0 + t} = \frac{1}{t_0 - \tau_\mathrm{m} \ln\{1 - (I_\mathrm{Rh}/I)\}}. \tag{11.5}$$

The action potential duration may be regarded as equivalent to the absolute refractory period. To make the model more realistic we may also wish to include a relative refractory period in which the voltage threshold is temporarily higher for a period of time following an action potential. We need to have a description of the time course of the relative refractory period. The one which Stein (1967a) selected was

$$(V_\mathrm{Th})_t = \frac{(V_\mathrm{Th})_\mathrm{rest}}{1 - \exp(-t/\tau_\mathrm{R})}, \tag{11.6}$$

where τ_R is the time constant of recovery from refractoriness. The resulting equation for the interspike interval is then

$$\{1 - \exp(-t/\tau_\mathrm{m})\}\{1 - \exp(-\tau/\tau_\mathrm{R})\} = I_\mathrm{Rh}/I. \tag{11.7}$$

An exact solution for t can be obtained when $\tau_\mathrm{R} = \tau_\mathrm{m}$

$$t = -\tau_\mathrm{m} \ln\{1 - (I_\mathrm{Rh}/I)^{\frac{1}{2}}\}. \tag{11.8}$$

For other cases it is easiest to treat I/I_Rh as the dependent variable in making the calculations.

Fig. 11.2 shows some calculations made with these models. Both abscissa and ordinate are scaled in normalized coordinates. Notice that when the current just exceeds the rheobasic value the frequency of firing jumps fairly abruptly up to a value of $1/4\tau_\mathrm{m}$. The curves then display slight downward concavity before reaching a final region of fairly linear behaviour. The slope of the linear region is very sensitive to the duration of the absolute and relative refractory periods.

It is very uncommon in either experimental or natural circumstances to have uniform polarization of an excitable cell. Generally, current passes into the cell at one point. In the case of receptors the current enters at the end of the axon near where it terminates on the receptive surface. The only problem about extending the above type of model to this circumstance is to decide an appropriate criterion for threshold. From the discussion already given in Chapter 9 it will be clear that exceeding the voltage threshold for inward current at the point of excitation would be an inadequate criterion, since the net inward current would be much less than the net outward current in other regions of the cable. In Chapter 12 we discuss Rushton's (1937) concept of a liminal length for excitation. This is the minimum length of cable which must have its potential displaced beyond the uniform voltage threshold in order that excitation is achieved. (It will be designated by the term X_LL.) In extending the present model we will simply use this concept and leave the

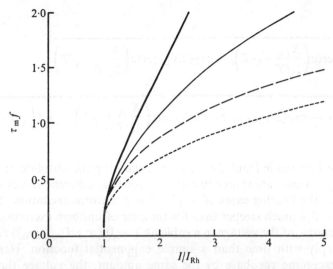

FIG. 11.2. Current–frequency plots for some simple examples of the uniformly polarized voltage-threshold model: thick continuous line, model with no refractoriness; thin continuous line, model with an absolute refractory period, $t_0 = \frac{1}{4}\tau_m$; dashed lines, models with both absolute and relative refractoriness: $t_0 = \frac{1}{4}\tau_m$, $\tau_R = \frac{1}{2}\tau_m$ (long dashes) and $\tau_R = \tau_m$ (short dashes). Note that both ordinate and abscissa are normalized.

reader to consult Chapter 12 for qualifications in the validity of such an assumption.

The voltage response of an infinite cable to a step current waveform is given by

$$V = \frac{r_a I \lambda}{4}\left\{\exp(-X)\mathrm{erfc}\left(\frac{X}{2\sqrt{T}}-\sqrt{T}\right) - \exp(X)\mathrm{erfc}\left(\frac{X}{2\sqrt{T}}+\sqrt{T}\right)\right\}.$$

The condition for excitation is that the voltage displacement at all points on the cable where $0 < X < X_{LL}$ must equal or exceed V_{Th}. The subscript LL stands for liminal length. This is achieved when the voltage at the point $X = X_{LL}$ is equal to V_{Th}.

$$V_{Th} = \frac{r_a I \lambda}{4}\left\{\exp(-X_{LL})\mathrm{erfc}\left(\frac{X_{LL}}{2\sqrt{T}}-\sqrt{T}\right) - \exp(X_{LL})\mathrm{erfc}\left(\frac{X_{LL}}{2\sqrt{T}}+\sqrt{T}\right)\right\}.$$

$$(11.9)$$

Hence

$$I = \frac{4V_{Th}}{r_a\lambda\left\{\exp(-X_{LL})\mathrm{erfc}\left(\frac{X_{LL}}{2\sqrt{T}}-\sqrt{T}\right) - \exp(X_{LL})\mathrm{erfc}\left(\frac{X_{LL}}{2\sqrt{T}}+\sqrt{T}\right)\right\}}.$$

The rheobasic current is given by

$$I_{Rh} = \frac{2V_{Th}}{r_a\lambda e^{-X_{LL}}},$$

so that

$$\frac{I}{I_{Rh}} = \frac{2}{\text{erfc}\left(\frac{X_{LL}}{2\sqrt{T}} - \sqrt{T}\right) - \exp(2X_{LL})\text{erfc}\left(\frac{X_{LL}}{2\sqrt{T}} + \sqrt{T}\right)},$$

$$= \frac{2}{1 - \exp(2X_{LL}) + \exp(2X_{LL})\text{erf}\left(\frac{X_{LL}}{2\sqrt{T}} + \sqrt{T}\right) + \text{erf}\left(\sqrt{T} - \frac{X_{LL}}{2\sqrt{T}}\right)}.$$

(11.10)

Calculations made from this equation, ignoring the absolute and relative refractory periods, are shown in Fig. 11.3 for four different values of X_{LL} as well as for the limiting cases of $X_{LL} = 0$ and uniform excitation. The curve for $X_{LL} = 0$ is much steeper than for the case of uniform excitation because the time course of the voltage in a cable at $X = 0$ ($\propto \text{erf}\{\sqrt{(t/\tau_m)}\}$ rises much more steeply with time than a simple exponential function. Hence for a current exceeding rheobase by the same amount, the voltage threshold is reached much more quickly. The time course of the voltage in a cable at a distance from the point of current injection rises more slowly, however, and the curve for a liminal length of 0·75 space constants is fairly similar to the uniformly polarized case.

The effects of absolute and relative refractory periods can be introduced

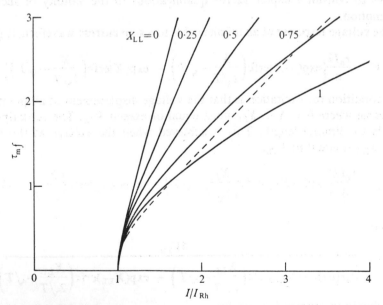

FIG. 11.3. Examples of the current–frequency plot for the point-polarized 'voltage-threshold' model. The continuous lines are, from left to right, for $X_{LL} = 0, 0.25, 0.5, 0.75$, and 1·0 respectively. Refractoriness has not been included. The dashed line is the comparable curve for the uniformly polarized model.

in this model in exactly the same way as shown above for the uniformly polarized model. Once again, their main effect is to reduce the steepness of the slope of the relation between frequency and current.

The question which naturally arises is whether these very simple models are an adequate description of the response of excitable cells. Unfortunately there are insufficient data in the literature to make a careful quantitative comparison, but it does appear that they are only in qualitative agreement with the behaviour of squid and *Carcinus* axons. These comparisons will be presented in a later section (see *Numerical computations of repetitive activity*).

Voltage-threshold model with a time-dependent conductance

In the preceeding section we attempted to make the simple voltage-threshold model more realistic by incorporating a relative refractory period. This was done by assuming that the voltage threshold varied with time following an action potential. In terms of the Hodgkin–Huxley theory this change in value of the voltage threshold will depend on both the return of the potassium conductance to a normal level and the recovery from sodium inactivation. This representation is therefore not very good because the increased potassium conductance following an action potential will not only raise the voltage threshold but also will alter the slope conductance of the membrane current–voltage relation in the linear region (below V_{Th}). In some nerve cells, such as the motoneurone, an action potential in the cell soma is followed by a prolonged after-hyperpolarization which lasts for 50–200 ms. This after-hyperpolarization seems to be the result of a very slow decline of the potassium conductance following its activation during the action potential. If the recovery from sodium inactivation is as rapid as for other excitable membranes it would be reasonable to neglect it, since it would only be significant at interspike intervals much shorter than those affected by the potassium conductance.

In the motoneurone, the time course of recovery of the potassium conductance is described quite accurately by an exponential function (Coombs, Eccles, and Fatt 1955a; Baldissera and Gustafsson 1971a,b) although the membrane voltage is changing in concordance with the conductance. There is no detailed analysis of the voltage and time-dependence of this conductance but such simple behaviour would be expected of a Hodgkin–Huxley potassium conductance if the membrane potential variation remained within the range in which the rate constant was at a minimum and hence also least sensitive to small voltage changes.

Kernell (1968, 1969, 1970, 1971) has presented two models based on these properties of the motoneurone. In his first model, described in the first three papers, he made two additional assumptions: that the neurone could be represented by a simple lumped circuit and that the effect of the membrane capacitance could be neglected. The type of membrane circuit assumed, and the form of the membrane current–voltage relation, are illustrated in Fig. 11.4.

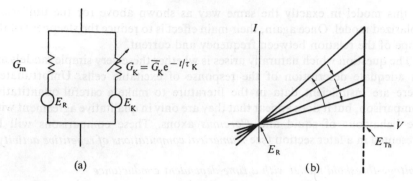

FIG. 11.4. (a) Circuit diagram of the model. The membrane capacitance is neglected. (b) Current–voltage relations displayed by the model. The line of lowest slope represents the resting conductance ($=G_m$). After an action potential has been generated the membrane conductance jumps up to the line of greatest slope ($= G_m + \bar{G}_K$) and subsequently declines exponentially with time back to the resting conductance (provided no further action potential is generated). Note that the voltage threshold remains constant, but the current threshold changes with time following an action potential.

The current required to displace the membrane potential to E_{Th} is given by

$$I = G_m(E_{Th} - E_R) + \bar{G}_K(E_{Th} - E_K)\exp(-t/\tau_K)$$

from which we obtain the interspike interval

$$t = \tau_K \ln\left\{\frac{\bar{G}_K(E_{Th} - E_K)}{I - G_m(E_{Th} - E_R)}\right\}. \tag{11.11}$$

The rheobasic current is given by

$$I_{Rh} = G_m(E_{Th} - E_R)$$

so that

$$t = \tau_K \ln\left\{\frac{\bar{G}_K(E_{Th} - E_K)}{I - I_{Rh}}\right\}$$

$$= \tau_K \ln\left\{\frac{I_K/I_{Rh}}{(I/I_{Rh} - 1)}\right\}, \tag{11.12}$$

where $I_K = \bar{G}_K(E_{Th} - E_K)$.

Calculations of the frequency of firing of such a model (neglecting the absolute refractory period) are shown in Fig. 11.5 for four different values of the I_K/I_{Rh} ratio (3, 6, 10, and 16). Once again the results are plotted on normalized coordinates. As in the previous model, the frequency of firing jumps up quite abruptly to a value of about $1/4\tau_K$. It then has a linear region before curving upwards. The basic form of the curves is to display an upward concavity (unlike the previous model which displayed a downward concavity). The slope of the linear region is dependent on the I_K/I_{Rh} ratio, being proportional to $2I_{Rh}/I_K$.

Both Kernell (1968, 1969, 1970) and Baldissera and Gustafsson (1971a,b) have compared the predictions of this model with experimental observations

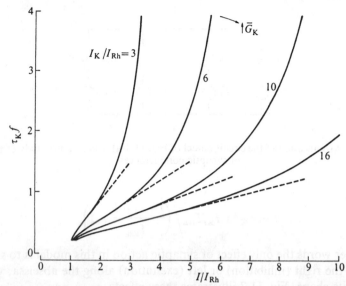

FIG. 11.5. The firing behaviour of the Kernell model of the motoneurone. The ordinate is normalized frequency ($\tau_K f$) and the abscissa is current normalized by the rheobasic current I_{Rh}. Four different values of the I_K/I_{Rh} ratio are shown (from left to right, 3, 6, 10, and 16). If the voltage threshold and E_K remain constant these curves also illustrate the effect of an increase in the maximal potassium conductance at the end of the spike ($\bar{G}_K$.)

of the frequency of firing of motoneurones in response to steady injected current. There is very good agreement except at higher current strengths, where the main deviation is that even higher firing frequencies than predicted can be observed. This is exactly the opposite result to that expected if the model were modified to include an absolute refractory period and, in particular, the effect of the membrane capacitance. Baldissera and Gustafsson (1971b) have presented evidence that the deviation is due to the fact that the conductance does not decline exponentially over the early part of its time course. The explanation of this more complicated time course is not clear but it might be either the result of the voltage sensitivity of the potassium conductance rate constant or else the result of a delayed activation of sodium and/or potassium conductance in the dendritic tree.

Kernell's model has also made an important contribution to the understanding of the effects of synaptic inputs on the firing frequency of the motoneurone. In order to include synaptic action the circuit diagram of the model is modified as in Fig. 11.6. The current required for firing is then

$$I = G_m(E_{Th}-E_R)+G_I(E_{Th}-E_I)+G_E(E_{Th}-E_E)+\bar{G}_K(E_{Th}-E_K)\exp(-t/\tau_K)$$

$$= I_{Rh}+I_I-I_E+I_K.\exp(-t/\tau_K),\dagger$$

where subscript I stands for inhibition and E stands for excitation.

† The sign for I_E represents the fact that it will be in the opposite direction to the other currents since E_E is further from the resting potential than E_{Th}.

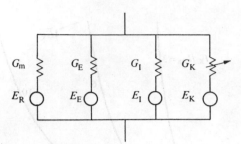

FIG. 11.6. Modification of the circuit model of Fig. 11.4(a) which allows study of the effect of synaptic conductances.

Hence

$$t = \tau_K \ln I_K/I_{Rh} \bigg/ \left(\frac{I+I_E-I_I}{I_{Rh}}-1\right). \qquad (11.13)$$

In other words the only effect of synaptic action in this model is to shift the curve to the right (inhibition) or left (excitation) along the abscissa, without altering its shape. Fig. 11.7 illustrates these effects.

Exactly this result has been observed experimentally for many synaptic actions on the motoneurone (see Granit, Kernell, and Lamarre 1966; Kernell 1969). Sometimes, however, synaptic effects do not produce a parallel shift. In order to simulate these results Kernell (1971) has introduced a more complicated model which includes a membrane capacitance and represents the motoneurone as a distributed structure by means of a compartmental model (see Rall 1964 and Chapter 7). Kernell found that when it was assumed that the potassium conductance occurred only in the soma compartment following an action potential the slope of the frequency–current relation was relatively unaffected by synaptic action. The same result was obtained if the potassium conductance increase also occurred in some of the dendritic compartments but the synaptic action was restricted to the soma. It was only when the synaptic input acted on dendritic compartments that a change of slope, comparable to those observed experimentally, was obtained. It will be of some interest therefore to determine experimentally whether the particular synaptic inputs which alter the slope of the frequency–current relation do in fact terminate on the dendrites.

The models described in this section have been remarkably successful in predicting some of the behaviour of motoneurones. It is possible that these models could also be applied to pyramidal tract cells, since they display a similar frequency–current relation (Koike, Mano, Okada, and Oshima 1970). A further check on the adequacy of the models would be to compare the time course of the voltage (in the intervals between each action potential) that they predict with those observed experimentally (see e.g. Calvin and Schwindt 1972; Schwindt and Calvin 1972).

It is unlikely, however, that the above models will serve as a general

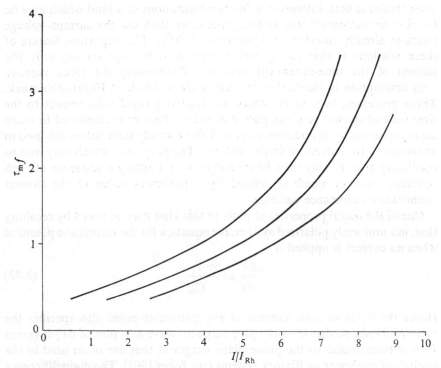

FIG. 11.7. The effect of synaptic conductance on the Kernell model of the motoneurone. The curve in the middle is for an I_K/I_{Rh} ratio of 10. The curve to the left shows the effect of a synaptic excitatory conductance whose magnitude is given by

$$G_E = I_{Rh}/(E_E - E_{Th}),$$

and similarly the curve to the right is for a synaptic inhibitory conductance whose magnitude is given by

$$G_I = I_{Rh}/(E_I - E_{Th}).$$

description of nerve cell firing since many of them do not show the characteristic upward concavity in the current–frequency relation (see e.g. Terzuolo and Washizu 1962; Kuno and Miyahara 1968; Eide, Fedina, Jansen, Lundberg, and Vyklicky 1969a). Some insight into such differences may be obtained by considering another nerve cell where a detailed analysis (Connor and Stevens 1971a,b,c) has been made (see p. 344).

Oscillator theory

It has been mentioned already that excitable membranes may be compared with electrical oscillators; Cole (1968) and FitzHugh (1969) have reviewed the development of these ideas. As FitzHugh has shown, the mathematical description of such physical systems may be used as a simplified way of representing the Hodgkin–Huxley system. The account we will give in this section is similar in many respects to that of FitzHugh. One difference in the presentation is that, instead of using representations of a kind which may be familiar to mathematicians and engineers, we shall use the current–voltage relations already described in Chapter 8 (p. 250). The important feature of these relations is that the system is simplified by representing only the slowest of the time-dependent reactions determining the ionic current. This assumption is identical with that made in much of FitzHugh's work. Those processes, such as m, which are relatively rapid with respect to the time scale of interest (e.g. the period of the oscillation) are assumed to reach steady state values instantaneously, and these steady state values are used in constructing the current–voltage relations. The properties which may lead to oscillatory activity may then be investigated by plotting a sequence of such relations, each of which is defined by a particular value of the slowest membrane conductance variable.

One of the useful properties of plots of this kind may be noted by recalling that, in a uniformly polarized system, the equation for the membrane potential when no current is applied is

$$\frac{dV}{dt} = -\frac{I_i}{C_m}. \tag{8.43}$$

Hence the value of ionic current at any particular point also specifies the value of dV/dt, so that the $i_i(V)$ plot may also serve as a plot of dV/dt versus V. It is then similar to the phase-plane diagrams that are often used in the analysis of nonlinear oscillatory systems (see Jones 1961). The main difference lies in the fact that, since dV/dt is proportional to $-I_i$, points above the voltage axis correspond to negative values of dV/dt, which is opposite to the phase-plane convention. The justification for adopting a different convention is that we think it is more helpful to physiologists for the current–voltage relations to be plotted using the conventions normally adopted in voltage-clamp work.

Given any initial point, that is, initial values of V and dV/dt, we may compute a trajectory describing how V and I_i change as the time varies. In general this will be given as a solution to the equation

$$\frac{dI_i}{dV} = \frac{(dI_i/dt)}{(dV/dt)} = -C_m \frac{(dI_i/dt)}{I_i}, \tag{11.14}$$

where I_i is a function of V and t, and is given by a simplified version of the Hodgkin–Huxley equations. Unless the simplification is great enough to allow analytical solutions to be obtained (see the section *van der Pol's equation*, p. 325), it will be necessary to integrate numerically, in which case it is simpler to use eqn (8.43) to obtain curves for V and I_i as functions of time. The trajectory in the phase-plane may then be plotted using these curves. We shall illustrate the use of these phase-plane diagrams in the sections which follow.

Qualitative analysis of repetitive activity. The curves drawn in Fig. 11.8 are a family of current–voltage relations similar to those already shown in Figs. 8.11 and 8.12. Each curve refers to the state of the membrane achieved momentarily when the fastest permeability changes have occurred. Therefore we have called the relations 'momentary current–voltage relations'. In this case it is assumed that the fastest reaction m which determines the activation of the sodium current, is instantaneous. The curve labelled y_1 represents the 'resting' state, that is, the curve which would be obtained when the potential of a resting cell is varied sufficiently to change m but not to change h and n. Such curves may be obtained experimentally, to a reasonable degree of approximation, by using ramp voltage-clamps (see Palti and Adelmam 1969). The curves labelled y_2, y_3, etc. are the relations obtained for various degrees of sodium inactivation and potassium activation. For simplicity, we shall assume in this section that the decline in h (sodium inactivation) and the increase in n (potassium activation) may be represented as a single process. This is not an unreasonable approximation for some purposes since these changes both occur at similar speeds in some nerve membranes, they have approximately the same net effect on the current–voltage relation and they both underlie the phenomena of accommodation and refractoriness. In cardiac membranes, it would be more realistic to allow both m and h reactions to occur instantaneously. The remaining time-dependent processes would then be the potassium activation processes (Noble and Tsien 1968, 1969a) and the slower sodium or calcium inactivation processes (Beeler and Reuter 1970b; Vitek and Trautwein 1971). The principles of the analysis, however, remain the same. In Fig. 11.8, the curves labelled y_2, y_3, etc. represent membrane states corresponding to different magnitudes of the single process y assumed to represent the changes in h and n.

In the resting state, $dV/dt = 0$, and the voltage will be at its resting value E_r. Now let an outward current I_1 be applied. This may be represented on the diagram by shifting the abscissa in the outward current direction by an amount equal to I_1. The voltage will now move to a new stable point E_I. This corresponds to the steady depolarization produced by a subthreshold current.

Now let the applied current be increased to I_2. This current is larger than the peak outward current reached by the curve y_1. The potential will therefore move into a range in which the net current is inward so that the membrane will depolarize at an increasing rate and the trajectory will initially follow the curve y_1. As it does so, the depolarization will initiate the accommodative processes which shift the current-voltage relation towards y_2, y_3, etc. The trajectory will therefore shift from one curve to another until it reaches a point at which the net current (including the steady current I_2) becomes a repolarizing current. dV/dt now becomes negative, and the system repolarizes along one of the curves corresponding to a large y, e.g. y_5. As the repolarization occurs, the system recovers and the current–voltage relation shifts back towards y_1. If the applied current is maintained, the potential will once again

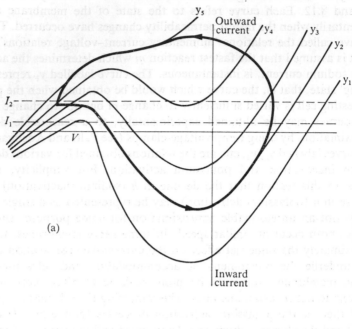

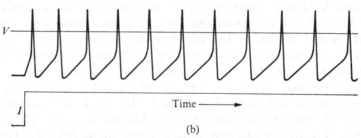

FIG. 11.8.(a) General form of current–voltage relations in a cell capable of generating repetitive activity. The curve labelled y_1 is the 'resting' relation, that is, the curve which would be obtained in response to brief voltage changes applied to a resting cell. The curves labelled y_2 to y_5 correspond to relations obtained when the slower permeability changes (e.g. sodium inactivation, potassium activation) are given time to occur. I_1 corresponds to a subthreshold current which depolarizes the cell from the resting potential E_r to E_1. I_2 corresponds to a suprathreshold current capable of generating the repetitive firing. The trajectory followed during repetitive activity is indicated by the thick line.

These diagrams are loosely based on experimental current–voltage relations found in excitable cells (see Chapter 8), but are not accurate representations of any particular experimental or theoretical results.

(b) Form of voltage as a function of time corresponding to trajectory shown in (a). (Taken from Fig. 11.22).

change in the depolarizing direction. Assuming that no processes occur which are not represented in this diagram, the trajectory will be a closed loop which corresponds to an indefinite train of oscillation. If the increase in y on depolarization occurs faster than the decrease in y on repolarization, the $V(t)$ curve will resemble a train of nerve action potentials similar to the experimental record shown at the bottom of Fig. 11.8.

According to this account, the critical level of depolarizing current which is just sufficient to produce repetitive activity will be equal to the peak outward current on the curve y_1. Moreover, any current which may excite may also produce repetitive firing. Experimentally, however, a level of current which is just sufficient to excite does not usually produce repetitive firing, that is, the threshold for repetitive firing is usually larger than the threshold for initiating a single action potential. The reason for this is that the accommodative processes may occur at subthreshold voltages so that whether or not the curve stays at y_1 while excitation occurs depends on how quickly the membrane is depolarized towards the threshold. If the approach to threshold is slow, the curve will shift towards y_2, etc. before firing occurs and even if a single response is initiated, the accommodative processes may have progressed too far for a second response to occur. Clearly the accommodative process will determine the threshold for repetitive firing and will also set a lower limit on the steady firing frequency which may occur in response to steady current flow. If the accommodative processes are sufficiently rapid they may prevent repetitive firing from occurring or limit it to a very narrow range of frequencies (see p. 331).

The steady state current–voltage relation. If the approach towards threshold is made indefinitely slow, the accommodative processes will have time to reach steady state at each potential. The current–voltage relation which is then followed will be the steady state current–voltage relation. In some excitable cells, it is possible to determine this relation simply by varying the applied current extremely slowly. In others, the system is not sufficiently stable at all potentials and it is then necesary to force the membrane to stay at various voltages by using the voltage-clamp technique (see Chapter 2, p. 7)

Fig. 11.9 illustrates how the steady state current–voltage relation (interrupted line) is related to the curves already shown in Fig. 11.8. The line has been drawn with a positive slope at all potentials, which is in accordance with the steady state current–voltage curve given by the Hodgkin–Huxley equations (Cooley, Dodge, and Cohen 1965, Fig. 7) and with the curves obtained experimentally in many excitable cells. There are, however, some exceptions, e.g. cardiac Purkinje fibres (see Noble and Tsien 1969a, Fig. 10). The steady state relation plotted in Fig. 11.9 intersects with the level of applied current at the point labelled 2. This is a point at which dV/dt is zero and, since the steady state slope is positive, the point might be stable. On the other hand, the slope conductance for rapid changes in potential is negative.

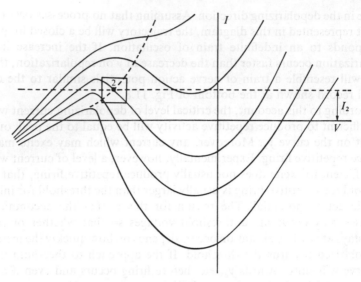

FIG. 11.9. First stage in simplifying system illustrated in Fig. 11.8 to determine stability conditions. The continuous curves are same relations as shown in Fig. 11.8. The interrupted curve is the steady state current–voltage relation. Point 2 is the intersection of this relation with the voltage axis after shifting by amount I_2. When current I_2 is applied the point 2 is a possible stable point. The area shown in square is linearized as shown in Fig. 11.10.

Perturbations about point 2, e.g. due to noise, may occur too quickly for their effects to be determined by the steady state relation. Clearly the stability of the system in the vicinity of this point requires a quantitative analysis. In order to do this we shall consider a small region in which the current–voltage relations may be treated as linear. This is illustrated in Fig. 11.10.

Conditions for instability: quantitative analysis. Since we are assuming the current–voltage relations to be linear, the effects of rapid changes in potential may be represented by a conductance g_1, which is the slope of the momentary current–voltage curve. This slope may be either positive or negative. The accommodative properties of the system may be represented by a time constant τ_y which is the time constant of the process y which changes the 'momentary' curve from one y value to another. This time constant may be called the time constant of accommodation. Its reciprocal, the rate constant of accommodation, will be called k. The steady state curve I_∞, may be characterized by a conductance g_2. This may be either positive or negative. In Fig. 11.10, we have assumed that g_2 is positive and that g_1 is negative, but the derivation of the equation does not require these assumptions.

In order to determine the conditions for stability, we first derive a general differential equation for the system. We may then use the results of nonlinear oscillation theory (see Jones 1961, Chapter 5) to obtain the properties of the system from the coefficients of the differential equation. For the sake of

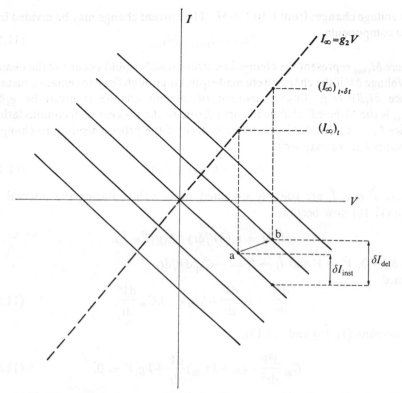

FIG. 11.10. Linearized region around point 2, which was used as the zero for the ordinate and abscissa in this diagram. The full lines show current–voltage relations for various values of y. The interrupted line shows the steady state current–voltage relation. Equations are obtained by considering the trajectory from an arbitrary point (a). The current change as the locus moves to point b is divided into an instantaneous component and a delayed component. See text for further explanation.

convenience we will define $V = 0$ as the point at which the steady state relation intersects the zero current axis (this assumption is arbitrary but trivial since, if V were not zero at this point that is, if we continued to use the convention that V is zero only at a quiescent point—see the section *Notation and definitions*, p. xiii—we could simply define a new variable on the voltage axis which is zero at the intersection point). We will assume also either that there is no applied current (in which case the voltage axis is a true zero-current axis and any instability obtained will be spontaneous instability) or that the presence of a steady applied current has already been allowed for by the device of shifting the voltage axis—see Fig. 11.8.

Differentiating eqn (8.43) we obtain

$$C_m \, d^2V/dt^2 + dI_i/dt = 0. \qquad (11.15)$$

We now obtain an equation for dI_i/dt using the model illustrated in Fig. 11.10. Consider the change in current which occurs between point a and point b as the voltage changes from V to $V+\delta V$. The current change may be divided into two components:

$$\delta I_i = \delta I_{inst} + \delta I_{del}, \tag{11.16}$$

where δI_{inst} represent the change in current which would occur for the change in voltage δV if the change were made quickly enough for y to remain constant. Since $\partial I/\partial V = g_1$ this component of current change is given by $g_1 \delta V$. δI_{del} is the 'delayed' change in current due to the process of accommodation. Since $I_\infty = g_2 V$ and k is the rate constant of the process tending to change I towards I_∞, we may write

$$\delta I_{del} = k(g_2 \hat{V} - \hat{I}_i)\delta t, \tag{11.17}$$

where $\hat{V}$ and $\hat{I}_i$ are suitably weighted mean values during the interval δt. Eqn (11.16) now becomes

$$\delta I_i/\delta t = g_1(\delta V/\delta t) + k(g_2 \hat{V} - \hat{I}_i).$$

As $\delta t \to 0$, $\hat{V} \to V$ and $\hat{I}_i \to I_i = -C_m \, dV/dt$.
Hence

$$\frac{dI_i}{dt} = g_1 \frac{dV}{dt} + kg_2 V + kC_m \frac{dV}{dt}. \tag{11.18}$$

From eqns (11.15) and (11.18),

$$C_m \frac{d^2 V}{dt^2} + (g_1 + kC_m)\frac{dV}{dt} + kg_2 V = 0. \tag{11.19}$$

The conditions for stability and instability may now be obtained by investigating solutions to eqn (11.19) for various cases. These solutions are well known and may be obtained from standard texts on the theory of oscillations (e.g. Andronow and Chaikin 1949; Jones 1961). The account given by Jones (1961, Chapter 5) is a relatively simple one, and may be useful to many physiologists since it requires relatively little mathematical knowledge. For a more advanced mathematical analysis of nonlinear systems the reader is referred to Cunningham (1958).

The general solution to eqn (11.19) is

$$V = A \exp(r_1 t) + B \exp(r_2 t),$$

where r_1 and r_2 are the roots of the auxiliary equation

$$C_m r^2 + (g_1 + kC_m)r + kg_2 = 0,$$

which are given by

$$r = \frac{-(g_1 + kC_m) \pm \sqrt{\{(g_1 + kC_m)^2 - 4kC_m g_2\}}}{2C_m}.$$

When g_2 is positive, five cases may be distinguished.

1. (g_1+kC_m) positive; $(g_1+kC_m)^2 > 4kC_mg_2$. r_1 and r_2 are then negative and real so that the solution will consist of a pair of decaying exponentials and the system will be stable at the origin.
2. (g_1+kC_m) positive; $(g_1+kC_m)^2 < 4kC_mg_2$. r_1 and r_2 are complex and their real parts are negative. The solution will then be oscillatory, but the magnitude of the oscillation will decay exponentially. Once again the system will be stable at the origin.
3. $(g_1+kC_m) = 0$. This is a special case when eqn (11.19) reduces to the equation for a simple harmonic oscillator. However, this condition is unlikely to hold over any substantial range of potentials in excitable cells and is of little physiological interest.
4. (g_1+kC_m) negative; $(g_1+kC_m)^2 < 4kC_mg_2$. The roots are complex but their real parts are positive. The solution will be oscillatory and the oscillation amplitude will increase exponentially. The system is now unstable.
5. (g_1+kC_m) negative; $(g_1+kC_m)^2 > 4kC_mg_2$. The roots are real and positive. The solution is an increasing exponential away from the origin and the system is unstable.

These results are summarized in Fig. 11.11 which shows the forms of the trajectories and $V(t)$ curves. It can be seen that the condition for instability is

$$-g_1 > kC_m. \tag{11.20}$$

This requires that the negative slope conductance for rapid changes should be larger in magnitude than the product of the membrane capacitance and the rate constant of the accommodative process. In general, the faster the accommodative process, the more stable the system will be. This 'stabilizing' effect of accommodation will be discussed further below (p. 352). It is also clear that the existence of a negative slope conductance for rapid change is not by itself a sufficient condition for instability. The negative slope conductance must be large enough for condition (11.20) to hold. It may be seen that, when g_2 is positive, it does not determine the condition for instability. However, it does determine the form of the solutions obtained. When g_2 is large, the term in column 2 of Fig. 11.11 will tend to be negative and the trajectory will be a spiral, corresponding to damped or increasing oscillations. When g_2 is sufficiently small, the intersection point will form a node and the system will move towards or away from the intersection point with little or no oscillation.

Applications of linearization theory. Chandler, FitzHugh, and Cole (1962) used an analysis of the kind described in the preceding section to study the stability of the squid axon membrane under incomplete voltage-clamp control. Cooley, Dodge, and Cohen (1965) have extended the approach to analyse the stability properties of the Hodgkin–Huxley equations as a function of the magnitude of applied current. Their analytical results were

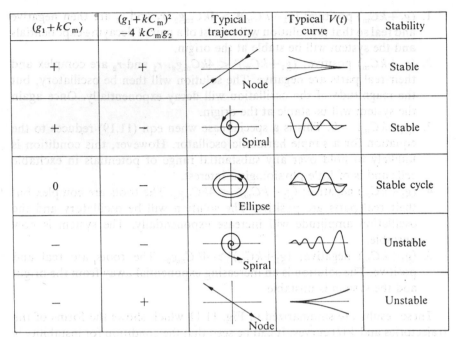

$(g_1 + kC_m)$	$\dfrac{(g_1 + kC_m)^2}{-4\,kC_m g_2}$	Typical trajectory	Typical $V(t)$ curve	Stability
$+$	$+$	Node		Stable
$+$	$-$	Spiral		Stable
0	$-$	Ellipse		Stable cycle
$-$	$-$	Spiral		Unstable
$-$	$+$	Node		Unstable

FIG. 11.11. Summary of types of solution to eqn (11.19) when g_2 is positive. Note that stability depends on the sign of the term $(g_1 + kC_m)$. Whether the solutions are oscillatory or monotonic is determined by the sign of the expression given in the second column. In the trajectory graphs, the ordinate is $-\dot V$ and the abscissa is V.

compared with numerical computations. For the most part, the results were consistent with those expected from analysis of linearized systems. Near the resting potential, the linearized model predicts damped oscillations (cf. Chapter 10, p. 303). On the other hand, relatively strong currents (at least 3 times rheobase) produce an unstable point in the linearized system corresponding to the maintained oscillatory activity which was computed.

However, a rather unexpected result appeared in the numerical calculations over an intermediate range of currents (between 1·5 and 3 times rheobase). The membrane characteristics near the singular point are stable and, as expected, *small* perturbations produced slowly decaying oscillations; but *larger* perturbations gave rise to a series of oscillations which grow to a continuous train of impulses. In this instance, a stable limit cycle surrounds the stable quiescent potential, like a moat surrounding a well.

More detailed applications of linearization theory to nerve axons have been described by Sabah and Liebovic (1969) and Mauro, Conti, Dodge, and Scher (1970).

Eqn (11.20) may be used to produce an interesting comparison between the conditions for repetitive firing in nerve and cardiac membranes. In nerve, the rate constant of accommodation is of the order of 1 ms^{-1} and the capacitance is about 1 μF cm^{-2}. The negative slope conductance must therefore be of the order of 1 mS cm^{-2} or more for instability to occur. In cardiac cells, the rate constant is of the order of 1 s^{-1}. If the capacitance is assumed to be 10 μF cm^{-2} (which is the case in Purkinje fibres) the minimum slope conductance is only about 10 μS cm^{-2} (cf. Hauswirth, Noble, and Tsien 1969, p. 259). Repetitive activity may therefore occur in cardiac membranes with negative slope conductances which are about two orders of magnitude smaller than required for repetitive firing in nerve cells.

van der Pol's equation. Although linearization may be sufficient for determining the stability of a given point, the analysis does not apply when the trajectory leaves the region over which the current–voltage relations may be assumed to be at least approximately linear. We then require a system which includes nonlinear current–voltage relations. The simplest system of this kind was first developed by van der Pol (1926), who obtained an equation for a valve oscillator with a nonlinear relation between anode current and grid voltage. We shall first show that van der Pol's equation may also be derived from a simple extension of the model we have described so far. As in van der Pol's model, we assume that the current–voltage relation for rapid changes in potential may be represented by a cubic equation (see Fig. 11.12):

$$I_i(V, t = \text{constant}) = g_1 V + \alpha V^3 + I_{\text{del}}, \qquad (11.21)$$

where g_1 is a negative constant giving the negative slope conductance near $V = 0$, α is a positive constant, and i_{del} is the amount of shift in the $i(V)$ curve due to accommodation. As before, $i_\infty = g_2 V$. These equations give a system with N-shaped $i(V)$ relations which shift upwards when the membrane is depolarized and downwards when the membrane is hyperpolarized from the steady state intersection point. The system therefore possesses all the major characteristics of excitable cells. We shall discuss later (see p. 328) the importance of the simplifications which are made in obtaining van der Pol's equation.

Differentiating eqn (11.21):

$$\left(\frac{dI_i}{dt}\right)_{\text{inst}} = (g_1 + 3\alpha V^2)\frac{dV}{dt}. \qquad (11.22)$$

The delayed component of current change is given by eqn (11.17) as before. Hence

$$\frac{dI_i}{dt} = (g_1 + kC_\text{m} + 3\alpha V^2)\frac{dV}{dt} + kg_2 V, \qquad (11.23)$$

and we obtain

$$C_\text{m}\frac{d^2V}{dt^2} + (g_1 + kC_\text{m} + 3\alpha V^2)\frac{dV}{dt} + kg_2 V = 0. \qquad (11.24)$$

This is identical with eqn (11.19) except for the additional term $3\alpha V^2$ in the coefficient of dV/dt. However, it is important to note that the conditions of

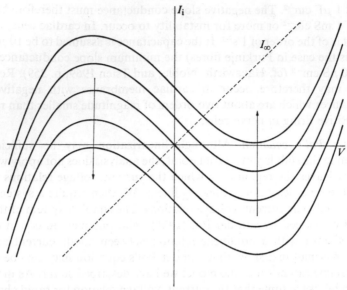

FIG. 11.12. Representation of van der Pol's model in terms of current–voltage diagrams. The continuous curves are solutions of eqn (11.21) for various values of I_{del}. The interrupted line is the steady state current–voltage relation. Since I_{del} changes so as to shift I_i towards I_∞, the current–voltage curves shift upwards when the voltage is positive and downwards when the voltage is negative (as indicated by arrows).

application of the two equations are quite different. Eqn (11.19) applies only to the vicinity of one point in the phase-plane, whereas eqn (11.24) may apply to the whole phase-plane. Before discussing solutions, it is convenient to rearrange eqn (11.24) into a more familiar form. Let $T = \psi t$, where $\psi = \sqrt{(kg_2/C_m)}$. We will call ψ the 'accommodation factor' since it is given by the variables which determine the rate and magnitude of accommodation. Eqn (11.24) becomes

$$\frac{d^2V}{dT^2} + \frac{1}{\psi}\left(\frac{g_1}{C_m} + k + \frac{3\alpha V^2}{C_m}\right)\frac{dV}{dT} + V = 0. \tag{11.25}$$

Now make the substitution

$$w = V \Big/ \sqrt{\left\{-\left(\frac{g_1}{C_m} + k\right)C_m/3\alpha\right\}}.$$

We then obtain

$$w'' - \varepsilon(1 - w^2)w' + w = 0, \tag{11.26}$$

where $\varepsilon = -(g_1/C_m + k)/\psi$ and the prime notation denotes differentiation with respect to T. When ε is positive, eqn (11.26) is van der Pol's equation. Solutions to this equation are well known and we shall simply summarize some of the important results before discussing their application to excitable cells.

1. When ε is very small (that is, for large values of k, g_2, C_m, or small values of g_1—provided the latter is negative and sufficiently large for ε to be positive) the equation is that for a simple harmonic oscillator with a very small nonlinear damping term. The solution is then approximately

$$w = 2 \cos T = 2 \cos(\psi t). \qquad (11.27)$$

The frequency is proportional to the accommodation factor, that is,

$$f \propto \sqrt{(kg_2/C_m)} \simeq \sqrt{(-g_1 g_2/C_m^2)}. \qquad (11.28)$$

Thus f is proportional to the square root of the rate and magnitude of accommodation, and since for ε to be very small, $k \simeq -g_1/C_m$, the frequency will also be proportional to $\sqrt{(-g_1)}$. Moreover, $-g_1$ must exceed kC_m for oscillatory activity to occur. The error in this estimate of f is of the order ε^2 (Jones 1961, p. 85).

2. When ε is large (i.e. for small values of k, C, and g_2 or large values of g_1), the trajectory is more complex. The equation for the trajectory may be obtained directly from eqn (11.26) and is

$$-dw'/dw(\propto dI_i/dV) = (w/w') - \varepsilon(1 - w^2). \qquad (11.29)$$

On integration it may be shown (see Jones 1961, pp. 90–3) that the trajectory has two regions in which its slope and w' are very small and two regions in which it closely follows the current–voltage relation given by eqn (11.21) (with I_{del} equal to a constant). The relation of the trajectory to the current–voltage relation is shown in Fig. 11.13(a). Between $w = 2$ and $w = 1$, the curve closely follows the intersection of the momentary current–voltage curve as it shifts upwards owing to accommodation. I_i and dV/dt are then very small and the result is the formation of a 'plateau' (see Fig. 11.13(b)). Below $w = 1$, the trajectory follows the momentary current–voltage curve and the system shifts rapidly towards $w = -2$. The trajectory in the positive direction then follows the intersection point up to $w = -1$ (this phase corresponds to the 'pacemaker depolarization'—see Fig. 11.13(b)). Above $w = -1$, the potential changes rapidly again towards $w = 2$. This trajectory is a quantitative example of the qualitative description of oscillatory activity given earlier in this chapter (see p. 317).

The period of the oscillation is of the order of $2\varepsilon/\psi$. Since g_1 is assumed to be large, we obtain

$$f \to -kg_2/2g_1. \qquad (11.30)$$

In contrast to eqn (11.28), the frequency is now directly proportional to the rate and magnitude of accommodation, and C_m is unimportant.

Applications of van der Pol's equation and related models to excitable cells

1. *Cardiac pacemaker activity.* The curve shown in Fig. 11.13(b) bears a strong resemblance to records of electrical activity in cardiac muscle cells. van der Pol's equation has therefore been used as a model of the cardiac

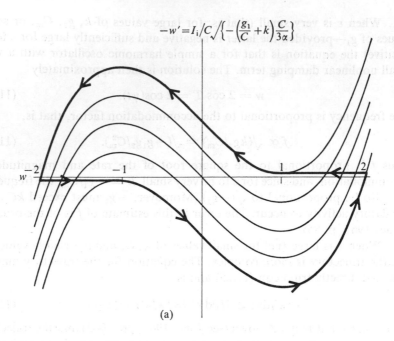

$$-w' = I_i/C \bigg/ \left\{ -\left(\frac{g_1}{C} + k\right)\frac{C}{3\alpha} \right\}$$

(a)

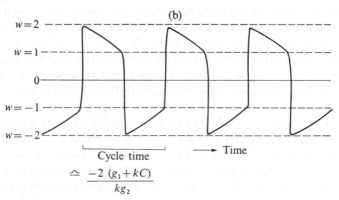

(b)

$$\text{Cycle time} \simeq \frac{-2\,(g_1 + kC)}{kg_2}$$

FIG. 11.13. (a) Trajectory given by a solution to van der Pol's equation when ε is large. The trajectory shown is the stable limit cycle towards which all other trajectories converge. (b) Voltage–time curve corresponding to trajectory shown in (a). Note the strong resemblance to action potentials and pacemaker potentials observed in cardiac muscle fibres.

pacemaker (e.g. van der Pol and van der Mark 1928). Since the mechanism of the pacemaker potential in cardiac Purkinje fibres is now known in terms of ionic current changes (see p. 351; Noble and Tsien 1968; McAllister, Noble, and Tsien 1975), it is possible to assess the usefulness and limitations of the van der Pol model.

In general, the model is correct in attributing the oscillations to approximately vertical shifts in an N-shaped current–voltage relation. Moreover, the constants g_1, g_2, and C_m may all be given at least approximate experimental values. g_1 is given by the slope of the negative-slope region of the current–voltage relations which are obtained when time is allowed for m and h to reach steady state values, g_2 is given by the slope of the steady state current–voltage relation (or an approximate linear equivalent), k is given by the rate coefficient of one of the outward current systems (since these change exponentially with time in Purkinje fibres, there is no difficulty arising from power function kinetics), and C_m is the observed membrane capacitance. Nevertheless, there are some important differences between the van der Pol model and a numerical model based on voltage-clamp data. In the latter, g_2 is not a constant (the steady state current–voltage relation is nonlinear) and the shape of the momentary $I(V)$ curves is not independent of the process of accommodation. However, these differences are of relatively minor importance and would lead only to quantitative errors. Of more importance is the fact that experimentally it is found that k is a very steep U-shaped function of the membrane potential, as expected for a system obeying equations of the Hodgkin–Huxley type, where $k = \alpha + \beta$ (see Chapter 8, Fig. 8.7). Moreover, in the case of Purkinje fibres there is more than one process which is responsible for shifting the momentary current–voltage curve. In particular, there are two functionally separate processes (with quite different values of k at each potential) which control the 'plateau' and 'pacemaker' phases of the oscillation.

The importance of the nonlinear dependence of k on potential may be illustrated by comparing the way in which the models might account for the increase in frequency produced by adrenaline. An obvious way of producing this effect in the van der Pol model is to increase k (see eqn (11.26) and (11.30)). In practice, this is how the acceleration is achieved (Hauswirth, Noble and Tsien 1968). However, the van der Pol model would be misleading if used to explain this effect. The experimental action of adrenaline is to shift the $k(V)$ curve on the voltage axis (see p. 351), so that at some potentials k is larger than before while at other potentials it is smaller. The effect, therefore, is not equivalent to a uniform increase in k at all potentials. Moreover, adrenaline increases the equivalent of g_2, which also has the effect of increasing f in the van der Pol model. This must be represented as an additional action in the van der Pol system, whereas, in the Hodgkin–Huxley equations a shift in the voltage dependence of the rate constants automatically produces a corresponding shift in the steady state relations since the latter are also functions of the rate constants (Chapter 8, eqns (8.35) and (8.36)).

We may conclude, therefore, that, although the van der Pol model is a surprisingly simple mathematical description of a process similar to the cardiac pacemaker, it could be misleading if used without reference to numerical reconstructions based on measurements of voltage-clamp currents. It is best regarded as an extreme simplification of the numerical model of the

Hodgkin–Huxley type which may be used to obtain general conditions for oscillatory activity.

2. *Repetitive activity in nerve: FitzHugh's BVP model.* In normal conditions, isolated nerve axons do not show spontaneous repetitive activity. Van der Pol's equation must therefore be modified if it is to serve as a simplified model. FitzHugh (1960, 1961, 1969) has done this by modifying the equation for the time-dependent component of current change. In place of

$$(\mathrm{d}I_{\mathrm{i}}/\mathrm{d}t)_{\mathrm{del}} = k(g_2 V + C\,\mathrm{d}V/\mathrm{d}t), \tag{11.31}$$

which is given by subtracting the instantaneous current change from eqn (11.18), FitzHugh's model uses an equation of the form

$$(\mathrm{d}I_{\mathrm{i}}/\mathrm{d}t)_{\mathrm{del}} = k\{g_2 V + A - B(I_{\mathrm{i}})_{\mathrm{del}}\}, \tag{11.32}$$

where A and B are constants. By replacing $C_{\mathrm{m}}\,\mathrm{d}V/\mathrm{d}t$ by $A - B(I_{\mathrm{i}})_{\mathrm{del}}$, the steady state current–voltage relation becomes nonlinear. The addition of the constant A ensures that the steady state intersection with the voltage axis is moved to a stable point which corresponds to a quiescent (resting) potential. Since the new equations bear some resemblance to a model developed by Bonhoeffer (1948), FitzHugh has called his model the Bonhoeffer–van der Pol (BVP) model.

The BVP model reproduces many of the features of nerve excitation, including threshold phenomena and repetitive firing to constant depolarizing currents. Like the van der Pol model, however, it does not represent the voltage dependence of the rate constants and steady state values of current change. One consequence of the lack of voltage dependence of the rate constants is that the duration of the action potential is comparable to the interval between action potentials—hence the model is not a very realistic simulation of low-frequency firing in nerve (see Fig. 11.13). There is also no simple way or reproducing the action of divalent ions (see p. 348) which, like adrenaline on cardiac membranes (see p. 351), change the voltage dependence of the Hodgkin–Huxley rate coefficients. Nevertheless, the relative simplicity of the model, compared to the Hodgkin–Huxley equations, allows some general properties to be investigated more easily. Readers interested in the analysis of models of this kind should consult FitzHugh's papers. (For the purpose of comparison with FitzHugh's work, it may be helpful to have the following conversions to FitzHugh's variables: $(I_{\mathrm{i}})_{\mathrm{del}} \propto W$; $w \propto V$; $kg_2 \propto \phi$; $A \propto a$; $B \propto b$; where the second symbol in each case is a variable or constant in FitzHugh's formulation of the BVP model. We should also explain that FitzHugh deliberately uses dimensionless variables where possible and does not directly interpret the constants in terms of conductances, rate coefficients of current change, and capacitances. The particular physical interpretation implied by the use of g_1, g_2, k, and C_{m} in our notation is not required by FitzHugh's equations.)

Numerical computation of repetitive activity

Although the simplified models discussed in the previous sections are useful for explaining the mechanism and general properties of repetitive activity they need to be tested to determine whether they are accurate descriptions of particular excitable cells. As a first step, this test can be performed by obtaining the predicted behaviour of the cell from full numerical computations which include all the ionic currents measured by a voltage-clamp analysis. This more sophisticated model can also be compared with the actual behaviour of the cell.

As yet there are very few cells for which this ideal procedure can be followed. Most of the following discussion therefore will be confined to the large unmyelinated axons of squid and crab, frog myelinated nerve fibre, and a gastropod nerve cell.

It has already been noted at the beginning of this chapter that the firing pattern in nerves encodes the information that they transmit. The relation between frequency of firing and stimulus strength is therefore a very important one.

Current–frequency relation

Axons of squid and crab. Several workers have made calculations for repetitive firing of the Hodgkin–Huxley model of the squid axon (Cole, Antoziewicz, and Rabinowitz 1955; FitzHugh and Antoziewicz 1959; Huxley 1959; FitzHugh 1961; Agin 1964; Cooley, Dodge, and Cohen 1965; Wright and Tomita 1965; Cooley and Dodge 1966; Stein 1967a; Guttman and Barnhill 1970; Dodge 1972). Most of these authors have made the calculations for the case of a uniformly polarized axon.

Fig. 11.14 is taken from Stein's (1967a) paper and shows the relation between current and frequency. Notice that above a critical value of 6·5 μA cm^{-2} (the 'rheobasic' current) the frequency jumps up to a value of about 50 Hz, and then the curve shows an upward convexity. There is a fairly limited range of frequencies at which the membrane fires—even with a current 10 times 'rheobasic,' the frequency is less than 125 Hz. At even stronger currents the membrane potential shows oscillations but does not generate negative current. In the range of frequencies 50–125 Hz the curve is accurately fitted by the equation $f = 27 \ln (I+1)$, where I is the current applied to a unit membrane area (Agin 1964).

Another feature of Stein's calculations is that the minimum current required to produce prolonged repetitive firing (the 'rheobasic' current, = 6·5 μA cm^{-2}) is much larger than that required to produce a single spike (2·24 μA cm^{-2}, Noble and Stein 1966). This nearly three-fold difference in current strength shows that any simple model which attempts to predict the interval between action potentials by means of a strength–duration relationship (which is the main principle of the procedure adopted in the first models described in this chapter) is unlikely to be satisfactory. The difference between

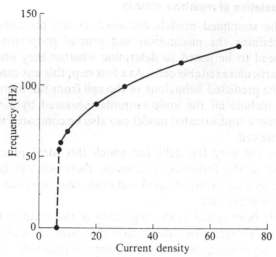

FIG. 11.14. Frequency–current curve for the uniformly polarized Hodgkin–Huxley model of the squid axon. (From Stein 1967*a*.)

the two current strengths (true rheobase and 'rheobase' for prolonged repetitive firing) may be attributed to a steady level of accommodation produced by the impulse train (cf. p. 353). The current strengths intermediate between these two values could initiate repetitive firing, but the train terminates after a variable number of impulses. Stein (1967*a*) noted, however, that the frequency during the longer of such transient periods of firing was similar to that maintained in a prolonged train by a slightly stronger stimulus.

Hagiwara and Oomura (1958) described experiments showing that there was only a limited band of frequencies at which the squid axon would fire. They were unable to produce maintained firing. It is possible that their axons were not healthy, since Tasaki (quoted in FitzHugh 1961) and Chapman (1966) have reported that it is possible to produce such maintained firing.

Guttman and Barnhill (1970) have compared experimental observations with numerical calculations for the uniformly polarized squid axon membrane in the circumstance of lowered external calcium concentration. The observations were made at several temperatures. Fig. 11.15 shows these results and some matching calculations. The left-hand side of the graph shows the experimental current–frequency relations and the theoretical calculations are on the right-hand side; there is quite close agreement between them. Guttman and Barnhill always used fairly long (300 ms) current pulses, so that it is likely that the frequencies they recorded would be maintained.

With higher current strengths their experiments showed 'skip-runs' in the frequency (a sequence of oscillations with every second, third, or fourth oscillatory peak giving rise to an action potential) and they noted that this phenomenon could not be reproduced with the numerical calculations for

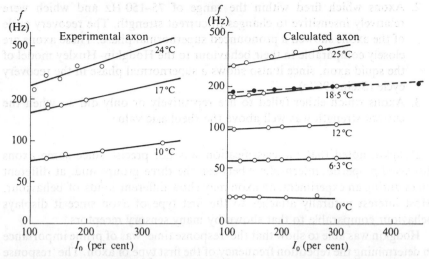

FIG. 11.15. Frequency–current curves for both the squid axon (experimental axon) and the Hodgkin–Huxley model of the squid axon (calculated axon) at different temperatures. Note that the frequency scale is logarithmic. The current density I_0 is expressed as a percentage of the rheobasic value. The broken line and filled circles illustrate the effect of increasing all the conductances by a factor of 1·5 when going from the 'standard' temperature of 6·3 °C to 18·5 °C (From Guttman and Barnhill 1970.)

uniform polarization, but Stein (1967*a*, Fig. 4) has reported such a phenomenon for the point-polarized cable calculations.

As mentioned above, Stein (1967*a*) has also illustrated the current–frequency relation calculated for a point-polarized Hodgkin–Huxley model of the squid axon. The frequency range is even more narrow, being restricted to 50–100 Hz. As far as we know there are no experimental observations with which this result can be compared.

The Hodgkin–Huxley model of the squid axon therefore does make reasonably satisfactory predictions of the behaviour of that cell. More extensive comparisons might still be needed however, particularly in the light of recent suggestions that the Hodgkin–Huxley description of the sodium current is inadequate (Goldman and Schauf 1972, 1973). One important property of the model is that it does not predict any adaptation in the firing rate with time (see p. 352). The model will certainly not serve as a general description of the current–frequency relation of nerve axons, since some can maintain a much lower firing rate than 50 Hz in response to a steady stimulus. The classical description of such behaviour in certain crab axons was given by Hodgkin (1948). Hodgkin classified the crab axons he studied into three main groups.

1. Those axons which could be made to fire over a range of about 5–150 Hz. He noted that the recovery cycle of such axons showed no measurable supernormal phase.

2. Axons which fired within the range of 75–150 Hz and which were relatively insensitive to changes in current strength. The recovery cycle of the axons showed a pronounced supernormal phase. These axons are closely comparable in their behaviour to the Hodgkin–Huxley model of the squid axon, since it also shows a supernormal phase in the recovery cycle (Stein 1967a).
3. Axons which either failed to fire repetitively or only did so when the current strength was well above the rheobasic value.

Hodgkin noted that the classification was not precise since some axons displayed properties intermediate between the three groups and, at different times during an experiment, an axon may show different kinds of behaviour. Most interest naturally attaches to the first type of axon since it displays behaviour comparable to that shown by many sensory receptors.

Hodgkin was able to show that the 'response time' was of prime importance in determining the repetition frequency of the first type of axon. The 'response time' is the time between the make of a constant current and the first action potential. A simple way to display the correlation between the response time and the repetition interval is to plot them both against the current strength. Fig. 11.16 is taken from Hodgkin's (1948) paper and shows that, although the two curves are not identical, they are very similar. The main deviations are near rheobase and at high current strengths; presumably the explanation of the latter deviation is that refractoriness becomes important at higher firing frequencies.

Hodgkin pointed out two factors that affected the response time: first there was the rate of development of a passive depolarization of the membrane time constant; secondly, superimposed on this passive potential there is a local response and this subthreshold response grows at a rate which is partly dependent on the strength of the applied current. A similar mechanism operates for subsequent action potentials in a repetitive train.

Chapman (1966) confirmed and extended Hodgkin's observations. He also found a fairly close correlation between the response time and the mean spike interval for the type-1 axon, but not for the other two types. This is illustrated in Fig. 11.17, taken from Chapman's paper. The behaviour of all three types of axon is illustrated. Notice that there is a difference between the curves relating the reciprocal mean interval and the reciprocal last interval to current strength for the type-1 axon; in other words, the frequency of firing is displaying adaption. The second type of axon displays an interesting piece of behaviour: for a given current strength the reciprocal mean interval is less than the reciprocal of the response time (see later). The relationship between mean frequency and current for the third type of axon is rather like that displayed by the Hodgkin–Huxley model of the squid axon, but unlike the latter will not give a maintained train of impulses.

How can the behaviour of the type-1 crab axon be explained? Hodgkin

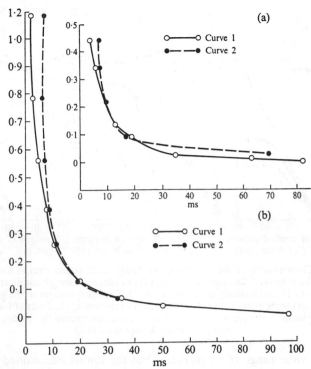

FIG. 11.16. Comparisons of the relation between current strength (expressed as log (current/rheobase)) and 'response time' (Curve 1) and of current strength and the interval between the first and second action potentials (Curve 2). (a) and (b) are different axons. (From Hodgkin 1948.)

(1964) suggested that relatively small modifications to some of the parameters of the Hodgkin–Huxley model of the axon would be sufficient to simulate this behaviour; he had observed that these axons differed from the squid axon in their steady state current–voltage relation, which showed a much higher membrane resistance for small depolarizations (Hodgkin, quoted by Dodge, 1972; see also Chapman 1966). This is reflected in the much longer membrane time constant for crab nerve (5–15 ms, Hodgkin 1948) than for squid axon (about 1·0 ms, calculated from the data of Hodgkin and Huxley (1952d)).

Dodge (1972) tested Hodgkin's prediction by modifying two parameters of the Hodgkin–Huxley model; he reduced the leakage conductance to half its normal value and also shifted the voltage sensitivity of the potassium activation rate constants by 8 mV in the depolarizing direction. Fig.11.18 shows the results he obtained; at current strengths near 'rheobase' the frequency was about 10 Hz. Stein (1967a) made a slightly different modification; he reduced the leakage conductance to one-fortieth of its normal value and shifted the sodium h variable by 20 mV in the depolarizing direction. He

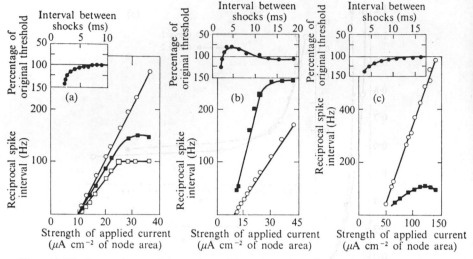

FIG. 11.17. Comparisons of the reciprocal 'response time' (open circles) and reciprocal mean interspike interval (filled squares) in relation to current strength for all three types of crab axon (types 1–3 respectively from left to right). The open squares in the graph for the type-1 axon plot the reciprocal of the last interval. The current applied was of 1 sec duration. The inset graphs show the recovery cycle for each axon determined by the double stimulus technique. (From Chapman 1966.)

obtained a wider range of frequencies, but the lowest maintained frequency was 30 Hz. Neither Stein's nor Dodge's frequency–current curves are very satisfactory simulations of either Hodgkin's or Chapman's observations (see Figs. 11.17 and 11.20; Dodge's curve, for example, is much flatter and required a current strength more than eight times 'rheobase' to reach a frequency of 70 Hz) but they do suggest that a suitably modified version of the Hodgkin–Huxley equations could match the observed current–frequency behaviour.†

How well do the simplified models described at the beginning of this chapter predict the current–frequency relations of squid and crab axon? Attempts to fit either the uniformly polarized or point-polarized data of Stein (1967a) with the appropriate simple model (including absolute and relative refractory periods) do not yield a satisfactory agreement. The explanation for this could, of course, be simply the result of not allowing for a supernormal phase in the recovery of excitability following an action potential. It is easy to introduce a modification in the model so that the supernormal period is simulated. A mathematical function of the form

$$V_{th}(t) = \frac{V_{th}(\text{rest})}{1 - \exp(-t/\tau_R) + \alpha t^\beta \exp(-\gamma t)}$$

† A detailed investigation of the effect of various modifications to the Hodgkin–Huxley equations on the frequency–current relation has recently been reported by Shapiro and Lenherr (1972).

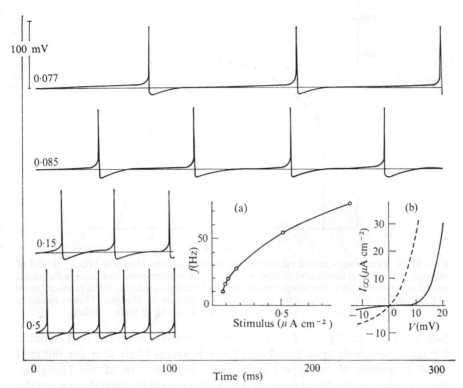

FIG. 11.18. Simulation of the firing frequency behaviour of the type-1 crab axon. (a) Current–frequency curve obtained.   (b) The solid line shows the form of steady state current–voltage relation assumed. Note the very low slope conductance for the first 10 mV. depolarizing displacement (cf. the 'standard' current–voltage relation indicated by the dashed line). The rest of the figure illustrates examples of the computed time course of the membrane potential. This figure was kindly provided by F. A. Dodge (see also Dodge (1972)).

could be fitted to the recovery cycle, or else direct measurement of the experimental curve for excitability can be employed. Such measurements have been made from Stein's (1967a) Fig. 10 and incorporated into a simple model for point polarization. The membrane time constant has been taken as 1·5 ms. The liminal length was assumed to be 0·77λ (see Chapter 12). The continuous line in Fig. 11.19 shows the result of such calculations, whereas the filled circles represent Stein's numerical calculations. Both calculations reveal the property of the frequency jumping abruptly up to the value of about 50 Hz but for all stronger currents the simple model predicts much higher firing frequencies. The abrupt jump in the simple model is explained by the fact that, because of the supernormal phase, the excitability goes through a maximum at 20 ms. In this model, once the first action potential is initiated, a current lower than rheobase would sustain repetitive firing whereas exactly the opposite effects holds for the Hodgkin–Huxley calculations

12

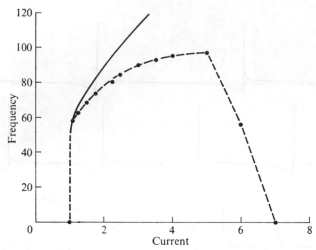

FIG. 11.19. Frequency–current curves for the point-polarized Hodgkin–Huxley model of the squid axon (filled circles and dashed line) and for the point-polarized voltage-threshold model (continuous line). The simple model has had incorporated into it the recovery cycle of the Hodgkin–Huxley model of the squid axon. The data on the Hodgkin–Huxley model of the squid axon was obtained from Figs. 5 and 10 of Stein (1967a).

(see Stein 1967a, Fig. 4). This result was, however, observed by Chapman for the type-2 crab axon (see Fig. 11.17). It seems likely that the minimal firing frequency of the type-2 crab axon, and perhaps of the Hodgkin–Huxley model of the squid axon, is set by the interval between the start of the action potential and the peak of a supernormal phase. This interval, however, may not be identical when a condition of constant current stimulation is compared with the recovery cycle measured by a double shock technique. Hodgkin (1948) reported that there was no correlation between the minimal firing and the time to peak of the supernormal phase when measured with a double shock technique, and the same result can be seen in Chapman's observations (see Fig. 11.17). Thus even if the explanation of the minimal firing frequency is correct, the fact that the recovery cycle may be influenced in its time course by a steady current makes any attempt at simulation by a simple model inappropriate.

Can nerve axons which do not have a significant supernormal phase be simulated with a simple model? Type-1 crab axons are of this kind and Fig. 11.20 shows the frequency–current relation observed by Chapman (filled circles, measured from Fig. 11.17) and one of Hodgkin's results (open circles, measured from Fig. 11.16). Both Hodgkin and Chapman show the time course of recovery of this type of axon, from which estimates of the absolute refractory period (2·5 ms) and the time constant of the relative refractory period ($\simeq$ 1·0 ms) can be derived. Unfortunately neither author provides a measurement of the membrane time constant for these particular axons, but Hodgkin mentions that his axons displayed time constants in the range 5–15 ms; the membrane resistance value quoted by Chapman would imply a membrane time constant of about 10 ms (assuming the membrane capacitance is slightly greater than 1 μF cm^{-2}). Therefore, a time constant of 10 ms has been assumed.

For the simulation by a simple model it is also necessary to know the geometry of current spread. Both Hodgkin and Chapman used extracellular electrodes only; Chapman's experimental design is an approximation to uniform polarization whereas Hodgkin's design is more like that of a point polarization. The right-hand line in Fig. 11.20 shows the predictions of a simple model with uniform polarization, using the above values of time constant, absolute and relative refractory periods. It is not a good fit to Chapman's points (filled circles).

In order to simulate Hodgkin's data it is also necessary to know the value of the liminal length. No information is available for these axons about the slope of the momentary current–voltage relation in the regions of uniform threshold, but if it is similar to that for the squid axon then the liminal length for the crab axon will be shorter than for the squid axon by a factor given by the square root of the ratio of the membrane slope conductances at the resting potential (see Chapter 12). A crab axon with a membrane time constant of 10 ms therefore should have a liminal length of

$$0.77 \times \sqrt{(1.5/10)} \lambda \simeq 0.3\lambda.$$

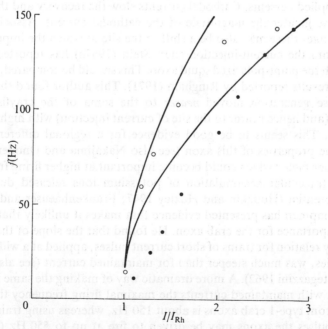

FIG. 11.20. Frequency–current relations for type-1 crab axons (filled and open circles) and for the corresponding simple models. See text for further details.

The left-hand line in Fig. 11.20 shows the frequency–current relation predicted by a simple model with this liminal length. The fit to Hodgkin's data (open circles) is only approximate, but the main deviation occurs at higher frequencies.

Considering the uncertainties in the values to be used in the simple model, it is quite encouraging that there can be an approximate fit, at least for the lower firing frequencies. This may not be a coincidence since the membrane properties of the type-1 crab axon make it more suitable than the squid axon to be simulated in this way. It will be remembered that the prime assumption of the simple models is that the membrane time constant is much longer than the time constants of the parameters controlling the action potential. Not only is the membrane time constant of the crab axon much longer, but Chapman has also presented evidence that the membrane resistance recovers much more rapidly after an action potential than for the squid axon. This is also manifest in the time course of recovery from refractoriness.

The deviations that the theoretical curves can display at higher current strengths, when compared with the experimental observations, is hardly surprising. The steady depolarizing current maintained between each action potential will tend to prevent recovery from sodium inactivation and lead to a maintained activation of the potassium channel (cf. Fig. 11.21); both these effects will become greater with larger currents. Chapman (1966) has shown that the process of membrane recovery following an action potential is sensitive to applied currents. Cathodal currents slow the recovery and the effect is greater the greater the magnitude of the cathodal current. Associated with these changes there may also be a shift in the site at which the impulses arise (away from the current-injection site). Stein (1967a) has reported such an effect with the point-polarized squid axon. This should be compared, however, with the results reported by Ringham (1971). This author found that the site of impulse generation moved *nearer* to the soma of the crayfish stretch receptor (and hence nearer to the site of current injection) with higher current strengths. This seems to be good evidence for a regional difference in the membrane properties of this axon (see also Nakajima and Onodera 1969a).

A further factor which could become important at higher firing frequencies is the extracellular accumulation of potassium ions released during each action potential (Hodgkin and Huxley 1947; Frankenhaeuser and Hodgkin 1956). Chapman has presented evidence that makes it unlikely that this is of major importance for the crab axon. He found that the slope of the current–frequency relation for trains of short current pulses, applied at a wide range of frequencies, was much steeper than for maintained current (see also Fuortes and Mantegazzini 1962). A more dramatic way of making the same point is to note that with maintained currents the maximal firing frequency that can be elicited from type-1 crab axons is about 150 Hz, whereas using trains of brief current pulses the axons may be driven to fire at up to 550 Hz (Chapman 1966).

It seems likely, therefore, that the main reason why the experimental observations in the type-1 crab axon differ from the simple model at higher firing frequencies is a result of the maintained current causing a greater delay in the recovery from an action potential, and this effect is only manifest when the interval between action potentials approach the recovery time. The two possible mechanisms mentioned were maintained activation of the potassium conductance and a slower recovery from sodium inactivation. Qualitatively the first mechanism will lead to a decreased membrane time constant, an increase in the liminal length, a decrease in the input resistance, and a rise in the voltage threshold, whereas the sodium inactivation mechanism would lead to an increase in the liminal length and a higher voltage threshold. A decrease in the membrane time constant would tend to steepen the slope of the frequency–current curve but the increase in liminal length, decrease in input resistance, and the increase in voltage threshold would have an opposite action. Obviously the net result would depend on the relative magnitude of the effects on sodium inactivation as compared with that of maintained potassium activation. Confining the interpretation to the simple model, it is obvious that the experimental results suggest that the effect of a decreased membrane time constant is outweighed by the other factors.

It may be in order, at this point, to explain why we have given such a detailed discussion of the appropriateness of simple models for the simulation of repetitive firing in nerve. As will be discussed briefly later, the models used by various workers to predict the firing patterns of nerve cells when the synaptic (or receptor) current is varying with time have been of the simple kind. This is a natural consequence of either attempting to obtain mathematical formulae which describe the firing behaviour or of making computer simulations of the firing patterns. Using a more complicated model, such as the Hodgkin–Huxley description, would not yield analytical solutions and would demand a possibly prohibitive amount of computing time. It is fortunate therefore that the simple models do seem to be reasonably appropriate for some, although not all, nerve axons in the lower firing-frequency range. At least, they do render less mysterious the fact that some axons can fire at much lower maintained frequencies than the squid axon.

A rather different way of interpreting the minimal firing frequency of a type-1 crab axon is in terms of oscillation theory. The essence of the successful modification (either of the Hodgkin–Huxley equations or of the simple model) is to reduce the slope of the steady state current–voltage relation. In terms of the van der Pol model this is equivalent to decreasing g_2, and this will decrease the frequency (eqn (11.30)) at which firing may occur. Qualitatively this effect arises in the van der Pol model because, if the steady state relation is fairly flat in the voltage range of interest, the momentary current–voltage relation shifts, in absolute terms, less rapidly than it would if it were shifting towards a steady state relation with a large slope.

Finally, it should be mentioned that there are other ways in which the

Hodgkin–Huxley equations may be modified in order to produce low firing frequencies.† One way is to decrease the rate of accommodation (cf. eqn (11.30)). Stein (1967a) performed calculations for the point-polarized Hodgkin–Huxley model of the squid axon in which the potassium rate constants were slowed by a factor of three. The minimal firing frequency was reduced from about 60 Hz to about 25 Hz. The slow rate of firing of cardiac muscle seems to be achieved in both the above mentioned ways. The steady state current–voltage relation is very flat (it may even have a region of negative slope, but positive chord, conductance) and the rate of change of potassium current is very much smaller than in nerve fibres.

Frog myelinated axon. Bromm and Frankenhaeuser (1972) have described recently numerical calculations for the current–frequency relation of a node of Ranvier (see also Kernell and Sjöholm 1973). The general form of their results are very similar to those obtained for the squid axon. Maintained firing could only be produced by currents well above the true rheobasic value ($>$ 1·7 times rheobase). The frequency jumped from zero to 155 Hz and the curve then displayed an upward convexity, the frequency being fairly insensitive to increases in current strength. At 2·7 times rheobase the frequency was 290 Hz, and above 3 times rheobase only damped oscillations were observed. Bromm and Frankenhaeuser mention that their current–frequency curve is linear on a semilogarithmic plot (i.e. $f \propto \ln{(I)}$).

These calculations were performed for a model of the myelinated axon (see Frankenhaeuser and Huxley 1964) assumed to be at 20 °C, whereas the frequencies quoted earlier from Stein's (1967a) paper were for the model squid axon at 6·3 °C. However, Stein did also perform calculations for a temperature of 20 °C; the Hodgkin–Huxley model of the squid axon then had a very narrow frequency range of 200–280 Hz.

One interesting figure presented in Bromm and Frankenhaeuser (1972) is included as Fig. 11.21. It shows the values of P_{Na} and P_K at various times during maintained current of two different strengths. At the higher stimulus strength, damped oscillations occur after the first action potential, and this is reflected in the time course of the two permeabilities; at the lower stimulus strength, there is a sustained train of action potentials and between each action potential a maintained potassium permeability can be seen. It is obvious also from the smaller size of the peak sodium permeabilities during each action potential after the first that there is a sustained sodium inactivation.

Bromm and Frankenhaeuser (1972) do not compare their computations with any experimental results for the behaviour of the frog axon, and we have not found any quantitative documentation of such experiments in the literature. Sato (1952) has shown that there is a fairly linear relationship between frequency and the logarithm of the stimulus strength (a constant voltage stimulus) for a single toad myelinated axon. The lowest frequency

† See also Shapiro and Lenherr (1972).

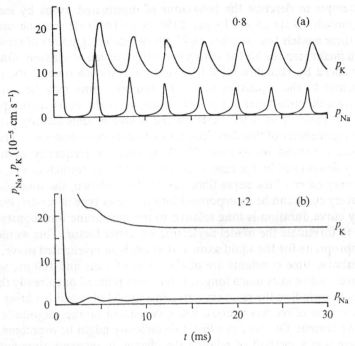

FIG. 11.21.(a) and (b). Computed values of the change in both p_K and p_{Na} with time in response to prolonged current of 0·8 (a) and 1·2 mA cm^{-2} (b). Frankenhaeuser–Huxley model of the frog node of Ranvier. (From Bromm and Frankenhaeuser 1972.)

observed was about 30 Hz at a temperature of 12–14 °C. It is well known, however, that frog myelinated axons differ in their ability to maintain a steady firing frequency; the nodes of Ranvier of frog motor fibres rarely exhibit repetitive activity, in contrast to the sensory fibres of the same species (Erlanger and Blair 1938; Schmidt and Stämpfli 1964). This seems to be due mainly to the fact that there is a much higher potassium conductance, both at the resting potential and during steady depolarization, in the nodes of the motor axons (Bergman and Stämpfli 1966). Since the potassium conductance of myelinated axons may be greatly reduced by applying tetraethylammonium ions (Bergman and Stampfli 1966; Hille 1967), it is not surprising that this treatment, together with removal of most of the external calcium ions (see later), produces sustained repetitive firing in motor nerve nodes (Bergman, Nonner, and Stämpfli 1968). The latter authors present evidence that the 'pacemaker' potential in this preparation is determined largely by the recovery of h so that the firing frequency is directly related to the rate constants of h.

The simple models described at the beginning of this chapter are not appropriate for simulating the behaviour of myelinated axons, for much the same reasons as have been given for the squid axon. However, there have been attempts to describe the behaviour of myelinated axons by means of simple models (Katz 1936; Tasaki 1950; Sato 1952). It may be useful to discuss these models because they highlight two differing points of view about the main factor responsible for setting the firing frequency of axons. One view, first advanced by Hodgkin (1948), is that the response time of the axon is fundamental to the repetition rate. The response time may be correlated loosely with the time taken for passive charging of the nerve axon up to the voltage threshold. This is the basis for the simple models already presented and the applicability of this description has already been discussed. The other view, first advanced by Adrian (1928), is that the frequency of firing is primarily determined by the time course of the relative refractoriness, that is, the recovery curve of the nerve fibre. As has been shown, the time course of the recovery curve can be incorporated into the other type of model; but if the recovery curve duration is long relative to the membrane time constant it is simpler to formulate the model neglecting the latter factor. This would seem to be appropriate for the squid axon and amphibian myelinated nerve, where the membrane time constants are of the order of 1 ms and 100 μs, whereas the recovery curve lasts much longer. It has been pointed out already that one difficulty in including the recovery curve in a model of repetitive firing is that the time course of recovery varies, being dependent on the magnitude of the polarizing current. One way in which this difficulty might be overcome would be if there was a method of relating the change in recovery time course to some other independently measurable response of the nerve to a constant current. An obvious possibility is Hill's time constant of accommodation (Hill 1936). Katz (1936) shows that, on Hill's general theory of excitation and accommodation in nerves, there should be a relation between the time for which repetitive firing can be maintained for a given accommodation time constant, and strength of applied stimulus. Although there was qualitative support for this prediction it proved to apply over a limited range of stimulus strengths (Sato 1952), probably because the value of the time constant of accommodation is itself sensitive to the stimulus strength (see Sato 1952). At the moment, therefore, it appears that any attempt to formulate a simple model for squid axon and amphibian myelinated nerve is bound to be an arbitrary exercise (see comment in Frankenhaeuser and Vallbo 1965); however, it does seem that the response time is a much less important factor in their repetitive firing than for the crab axon.

Gastropod nerve cell. Connor and Stevens (1971a,b,c) (see also Neher and Lux 1971; Neher 1971) have made a voltage-clamp analysis of a Gastropod nerve cell which responds repetitively to steady current injection. Fig. 11.22 shows examples of repetitive firing induced by constant depolaring currents. The current–frequency relation is almost linear over a fairly large range and, unlike the relation obtained in squid nerve, extends down to very low frequencies.

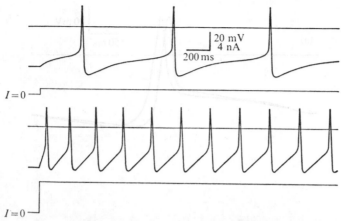

FIG. 11.22. Repetitive firing induced in a gastropod nerve-cell soma by two different intensities of current. The upper traces are membrane voltage and the lower traces stimulating current. (From Connor and Stevens 1971a.)

Voltage-clamp analysis of the ionic currents showed that, in addition to the transient sodium conductance and maintained potassium conductance characteristic of squid nerve membrane there is also a transient potassium conductance whose kinetic properties are very important in controlling the repetitive activity at low frequencies. At potentials positive to about −50 mV this conductance is inactivated and it is, therefore, necessary to hyperpolarize the membrane to more negative potentials before this third conductance can be activated. During natural activity this is precisely what happens. At the end of an action potential the conventional potassium conductance hyperpolarizes the membrane towards the potassium equilibrium potential, so removing the inactivation of the second potassium conductance. As the conventional potassium conductance decays, the membrane begins to depolarize again and, since the decay of this potassium conductance is fairly rapid, the threshold for activating the sodium conductance would be approached fairly quickly, thus generating a high frequency of firing. When large depolarizing currents are applied to the membrane this is, in fact, what occurs. At low current strengths, however, there is time for the more slowly developing depolarization to activate the second potassium conductance g_A. The current flow (I_A in Fig. 11.23) produced by this conductance then prevents the membrane potential from reaching threshold. Threshold is approached only very slowly as the conductance g_A is inactivated. Fig. 11.23 shows a reconstruction of these events calculated by Connor and Stevens (1971c) on the basis of their voltage-clamp measurements. The top records show experimental and computed potentials, and the lower records show the computed changes in the first g_K and second g_A potassium conductances. There is close agreement between the results of the computations and the experimentally recorded activity, as is also shown in Fig. 11.24, where the

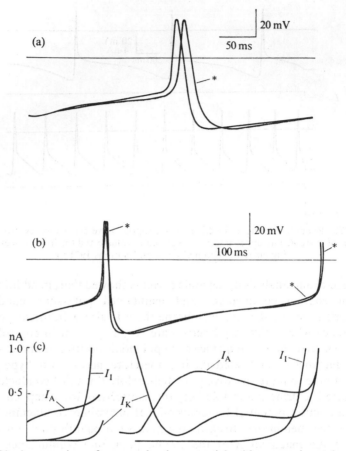

FIG. 11.23. A comparison of computed action potentials with an experimental example of the gastropod nerve cell. In (a) and (b) the computed trace is indicated by the asterisks. (c) shows the computed membrane currents associated with the computed membrane potential behaviour in (b). (From Connor and Stevens 1971c.)

firing frequencies predicted are compared with those observed at different current strengths.

Connor and Stevens's analysis is of considerable importance because it shows that there is at least one example of repetitive firing in a nerve cell which is qualitatively quite different in its mechanism (at low frequencies) from that of the conventional Hodgkin–Huxley model of the action potential. This is, of course, not the only excitable cell in which more than one voltage-sensitive potassium conductance has been demonstrated. Another example is the cardiac Purkinje fibre. In some respects the mechanism of firing in the Gastropod nerve cell, at low frequencies, resembles the mechanism of pacemaker activity in Purkinje fibres. In both there is a different conductance mechanism responsible for controlling the rate of firing to that for action potential repolarization. The resemblance does not extend much further than this, however, since in Purkinje fibres it is the decay in the activation of a potassium conductance, which shows no inactivation properties, which is responsible for the pacemaker depolarization. Moreover, although this conductance is not primarily involved in repolarization, it is nevertheless activated during the action potential. By contrast, in the Gastropod nerve cell it is the inactivation process which is the dominant factor controlling the interspike membrane potential, and the conductance involved is activated in the period just following the previous action potential.

Voltage–frequency relation

It has been a common experimental technique in receptor physiology to record the relationship between stimulus strength and firing frequency; this commonly takes the form frequency $\propto$ logarithm (stimulus strength) (Matthews 1931; Hartline and Graham 1932; Katz 1950; Terzuolo and Washizu 1962; Gillary 1966). Does this mean that the different kinds of transducer mechanism have similar properties, or is this result due to a common property of the response of the nerve fibres to the electrical result of the transduction process? As already shown, many (though not all), nerve fibres have a logarithmic relationship between firing frequency and injected current; this would suggest that it is the latter possibility which explains the result, and that the transduction process is fairly linear. On the other hand, Chapman's (1966) results for the type-1 and type-2 crab axon show a fairly linear relation between current and frequency, suggesting that the logarithmic nature of the over-all process might occur in the transduction mechanism. It is conventional to hold the latter conclusion, on the basis of a direct study of certain receptors; several investigators have reported a linear relationship between an electrical

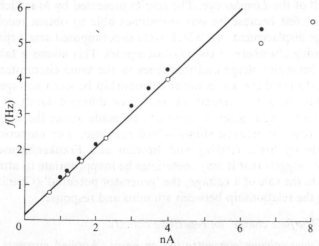

FIG. 11.24. Current–frequency behaviour of a gastropod nerve cell (open circles) and the computed model (filled circles). (From Connor and Stevens 1971c.)

result of transduction and the firing frequency of the receptor (e.g. Katz 1950; MacNichol 1958; Wolbarsht 1960; Terzuolo and Washizu 1962). During the firing of a receptor there is often no obvious electrical parameter which is suitable for measuring (as a steady state value), and so many of the experiments have involved the measurement of voltage following the blocking of nerve firing by local anaesthetics or tetrodotoxin. Both these compounds at low concentrations produce a more or less specific block of the sodium conductance mechanism, leaving the potassium conductance relatively unaffected.

Clearly, the measurements of voltage produced (in the presence of a blocking compound) and frequency of firing (without any blocking compound) are made in different circumstances, so the linear relation between voltage and frequency may be fortuitous. When the sodium conductance mechanism is blocked the over-all membrane current–voltage relation will be affected by the properties of the 'leak' conductance and the Hodgkin–Huxley potassium conductance. If the latter is quantitatively dominant, the current–voltage relation will be approximately described by $I \propto \exp(V)$ (see Stein 1967a). If the unblocked axon has a logarithmic relation between frequency and applied current (i.e. $f \propto \ln (I)$ or $I \propto \exp(f)$) it is obvious that these two quite different forms of nonlinearity could cancel out when a comparison was made, leading to a fortuitously linear relation between voltage and frequency (Stein 1967a; Bromm and Frankenhaeuser 1972).

For most of the receptors studied it is not known whether the region of impulse initiation does display a logarithmic or a linear relation between current and frequency. Terzuolo and Washizu (1962) found a logarithmic relationship between current and frequency for the crayfish stretch receptor neurone, whereas MacNichol (1958) described a linear relationship for the eccentric cell of the *Limulus* eye. The results presented by MacNichol are of particular interest because he was sometimes able to obtain recordings of large voltage displacements on which were superimposed small spikes (presumably arising elsewhere in the photoreceptor). This allows a fairly direct correlation between voltage and frequency in the same circumstance. Over the range 2–15 Hz there was a linear relationship between voltage and frequency. Thus these experiments on two very different kinds of receptor suggest that no simple generalization can be made about the origin of the logarithmic relation between stimulus and response. The numerical calculations made by Stein (1967a) and Bromm and Frankenhaeuser (1972) however, do, suggest that it may sometimes be inappropriate to attach much importance to the role of a voltage, the 'generator potential' (Granit 1955) in considering the relationship between stimulus and response.

Influence of 'surface charge' on repetitive activity

Action of low calcium concentration on nerve. Applied currents shift the membrane potential towards threshold and so induce firing. This may also be achieved by the converse procedure of shifting the threshold towards the resting potential. Experimentally, one way of shifting the threshold is to reduce the divalent-ion concentration (Frankenhaeuser and Hodgkin 1957; Blaustein and Goldman 1968; Hille 1968). This is thought to reduce the number of cations absorbed at surface negative sites, and so increase the surface negative charge of the membrane. If the potential at the outside surface is shifted in a negative direction, without any concomitant change in the potential at the inside surface, the electric field across the membrane will be reduced. So far as the voltage-dependent permeability reactions are concerned, this will be equivalent to the effect of depolarization. Less depolarization will be required, therefore, to produce a given degree of activation of the m reaction of the sodium conductance. Therefore a sufficiently large shift in threshold will produce a net inward current at a level of potential at which the membrane was previously quiescent. Like the application of constant currents, this may give rise to sustained membrane oscillations. Huxley (1959a) has reproduced the effects of reduced calcium by computing solutions to the squid axon equations for various amounts of threshold shift. Examples of some of his results are shown in Fig. 11.25. The results show a remarkable range of behaviour. As expected, it is possible to produce damped subthreshold oscillations, single action potentials, and sustained repetitive firing. The results also include more unusual activity. Thus it is possible to calculate, as the response to single shocks of slightly different magnitudes, membrane-potential oscillations which gradually increase or decrease in amplitude over several cycles. This kind of behaviour corresponds to the spiral trajectories in the phase-plane (see Fig. 11.11). Like the propagating subthreshold wave in the Hodgkin–Huxley cable model (see p. 301), these unusual responses are elicited only within a rather restricted range of stimulus strengths and calcium concentrations, and may be difficult to observe experimentally.

Although the effect of low calcium-ion concentration is well known experimentally, there is very little of such work which may be compared quantitatively with Huxley's computation. Huxley (1959a) shows examples of repetitive firing in low calcium concentrations recorded by Hodgkin, Frankenhaeuser, and Keynes. Recently, Guttman and Barnhill (1970) have compared the effects of applied currents and temperature on computed and experimental records for squid axons in low-calcium solutions (see Fig. 11.15). Guttman (1971) has also reviewed the literature on oscillations and repetitive firing in nerve axons.

Action of adrenaline on a cardiac pacemaker. Changes in calcium concentration shift the voltage dependence of both the sodium and the potassium permeability variables. The effect on membrane surface charge, therefore, may be relatively non-specific. An example of a more specific effect is provided by the action of adrenaline on cardiac Purkinje fibres, which shifts the voltage dependence of one of the potassium permeability variables (Hauswirth,

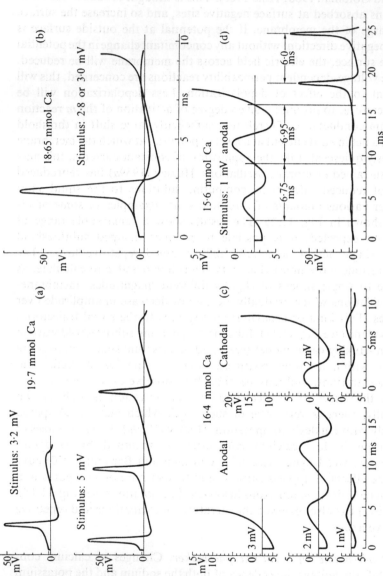

Fig. 11.25. Computed responses of the Hodgkin–Huxley equations for various values of extracellular calcium concentration. The normal concentration is 44 mmol, at which the solution is always a single action potential. At 19·7 mmol (a), the solution is a single response or a train of responses depending on the size of the stimulus. At 18·65 mmol (b), trains or subthreshold damped oscillations occur. At 16·4 mmol (c), the subthreshold oscillations to some stimuli gradually increase until excitation occurs (cf. Fig. 11.11); the response to a 1 mV deflection is almost the sinusoidal response described in Fig. 11.11. At 15·6 mmol (d), all stimuli produce responses that grow into a full action potential train.
(From Huxley 1959a.)

Noble, and Tsien 1968) while having no effect on the sodium permeability variables (Trautwein and Schmidt 1960). The effect of adrenaline seems to be to change the surface charge near the potassium channels so that the field in these regions is increased. The same depolarization therefore activates less of the potassium conductance and, furthermore, the conductance decreases more rapidly when the membrane is repolarized. Since this decay of the potassium conductance is responsible for controlling the depolarization during the pacemaker phase of oscillatory activity (Noble and Tsien 1968) the action of adrenaline is to increase this rate of depolarization. Computations, using a model of the Hodgkin–Huxley type, have been performed to reproduce this action of adrenaline (Hauswirth, McAllister, Noble, and Tsien 1969), and the results are shown in Fig. 11.26.

It can be seen that quite small voltage shifts (corresponding to small

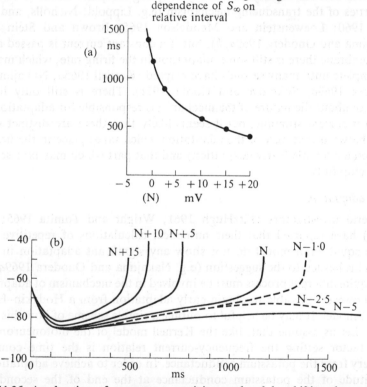

Fig. 11.26. Acceleration of spontaneous activity produced by shifting the potassium current activation curve in a model of the Purkinje fibre. The curve labelled N is the normal case. Depolarizing shifts are indicated (in millivolts) by positive values. Hyperpolarizing shifts are negative values. Thus, N+10 means that the activation curve was shifted by 10 mV in the depolarizing direction. (a) shows the dependence of the duration of the 'pacemaker' potential on the shift imposed. (b) shows the computed potential changes as a function of time. The curves start at the end of an action potential.

changes in 'surface charge') produce large increases in the rate of depolarization during the pacemaker potential. The precise concentrations of adrenaline to which these shifts may correspond are not yet known, but since much larger shifts may be observed experimentally at large adrenaline concentrations (Hauswirth, Noble, and Tsien 1968), it seems likely that the effects are adequate to account for the accelerating action of adrenaline. As already noted above (p. 329), the accelerating effect might also be reproduced in a model of the van der Pol type, but it is not possible to give any simple physical interpretation (such as the surface charge hypothesis discussed here) to the changes in k and g_2 which would be required.

Adaptation

Almost all nerves or nerve cells which are capable of maintained firing show some decline in the firing frequency with time. A large part of the difference between rapidly and slowly adapting receptors seems to be in the properties of the transducing mechanism (e.g. Lippold, Nicholls, and Redfearn 1960; Loewenstein and Mendelson 1965; Brown and Stein 1966; Nakajima and Onodera 1969a, b), but if a constant current is passed across the membrane there is still some adaptation in the firing rate, which may not be complete until many seconds have elapsed (Kernell 1965a; Nakajima and Onodera 1969a; Sokolove and Cooke 1971). There is still only limited evidence about the nature of the mechanisms responsible for adaptation to a constant current stimulus, but it seems likely that there are distinct differences between that part of the adaptation which takes place in the first few interspike intervals (early adaptation) and that part which may take seconds to develop fully.

Early adaptation

Several investigators (FitzHugh 1961; Wright and Tomita 1965; Stein 1967a) have reported that their numerical calculations of repetitive firing in the squid axon model do not show any significant adaptation in firing rate. This has led to the suggestion (e.g. Nakajima and Onodera 1969a) that some quite different process must be involved in the mechanism of adaptation. It is, however, possible to obtain early adaptation from a Hodgkin–Huxley type of model. It may be helpful to outline a simple scheme of how this could occur. Let us assume that, like the Kernell model of the motoneurone, the main factor setting the frequency-current relation is the time course of recovery from the potassium conductance. In order to achieve adaptation the magnitude of the potassium conductance at the end of the second (and subsequent) action potential must be greater than it is after the first. In a Hodgkin–Huxley system this potassium conductance is proportional to n^γ ($\gamma = 2$, in the myelinated nerve model and 4 in the squid axon model). If the time course and amplitude of the depolarizing part of the action potential waveform is identical for the first and second impulses, this means that n^γ

must be larger just before the second action potential than it is before the first. This would be very likely because the membrane potential trajectory before the second action potential is quite different from that before the first. When a constant current is first applied, the response, as a first approximation is a passive charging of the membrane up towards the voltage threshold. If the steady state voltage dependence of the potassium system is placed sufficiently far along the voltage axis, so that it is not significantly activated by depolarizations below the voltage threshold, the value of n (and hence n^γ) just before the first action potential will be near zero. During the first action potential n will increase its value and subsequently slowly decline.

In general, n will not reach its maximal value of $1\cdot0$, owing to the brevity of the depolarizing phase of the action potential; the peak value it reaches during the action potential will be dependent, of course, on its initial value being least if $n = 0$. (This can be most easily seen by the equation for n at a constant voltage. It is, $n = n_\infty - (n_\infty - n_0)\exp(-t/\tau_n)$, where n_0 is the initial value, n_∞ the steady state value to which n tends for a given voltage, and τ_n the voltage-dependent time constant.) Let us assume that n reaches a peak value of $0\cdot6$ (cf. Frankenhaeuser and Huxley 1964), so that if $g_K \propto n^2$ the potassium conductance will reach a peak value which is just over one-third ($0\cdot36$) of its maximal value. After the depolarizing part of the action potential is over the value of n will decline with an approximately exponential time course (assuming that the voltage remains within a range in which τ_n does not change very much), and hence g_K will decline exponentially with a time constant equal to τ_n/γ. For a given value of current injection the cell will fire a second impulse when

$$I = g_m(E_R - E_{Th}) + g_K(t)(E_K - E_{Th}), \tag{11.33}$$

where g_m is the resting membrane conductance of the cell. Let us assume the current is slightly above 'rheobase' and this condition is met when $g_K(t)$ has reached $\frac{1}{100}$ of its maximal value. Then, if $\gamma = 2, n = 0\cdot1$ (since $n^2 = 0\cdot01$). The next action potential therefore starts with a large initial value of n ($0\cdot1$ instead of 0) so that, if it has a very similar time course to the first action potential, n will reach a peak value of, say, $0\cdot65$ instead of $0\cdot6$. The peak value of the potassium conductance during the action potential is now approximately $0\cdot42$, instead of $0\cdot36$, and, if there is the same value of the time constant for decline in n after the action potential, it will take longer for it to decline from its peak value to a value of $0\cdot1$ than it did after the first action potential. On this very simple hypothesis the third and subsequent intervals will all have identical values to the second interspike interval. It would also be a prediction of this scheme that, if n reaches a value of $1\cdot0$ (or near to $1\cdot0$) during the first action potential, no adaptation would occur.

This particular explanation of the way in which early adaptation may occur makes two crucial assumptions which are unlikely to be strictly correct. The objections to them may be couched as follows: (1) if the second action potential starts with a higher initial value of n there will be a faster development of a substantial potassium conductance and hence the waveform of the depolarizing part of the action potential may be smaller in amplitude and curtailed in its time course. There might therefore be less of an increment in n when it starts at a higher initial value; (2) even if the peak value of n is higher after the second action potential the higher potassium conductance will lead to a larger postspike hyperpolarization. The voltage dependence of τ_n takes the form of an inverted U-shape, so that the larger repolarization may tend to shift τ_n to a smaller value. Consequently, although the peak value of n is larger, it may decline back to its resting level at a faster rate; this effect therefore would tend to cancel out the effect of a large peak conductance, and the first and second interspike intervals could be very similar. It is clearly inappropriate to explore these objections any further in a qualitative manner since the only secure way of assessing their significance is by computation of the Hodgkin–Huxley type of equations.

Examples of such computations have been performed by Kernell and Sjöholm (1972, 1973)[†]. They wished to model the mammalian motoneurone and therefore chose the Frankenhaeuser–Huxley description of the vertebrate axon as their basic model. Modifications to this description had to be introduced before they were satisfied that the model displayed properties comparable to the motoneurone: the main ones being a shift of the sodium inactivation variable in the depolarizing direction and the addition of a second, slow potassium conductance system (see Kernell and Sjöholm 1972).

Kernell and Sjöholm (1973) reported that early adaptation in their model was dependent on the presence of the slow potassium system. Accompanying the adaptation the voltage time course revealed an increase in the size of successive after-hyperpolarizations. They were able to vary the amount of early adaptation (and the degree of 'summation' of the after-hyperpolarizations) simply by changing the voltage sensitivity of one of the rate constants α of their slow potassium system. As would be expected, there was no adaptation when this conductance reached its maximal value during the first spike.

The simple scheme outlined above predicted that all the early adaptation would take place between the first and second intervals. In Kernell and Sjöholm's models this was only true for lower firing frequencies; at higher frequencies further adaptation took place over the next few intervals. In Fig. 20 of Kernell and Sjoholm (1973) it can be deduced from the time course of the membrane potential that there is a progressive increase in potassium conductance following the first eight action potentials. Accompanying this increase there is also a progressive increase in the peak height of the second to eighth action potentials so that it is natural to assume that the mechanism is as follows. After the second action potential the increased potassium

---

[†] We are grateful to Drs Kernell and Sjöholm for allowing us to see their manuscripts before publication.

conductance (and hence relative repolarization) leads to greater recovery from sodium inactivation than after the first action potential. This leads to the peak height of the third action potential being greater than that of the second action potential—in turn leading to a greater activation of potassium conductance during the third action potential than during the second, etc. In other respects the assumptions of the simple scheme seem to be largely justified.‡

Kernell and Sjöholm (1972, 1973) discuss the resemblance between their models and the properties of the mammalian motoneurone. It has been known for some time from experimental investigations that the motoneurone after-hyperpolarization does show temporal summation (Ito and Oshima 1962), so this theoretical study of Kernell and Sjöholm adds further weight to the suggestion that the explanation of the early adaptation is such temporal summation (Kernell 1968; Baldissera and Gustafsson 1971b; Calvin and Schwindt 1972; Schwindt and Calvin 1972; Kernell 1972).

Although the discussion of early adaptation has concentrated on one particular model it is possible that quite different modifications of the parameters of the Hodgkin–Huxley equations might also lead to an initial reduction in firing rate. For example, if the repetition frequency was pre-dominantly set by the behaviour of the h variable, that is by the recovery from sodium inactivation (cf. Bergman, Nonner, and Stämpfli 1968) a comparable behaviour could be ascribed to it as has been to the n variable above. Two kinds of experimental observation would point to this being an appropriate model: an increase in the voltage threshold between the first and second spikes and a reduction in the peak amplitude of the action potential in the second and subsequent spikes.

As discussed above, it is very likely that many examples of early adaptation are a result of a combination of mechanisms, so that they could not be described by solely discussing a single Hodgkin–Huxley variable.

Late adaptation

The discussion of early adaptation concentrated on the difference between the first two interspike intervals. It has been noted already that the increase in interspike intervals can continue for several seconds before adaptation is complete. Although it is possible that a complicated mechanism, involving more than one Hodgkin–Huxley variable, might explain the decrease in frequency over several interspike intervals, it does seem unlikely that the

‡ Baldissera, Gustaffsson, and Parmiggiani (1973) have shown recently that a simple modification to such a scheme, in which the time course of the potassium conductance is simulated by a function of the form $G_K(t) = A \exp(-\alpha t) + Bt^\beta \exp(-\gamma t)$, will give adaptation between the second and third intervals as well, if the potassium conductance change after each action potential is assumed to summate with that generated by the previous spike. (This is equivalent to assuming that n^γ after an action potential is incremented by the same amount, whatever the value of n^γ just before the action potential is generated. This would not be expected, in general, for a Hodgkin–Huxley variable, particularly if $\gamma \neq 1$, but might hold over a certain range.)

effect would last much longer. Various suggestions have been made about the kind of mechanisms which may underlie this slower phase of adaptation. They can be classified into three broad categories: (1) changes in conductance with a very slow time course, (2) changes in ionic concentrations on either side of the membrane, and (3) a slow increase in a hyperpolarizing (or repolarizing) current produced by an electrogenic pump.

Changes in conductance with a slow time course. One obvious possibility is that there is a further voltage-sensitive conductance, with a reversal potential on the hyperpolarizing side of the voltage threshold, which has a very long time constant for activation. One possible example of such a mechanism is reported by Connor and Stevens (1971a) for a Gastropod nerve cell. They reported that their model, based on voltage-clamp data (see p. 345), generally showed no change in firing frequency with time (except in special circumstances, when the first interval could be longer). Experimental observations however, did show adaptation, and associated with this they reported evidence for a potassium conductance with a much longer time constant than the conventional g_K (see also, Meves 1961).

Another phenomenon reported by Connor and Stevens (1971a) was that of a slow inactivation of g_K. This has also been reported for many other cells, including the squid axon (Frankenhaeuser and Waltman 1959; Nakamura, Nakajima, and Grundfest 1965; Ehrenstein and Gilbert 1966; Nakajima and Kusano 1966; Adrian, Chandler, and Hodgkin 1970). If the major factor affecting the firing frequency of a cell was the 'response time' and the stimulating current strength was large enough to produce a maintained potassium conductance between each spike, it is obvious that a slow reduction in this maintained conductance could lengthen the effective membrane time constant. On the other hand, since the membrane resistance has increased, less current would be required to reach the voltage threshold. It is possible to show that, in the simple models described at the beginning of this chapter, the net effect would be an *increase* in frequency. This is easy to show for uniform polarization, and it also holds for the point polarization model, at least within the expected physiological range (i.e. liminal length between 0·2 and 2·0 space constants, current less than 20 times 'rheobase').

It is possible, however, that potassium inactivation could lead to a slow adaptation. A slow decline in the potassium conductance could lead to a slowing of the repolarizing phase of the action potential. If the spike frequency was high then there might be less and less recovery from sodium inactivation which in turn led to an increase in the voltage threshold and consequent decrease in firing frequency (Nakajima and Onodera 1969a).

A final possibility is a mechanism for a slow reduction in the maximum available sodium conductance; either by a very slowly developing sodium inactivation or by a slow process of recovery from sodium inactivation (cf. Haas, Kern, Einwächter, and Tarr 1971).

Changes in ionic concentration. During a prolonged train of action potentials there is a net movement of sodium (and perhaps chloride) ions into the cell and a net loss of potassium ions. These net transfers of ions can lead to marked changes in the intracellular and extracellular ion concentrations. Such changes could affect the firing frequency of the cell as a direct result of a change in the equilibrium potential of the particular ion concerned or because the changed concentration of the ion has an effect on one of the membrane conductances.

The accumulation of sodium ions inside the cell will lead to a reduction in E_{Na} and hence, if everything else remains the same, to an increase in the voltage threshold. In most circumstances this is unlikely to be a major effect because the volume-to-surface ratio of the cell and/or the activity of a sodium extruding mechanism tend to limit the magnitude of the increase in $[Na^+]_i$. Bergman (1970), however, has presented evidence that large changes in E_{Na} can result from quite brief periods of activity in frog myelinated nerve fibres; he suggests that this could be the result of a very small effective volume-to-surface ratio for the period immediately after activity—perhaps because diffusion of the sodium ions from the axoplasm near the node into the internodal axoplasm is fairly slow. It is not yet clear whether this internal accumulation of sodium ion near the nodal membrane is of sufficient magnitude in normal firing behaviour to account for, or contribute to, adaptation.

Bergman (1970) also made the interesting observation that the internal accumulation of sodium had an effect on the voltage-sensitive potassium conductance. The effect, a reduction, was only quantitatively significant when the membrane voltage was more depolarized than the presiding value of E_{Na} (i.e. when both sodium and potassium ions were tending to move outward) so that it would not be important during normal firing.

Since the classical study of Frankenhaeuser and Hodgkin (1956) it has been realized that there may be an external barrier to diffusion surrounding excitable cells. Potassium ions which have travelled out across the membrane may then accumulate in a small volume of extracellular space, leading to a significant reduction in E_K. Such an increase in $[K^+]_0$ will lead generally to a membrane depolarization (see, however, Noble 1965; Gorman and Marmor 1970) and therefore, if there are no other effects, to a decrease in the amount of current required to reach threshold for firing. It is possible that this process is responsible for the effect of 'warming-up', seen in some crab axons (Hodgkin 1948; Chapman 1963); 'warming-up' is exactly the opposite process to adaptation since there is a decrease in interspike-interval with time.

Large increases in $[K^+]_0$ might have the opposite effect, however. Large depolarizations, associated with greater reductions in E_K, may lead to the development of sodium inactivation following an action potential. Furthermore, the increased $[K^+]_0$ may lead to an increase in the resting membrane conductance (cf. Hodgkin and Huxley 1947). Either of these two effects, if dominant, could lead to adaptation.

Both Chapman (1966) and Nakajima and Onodera (1969a) have sought evidence for the accumulation of extracellular potassium as a significant factor in repetitive firing. Chapman's technique was to compare the response of a crab axon to steady current and to trains of short current pulses. He was concerned primarily with the factors affecting the slope of the current–frequency relation (see p. 340), so that these results do not bear directly on the problem of adaptation. Nakajima and Onodera (1969a) tested for the effects of ion accumulation by introducing a brief interruption in the passage of a constant current. The results they obtained for both a slowly and rapidly adapting stretch receptor of the crayfish are reproduced in Fig. 11.27. They concluded that changes in ionic concentrations were not a primary factor in the adaptation since they thought that the brief cessation of the depolarizing current would have little effect on any ion accumulation. This conclusion depends on the assumption that the time course of diffusion of ions out of an extracellular space is longer than the period of current cessation. The pause in Nakajima and Onodera's record is of the order of 150 ms. The time constant for the recovery from extracellular potassium accumulation in Frankenhaeuser and Hodgkins' experiments (on the squid axon) varied between about 20 ms and 140 ms, which means that, if the crayfish stretch receptor had an extracellular space with similar properties to that of the squid axon, 150 ms might be sufficient time for substantial recovery to occur. It is also evident from inspection of Nakajima and Onodera's figure (Fig. 11.27) that, at least for the rapidly adapting receptor, the firing frequency after the brief pause is significantly lower than at the start. In other words, there is some process, presumably contributing to spike adaptation, which does not fully recover in 150 ms. This could be any one of the three mechanisms suggested as underlying adaptation. Experimental evidence pertaining to this will be discussed in the next section.

Activity of an electrogenic pump. It is now well established in a variety of tissues that the ion pump which moves sodium out of the cell, and potassium into the cell, can be electrogenic in its action, producing a net outward movement of charge (Kerkut and Thomas 1965; Adrian and Slayman 1966; Nakajima and Takahashi 1966; Rang and Ritchie 1968; Thomas 1969, 1972; Gorman and Marmor 1970). The pump is increased in its activity by a rise in internal sodium concentration and also by an increase in external potassium concentration, although the activation by external potassium may not occur in the physiological range (Sokolove and Cooke 1971; Thomas 1972). Therefore, the increase in $[Na^+]_i$ accompanying firing of a cell would lead to the slow development of a hyperpolarizing current which, by reducing the magnitude of the effective stimulating current, would reduce the rate of firing of the cell.

Good evidence that the sodium pump does play a significant quantitative role in adaptation to constant current stimulation has been provided by Sokolove and Cooke (1971). They studied the slowly-adapting crayfish

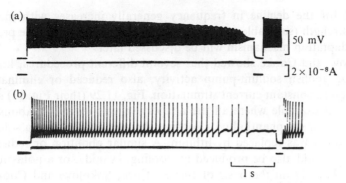

FIG. 11.27. Spike adaptation in both slowly adapting (a) and rapidly adapting (b) crayfish stretch receptors. Note the considerable recovery in both examples when the current is turned off briefly. (From Nakajima and Onodera 1969.)

stretch receptor and found that, in response to constant currents of various strengths, the frequency of firing jumped up to an initial value and then slowly declined along an exponential time course to a lower steady firing frequency. This behaviour is illustrated in Fig. 11.28 (Fig. 2 of Sokolove and Cooke 1971). Notice that with the lowest current strength the frequency eventually declines to zero, and that with increasing current strength the ratio of steady frequency to initial frequency becomes larger. Another feature of the firing behaviour of these cells, not very obvious in Fig. 11.28, is that the time

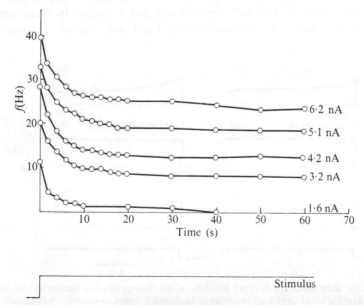

FIG. 11.28. Behaviour of a slowly adapting crayfish stretch receptor at different current strengths. 'Instantaneous frequency' is plotted against time in response to steps of current. (From Sokolove and Cooke 1971.)

constant for the decline in frequency generally increases with increasing current strength (see Table 1 of Sokolove and Cooke 1971). These properties of the adaptation mechanism will be discussed later.

Sokolove and Cooke showed that several different procedures, known to reduce or abolish sodium-pump activity, also reduced or eliminated the adaptation to constant current stimulation. Fig. 11.29 (their Fig. 11) shows a very elegant example where all the adaptation was reversibly abolished by treatment with strophanthidin or bathing the preparation with a solution in which sodium was replaced by lithium. A similar abolition or reduction in adaptation could also be produced by cooling, cyanide, or a potassium-free solution. Largely on the basis of this evidence, Sokolove and Cooke concluded that electrogenic sodium pumping was the sole important mechanism in producing adaptation. However, they did point out that other factors, such as potassium inactivation, may be important at the higher current strengths used by Nakajima and Onodera (1969*a*). One important technical point that Sokolove and Cooke make is that the use of potassium citrate electrodes (favoured because they pass steady current fairly readily) may interfere with the sodium pump as a result of a reduction in intracellular magnesium concentration (see Baker, Foster, Gilbert, and Shaw 1971), hence reducing the magnitude of observed adaptation.

Although Sokolove and Cooke's data strongly favours the sodium pump as being of major quantitative significance in adaptation, there is one result that they report which indicates another process at current strengths just above 'rheobase'; some of their cells showed not just a decline in frequency to a lower steady value, but cessation of firing. Obviously a complete cessation of

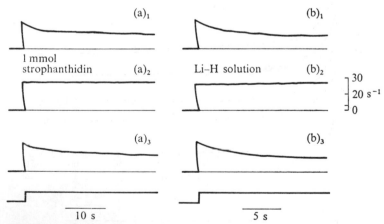

FIG. 11.29. Smoothed (penwriter) records of the 'instantaneous frequency' of firing of slowly adapting crayfish stretch receptor in response to steps of current. Adaptation is seen before (a)$_1$ and after (a)$_3$ application of 1 mmol strophanthidin (a)$_2$ and, in another cell, before (b)$_1$ and after (b)$_3$ perfusion in Li–van Harreveld's solution (b)$_2$. (From Sokolove and Cooke 1971.)

firing could not be easily attributed to a maintained contribution of hyper-polarizing current from an electrogenic sodium pump since, once firing has ceased, there is no net inward movement of sodium ions across the membrane to stimulate its activity above the resting level (unless the depolarizing current produced a maintained subthreshold increase in sodium conductance). Sokolove and Cooke also suggested that some of the adaptation they observed at low temperatures might be due to a slow increase of sodium inactivation. These considerations, however, do not modify the conclusion that sodium-pump activity is of major quantitative significance in adaptation.

It is a straightforward matter to give a theoretical account of sodium-pump activity and apply it to the data of Sokolove and Cooke. A similar, but more extensive, theoretical investigation has recently been reported by Sokolove (1972).

Let us assume that the activity of the sodium pump is directly proportional to the internal sodium-ion concentration (cf. Hodgkin and Keynes 1956; Mullins and Brinley, 1969; Baker, Blaustein, Keynes, Manil, Shaw, and Steinhardt 1969; Thomas 1972). The equation for a single sodium-pump site would be given by

$$\frac{d\{Na_p^+\}}{dt} = k[Na^+]_i$$

and for an area of membrane, it is given by

$$\frac{d\{Na_p^+\}}{dt} = knS[Na^+]_i$$

where $\{Na_p^+\}$ is the amount of pumped sodium, k has dimensions of volume per second, n the number of sites per unit area of membrane, and S the total area of membrane. It can be readily shown that the response following a sudden increase in $[Na^+]_i$ (neglecting the resting values of $[Na^+]_i$ and Na^+ influx) is given by:

$$\frac{d\{Na_p^+\}}{dt} = knS[Na^+]_{i,t=0} \exp\left(-\frac{knS}{V}t\right),$$

$$\left(\text{since } \frac{d[Na^+]_i}{dt} = -\frac{1}{V}\frac{d\{Na_p^+\}}{dt}\right)$$

or, expressing pump activity as a current,

$$i_p = cknS[Na^+]_{i,t=0} \exp\left(-\frac{knS}{V}t\right). \tag{11.34}$$

where i_p is the sodium-pump current, c is the proportion of the total sodium pumped as a net charge (that is, unaccompanied by inward potassium move-ment or movement of some other ion by the pump), and V is the cell volume.

To model Sokolove and Cooke's data we need to know the response of the pump for an increase in sodium influx which follows the time course of the frequency during a constant current stimulation, that is

$$f(t) = f_{ss} + (f_0 - f_{ss})\exp(-t/\tau_a), \tag{11.35}$$

where f_{ss} is the steady state frequency, f_0 the initial frequency, and τ_a the time constant of the adaptation process.

It is convenient to ignore the discrete temporal character of the sodium influx associated with each impulse, and treat the net sodium influx as if it had a smoothly changing time course. In this case, if g is the total net charge of sodium ion which moves in with each impulse (assumed to be constant, that is, ignoring the possibility of increasing sodium inactivation, etc.), then the sodium influx, expressed as a current, is given by

$$i_{Na}(t) = g\{f_{ss} + (f_0 - f_{ss})\exp(-t/\tau_a)\}.$$

The pump current in response to this time course of sodium influx can be shown to be

$$i_p = cgf_{ss}\left[1 - \exp\left(-\frac{knS}{V}t\right) + \left(\frac{f_0}{f_{ss}} - 1\right)\left\{\frac{\exp\left(\frac{-knS}{V}t\right) - \exp\left(\frac{-t}{\tau_a}\right)}{(V/knS\tau_a) - 1}\right\}\right] \tag{11.36}$$

$$\left(\text{providing } \tau_a \neq \frac{V}{knS}\right).$$

The final problem is to translate the time course of the pump current into a time course for frequency decline. If the current–frequency curve can be represented as a linear segment over the frequency-range considered (that is, f_{ss} to f_0), this is straightforward. Since the problem was formulated with the assumption that the frequency decline was exponential (with a time constant τ_a) it follows that the sodium-pump current must have the following time course

$$i_p(t) = i_{p(ss)}\{1 - \exp(-t/\tau_a)\}, \tag{11.37}$$

where $i_{p(ss)}$ is the steady state pump current, which must in turn be equal to c times the steady state inward leak of sodium (i.e. $i_{p(ss)} = cgf_{ss}$).

By inspection it can be seen that the right-hand side of eqn (11.37) is identical to the right-hand side of eqn (11.36) when

$$(f_0/f_{ss} - 1)\bigg/\left(\frac{V}{knS\tau_a} - 1\right) = 1,$$

that is,

$$\tau_a = \frac{f_{ss}}{f_0}\frac{V}{knS}. \tag{11.38}$$

Now V/knS is the time constant for rate of activity of the pump following an increase in $[Na^+]_i$ (see eqn (11.34)). Eqn (11.38) demonstrates that τ_a will be briefer than this time constant by a factor determined by the ratio of the steady state to initial frequencies.

How can this analysis be applied to Sokolove and Cooke's data? The first question is to establish whether sodium-pump activity is directly proportional to the internal sodium concentration. Sokolove and Cooke observed that the decline in the hyperpolarization following a brief tetanus (which is a sign of electrogenic pump activity) is usually exponential in time course (cf. eqn (11.34)). The time constant ($= V/knS$ in the above analysis) estimated by Sokolove and Cooke varied in different cells, with a mean value of 11·1 s and a range of 5·8–17 s. There were some exceptions to the exponential time course noted by Sokolove and Cooke, particularly with larger post-tetanic hyper-polarizations, but this may be because of some potential-dependence of the pump rate (see Kostyuk, Krishtal, and Pidoplichko 1972) or a nonlinearity between current and voltage. If these exceptions are put aside, the usual exponential behaviour of the post-tetanic hyperpolarization can be taken as strong evidence for a linear relation between electrogenic pump activity and $[\mathrm{Na}]_i^+$.

Another assumption of the theoretical account which can be tested is that the amount of sodium entering with each impulse remains roughly constant whatever the frequency. In order to check this, one can use the fact that in the steady state the electrogenic pump current should bear a constant relationship to the inward sodium current (ignoring the possibility that there is a change in the amount of bound intracellular sodium) such that

$$i_p = ci_{\mathrm{Na}}. \tag{11.39}$$

If the amount of sodium ion entering with each impulse is constant then eqn (11.39) may be written as

$$i_p = cgf_{ss}$$

so that a plot of i_p against f_{ss} should be linear, through the origin, with a slope equal to $1/cg$.

Sokolove and Cooke give data for four cells in their Table 1 of which their cell D has been studied at the largest number (8) of different constant current strengths. Fig. 11.30 illustrates the current–frequency curves for this cell, with the open circles indicating the initial frequency and the filled circles the steady state frequency. From these two curves the magnitude of the sodium-pump current, for each steady state frequency, may be calculated; it is the horizontal distance separating the two curves. This calculation assumes, of course, that the only factor in adaptation is the slow development of pump activity in response to the increased external sodium concentration.

Fig. 11.31(a) illustrates the results of such a calculation for Sokolove and Cooke's cell D. It can be seen that there is a roughly linear relation between pump current and steady state frequency, with a tendency for a decrease at

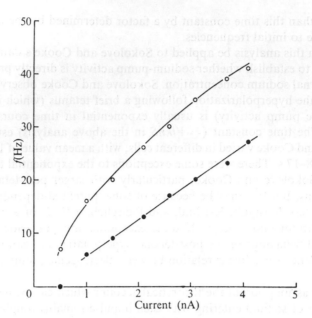

FIG. 11.30. Plots of the initial (open circles) and steady (filled circles) frequency of firing of a crayfish stretch receptor in response to steps of current. Data obtained from Table 1 (cell D) of Sokolove and Cooke (1971).

higher frequencies. This tendency is revealed more clearly in Fig. 11.31(b) where the total amount of sodium ion pumped out electrogenically per impulse is plotted against the steady state frequency. If, as in other tissues (Hodgkin and Keynes 1956; Thomas 1969), the pump current is one-third of the total sodium transported the value of the ordinate of Fig. 11.31(b) should be multiplied by 3 to determine the amount of sodium ion entering per impulse. Similar calculations on the data for the other three cells provide only two or three points per cell; they do not show the same clear relationship, but this may simply be a result of experimental scatter.

The value of the time constant V/knS is not given by Sokolove and Cooke for the four cells in their Table 1, so that it is not possible to make a watertight check of the prediction of the above model.† A partial check can be performed, however, because for each current strength the values of f_0, f_{ss}, and τ_a are given. In a single cell the value of V/knS should remain constant, so that we can check the predicted value of V/knS for each current strength in a single cell and see how closely they resemble each other. Fig. 11.32 shows such calculations for cell D. The mean value for $(f_{ss}/f_0)\tau_a$ (plotted as open circles) is 6·85 s, with a standard deviation of 0·37 s. Notice that there is a definite tendency for the estimate to increase with higher current strengths.

† Sokolove (1972) does report such a comparison for his cell E. The calculated value of the predicted time constant was 5·5 s, whereas the observed time constant for post-tetanic hyperpolarization was 4·8 s.

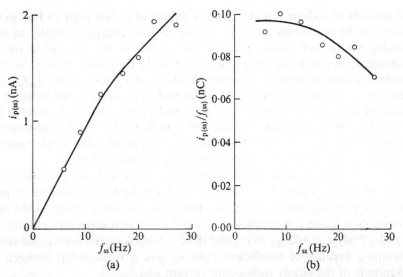

FIG. 11.31(a). A plot of the estimated steady sodium-pump current $i_{p(ss)}$ against steady firing frequency for the same cell as in Fig. 11.30. (b) The estimated net charge transferred by the sodium pump per impulse is plotted against the steady firing frequency. Same data as in (a).

The explanation for this may be simply that no correction has been applied for the smaller amount of sodium ion entering per impulse at higher steady frequencies (see Fig. 11.31(b)). It may be presumed that at the beginning of constant current stimulation, whatever its strength, the same amount of sodium ion will enter with each impulse, but that this amount will decline to a variable degree with the final value depending on stimulating current strength and hence f_{ss}. Inspection of Fig. 11.31(b) suggests that there is no decline in

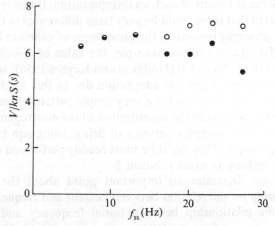

FIG. 11.32. Calculations of the time constant V/knS for the same cell as Figs. 11.30 and 11.31. See text for further description.

the amount of sodium ion entering for values of f_{ss} less than 15 Hz; so this value can be tentatively taken to be the amount entering initially as well. Making the further assumption that the decline in the amount of sodium entering occurs with the same time course as the frequency decline, an exact correction can be made. The correct estimate of V/knS is now $(g_{ss}f_{ss}/g_0f_0)$, where g_{ss} is obtained from Fig. 11.31(b) and g_0 is taken as the mean of the values at steady state frequencies of 6, 9, and 13 Hz. The filled circles in Fig. 11.32 show these modified estimates for V/knS. There is now a tendency for the estimate to decrease with the value of f_{ss}. The mean value for the modified estimate of V/knS is 6·16 s, with a standard deviation of 0·51 s.

Both the original and modified estimates for V/knS are just within the range of values reported by Sokolove and Cooke for the decline of post-tetanic hyperpolarization studied in other cells. Similar analyses of the three other cells in Table 1 of Sokolove and Cooke lead to uncorrected estimates of $(12·43\pm2·54)$ s, $(10·72\pm1·47)$ s, and $(8·65\pm1·02)$ s. Corrections could not be attempted because of insufficient data to give a relationship between the magnitude of the steady state pump current and f_{ss}.

As mentioned above, the test of the model made above is less satisfactory than making a direct comparison between the value of V/knS estimated by the above method and that obtained by other means.† Nevertheless, the fairly close agreement between the values of V/knS estimated for the same cell at different current strengths does suggest that the model gives a satisfactory explanation of the experimental results. This analysis therefore lends circumstantial support to Sokolove and Cooke's conclusion that adaption in the slowly adapting crayfish stretch receptor is mainly a result of electrogenic pump activity.

It is natural to wonder whether late adaption in other cells has a similar explanation. Sokolove and Cooke (1971) mention several examples where there is indirect evidence in favour of such an interpretation. If this is correct then it would be expected that there could be very large differences in the time course of adaptation, primarily because of the great range of values in the volume-to-surface ratio. To take an extreme example, the value of V/knS for the giant squid axon is of the order of 5 h (Hodgkin and Keynes 1956), so that it would be very difficult to detect any late adaptation due to this mechanism.

The model presented above is for a very simple pattern of impulse activity; since the equations are linear the contribution of the electrogenic pump could be assessed with more complex patterns of firing, using eqn (11.34) and the superposition principle. This might be most readily performed on a computer rather than by seeking an exact solution.‡

Fig. 11.30 also illustrates an important point about the effect of the electrogenic pump on the relation between current and frequency. The open circles show the relationship between initial frequency and current. The

---

† See footnote on p. 364.
‡ See Sokolove (1972).

shape of this curve is slightly convex upwards, the steady state frequency—current curve is linear. Any difference in shape must be attributed to the action of the electrogenic pump. This therefore introduces a further factor affecting the shape of frequency–current curve, not discussed in the earlier sections, and reinforces the conclusion reached there that simple models of firing frequency are unlikely to be quantitatively adequate accounts of the response of excitable cells to constant current stimulation.

Response to a time-varying input

The bulk of this chapter has been devoted to models of repetitive firing under conditions of a steady excitation. Apart from the process of adaptation there has been no consideration of differences between individual interspike intervals, that is, once a steady firing frequency is attained it has been assumed to be completely regular. This is a very inadequate simulation of the actual behaviour of excitable cells, for regular firing is the exception rather than the rule. Some part of the observed variability in interspike intervals may be a result of fluctuations in the voltage threshold or membrane potential of the firing cell, but commonly the most important cause is that the current causing excitation of the cell is not constant but varies with time.

There is now an extensive theoretical literature concerned with giving a complete description of the pattern of firing frequency of cells excited in such a way. It is beyond the scope of this book to attempt a detailed review of this work. Instead we will give a brief pastiche, in the hope that interested readers will be given a start in the field by way of the references quoted.

The model of the excitable cell generally adopted is of a simple form. Uniform excitation is almost invariably assumed and the passive membrane is taken to be either a simple capacitor (the 'linear integrator' model) or a resistance and capacitance in parallel (the 'leaky integrator' model). The criterion for firing is that a voltage threshold be reached, and after the action potential the membrane potential is reset to its resting level. A few models have included absolute and relative refractory periods. The 'leaky integrator' model, with voltage threshold, is identical to the simple model described at the beginning of this chapter; for constant current stimulation it gave, at best, a qualitative fit to the actual firing behaviour of cells (see the section *Numerical computations of repetitive activity*, p. 331). The 'linear integrator' is, in general, even less realistic as a model and will not be discussed, although it is mathematically more easily solved (see e.g. Gerstein and Mandelbrot 1964; Stein 1965, 1967b; Johannesma 1969; Sugiyama, Moore, and Perkel 1970; Knight 1972a).

The form of the time-varying input has also taken different forms, either continuous or discrete. A continuous time-varying input is particularly pertinent to the modelling of the firing of receptors—e.g. to predict their response to a sinusoidal stimulus—or to nerve cells which receive synaptic action from a large number of different sources, each of which produces a very small effect. Discrete inputs, such as a series of current pulses, are a better model of the synaptic firing of nerve cells when individual fibre inputs produce a large effect (relative to the threshold value). Some workers have modelled synaptic inhibition as well as synaptic excitation with the discrete input model.

Continuous input

As already mentioned, one special form of continuous input often employed experimentally is that of a sinusoidal stimulus. A detailed mathematical analysis of the 'leaky integrator' model stimulated by a current of the form $I = I_0 + I_i' \cos(\omega t + \phi)$ is given by Rescigno, Stein, Purple, and Poppele (1970). The timing of firing in relation to the phase of the modulating current was derived, and it was also shown that the model tends to a stable pattern of firing which repeats periodically. This may mean that the output frequency pattern is a very poor 'copy' of the input current (see also Stein and French 1970; Knight 1972a). However, the presence of some neuronal variability (such as a 'noisy' membrane potential) tends to improve the quality of the 'copy' (Stein and French 1970; Knight 1972a). Stein (1970) has concluded that for a given range of stimulus frequencies and amplitudes there is an optimal level of neuronal variability to obtain minimal distortion, and has drawn attention to the fact that actual receptors seem to behave accordingly. Stein (1970) also pointed out that an additional method of reducing distortion was to perform a 'spatial' average over more than one unit. Knight (1972a) has considered the relationship between unit firing patterns and those of a population in some detail and compared the results of his theoretical predictions with the behaviour of a neurone in the eye of *Limulus* (Knight 1972b).

The other common form of continuous input which has been used is to assume a current which has a steady mean level, but fluctuates about the mean value with a Gaussian amplitude distribution. Although a complete set of explicit analytic solutions has not yet been obtained for this model there has been considerable success (see Gluss 1967; Johannesma 1968, 1969; Sugiyama, Moore, and Perkel 1970).[†] Sugiyama, Moore, and Perkel have given a lucid and brief review of the various approaches adopted.

Johannesma (1969) has computed the mean-frequency–average-current relation for different amounts of variability in the input. He took three examples: one where the variance in the input remained constant and two others where the amount of variability is related to the average amount of current (with either the standard deviation or the variance of the variability being proportional to the average amount of input current). His frequency–current relations for a constant amount of variability, whatever the average value of current, is illustrated in Fig. 11.33 (his Fig. 3.31). The limiting case of no variability ($S = 0.00$) is, of course, identical to the frequency–current

---

† Clay and Goel (1973) have recently reported solutions for a continuous input with refractoriness modelled by a varying threshold.

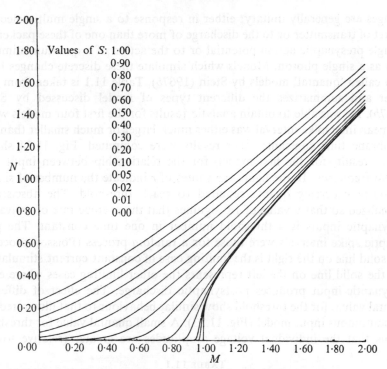

FIG. 11.33. Frequency–current curves for a model with continuous input with constant variability. The ordinate N represents frequency (normalized by the membrane time constant) and the abscissa M is the average value of input current divided by the rheobasic current. S indicates the amount of variability with increasing variability displacing the relationship to the left. (From Johannesma 1969.)

relation illustrated in Fig. 11.2 (with no absolute and relative refractory period). As one would expect intuitively, the effect of increasing variability is to increase the mean frequency for a given average value of current, with the greatest effect being evident with smaller currents. If the variability is sufficiently large the model fires even when the average value of current is zero. Another way of describing the effect of variability is that it tends to blur the frequency–threshold effect seen with constant current and to linearize the frequency–current curve. Qualitatively, similar effects are seen when the variability increases in proportion to the average current. Johannesma (1969) has also illustrated the behaviour of the membrane potential for subthreshold current strengths, but did not obtain interval histograms for any given mean level of firing frequency. Sugiyama, Moore, and Perkel (1970), however, do present two examples.

Discrete inputs

The mechanism by which a nerve cell or receptor is excited is usually in response to a conductance change to one or more ions. These conductance changes are generally unitary; either in response to a single multimolecular packet of transmitter or to the discharge of more than one of these packets to a single presynaptic action potential or to the action of a quantum stimulus such as a single photon. Models which simulate these discrete changes have been called 'quantal' models by Stein (1967b). Table 11.1 is taken from this paper and summarizes the different types of model discussed by Stein (1967b). He was able to obtain analytic results for the first four models when the mean interspike interval was either much larger or much smaller than the membrane time constant; other results were computed. Fig. 11.34 shows Stein's result (for the first model) for the relationship between input and output frequency. In the figure the values of r indicate the number of simultaneously occurring quanta required to reach threshold. The abscissa is normalized so that a value of 1·0 signifies that the average rate of arrival of presynaptic inputs is a threshold number in one time constant. The presynaptic spike intervals were those for a random process (Poisson process). The solid line on the right is the limiting case of constant current stimulation and the solid line on the left represents the other limiting cases where each presynaptic input produces postsynaptic discharge. The effect of different quantal values for the threshold show similar behaviour to that illustrated for the continuous input model (Fig. 11.33). A small quantal value for threshold means that the individual voltage amplitudes are greater (relative to the

TABLE 11.1

Model	Quantal effect	Subthreshold decay between inputs	Change upon reaching threshold
1. Exponential decay	Unit voltage change	Exponential with time constant τ	Voltage reset
2. Variable duration	Unit voltage change	At discrete times; duration of unit voltage changes variable with mean τ	Voltage reset
3. Variable size	Voltage change having amplitude density function $q(a)$ with unit mean and variance σ_p^2	Either model 1 or 2 above	Voltage reset
4. Single pore	Unit current change	Current decay at discrete times; duration of unit current changes random with mean t_p; voltage changes exponential with time constant τ	Voltage reset; current unaffected
5. Modified Hodgkin–Huxley	Unit current change	Current decay as in model 4 above with mean 1 ms; voltage changes according to Hodgkin–Huxley equations	Voltage changes according to Hodgkin–Huxley equations; current unaffected

Note: The excitatory input is assumed to occur at random with a mean rate of p_e. The current–voltage curves of models 1–4 are as for the voltage-threshold simple model discussed earlier in this chapter. From Stein (1967b).

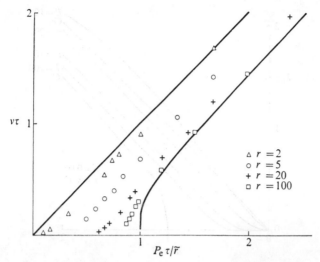

FIG. 11.34. Normalized input–output relation for the exponential decay model of Table 11.1. The ordinate is frequency-normalized by the membrane time constant. The abscissa expresses the mean rate of 'quanta' p_e normalized by both the membrane time constant and by the minimum number of quanta required to reach threshold, so that different threshold levels r may be compared. (From Stein 1967b.)

separation between the resting potential and the threshold level) and thus can be regarded as analogous to greater variability in the continuous input model.

Stein's computations with the Hodgkin–Huxley model of the squid axon assumed that the stimulating current changed randomly by small amounts. With this source of variability the abrupt jump up to a minimal steady frequency was not seen, as shown in Fig. 11.35. This is similar to the effect in simpler models, illustrated in Figs. 11.33, 11.34, and 11.36.

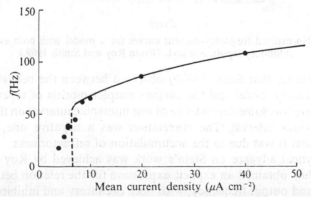

FIG. 11.35. Current–frequency curves for the uniformly polarized Hodgkin–Huxley model of the squid axon. The lines indicate the result with constant current (as in Fig. 11.14) and the filled circles when the current source fluctuates about a mean amplitude. (From Stein 1967b.)

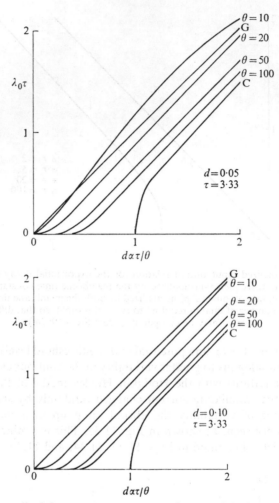

FIG. 11.36. Normalized frequency–current curves for a model with both excitatory and inhibitory inputs. See text. (From Roy and Smith 1969.)

One difference that Stein (1967b) observed between the behaviour of the Hodgkin–Huxley model and the various simpler models of repetitive firing was that there was some dependence of one interspike interval on the duration of the previous interval. The correlation was a negative one, and Stein presumed that it was due to the accumulation of refractoriness.

An analytical advance on Stein's work was achieved by Roy and Smith (1969).† They obtained an explicit expression for the relation between input frequency and output frequency, with both excitatory and inhibitory voltage displacements undergoing exponential decay. Fig. 11.36 is a rearrangement

† See also Fienberg (1970), Capocelli and Ricciardi (1971), Hochman and Fienberg (1971), Goel, Richter-Dyn, and Clay (1972), and Osaki and Vasudevan (1972).

of their Figs. 2 and 3 and shows similar effects to those computed by Stein (1967b). Their θ is equivalent to Stein's r. The curves labelled C and G are identical to the solid lines in Fig. 11.34. The parameter d in their model signifies the difference in the frequencies between excitatory and inhibitory inputs, divided by the total frequency. (A d equal to 1·0 would mean there was no inhibitory input and a d of 0 would mean that the excitatory and inhibitory inputs had equal frequencies.) Note that the abscissa is normalized so that an equal amount of net excitatory action, relative to threshold, is the independent variable.

The novel feature of Roy and Smith's curves is that when the input frequency for excitation and inhibition are very similar ($d = 0·05$, that is, the ratio of the two frequencies is 0·525 to 0·475) and a small number of simultaneously occurring excitatory quanta are required to reach threshold ($\theta = 10$), the curve crosses the line labelled G. It was pointed out earlier that this latter line represents an output pulse for each excitatory input pulse (that is, $\theta \leq 1$), so that this result may seem paradoxical. The paradox is less worrying if it is remembered that the line labelled G also represents the output that would be obtained if a threshold number of inputs occurred sufficiently close together in time that the exponential decay could be neglected. Thus if $\theta = 10$, an abscissal value of 1·0 is equal to a mean excitatory input frequency of $10/\tau_m$. If for each interval of one time constant all ten inputs occurred very close together, one output pulse would be produced per time constant (this is, of course, also the behaviour of the 'linear integrator' model where the 'time constant' for decay is infinite). In Roy and Smith's figures the abscissa has been normalized relative to threshold. For the above values ($\theta = 10, d = 0·05$ an abscissal value of 1·0 is equal to a mean excitatory input frequency of $105/\tau_m$ and a mean inhibitory input frequency of $95/\tau_m$. Both inputs are assumed to be random. Roy and Smith's result must mean that with such Poisson input patterns the near coincidence of a *net* number of excitatory inputs greater than 10 (e.g. 14 excitatory and 3 inhibitory) occurs slightly more frequently than once per time constant.

So far the discussion has concentrated on the relationship between mean input frequency and mean output frequency. The models described can also be used to compute other characteristics of the output frequency behaviour, such as the interval histogram and the variance of the frequency about the mean. These measures of the output behaviour represent only the most elementary of statistical measurements which can be made (Moore, Perkel, and Segundo 1966). Part of the purpose in making such measurements is that it may help, in experiments in which detailed methods of intracellular analysis are impractical, in deducing both the properties of the postsynaptic cell and the characteristics of the synaptic inputs impinging upon it. An ideal of this form of backwards deduction would be to find some statistical measurement of the output behaviour which uniquely specified one kind of model. Once the particular type of model had been so specified, it would be an elementary, if laborious, procedure to determine the parameters of the model which exactly fitted the data. The necessary background to such a possibility is to determine the behaviour of a wide variety of physiologically plausible models in order to see in what way their output behaviour differs. Perkel and his colleagues have already done a great deal of this background work (e.g. Segundo, Perkel, Wyman, Hegstad, and Moore 1968).

Perkel's computer model of the nerve cell (Perkel 1964) is extremely flexible. It can provide a large number of separate presynaptic inputs, excitatory or inhibitory, to the model postsynaptic nerve cell, with independent control of the frequency pattern and strength of each input, although in most of the computations reported by Segundo, Perkel, Wyman, Hegstad, and Moore all inputs had similar mean rates and excitatory potential amplitudes. The degree of interdependency on the firing of the inputs was an independent variable. The postsynaptic cell had a linear current–voltage relation with a voltage threshold; with both absolute and relative refractory periods reflected in the time course of the threshold level. After firing, the membrane potential could be either reset to a value more hyperpolarized than the resting potential or returned to the value of the membrane potential at which it would have been (that is, influenced by preceding synaptic inputs) if a spike had not occurred. Many of the features of Perkel's model are illustrated in Fig. 11.37.

Segundo, Perkel, Wyman, Hegstad, and Moore (1968) chose four different forms of presynaptic impulse interval distribution: Gaussian, gamma, bimodal, and exponential (Poisson). When each presynaptic input was independent there was little difference in the postsynaptic mean interval, for a constant number and strength of inputs, whatever the form of the presynaptic interval distribution; the standard deviation (and coefficient of variation) of the postsynaptic intervals is greatest for a Poisson form, and is successively smaller for bimodal, gamma, and Gaussian distributions. As the number of input channels increases (and their strengths decrease accordingly) the output tends to become regular for all forms. Three forms of output-interval histogram were observed:

1. *Asymmetrical, with positive skew.* Most of these could be fitted by a gamma function (cf. Stein 1965). Greatest skew was obtained with a smaller number of input channels and the distributions were, respectively more skewed by input forms in the order Poisson, gamma, bimodal, and Gaussian. Two types of autocorrelation histograms were associated with this type of postsynaptic interval distribution, the more highly periodic autocorrelogram being associated with a less-skewed and narrower interval histogram. The degree of skewness of the interval histogram depended not only on the number of input channels and input form but also on the 'quantal' sizes, membrane and relative refractory time constants, and the resetting mode.

2. *Symmetrical*. This is the limiting case of the first type of distribution and is produced by all forms of presynaptic input when the number of input channels is large (and the 'quantal' sizes accordingly small).

3. *Bimodal*. These are seen most commonly in postsynaptic cells whose potential is reset without undershoot following each spike.

Making the presynaptic inputs interdependent tends to make the activity of the postsynaptic cell mirror the impulse pattern of the inputs, with the added feature that the postsynaptic interval histogram may become bimodal even if the presynaptic inputs are fairly regular (i.e. Gaussian, with a small standard deviation).

These results led Segundo and his colleagues to three main conclusions. The first was that at junctions where there are a small number of powerful presynaptic inputs, the firing of the postsynaptic cell closely follows the form of the presynaptic input, whether the latter are completely independent or highly interdependent. The second conclusion is that, if there are a large number of weak and independent presynaptic inputs, the postsynaptic cell generates the same form, uninfluenced by the form common to all the inputs. This conclusion would hold also, of course, if the form varied (e.g. some Gaussian, some Poisson, and some bimodal) for different input channels, since the over-all interval histogram for a large number of independent input channels will always tend to a Poisson distribution. The third

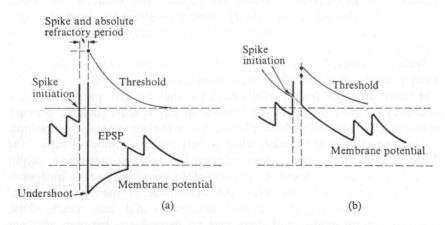

FIG. 11.37. An illustration of some of the features of Perkel's model nerve cell. The synaptic input is represented as discrete displacements of the membrane potential with subsequent exponential decay. Individual potentials summate linearly. Once threshold is exceeded there is an absolute and relative refractory period (represented by the increase in threshold). In the two variations in the model shown there is either a reset potential following the spike, with subsequent exponential decay to the resting potential (a) or the post-spike potential returns to the value it would have had if a spike has not intervened (b). EPSP: excitatory postsynaptic potential. The steady state value of threshold and potential are indicated by horizontal dashed lines. (From Segundo, Perkel, Wyman, Hegstad, and Moore 1968.)

conclusion was that, when the input channels are interdependent, the output will reflect the form of presynaptic input, even if there are numerous weak channels.

It will be evident from this presentation that the use of simple statistical measures for the description of the firing pattern of a postsynaptic cell will not uniquely specify the membrane properties of the postsynaptic cell and the character of the synaptic input. Nevertheless, the use of either more sophisticated statistical measurements and/or some independent experimental information about the properties of the input or of the postsynaptic cell may help to decide.

A notable example of such a careful matching between experiment and theory can be found in the work of Walløe and his colleagues (Jansen, Nicolaysen, and Rudjord 1966; Walløe 1968; Walløe, Jansen, and Nygaard 1969). They recorded from the axons of dorsal spino-cerebellar tract (DSCT) cells activated to fire by the primary endings from muscle spindles, which in turn were made to fire by muscle stretch. The relationship between DSCT firing frequency and muscle length was fairly linear. At any single muscle length the interval histograms were broad and unimodal, with a roughly constant coefficient of variation. It was found that there was a strong dependency between successive intervals in the postsynaptic firing. This was expressed quantitatively by measuring the first serial correlation coefficient R_{12}, which had a mean value of about -0.6 and increased slightly with increasing mean frequency. Jansen, Nicolaysen, and Rudjord (1966) had suggested that this strong dependency between neighbouring intervals might be a consequence of an accumulation of subnormality, when two intervals occurred close together.

Walløe attempted to reproduce these results by computer simulation using a nerve cell model whose main features are illustrated in Figs. 11.38. The main difference from Perkel's model is that the cell 'pacemakes' at a frequency of about 10 Hz in the absence of any specific (that is, primary muscle spindle) input. This feature is in accord with experimental observations.

The output from the model, when a Poisson distribution of input was assumed, showed a nearly linear relationship between input and output frequency and the coefficient of variation of the output was nearly independent of the input frequency when the magnitude of the input 'quanta' were large. However, the interval histograms did not match those obtained experimentally, and there was no dependency between adjacent intervals.

Walløe then introduced a 'memory' into the model by making the time constant of the exponential decay of threshold following a spike (that is, the relative refractory time constant) a function of the length of the preceding interval. Some serial dependency of the intervals was obtained ($0 > R_{12} > -0.3$) but not as great as in the experimental results. The interval histograms were also still a poor fit. Walløe therefore concluded that with this nerve

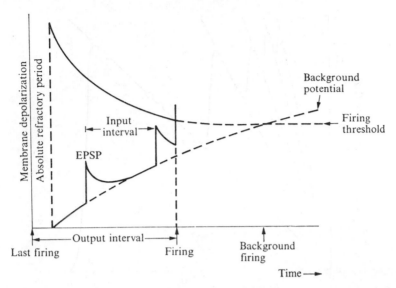

FIG. 11.38. An illustration of some of the properties of Walløe's nerve cell model. (From Walløe 1968.)

cell model it was not possible to simulate the behaviour of the DSCT cell when Poisson-distributed inputs were assumed.

Presumably guided by the known behaviour of muscle primary endings, Walløe then went on to determine the output behaviour of the same nerve-cell model when it received a number of independent excitatory inputs, with the input pattern of each taking the form of a Gaussian distribution (cf. Stein and Matthews 1965; Matthews and Stein 1969). As before, the input–output relation was linear, and the coefficient of variation of the output was roughly constant. In order to obtain a variability in the output discharge comparable to the DSCT cell, the size of the 'quantal' potentials had to be larger than 40 per cent of the difference between resting potential and threshold. This also set a limit on the total number of inputs, with a good fit being obtained by between 10 and 20 input fibres. With such a model an excellent fit was achieved, including the simulation of the interval histograms and the serial dependency.

Independent experimental work with intracellular recording from DSCT cells (Eide, Fedina, Jansen, Lundberg, and Vyklicky 1969a,b) provided striking confirmation for the assumptions of the modelling.

It is worth noting that the serial dependency in the output intervals could be produced in the model without any 'memory' in the model cell's refractoriness. It is the result of 'beating' between the different, fairly regular inputs (see Walløe 1968).

Another rather interesting result which has 'beating' as its underlying explanation is the effect of a regular input of excitation or inhibition to a neurone already firing regularly. Perkel, Schulman, Bullock, Moore, and

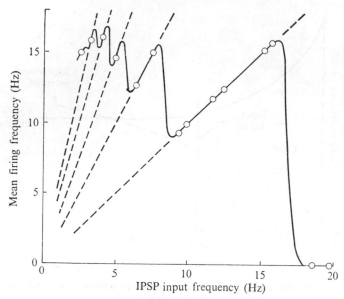

FIG. 11.39. Mean firing frequency of monosynaptically inhibited crayfish stretch receptor (open circles). The inhibiting impulses were regularly spaced in time. The dashed lines are drawn according to theory. IPSP: Inhibitory postsynaptic potential. (From Perkel, Schulman, Bullock, Moore, and Segundo 1964.)

Segundo (1964) have shown both experimentally and in matching computer-simulation studies that an increase in the frequency of the inhibition may increase the firing frequency of the cell, and vice versa for excitation. This effect is shown for inhibition in Fig. 11.39, it is not observed for a Poisson input. The possible functional significance of this paradoxical effect is discussed by Perkel, Schulman, Bullock, Moore, and Segundo (1964). Dong and Reitz (1970) have drawn attention to the fact that this effect may be important in the vagal control of heart-rate.

The importance of the modelling of nerve cell firing described in this section extends beyond the consideration of making adequate models. Both Stein (1967c) and Walløe (1968, 1970), amongst others, have discussed the importance of the input–output behaviour of nerve cells in setting limits on the information capacity of a synaptic relay, and also pointed out the significance of this work for making deductions about the nature of neural codes (see also, Perkel and Bullock 1968). It seems clear, however, that more sophisticated models will be required to do justice to the known complexity of some synaptic relays (see e.g. Porter and Muir 1971; and Chapter 7). A further problem is that it may be necessary to include the effects of other cells—such as those producing recurrent synaptic action—so that the problem of modelling a single relay expands into that of modelling a nerve cell network.

12. Nonlinear cable theory: analytical approaches using polynomial models

ONE of the major difficulties in the analysis of excitation and conduction in excitable cells is that the problems frequently require solutions of nonlinear partial differential equations such as eqn (9.7). The required solutions may be obtained by numerical methods of approximation, but such solutions are not readily applicable to cases other than the particular ones for which they have been computed. A more general analysis usually requires a combination of numerical results together with analytical treatment of simplified systems for which the equations are less formidable. In Chapters 9–11 we discussed several simplified systems in which the nonlinearities or spatial variations are either neglected or reduced. In the first parts of the present chapter we shall discuss systems in which time dependence is neglected. The equations which may then be obtained will be relevant primarily to physiological situations in which the steady state properties are investigated. Nevertheless, the theory also helps in other cases since time-dependent changes may be regarded as changes from one steady state condition towards another. To describe the steady state behaviour will contribute therefore to an understanding of the possible transient behaviour. Moreover, as we shall show, although it is often very difficult to obtain analytical results when time dependence is included, it is much easier to obtain results for nonlinear systems when only spatial variations in current and voltage are considered.

The chapter is divided into four parts. In Part I we will obtain equations for the current–voltage relations in nonlinear cables, assuming that the membrane current–voltage relation is known. In Part II we will discuss briefly the use of polynomial expressions to represent membrane current–voltage relations, and in Part III we will use such expressions to obtain equations for the spatial variations. Finally, in Part IV we shall reintroduce time dependence by discussing the applications of polynomial cable models to excitation and propagation.

Part I: Current–voltage relations

Cole's theorem and related equations

An experimental method which is often used to study the current–voltage relations of excitable cells is to insert two microelectrodes in the interior of a cable-like cell. Usually, the electrodes are placed close together and one is used to apply current while the other is used to record voltage changes. The voltage electrode records the voltage at only one point so that information cannot be obtained directly on the spatial variation in voltage. Moreover, it is important to note that the current flowing through the current electrode will flow nonuniformly through the membrane. These difficulties have already been noted in Chapter 3 where we derived an equation for the input resistance measured by this experimental procedure. There is, however, an additional difficulty which occurs when nonlinear systems are investigated. This arises from the fact that the pattern of current spread is a function of the current strength so that the membrane current corresponding to the voltage changes recorded will not show a simple relation to the total current applied through the current electrode.

In order to deal with this situation we require an analysis of the relation between the membrane current–voltage relation $i_i(V)$, and the input current–voltage relation $I(V_0)$. This problem has been already discussed in Chapter 9 in an analysis of the threshold conditions for excitation. The treatment we shall give here is a more general one and we shall illustrate it in greater detail.†

We assume that the cable is polarized at its midpoint, which is defined as $x = 0$. At this point the applied current I divides into two as it flows into the two halves of the cable, so that

$$(i_a)_{x=0} = I/2. \tag{12.1}$$

Hence

$$I = -\frac{2}{r_a}\left(\frac{dV}{dx}\right)_{x=0} \tag{12.2}$$

Provided that the membrane ionic current i_i is a unique function of V,

$$i_i = F(V), \tag{12.3}$$

the value of i_i at each point on the cable will be determined by the value of V at that point. The problem now is to use eqns (12.2) and (12.3) to relate I to i_i. Since I is proportional to dV/dx and i_i is proportional to d^2V/dx^2 we may relate the two parameters by making use of the mathematical identity:

$$\frac{d}{dx}\left\{\left(\frac{dV}{dx}\right)^2\right\} = 2\left(\frac{dV}{dx}\right)\left(\frac{d^2V}{dx^2}\right). \tag{12.4}$$

Integrating both sides with respect to x and making it a definite integral between the end of the cable l and the point x,

$$\left[\left(\frac{dV}{dx}\right)^2\right]_l^x = 2\int_{V_l}^{V_x}\left(\frac{d^2V}{dx^2}\right)dV. \tag{12.5}$$

Hence

$$\left(\frac{dV}{dx}\right)_x = \pm\left\{2\int_{V_l}^{V_x}\frac{d^2V}{dx^2}\,dV + \left(\frac{dV}{dx}\right)_l^2\right\}^{\frac{1}{2}}. \tag{12.6}$$

† A slightly different derivation but with an equally general approach, will be found in Adrian, Chandler, and Hodgkin (1972).

The choice of sign is arbitrary since it depends simply on which half of the cable we choose to consider. If we take only positive values of x, then for $V \to 0$ as $x \to \infty$ in an infinite cable, we require the negative sign. Using the cable equation (Chapter 3, eqn (3.2))

$$i_i = \frac{1}{r_a} \frac{d^2 V}{dx^2},$$
(12.7)

we obtain

$$\left(\frac{dV}{dx} \right)_x = -\left\{ 2 r_a \int_{V_l}^{V_x} i_i \, dV + \left(\frac{dV}{dx} \right)_l^2 \right\}^{\frac{1}{2}}.$$
(12.8)

The boundary conditions are determined by the nature of the cable. We will first consider the infinite-cable case, and then discuss the terminated-cable situation.

Infinite cable, $l \to \infty$. In this case $(dV/dx)_l \to 0$ and it follows that the membrane current i_i must also tend to zero. Hence V_∞ must have a value such that i_i is zero in eqn (12.3). In fact, there may be more than one voltage $V_{i=0}$ which satisfies this condition. Since the electrodes are placed at $x = 0$, we may replace V_x by $V_{x=0}$. Hence, using eqn (12.2),

$$I = 2 \left(\frac{2}{r_a} \int_{V_{i=0}}^{V_{x=0}} i_i \, dV \right)^{\frac{1}{2}}.$$
(12.9)

This relation between I and i_i is similar to that already given in Chapter 9 (eqn. (9.4)), but is more general in that $V_{i=0}$ is not restricted to the case where $V_\infty = 0$. In this form the equation may be used to obtain I when i_i is a known function of V. The inverse relation may be obtained as follows,

$$I^2 = \frac{8}{r_a} \int_{V_{i=0}}^{V_{x=0}} i_i \, dV.$$
(12.10)

This equation may be evaluated (see eqn (13.98)) by differentiating with respect to I and noting that $V_{x=0}$ is a function of I,

$$2I = \frac{8}{r_a} i_i \frac{dV_{x=0}}{dI}$$

or

$$i_i = \frac{r_a}{4} I \frac{dI}{dV_{x=0}}.$$
(12.11)

This equation is the Cole theorem and has already been obtained in Chapter 9 (eqn (9.1)) using a different derivation. We have given both derivations since they illustrate important but different properties of nonlinear cables. The derivation given in Chapter 9 illustrates why the equation is applicable only to infinite cables, by considering the conditions for eliminating dx in the differential equations. The derivation given here emphasizes the relation of the Cole theorem to the more general equation (eqn (12.8)). As we shall show later, this allows the theorem to be extended to deal with terminated cables.

1. *Applications using a cubic equation for i_i (V).* We will illustrate now the use of these equations by considering a particular example. Suppose that the membrane current–voltage relation is given by an N-shaped curve of the kind already discussed in Chapter 8. We will also assume that the membrane obeys this relation for a period of time which is longer than the time taken for charge redistribution on the cable to take place, so that this distribution may be supposed to approach the steady state distribution. In most real situations, of course, this assumption will be only approximately valid. Nevertheless, as we have noted already, the steady state solutions may be used to obtain some important properties of the system which also apply (though not necessarily quantitatively) to non-steady state conditions.

It is convenient to represent the $i_i(V)$ relation using a simple cubic equation:

$$i_i = \frac{1}{R}\left\{V - V^2 + \left(\frac{V}{2}\right)^3\right\}. \tag{12.12}$$

where R is a constant representing the resistance near $V = 0$. We have already used a cubic expression in Chapter 11, although not the same one as eqn (12.12). We will explain the convenience of using polynomial expressions for i_i in Part II of this chapter.

Eqn (12.12) is plotted as the continuous curve linking points V_A, V_B, and V_D in Fig. 12.1. Clearly, the current scale is arbitrarily determined by the parameter R. The voltage scale is also arbitrary, but it may be helpful to note that it can be related to the normal millivolt scale for an excitable cell by allowing each unit on the V axis to correspond to about 20 mV. The equation then describes a membrane with an excitation threshold for uniform polarization V_B about 23 mV from the resting potential V_A. The 'action potential' which the system could generate would be about 140 mV (that is, point V_D). As already discussed in Chapter 8, a current–voltage relation of this kind gives a reasonable description of the membrane when the sodium activation process (m) is allowed to occur while the processes underlying accommodation (n and h) are held constant.

There are three possible values for $V_{i=0}$; V_A, V_B, and V_D. However, the only solution for eqn (12.9) when $V_\infty = V_B$ occurs when $V_{x=0}$ also equals V_B This solution is trivial since it corresponds to a uniformly polarized cable at threshold. As already noted in Chapter 8, this condition is unstable. Moreover, as we will show later (see p. 384), it is a condition which may not be approached using point polarization in an infinite cable. We may assume therefore that, in all cases of interest, V_∞ is either V_A (that is, 0) or V_D.

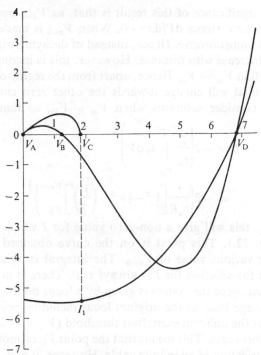

FIG. 12.1. Current–voltage relations for uniform and point polarization of a cable. The curve connecting points V_A, V_B, and V_D is given by eqn (12.12) and will serve in this chapter as a model for the current–voltage relation in an excitable cell. In subsequent diagrams we shall let each unit on the voltage axis equal 20 mV. The curves connecting points V_A and V_C, and connecting I_1 and V_D are the corresponding current–voltage relations for an infinite cable to which current is applied at one point.

When $V_\infty = V_A = 0$ we have

$$I = 2 \left\{ \frac{2}{r_a} \int_0^{V_{x=0}} i_1 \, dV \right\}^{\frac{1}{2}}$$

$$= 2 \left[\frac{1}{r_a R} \left\{ V^2 - \tfrac{2}{3} V^3 + \left(\frac{V}{2} \right)^4 \right\} \right]^{\frac{1}{2}}, \quad (12.13)$$

where V in each term in this equation is $V_{x=0}$. We will now assume that $V_{x=0}$ may be assigned any value. This is equivalent to assuming that the voltage at $x = 0$ is controlled by a voltage-clamp circuit.

The solution to eqn (12.13) has been computed and is shown as the curve linking points V_A and V_C. Up to the point $V_{x=0} = V_C$, the integral in eqn (12.13) has a positive value. However, beyond this point, the integral becomes negative and the square-root involved in computing I then gives an imaginary number. Real values for I do not therefore exist beyond this point when $V_\infty = V_A$.

The physical significance of this result is that, as $V_{x=0}$ approaches V_C, I and i_a approach zero. Hence $dV/dx \to 0$. When $V_{x=0}$ is made to exceed V_C, the sign of dV/dx must reverse. Hence, instead of decaying with distance, the potential must increase with distance. However, this is inconsistent with the boundary condition $V_\infty = V_A$. Hence, apart from the region of the controlled point, the potential will change towards the other zero current point V_D. We must then consider solutions when $V_\infty = V_D$, so that I is given by

$$I = -2\left\{\frac{2}{r_a}\int_{V_D}^{V_{x=0}} i_i\, dV\right\}^{\frac{1}{2}}$$

$$= -2\left\{\frac{1}{r_a R}\left[V^2 - \tfrac{2}{3}V^3 + \left(\frac{V}{2}\right)^4\right]_{V_D}^{V_{x=0}}\right\}^{\frac{1}{2}}. \tag{12.14}$$

For $V_{x=0} = V_C$, this will give a non-zero value for I which is, in fact, the point I_1 in Fig. 12.1. This point is on the curve obtained by integrating eqn (12.14) for various value of $V_{x=0}$. The integral in this case is always positive so that the solution for I is always real. There is no discontinuity, which means that, once the system is given by a locus on the curve including I_1, it cannot change back to the original locus including points V_A and V_C.

Note also that the uniform excitation threshold (V_B, $I = 0$) does not exist as a point on either curve. This means that the point V_B cannot be approached using point polarization of an infinite cable. However, it may be approached, to a reasonable degree of approximation, in terminated cables below a certain length (see below).

2. *Threshold phenomena during activity.* Eqn (12.14) describes a current–voltage curve with only one intersection V_D and no threshold point comparable to V_C. This property illustrates the more general result that, when the membrane current–voltage relation is such that

$$\left|\int_{V_B}^{V_D} i_i\, dV\right| > \left|\int_{V_A}^{V_B} i_i\, dV\right|, \tag{12.15}$$

there will exist a threshold for depolarization by point-applied currents, but there will be no threshold for repolarization. Similarly, when the inequality in expression (12.15) is reversed, there will exist a threshold for repolarization but none for depolarization. As we have noted already in Chapter 8, the current–voltage relations shown by excitable cells will generally satisfy expression (12.15) at the beginning of an action potential. Towards the end of an action potential, the inequality becomes reversed as the repolarization process occurs. Therefore we may expect a threshold for repolarization to be absent at the beginning of an action potential. This phenomenon is observed most clearly in the case of cardiac cells in which the action potential is relatively long. Fig. 12.2 summarizes the threshold phenomena which may

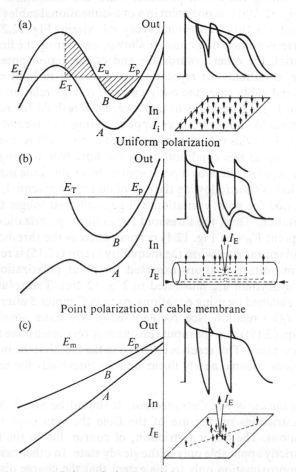

(a)

Out

E_r E_u E_p

E_T

B

A

In

I_i

Uniform polarization

(b)

Out

E_T E_p

B

In

I_E

A

I_E

Point polarization of cable membrane

(c)

Out

E_m E_p

B

A

In

I_E

I_E

Point polarization of sheet membrane

FIG. 12.2. Effect of cable and two-dimensional geometries on current–voltage relations and threshold phenomena during cardiac action potential. (a) Uniform polarization. A shows current–voltage relation during early part of action potential, B shows relation during late part of action potential. These relations are similar to those observed experimentally. Right-hand diagram shows responses expected for subthreshold and suprathreshold repolarizing currents. The threshold voltage is given by the intersection point E_u. E_T is the voltage at which the shaded areas are equal. This determines the intersection point for cable excitation (see (b)). (b) Cable polarization. The cable current–voltage diagram for curve A shows no threshold point. The diagram for curve B shows a threshold point at the voltage E_T. Right-hand diagrams illustrate responses expected. Hyperpolarizing currents applied early during action potential fail to initiate all-or-nothing repolarization. (Based on Hall and Noble (1963) and Noble and Hall (1963).) (c) Point polarization of syncytial sheet. Current-voltage relations become almost linear and no thresholds occur. (Noble 1962b.)

be observed during the action potential when the membrane is polarized uniformly (Fig. 12.2(a)), at one point in a one-dimensional cable (Fig. 12.2(b)) or at one point in a two-dimensional syncytial system (Fig. 12.2(c)). In each case, two current–voltage diagrams are shown, one during the first half of the action potential, the other towards the end of the action potential. In the uniform case, thresholds for repolarization E_u occur in both cases. In the one-dimensional cable case, the early current–voltage relation is similar to that shown in Fig. 12.1 and, as in this case, no threshold for repolarization exists. A threshold, however, does appear during the second half of the action potential. At this time the inequality in eqn (12.15) is reversed. Moreover, for so long as this condition holds the fibre will no longer display a threshold for depolarization by a point source. Since this state may persist for a certain period of time following the end of an action potential, the absolute refractory period for point excitation will generally last longer than that for uniform excitation. Thus, the threshold for uniform polarization appears as soon as the point V_B (see Fig. 12.1) occurs, whereas the threshold for point excitation will appear later when the inequality in eqn (12.15) is re-established.

The current–voltage relations expected for point polarization of a two-dimensional syncytium are illustrated in Fig. 12.2(c). The solution for this case may be obtained by using equations given in Chapter 5 after substituting a nonlinear $i_i(V)$ relation (see Noble 1962b). We have noted already in Chapter 5 (eqn (5.18)) that the input resistance is very insensitive to changes in membrane resistance. This result is reflected in the fact that the input current–voltage relations become nearly linear and no thresholds for repolarization are observed.

3. *Note on the use of the Cole theorem.* It should be noted that there are important restrictions on the use of the Cole theorem (eqn (12.11)) and related equations. The major restriction, of course, lies in the fact that the theorem is strictly applicable only in the steady state. In other cases, it may be used as an approximation only to the extent that the charge distribution on the cable approaches steady state conditions. The second restriction is that the membrane current–voltage relation should be stationary. This creates no difficulty if the relations are steady state ones since these are, by definition, time independent. However, it is worth emphasizing that it is not required that the membrane should show no time-dependent behaviour. The steady state relations may be investigated using the Cole theorem even if time is required for the steady state values of i_i to be established.

In cases where these conditions are not well met, it may still be possible to obtain useful qualitative information, but there may then be large quantitative differences between steady state results and those including time-dependent phenomena (see e.g. Hall and Noble 1963).

Terminated cables. 1. *Open-circuit termination: influence of fibre length on excitation threshold.* If the cell terminates in an open-circuit (as at the uncut end of a nerve or muscle fibre) then, as in the infinite cable, $(dV/dx)_l = 0$. However, the value of V_l is no longer restricted to V_A or V_D since the membrane current need not be zero at the termination. In place of eqn (12.9), therefore, we have

$$I = 2 \left(\frac{2}{r_a} \int_{V_l}^{V_0} i_i \, dV \right)^{\frac{1}{2}}, \tag{12.16}$$

where V_0 and V_l stand for $V_{x=0}$ and $V_{x=l}$. This simplification in notation is justified since there is no longer any ambiguity concerning whether V_0 denotes $V_{x=0}$ or $V_{i=0}$. This equation is more difficult to use than eqn (12.9) since V_l (or V_0, given V_e) must be estimated in each case. This may be done using equations for the spatial spread of current described in Part III. This difficulty also restricts the use of the Cole theorem. As already shown in Chapter 9, the theorem is not applicable in its simplest form to terminated cables. However, we may derive an equation analogous to the Cole theorem:

$$I^2 = \frac{8}{r_a} \int_{V_l}^{V_0} i_i \, dV.$$

Differentiating with respect to I, and noting that the functions dependent on I are the two limits on the integral, we obtain (eqn (13.103))

$$2I = \frac{8}{r_a} \left(i_i \frac{dV_0}{dI} - i_i \frac{dV_l}{dI} \right).$$

Hence

$$i_i = \frac{r_a}{4} I \left\{ \frac{d(V_0 - V_l)}{dI} \right\}^{-1}. \tag{12.17}$$

In this case, therefore, in order to obtain i_i we must measure the voltage at two points on the cable, V_0 and V_l. In practice, this severely limits the application of the equation since in many experimental situations information concerning the voltage at points other than the site of current injection is difficult or impossible to obtain.

The current–voltage relations for terminated cable of various lengths using eqn (12.12) for the ionic current are shown in Fig. 12.3. These relations were obtained using the numerical approximation method described at the end of this chapter (p. 435). Various values for V_l were chosen as initial conditions, and the cable equations were integrated with respect to x from the point $x = l$ with $(dV/dx)_l = 0$. Results for cables of various lengths could be obtained simply by terminating each computation at the appropriate length. The value of dV/dx at this length allowed I to be computed. This value was then plotted against the value of V at the same point.

The curve labelled ∞ is the infinite-cable current–voltage relation already shown in Fig. 12.1. The curve for a cable of length equal to one resting space constant is very nearly the same shape as the uniform current–voltage diagram

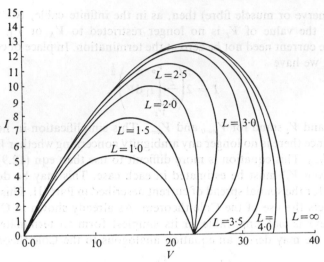

FIG. 12.3. Current–voltage relations for cables of various lengths. The membrane is assumed to obey eqn (12.12) and the voltage scale was obtained by allowing each unit in Fig. 12.1 to equal 20 mV. The current scale is arbitrary. The number on each curve gives the cable length L in terms of the number of resting space constants.

and the threshold point is exactly equal to V_B. The curves for other cable lengths are more complex and some further explanation is required. Thus the relation for $L = 2$ initially follows that for $L = \infty$. This is expected since, for small polarizations, the spatial spread is determined by the resting space constant, that is $\lambda = \sqrt{(R/r_a)}$, and in two space constants the voltage decrements to only about 10 percent of its initial value. However, as the depolarization is increased, the membrane chord resistance increases. The 'effective space constant' increases and the membrane becomes more uniformly polarized. In fact, as the threshold is approached the cable becomes highly uniform, and the threshold potential is nearly equal to the uniform threshold V_B. Hence a cable which is sufficiently long to behave as an infinite cable in response to small currents may behave as a uniform cable near threshold.

The curves for $L > 2$ are even more complex. First, it should be noted that for all lengths up to about $L = 3\cdot45$, the value of V at which the computed value of I is zero is *exactly* equal to V_B. The reason for this result will be discussed later (see eqn (12.18)). However, it does not follow from this that the voltage threshold should equal V_B since for values of L greater than about 2 the curves first reach a larger value of V and then turn back towards V_B. This is very evident in the curve for $L = 3$ (see Fig. 12.4). The voltage threshold in these cases will be given by the maximum voltage reached since stable solutions exist up to but not beyond this voltage.

Fig. 12.4(a) illustrates how the values for the voltage threshold and the rheobasic (that is, long-duration) current threshold are obtained from the computed current–voltage diagram for the case $L = 3 \cdot 0$. The effects of cable length on voltage threshold, current threshold, and the value of $V_{I=0}$ are shown in Fig. 12.4(b). The voltage threshold remains close to the uniform threshold up to $L = 2 \cdot 0$ and is nearly equal to the infinite cable threshold beyond $L = 4 \cdot 0$. The rheobasic current initially increases linearly with cable length (as expected, since the amount of current required to polarize the membrane uniformly to a particular voltage is proportional to its area) and then reaches a maximum at about $L = 4 \cdot 0$. Since in many physiological situations a fibre is usually polarized at or near its midpoint, these results

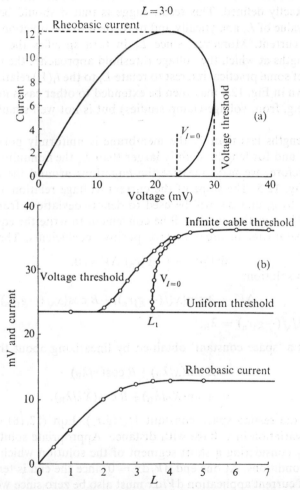

FIG. 12.4. (a) Diagram illustrating how the rheobasic current, voltage threshold, and value of $V_{I=0}$ are obtained from the current–voltage diagram for $L = 3 \cdot 0$. (b) Variation in rheobasic current, voltage threshold and $V_{I=0}$ with cable length. The current and voltage scales are the same as in Fig. 12.3. L_1 is the length at which the curve for $V_{I=0}$ meets the uniform threshold line.

show that for a fibre whose current–voltage relation is well simulated by eqn (12.12), the fibre must be of the order of 8 space constants in length to behave as an infinite cable near threshold. Thus, a skeletal muscle fibre with a space constant of the order of 2 mm would need to be about 2 cm in length. The way in which this estimate should be changed when the current–voltage relation is not well simulated by eqn (12.12) will be discussed below.

The most surprising result shown in Figs. 12.3 and 12.4 is that the value of $V_{I=0}$ remains equal to the uniform threshold for values of L up to 3·45 and then rapidly increases towards the infinite-cable threshold. In fact the curve for $V_{I=0}$ is vertical as it leaves the uniform threshold line and the length L_1 is therefore exactly defined. This result suggests that it should be possible to obtain the value of L_1 analytically and to relate it to the function used for the membrane current. Moreover, since L_1 in turn specifies the approximate range of lengths at which the voltage threshold approaches the infinite-cable value, it is of some practical interest to relate L_1 to the $i_i(V)$ relation, since the results shown in Fig. 12.4 may then be extended to other cases in which $i_i(V)$ is known (e.g. from voltage-clamp studies) but is not well simulated by eqn (12.12).

For all lengths less than L_1 the membrane is uniformly polarized at the point $V_{I=0}$, and for lengths slightly larger than L_1 the nonuniformity will be small. Therefore, we may linearize the equations around the point V_B as shown in Fig. 12.5. The slope of the current–voltage relation at V_B will be represented by g_1 and ΔV will be used to denote deviations from V_B. g_1, of course, will be negative and it will be convenient to write the equations with $-g_1$ as a parameter in order to give positive coefficients. The differential equation is

$$d^2V/dx^2 + (-g_1 r_a)\, \Delta V = 0,$$

which has a solution†

$$\Delta V = A \sin\{x\sqrt{(-g_1 r_a)}\} + B \cos\{x\sqrt{(-g_1 r_a)}\}.$$

Let

$$1/\sqrt{(-g_1 r_a)} = \lambda_B,$$

where λ_B is a 'space constant' obtained by linearizing about the point V_B. Then

$$\Delta V = A \sin(x/\lambda_B) + B \cos(x/\lambda_B)$$

$$= A \sin(X\lambda/\lambda_B) + B \cos(X\lambda/\lambda_B), \tag{12.18}$$

where λ is the resting space constant $1/\sqrt{(g_r r_a)}$. Eqn (12.18) describes an oscillating variation in voltage with distance. Appropriate solutions may be obtained by considering a short segment of the solution which satisfies the boundary conditions. At one end $dV/dx = 0$ since the end is terminated. At the point of current application dV/dx must also be zero since we are dealing

---

† Costantin (personal communication) has also obtained this solution using a membrane current–voltage relation consisting of three linear segments instead of the polynomial relation used here. In the region of threshold, linearization of our current–voltage relation makes the system identical with that considered by Costantin.

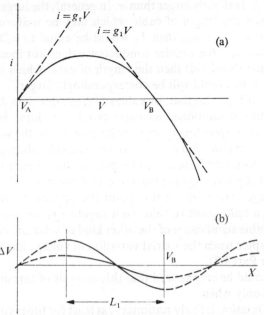

FIG. 12.5. (a) Linearization of current–voltage diagram to give lines for the resting slope conductance g_r and the slope conductance at threshold g_1. (b) Oscillating solutions for ΔV as a function of X. L_1 is given by distance between two points at which $dV/dX = 0$, that is L_1 equals one half cycle.

with the case when $I = 0$. Hence, $A = 0$ and $\Delta V = \Delta V_0 \cos(x\lambda/\lambda_\mathrm{B})$. Thus the appropriate solution is a half cycle separating two values of x at which $dV/dx = 0$. The length of this half cycle must be equal to L_1 since it will be impossible to obtain solutions satisfying the boundary conditions for lengths less than L_1 without allowing the fibre to be uniformly polarized at the voltage V_B. Hence,

$$L_1 = \pi\lambda_\mathrm{B}/\lambda = \pi\sqrt{\left(\frac{1}{-Rg_1}\right)} = \frac{\pi}{\sqrt{-g_1}}$$

since $\lambda = 1$ by definition (we are assuming $R = 1$ in eqn (12.12)). Now differentiating eqn (12.12) gives

$$di_i/dV = 1 - 2V + \tfrac{3}{8}V^2$$

and, since $V_\mathrm{B} = 1\cdot171$ (or more exactly $4/(2 + \sqrt{2})$),

$$(di_i/dV)_{V_\mathrm{B}} = g_1 = -0\cdot828.$$

Hence $L_1 = \pi/\sqrt{0\cdot828} = 3\cdot443$. This value is identical (within the limits of computational error) with the computed value of L_1 shown in Fig. 12.4.

It can now be seen that L_1 is determined by the magnitude of the slope conductance near threshold compared to the magnitude of the slope conductance at rest. In this case, the slope conductance g_1 is only a little smaller than g_r, and so L_1 is slightly larger than π. In general, the larger the magnitude of g_1, the shorter the length of cable which may be uniformly polarized at threshold. Thus if $|g_1| = 4g_r$ then L_1 would be equal to $\pi/2$. Conversely, if $|g_1|$ is small (that is, if the sodium conductance does not increase too rapidly with voltage near threshold) then the length of cable which may behave as a uniform cable at threshold will be correspondingly larger.

It is clear from Fig. 12.4 that for values of L greater than L_1 there are two possible nonuniform solutions, one corresponding to input voltages equal to $V_{I=0}$, the other corresponding to input voltages equal to the voltage threshold. In normal experimental situations it is the second solution which will be obtained, since this may be achieved by progressively increasing the value of V (preferably using a voltage control circuit) at the point of current injection up to the voltage threshold. At this point the applied current will change suddenly from a finite positive value to a negative value, and excitation will occur. The possible significance of the other kind of solution will be considered later in this chapter when the spatial variation in voltage has been considered more fully (see Part III).

Finally, it should be emphasized that this analysis of threshold phenomena strictly applies only when

(a) the $i_i(V)$ relation is fairly stationary, at least for times comparable to the time for redistribution of charge on the cable, and

(b) excitation is achieved either by prolonged currents (that is, rheobasic currents) or by using voltage control at the point of current application.

The analysis does not apply to excitation using fairly brief current pulses when the charge distribution does not even approach steady state conditions. Some of the complications which then arise have been discussed in Chapter 9. We shall discuss these problems further in a later part of this chapter (see p. 413).

2. *Short-circuit termination.* This condition applies either when the fibre ends in an unhealed cut or, approximately, when there is a sudden branching point which reduces the resistance to current flow towards zero. Examples of the latter are a profusely branching dendritic tree at the end of an otherwise unbranched dendrite and the case of a Purkinje fibre entering and making electrical contact with the ventricle muscle (cf. Weidmann 1952). In these cases the boundary condition becomes $E_l = 0$. We then obtain

$$I = 2\left\{\frac{2}{r_a} \int_{V_l=-E_r}^{V_{x=0}} i_i \, dV + (i_a)_i^2\right\}^{\frac{1}{2}} \tag{12.19a}$$

and

$$i_i = r_a \frac{dI}{dV_{x=0}}\left\{\frac{I}{4} - (i_a)_i \frac{d}{dI}(i_a)_i\right\}, \tag{12.19b}$$

where E_r is the *absolute* resting potential. This assumes that V is measured with respect to the resting potential, that is, $V = 0$ when $E = E_r$, so that the value of V when $E = 0$ must be $-E_r$. These equations are even more impractical than those for the open-circuit termination since the measurement of the short-circuit current $(i_a)_l$ requires two additional electrodes to measure dV/dx near the end of the cable.

Part II: The use of polynomials to represent membrane current–voltage relations

In Chapter 8 we discussed a variety of nonlinear current–voltage relations, many of which may be described by equations based on simple physicochemical models of ion transport. In general, these equations are too complex to be usefully treated in nonlinear cable theory, apart from their use in numerical computations. Moreover, most excitable membranes display current–voltage relations which are complex mixtures of the various kinds of nonlinearity.

A more generally useful way of representing the current–voltage relations is to expand the appropriate function in a power series. Thus if

$$i_i = F(V) \tag{12.3}$$

then, using MacLaurin's expansion, eqn (13.108),

$$i_i = F(0) + F'(0)V + \frac{F''(0)}{2!}V^2 + \frac{F'''(0)}{3!}V^3 + \dots. \tag{12.20}$$

In any particular case a power series may be obtained in either of two ways.
(1) If eqn (12.3) is defined analytically and is easily differentiated then eqn (12.20) may be obtained directly by differentiating eqn (12.3).
(2) If eqn (12.3) is given numerically (e.g. from experimental measurements of a current–voltage relation or from tabulation of a function which would be too cumbersome to be expanded directly) then the problem becomes one of curve fitting. Coefficients in V, V^2, V^3, etc, are chosen so that the numerical results are fitted to a desired degree of approximation.

In either case, it is usually found that a reasonable approximation may be obtained by terminating the series after only a few terms, e.g. up to V^3 or V^4. The resulting expression is known as a polynomial. The properties of polynomials have been treated extensively so that this method of representation readily allows further analytical treatment. Moreover, even for purely numerical purposes a polynomial representation is one of the most useful.

We shall use various polynomials in this chapter to represent current–voltage relations similar to those observed in excitable cells. Since the purpose of this chapter is to illustrate general properties, we shall restrict the treatment to a third-order polynomial of the form

$$i_i = \frac{1}{R}(V + pV^2 + qV^3), \tag{12.21}$$

where R has the dimensions of resistance per unit length, p has the dimensions of (voltage)$^{-1}$, and q has the dimensions of (voltage)$^{-2}$. There is no zero power of V since we are taking V as a displacement from the resting potential. Thus when $V = 0$, $i_i = 0$.

The eqn (12.12) used in Part I is a particular example of eqn (12.21), with $p = -1$ and $q = (2)^{-3}$. In Part III of this chapter, we shall consider two other examples.

In order to represent outward-going rectification (see Chapter 8) we will set $p = 0$ and $q = 1$. This allows the steady state current–voltage relation in nerve cells to be simulated. In the case of some muscle cells, the steady state relation shows a mixture of inward- and outward-going rectification. In this case, a useful simulation is given with $p = -3$ and $q = 2.5$. These relations are plotted in Fig. 12.6.

It may be worth noting also at this stage that more accurate simulations of particular cases may be obtained using higher-order polynomials. The equations for the infinite cable discussed in this chapter may be solved without too much difficulty using polynomials up to the fifth order. The equations for terminated cables, however, become difficult for polynomials above the third order.

Part III: Spatial distribution of charge

The spatial spread of charge in the steady state may be obtained using eqn (12.8) derived in Part I. Putting $(dV/dx)_t = 0$, as in the infinite and open-circuit cables, and using eqn (12.21) for i_i:

$$\left(\frac{dV}{dx}\right)_x = -\left\{\frac{2r_a}{R}\int_{V_t}^{V_x}(V+pV^2+qV^3)\,dV\right\}^{\frac{1}{2}}$$

$$= -\left\{\frac{r_a}{R}(V^2+\tfrac{2}{3}pV^3+\tfrac{1}{2}qV^4+C)\right\}^{\frac{1}{2}}, \qquad (12.22)$$

where C is a constant given by

$$C = -(V_i^2+\tfrac{2}{3}pV_i^3+\tfrac{1}{2}qV_i^4).$$

For the special case of the infinite cable in which $V_\infty = 0$ (but not for other values of V_∞ in this case) C becomes zero.

Eqn (12.22) may be transformed to give

$$\int_{V_x}^{V_0}\frac{dV}{(V^2+\tfrac{2}{3}pV^3+\tfrac{1}{2}qV^4+C)^{\frac{1}{2}}} = -\int_x^0\frac{dx}{(R/r_a)^{\frac{1}{2}}}. \qquad (12.23)$$

The general form of the left-hand side of eqn (12.23) is that of an elliptic integral of the first kind. An elliptic integral is any integral of the form

$$\int W[t, \sqrt{\{P(t)\}}]\,dt,$$

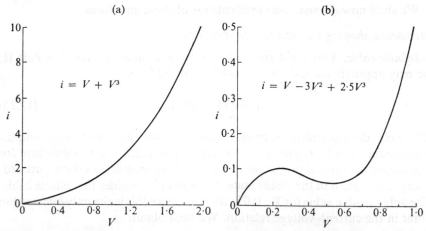

FIG. 12.6. Polynomial functions used to represent membrane current–voltage relations. (a) relation used to represent steady state outward-going rectification observed in nerve membranes. Given by $p = 0$, $q = 1$. (b) Relation used to represent mixture of inward-going and outward-going rectification observed in some muscle membranes. Given by

$$p = -3 \text{ and } q = 2 \cdot 5.$$

where $P(t)$ is a polynomial of the third or fourth degree and W is a rational function. A full treatment of these integrals may be found in many standard works (Greenhill 1892; Hancock 1917; Whittaker and Watson 1944; Erdelyi, Magnus, Oberhettinger, and Soni 1966), and they are extensively tabulated (Pearson 1934; Jahnke and Emde 1945; Byrd and Friedman 1954; Belyakov, Kravtsova, and Rappoport 1965; Abramowitz and Stegun 1964).

When $C = 0$ (infinite cable with $V_\infty = 0$) the left-hand side of eqn (12.23) simplifies further and may be expressed in terms of a small number of elementary functions. The reason for this is that a factor of $P(t)$ is V^2, so that this may be taken outside the square root to give

$$\int_{V_x}^{V_0} \frac{\mathrm{d}V}{V(1 + \tfrac{2}{3}pV + \tfrac{1}{2}qV^2)^{\frac{1}{2}}} = -\int_{x}^{0} \frac{\mathrm{d}x}{(R/r_a)^{\frac{1}{2}}}. \tag{12.24}$$

This gives a much simpler algebraic integral which may be handled by conventional methods (e.g. Hardy 1916; Durrell and Robson 1932; Dwight 1962). Moreover, since we may now take integrals containing terms up to V^6 (which reduces to V^4 when V^2 is taken out as a factor), we may use polynomials for i_1 containing terms up to V^5. In general, therefore, it will be possible to obtain solutions for the case $V_\infty = 0$ when the membrane current–voltage relation is well simulated by polynomials of the fifth degree. If accurate simulation requires higher powers (> 5 for this case; > 3 for cases in which $C \neq 0$) it may still be possible to obtain a solution, either as a hyperelliptic integral or, in special cases, the integral may be reduced to an elliptic integral or to a sum of elliptic integrals (see Byrd and Friedman 1954).

We shall now discuss some applications of these equations.

Membrane showing outward rectification only

Infinite cable. This is the simplest case to treat since, as noted in Part II, we may approximate i_i using an equation of the form

$$i_i = \frac{1}{R}(V + qV^3).$$ (12.25)

Of course, this equation may be inadequate in some cases, and more complex polynomials may be required. Nevertheless, eqn (12.25) is a useful form for illustrating the general properties of cables whose membranes show outward-going rectification in the steady state. We will first consider the infinite cable. The only possible value for V_∞ is 0 in this case since this is the only intersection point in the current–voltage relation. We then obtain

$$\frac{dV}{dx} = -\left\{ \frac{r_a}{R}(V^2 + \tfrac{1}{2}qV^4) \right\}^{\frac{1}{2}}.$$

Hence

$$\int_{V_a}^{V_0} \frac{dV}{V(\tfrac{1}{2}qV^2 + 1)^{\frac{1}{2}}} = -\left(\frac{r_a}{R} \right)^{\frac{1}{2}} \int_x^0 dx.$$

Since

$$\int \frac{dy}{y\sqrt{(y^2 + a^2)}} = -\frac{1}{a} \ln \left| \frac{a + \sqrt{(y^2 + a^2)}}{y} \right|$$

we obtain

$$\ln \left| \frac{V_0\{\sqrt{(qV^2 + 2)} + \sqrt{2}\}}{V\{\sqrt{(qV_0^2 + 2)} + \sqrt{2}\}} \right| = \frac{x}{\sqrt{(R/r_a)}},$$

which may be rearranged to give

$$V = \frac{2\sqrt{2}a}{1 - a^2 q}, \text{ where } a = \frac{V_0}{\sqrt{(qV_0^2 + 2)} + \sqrt{2}} \exp\{-x/\sqrt{(R/r_a)}\}.$$ (12.26)

Fig. 12.7 shows the results obtained using eqns (12.25) and (12.26). In Fig. 12.7(a) we have plotted the membrane current–voltage relation described by eqn (12.25) when $q = 1$, together with the cable input current–voltage relation given by inserting eqn (12.25) for i_i into eqn (12.9). It can be seen that the input current–voltage curve displays less outward-rectification than does the curve for i_i. This is an illustration of the rather large extent to which the cable properties tend to linearize the current-voltage relations of excitable cells. The curves in Fig. 12.7(b) show semilogarithmic plots of eqn (12.26) for various values of V_0 indicated by the figures on the current–voltage diagram. The results have been normalized in terms of V_0 (that is, we have plotted V/V_0) in order to facilitate comparisons of the *shapes* of the curves. For small values of V_0, the decay of voltage with distance is nearly exponential. As the voltage is increased, however, the initial decay becomes much steeper. This is

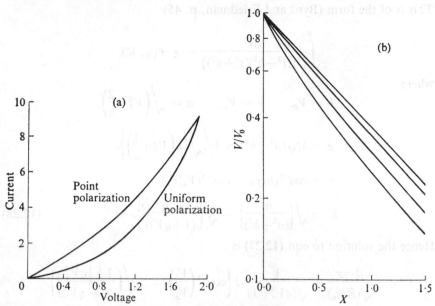

FIG. 12.7. (a) Current–voltage relation for uniform polarization and for point polarization of cable whose membrane shows outward-going rectification. (b) Semilogarithmic plots of $V(X)$ curves for V_0 from top to bottom = 0·15, 0·9, 1·7 and 2·6.

a consequence of the fact that, at stronger depolarizations, the membrane conductance increases. The apparent space constant therefore decreases as current escapes more easily through the fibre membrane. This effect provides a ready explanation for the linearization of the current–voltage relation. As more current is applied so the increment in polarization is restricted to a smaller area of membrane. The total applied current I therefore increases less steeply with voltage than does i_1.

Terminated cable. In this case V_t will not be zero. The constant C in eqn (12.23) then becomes $(-V_t^2 + \frac{1}{2} q V_t^2)$. We then obtain

$$\frac{d-x}{\sqrt{(R/r_a)}} = \int_{V_t}^{V_x} \frac{\sqrt{(2/q)}}{\sqrt{\left(V^4 + \dfrac{2V^2}{q} - \dfrac{2V_t^2}{q} - V_t^4\right)}}\, dV$$

$$= \int_{V_t}^{V_x} \frac{\sqrt{(2/q)}}{\sqrt{\left\{\left(V^2 + \dfrac{1}{q}\right)^2 - \left(V_t^2 + \dfrac{1}{q}\right)^2\right\}}}\, dV$$

$$= \sqrt{\left(\frac{2}{q}\right)} \int_{V_t}^{V_x} \frac{dV}{\sqrt{\left\{(V^2 - V_t^2)\left(V^2 + V_t^2 + \dfrac{2}{q}\right)\right\}}}. \tag{12.27}$$

This is of the form (Byrd and Friedman, p. 45)

$$\int_b^y \frac{dt}{\sqrt{\{(t^2-b^2)(t^2+a^2)\}}} = g \cdot F(\varphi, K),$$

where

$$b = V_d, \qquad y = V_x, \qquad a = \sqrt{\left(V_i^2 + \frac{2}{q}\right)},$$

$$g = 1/\sqrt{(a^2+b^2)} = 1 \Big/ \sqrt{\left\{2\left(V_i^2 + \frac{1}{q}\right)\right\}},$$

$$\varphi = \cos^{-1}(b/y) = \cos^{-1}(V_i/V_x),$$

$$K = \sqrt{\left\{\frac{a^2}{(a^2+b^2)}\right\}} = \sqrt{\left\{\frac{(1+\frac{1}{2}qV_i^2)}{(1+qV_i^2)}\right\}}. \tag{12.28}$$

Hence the solution to eqn (12.27) is

$$\frac{d-x}{\sqrt{(R/r_a)}} = \frac{1}{\sqrt{(qV_i^2+1)}} \cdot F\left\{\cos^{-1}\left(\frac{V_i}{V_x}\right), \quad \sqrt{\left(\frac{1+\frac{1}{2}qV_i^2}{1+qV_i^2}\right)}\right\}.$$

Since we wish to know V_x given V_i we can obtain an explicit solution, not in terms of the elliptic integral, but as an elliptic function (see Byrd and Friedman 1954):

$$(V_i/V_x) = \cos \varphi = \text{cn}(u_1, K).$$

Now

$$\frac{l-x}{\sqrt{(R/r_a)}} = g \cdot F(\varphi, K) = g \cdot u_1.$$

Therefore

$$u_1 = \frac{(l-x)\sqrt{(qV_i^2+1)}}{\sqrt{(R/r_a)}}. \tag{12.29}$$

We therefore obtain the solution

$$V_x = V_i/\text{cn}(u_1, K), \tag{12.30}$$

where K and u_1 are given by eqns (12.28) and (12.29).

Results obtained using a cable length equal to one 'resting' space constant (that is, $l = \lambda_r = \sqrt{(R/r_a)}$) are shown in Fig. 12.8. The method of computing the solutions is described in the *Note on numerical methods* at the end of this chapter.

Fig. 12.8(a) shows the input current–voltage relation for comparison with the infinite-cable and membrane relations. The curve follows the infinite-cable relation more closely than the uniform membrane relation, that is, the terminated cable in this case acts more closely like an infinite cable, as would be expected since the cable becomes less uniformly polarized as the strength of current is increased. This is shown in Fig. 12.8(b) where we have plotted the spatial decay of voltage for various values of V_0.

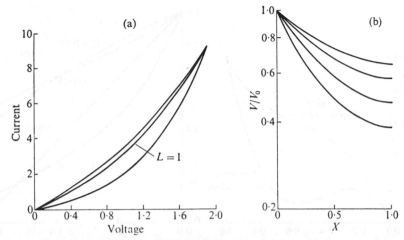

FIG. 12.8. (a) Same as in Fig. 12.7(a) but also including current–voltage relation for cable of length $L = 1$. (b) Semilogarithmic plots of $V(X)$ curves for values of V_0 from top to bottom $= 0.16, 0.88, 1.7$, and 2.6.

This result is particularly relevant to voltage-clamp techniques which involve polarizing a short cable, such as the sucrose gap technique and the use of microelectrodes for voltage-clamping short fibres. These methods require that the fibre length should be short enough to allow the polarization to be approximately uniform. It should be noted, however, that the degree of uniformity achieved in the resting state (that is, for very small polarizations) may not be a good guide to the behaviour of the system when large currents are applied. If a large degree of outward rectification occurs, the fibre will become less uniformly polarized as the current strength is increased.

Membranes showing inward-rectification but no excitation thresholds

Solution for infinite cable $(C = 0)$. For these cases, we require the full third-order polynomial (12.21) and the analysis is more difficult than for eqn (12.25). We will first obtain equations for $C = 0$ (that is, infinite cable with $V_\infty = 0$). Using a derivation analogous to that used to obtain eqn (12.26), we may obtain

$$\frac{V}{\{2\sqrt{(\tfrac{1}{2}qV^2 + \tfrac{2}{3}pV + 1)} + 2 + \tfrac{2}{3}pV\}}$$
$$= \frac{V_0}{\{2\sqrt{(\tfrac{1}{2}qV_0^2 + \tfrac{2}{3}pV_0 + 1)} + 2 + \tfrac{2}{3}pV_0\}} \exp\left\{\frac{-x}{\sqrt{(R/r_a)}}\right\}. \quad (12.31)$$

If we let $p = -3$ and $q = 2.5$, we obtain the membrane current–voltage relation shown in Fig. 12.9. It resembles the relations observed in cardiac membranes (cf. Deck and Trautwein 1964; Noble and Tsien 1969a, Fig. 10) in displaying a combination of inward-going and outward-going rectification, with the latter predominating at large values of V. We have also plotted the

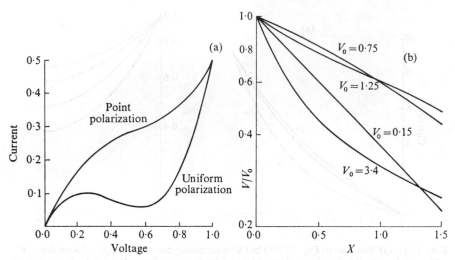

Fig. 12.9. (a) Current–voltage relations for uniform polarization and for point polarization of cable whose membrane shows inward-going and outward-going rectification. (b) Semi-logarithmic plots of $V(X)$ curves for values of V_0 indicated.

cable-input current–voltage relation using eqn (12.9). The 'linearization' referred to above is now very striking and the negative slope conductance observed at intermediate voltages in the $i_i(V)$ relation is completely absent in the $I(V_0)$ relation.

The spatial decays for various values of V_0 obtained from eqn (12.31), are plotted in Fig. 12.6(a). As before, for very small polarizations ($V_0 \ll 3/2q$), a simple exponential decay is obtained. With increasing values of V_0 the effects of the inward-rectification first dominate so that the voltage declines less rapidly with distance than in a linear cable. At large values of V_0 the opposite effect is observed.

Solutions for terminated cable ($C \neq 0$). Using a derivation similar to that used for eqn (12.27), we obtain

$$\frac{l-x}{\sqrt{(R/r_a)}} = \int_{V_l}^{V_x} \frac{\sqrt{(2/q)}\,\mathrm{d}V}{\sqrt{\left\{\left(V^4 + \frac{4}{3}\frac{pV_l^3}{q} + \frac{2}{q}V^2 - \left(V_l^4 + \frac{4}{3}\frac{pV_l^3}{q} + \frac{2}{q}V_l^2\right)\right\}}}. \tag{12.32}$$

One root of the polynomial is always V_l so that the integral may be written

$$\int_{V_l}^{V_x} \frac{\sqrt{(2/q)}\,\mathrm{d}V}{\sqrt{\left[(V-V_l)\left\{\left(V^3 + \frac{4}{3}\frac{pV^2}{q} + \frac{2}{q}V\right) + \left(V_l^3 + \frac{4}{3}\frac{pV_l^2}{q} + \frac{2}{q}V_l\right) + \right.}} \tag{12.33}$$
$$\left. + \left(V^2V_l + \frac{4p}{3q}VV_l + VV_l^2\right)\right\}\right].$$

Further solution of this integral involves finding the other three roots of the polynomial, and these will depend on the values of p, q, and V_l.

Once again we may use $p = -3$ and $q = 2 \cdot 5$ for this case. Since there is only one value of V (that is, 0) when $i_i = 0$, the only situation in which $V_l \neq 0$ is the terminated cable.

There is one additional real root b in this case, and the other two roots form a complex conjugate pair c and $\bar{c}$. The solution is therefore of the form

$$\int_a^y \frac{dt}{\sqrt{\{(t-a)(t-b)(t-c)(t-\bar{c})\}}} = g . F(\varphi, K),$$

where $b < a < y < \infty$; c, $\bar{c}$ are complex (see Byrd and Friedman, p. 135). If we denote the real part of the complex root by u and the complex part by v then the solution in elliptic function form is

$$V_x = \frac{bA - aB - \text{cn}(w, K)(aB + bA)}{A - B - (A+B)\text{cn}(w, K)}, \tag{12.34}$$

where $b = $ other real root of polynomial,

$$a = V_l$$

$$A = \sqrt{\{(V_l - u)^2 + v^2\}},$$

$$B = \sqrt{\{(b - u)^2 + v^2\}},$$

$$w = \sqrt{(\tfrac{5}{4}AB)} \frac{(d - x)}{\sqrt{(K/r_a)}},$$

and

$$K = \sqrt{\left\{ \frac{(A+B)^2 - (a-b)^2}{4AB} \right\}}.$$

Fig. 12.10(a) shows the current–voltage relation obtained for a cable length equal to one resting space constant. In contrast to the case of the membrane showing only outward rectification, the relation follows the $i_i(V)$ curve more closely than the infinite-cable curve. This results from the fact that the increase in resistance as V is increased causes the cable to become more uniformly polarized. This is shown in Fig. 12.10(b). For intermediate polarizations, the cable becomes very uniform indeed. As expected, at large values of V_0 the outward rectification predominates, and the system becomes highly non-uniform.

Current–voltage relations with excitation threshold

Infinite-cable solution. For this case we may use eqn (12.12) which is given when $p = -1$ and $q = 2^{-3}$. As we may expect, the results are generally more complex than in the preceding case. We will discuss first the infinite-cable situation. In this case $V_l = V_\infty$ and is given by either

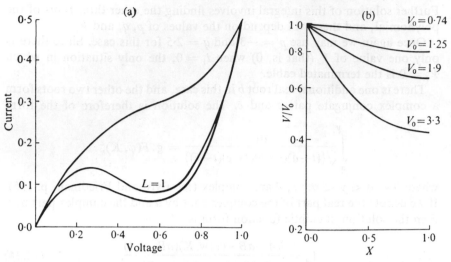

FIG. 12.10.(a) Same as in Fig. 12.9(a) but also including current–voltage relation for cable of length $D = 1$. (b) Semilogarithmic plots of $V(X)$ curves for values of V_0 indicated.

$V_A (= 0)$ or V_D (see Fig. 12.1). When $V_\infty = 0$, the solutions for $V(x)$ are similar to those obtained in the preceding case in that the voltage spread greatly increases in the region of negative slope conductance. However, it is important to note that real solutions will exist only for $V_0 < V_C$ (see Fig. 12.1). For more positive values of V_0, the solution becomes complex. The significance of this result has already been discussed in Part I (see p. 384).

When $V_\infty = V_D$, it is possible to obtain real solutions for all values of V_0.

The results in this section will enable us to investigate some excitation phenomena in more detail. In particular, we will use the results in later sections concerned with the spread of charge during the initiation of excitation (see p. 418).

The spatial variation in potential may be computed numerically direct from the cable differential equation

$$d^2V/dx^2 = r_a i_i, \qquad (12.35)$$

using the polynomial expression for i_i. The methods used are briefly described at the end of this chapter. (see the section *Note on numerical methods*, p. 433) and form an alternative approach to that using elliptic functions.

Results for various values of V_0 are shown in Fig. 12.11. As expected, there are two solutions for each value of V_0 when $V_0 < V_C$. When $V_0 > V_C$, there is only one solution. When V_0 approaches V_C, one of the two solutions is almost flat (that is, $dV/dx \to 0$) near $x = 0$. This corresponds to $I \to 0$, that is, the

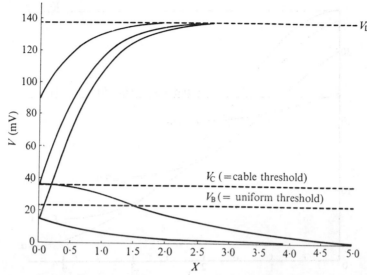

FIG. 12.11. $V(X)$ curves for infinite cable obeying membrane current–voltage relation given by eqn (12.12). For values of $V < V_C$ there are two solutions for each value of V_0. Above V_C there exists only one solution for each value of V_0.

cable threshold point in Fig. 12.1. In the other solution dV/dx is large and positive, which corresponds to a large negative value of I. This is the value I_1 in Fig. 12.1.

As the cable threshold is approached, the spatial spread of charge greatly increases. Thus the decay of potential in one space constant is less than 20 per cent. Hence the initiation of excitation will occur almost simultaneously over a fairly large area of membrane. This result is of importance in experiments concerned with the determination of conduction velocity. The relation between the distance from the stimulating electrode and the time of arrival of an action potential will not intersect the distance axis at zero but rather at some distance along the axis, since the excitation wave will be initiated over a substantial area of membrane near the electrode (Rushton 1937; see also p. 413).

Terminated-cable solution: influence of cable length on spatial distribution of charge and on current–voltage relations.. An important consequence of the extensive spatial spread of charge near threshold is that the effects of cable terminations will become apparent even in fairly long cables. This point has already been noted in Part I where it was shown that for a membrane current–voltage relation obeying eqn (12.12), the cable must be at least 4 space constants in length before behaving as an infinite cable near threshold (see Figs. 12.3 and 12.4). In this section we will discuss the spatial spread of charge in cables of various lengths in order to illustrate some of the properties of excitable cells which may be attributed to the presence of terminations.

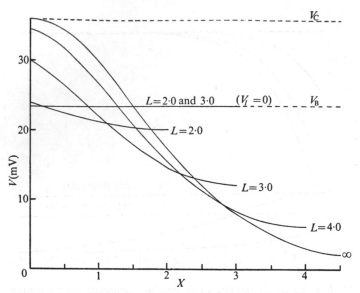

FIG. 12.12. $V(X)$ curves for cables of various lengths at threshold.

Fig. 12.12 shows the variation in V with X at threshold for cables of lengths $L = 2, 3, 4$, and ∞. When $L = 2$, the cable is nearly but not quite uniformly polarized. For all finite cables there are two solutions, one corresponding to the true voltage threshold and the other to a $V_{I=0}$ point (see Fig. 12.4). (Note that for $L_1 < L < \infty$ there are additional solutions since there are five, rather than three, points at which $I = 0$ on the cable current–voltage relation—see Fig. 12.15, $L = 4.0$). The first solution is non-uniform. The second is uniform; $V = V_B$ everywhere. As we have noted already in Part I (see p. 384), the nonuniform solution is the one which would normally occur experimentally. The uniform solution would be approached only in exceptional cases, e.g. where the end of the cable is initially polarized by a second current source or by an 'abortive' attempt at excitation. Moreover, this solution is unstable (see p. 382).

In most experimental situations, the variation in voltage with distance will not be available for direct investigation since this requires voltage recording from a number of points along the cable, although in some cases (e.g. the T system in skeletal muscle—see Chapter 6) indirect techniques may be used to assess the degree of nonuniformity near a threshold point. In general, the more useful relation, from the experimental point of view, is the $I(V)$ relation since this requires measurements of voltage and current at only one point.

Fig. 12.13(a) shows the current–voltage relations for cables of lengths $L = 0.5, 0.75, 1.0$, and 1.25, together with the uniform current–voltage relation (eqn (12.12)) scaled to compare with the case $L = 1.0$. The major interest in this family of curves lies in their application to voltage-clamp studies of the sodium and calcium inward currents in excitable cells. Most of the techniques used to study these currents in muscle cells involve using short cables, and it is therefore important to assess the extent to which the cable properties 'distort' the current measurements. The most obvious effects in Fig. 12.13(a) are that the relation becomes steeper in the region of threshold and the peak inward current occurs at less depolarized potentials. This is most evident in the case $L = 1.25$ when the relation becomes vertical just beyond threshold and the peak inward current occurs at 50 mV compared to a value of 95 mV for the uniform current–voltage relation. The distortions become less marked as the cable length is reduced but there are detectable differences even in a cable as short as $L = 0.5$ (the peak inward current occurs at about 90 mV in this case).

Cable lengths up to about 0.5 are often used in voltage-clamp experiments on muscle cells, particularly cardiac muscle. However, before the results are compared with those expected theoretically, it is very important to note that the curves for various values of L shown in Fig. 12.13(a) strictly apply only when the membrane current obeys eqn (12.12). This equation gives a peak inward conductance less than 10 times larger than the resting conductance g_r, and the slope conductance at threshold g_1 is actually slightly smaller than g_r (see p. 391). In the case of the sodium current in excitable cells, it is more typical for the peak inward chord conductance to be about 100 times larger than the resting chord conductance (see e.g. Hodgkin, Huxley, and

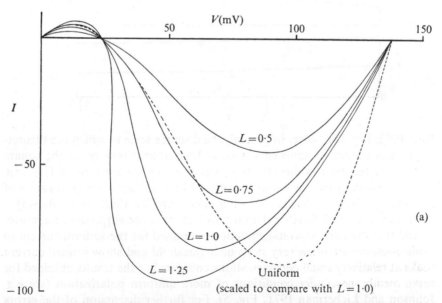

FIG. 12.13.(a) Current–voltage relation for cable of length $L = 0.5$, 0.75, 1.0, and 1.25. (b) $V(X)$ relations for $L = 1.0$. Note that there is only one solution for each value of V_0 and that, at threshold, the membrane is uniformly polarized.

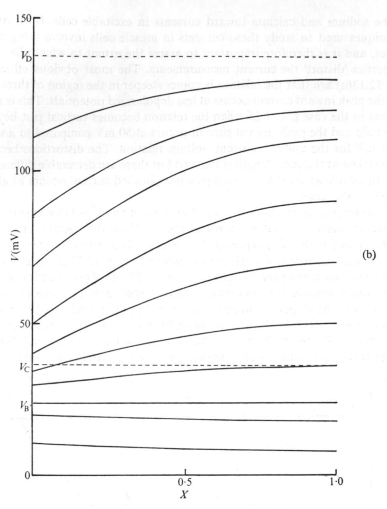

Katz 1952, Fig. 13). Since, in general, the distance scale to which the theoretical results should be adjusted varies, at least approximately, as the square root of the resistance values (thus L_1 varies as the square root of $1/g_1$—see p. 391) we may expect the curves shown in Fig. 12.13(a) to apply to values of L about $\sqrt{(100/10)}$ (that is, about 3 times smaller than those shown)—in the range 0·2 to 0·4 rather than 0·5 to 1·25. It is not surprising, therefore, to find that the current–voltage relations obtained for the sodium current in cardiac muscle are often very steep near threshold and show inward current peaks at relatively small depolarizations compared to the results obtained for nerve membranes under conditions of more uniform polarization (see e.g. Johnson and Lieberman 1971, Fig. 5). For further discussion of the errors involved in measuring the inward currents in cardiac muscle the reader is referred to Beeler and Reuter (1970a) and Johnson and Lieberman (1971).

In the case of the calcium (or mixed sodium and calcium) currents observed in cardiac muscle, the results of Fig. 12.13(a) may apply with relatively little scaling since the calcium currents are considerably smaller than the sodium currents and the current–voltage relations obtained are not very dissimilar from that given by eqn (12.12).

The $V(X)$ relations for the case $L = 1.0$ are shown in Fig. 12.13.(b) This diagram allows the degree of nonuniformity underlying the distortions in the current–voltage diagrams to be estimated. The cable is fairly uniformly polarized near threshold (in fact, the polarization is exactly uniform at V_B as expected—see p. 388), but becomes very nonuniform immediately above threshold when V_0 enters the steep region of the current–voltage diagram. Thus for a value of V_0 equal to 50 mV the end of the cable is at a potential 40 mV larger. $V(X)$ curves for cables shorter than 1.0 may be obtained readily from Fig. 12.13(b) by omitting part of the diagram. Thus the case $L = 0.5$ is given by omitting the curves between $X = 0$ and $X = 0.5$ and rearranging the rest of the distance axis to start at zero. The nonuniformity in this case is clearly considerably smaller (about 10 mV) and is, perhaps, tolerable in many experimental situations.

Another application of these results occurs in the case of the T system in skeletal muscle fibres.† As we have already noted in Chapter 6 this system appears to be excitable in the presence of sodium ions and, in response to voltage-clamp polarizations applied at the surface of the fibre, it should behave as a terminated cable clamped at one end, which is the condition for obtaining the solutions discussed in this chapter. Moreover, results obtained in the presence of tetrodotoxin suggest that the cable length is about one space constant (see Chapter 6, p. 122). Strictly speaking, computations of the spread of current should be done using a Bessel equation but since, as already discussed in Chapter 6, the current converges at the termination (that is, at the centre of the T system) rather than diverging from a point source, the results will be qualitatively fairly similar to those obtained using ordinary one-dimensional cable equations.

The important result in this case is that, in the region of threshold, there are two kinds of solution, one for which V decrements with distance (as for $V_0 < V_B$ in Fig. 12.13(b)) and another in which V increases with distance (as for $V_0 > V_B$). The latter kind of solution occurs below the infinite-cable threshold V_C. In fact, in rather longer cables it may also occur below V_B as an alternative solution to a decrementing one (see Fig. 12.14(b)). Hence, it is

---

† Costantin (personal communication) has obtained results similar to those discussed in this paragraph using a three-segment $i_1(V)$ relation (see footnote to p. 390), and we are indebted to him for letting us see an unpublished version of his theoretical work. In particular, he has also shown that the possibility that the centre of the T system may reach the mechanical threshold before the surface arises naturally from the cable solutions near threshold and does not require that a full action potential should occur at the centre, although it does require that some inward current should be generated (see Eisenberg and Costantin 1971).

possible for the T system to display a $V(X)$ curve which increases with distance from the clamped region while the fibre surface displays a $V(X)$ curve which decrements with distance away from the polarizing electrode since the fibre as a whole will behave as an infinite cable. The results obtained on recording contractile responses near threshold will then depend on the position of the contraction threshold compared to the electrical excitation threshold. If the contractile threshold is reached before the $V(X)$ curve in the T system changes to one which increases with X, the contractile response will be seen at the surface before the centre, as in the fibres investigated in the presence of tetrodotoxin (see p. 119). In other cases, the $V(X)$ curve may change to one in which V increases towards the centre of the T system before the contractile threshold is reached. The centre of the fibre will then respond before the surface does, as observed in Costantin's (1970) experiments. Note also that the depolarization at the centre of the T system may in fact be quite moderate. Thus the case $V_0 = 34$ mV in Fig. 12.13(b) gives a value for $V_L = 50$ mV. As Costantin has also noted, this emphasizes that the contractile response at the centre of the fibre need not indicate that the centre of the T system has escaped completely from the voltage-clamp control in the sense that an all-or-nothing depolarization has occurred. It may simply indicate that the nonuniformity has changed from a decrementing kind to an incrementing kind.

Fig. 12.14 shows that the solutions become more complex when cables longer than $L = 1.25$ are considered. To illustrate the important features we have given the $I(V)$ relation and $V(X)$ curves for $L = 1.5$. The $I(V)$ relations for a more extensive range of values of L will be discussed later (Fig. 12.15). The main difference between the $I(V)$ relation for $L = 1.5$ and those for lengths less than 1.25 is that the curve displays a range of voltages at which three current values occur as solutions at each voltage. This means that the fibre will display two discontinuities when its current–voltage relation is investigated using voltage control at one end. On increasing the voltage there will be a sudden jump (interrupted arrow) from point 1 to point 2. On changing the voltage in the opposite direction, there will be a similar discontinuity between points 3 and 4. Over a certain voltage range, therefore, there will be two stable $V(X)$ solutions for each voltage. This is shown in Fig. 12.14(b). The degree of nonuniformity is now very large at all voltages except for one class of solutions near V_B. The presence of a discontinuity in the current–voltage relation may therefore be taken as a fairly certain indication that the fibre is too long for voltage-clamp studies aimed at a quantitative analysis of the ionic currents.

Trautwein, Dudel, and Peper (1965) have described voltage-clamp experiments on cardiac Purkinje fibres in which, at some voltages, two steady state current values were obtained. Their results differ from the kind of result shown in Fig. 12.14 since the two current values they obtained were both positive (see Trautwein, Dudel, and Peper 1965, Fig. 5), whereas in

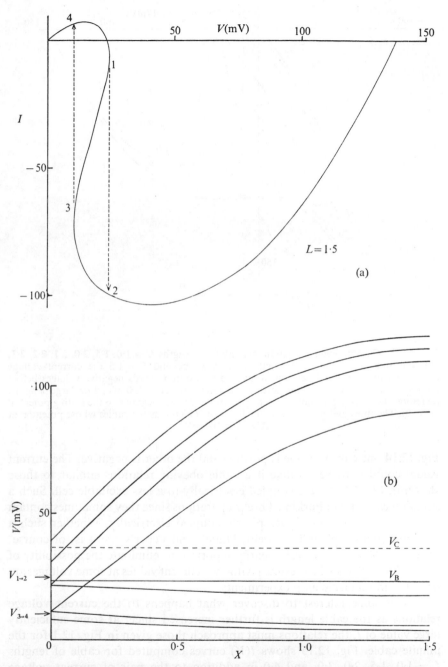

FIG. 12.14. (a) Current–voltage relation for cable of length $L = 1.5$. (b) $V(X)$ relations. Note that near threshold there are two solutions for each value of V_0.

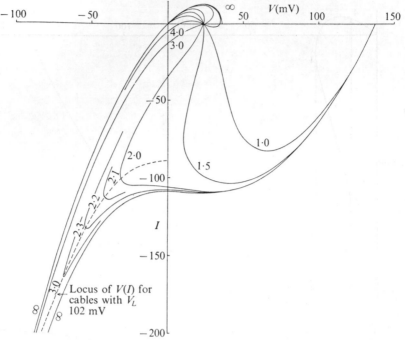

FIG. 12.15. Current–voltage relations for cables of lengths $L = 1\cdot0$, $1\cdot5$, $2\cdot0$, $2\cdot1$, $2\cdot2$, $2\cdot3$, $3\cdot0$, $4\cdot0$, and ∞. As the cable length is increased beyond $L = 1\cdot5$ the current–voltage relation beyond threshold develops a marked curvature towards negative voltages before joining the $L = \infty$ relation intersecting the V-axis at about 130 mV. For large lengths, therefore, the voltage threshold for repolarization becomes negative to the resting potential ($V = 0$). The threshold points lie on the locus of $I(V)$ points for cables whose potential at $X = L$ is 102 mV.

Fig. 12.14 one current value is positive and the other is negative. The current values could both be positive if a cable obeying relations similar to those shown in Fig. 12.14 were coupled electrically to a less excitable cell. Such a situation can arise in Purkinje fibre preparations since they sometimes contain two or more fibres which are poorly coupled electrically. Whether such a situation could explain Trautwein, Dudel, and Peper's result is, of course, uncertain. Nevertheless, it is clearly important to eliminate the possibility of nonuniform effects when results giving two current values at some voltages are obtained in voltage-clamp experiments.

It is of some interest to discover what happens to the current–voltage relations as the cable length is further increased since, at some sufficiently large value of l, the relations must approach those given in Fig. 12.1 for the infinite cable. Fig. 12.15 shows $I(V)$ curves computed for cable of lengths $L = 1\cdot0$, $1\cdot5$, $2\cdot0$, $3\cdot0$, and $4\cdot0$ in addition to the pair of current–voltage relations for infinite cables. Parts of the curves for $L = 2\cdot1$, $2\cdot2$, and $2\cdot3$ are also shown.

The set of curves for positive values of I is the set already shown in Fig. 12.3. It can now be seen that the curves for $L > 2{\cdot}0$ turn back towards V_B and then swing out to negative values of current and voltage before approaching the lower of the two infinite cable curves. This result means that for cables greater than a certain length (in fact for $L > 1{\cdot}65$ in this case—see caption to Fig. 12.16) the threshold for repolarization will be negative to the resting potential V_A. This phenomenon has been observed experimentally in cardiac muscle (Hoffman and Cranefield 1958; Woodbury 1961), and an explanation based on full computations using the cable differential equations, including the capacity current, has been given already by Hall and Noble (1963). Their explanation for thresholds negative to the resting potential depends on transient deviations from the steady state solutions occurring during charge redistribution. The results shown in Fig. 12.15 show that thresholds negative to V_A may occur in steady state solutions. An explanation of the experimental results does not therefore require charge-redistribution effects to be included.

A striking feature of Fig. 12.15 is that the curves for all cable lengths greater than $1{\cdot}65$, including cables longer than L_1, curve back towards negative voltages and eventually join the infinite-cable curve to intersect the voltage axis at V_D. For cable lengths greater than L_1, this is achieved by crossing the voltage axis at three points in the vicinity of V_B, as shown for the case $L = 4{\cdot}0$. The curve for $L = L_1$ (not shown in Fig. 12.15) simply becomes asymptotic to the voltage axis in the region of V_B. This result is expected since we have already shown (see p. 390) that the solutions near V_B in this case are oscillatory and that, over the range of voltages for which the $i_i(V)$ relation is approximately linear, $dV/dX = 0$ at both ends of the cable. Hence, there must exist a range of voltages for which $I \rightarrow 0$, which requires that the $I(V_0)$ curve should be asymptotic to the voltage axis.

The relations plotted in Fig. 12.15 are clearly very complex, and some further explanation is required in order to assess their applicability to physiological situations. Thus it is important to know how much of each $I(V_0)$ curve is, in fact, accessible to investigation using normal experimental techniques. This requires an analysis of the stability of the system in various regions of the $I(V)$-plane. We shall not attempt a full rigorous analysis of stability here. Instead, we shall give an intuitive account which enables the important features of Fig. 12.15 to be understood.

An important clue to what determines stability is obtained by asking what determines the threshold for repolarization, that is, the voltage at which the curve obtained by polarizing with negative currents from the point V_D swings back towards the voltage V_B. It is found that, for cables greater in length than $L = 1{\cdot}65$, all the threshold points lie on or very close to the locus of the $I(V_0)$ points for a particular class of cables: those in which $V_l = 102$ mV (see interrupted line in Fig. 12.15). Hence, although the repolarization threshold is a variable dependent on cable length when expressed in terms of the voltage at $X = 0$, its value becomes a constant when

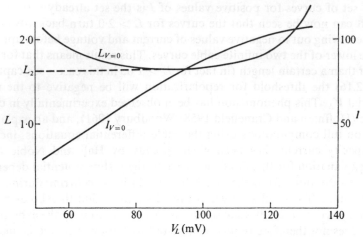

FIG. 12.16. Variation of the value of X at which $V = 0$ (that is, $L_{V=0}$) and the value of I at this point (that is, $I_{V=0}$) with the value of the end voltage V_D. Note that there is a minimum length L_2 of cable whose potential must lie at positive values.

expressed in terms of the voltage at the other end of the cable. This indicates that the conditions for stability are largely determined by the part of the cable lying at positive values of voltage. Now, the length of cable where $V > 0$ is a function of V_l. We have plotted this function in Fig. 12.16, and it can be seen that there is a minimum length ($L_2 = 1.65$) which occurs for values of V_l between about 85 mV and 100 mV. Outside this range the length of cable between V_l and $V = 0$ increases. The reason for this result is that when V_l approaches V_D (that is, about 137·5 mV) the solutions approach those for an infinite cable, and when V_l approaches V_B the solutions approach the oscillatory ones for a cable at or near the excitation threshold. In both cases the cable may extend for an indefinite distance before a point at which $V = 0$ occurs.

A full analysis of stability in this case would require a study of transient as well as steady state solutions. We have not done this. Nevertheless, it seems likely that cables with V_L less than about 100 mV will be unstable, since the length of cable at positive potentials may then increase while the current required to maintain the positive region of cable decreases as V_L falls. This is shown in Fig. 12.16 in the plot of $I_{V=0}$ against V_L. Any attempt by a voltage-control circuit at the negative end of the cable to control the positive end would tend to reinforce the tendency for the net charge to change in a negative-going direction since both the increase in $L_{V=0}$ (that is, length of cable at positive potentials) and the decrease in $I_{V=0}$ would tend to reduce the magnitude of the charge on the negative end of the cable. A voltage-control circuit would try to counteract this by applying even more negative charge to the cable.

It should be emphasized, however, that this instability applies under the condition that the cable is clamped at only one point. This *physical* condition is essential since there is no purely mathematical reason for expecting any of the steady state solutions to be unstable. This point may become clear in considering a physical situation for which all the steady state solutions would be stable. This would require a second voltage-control circuit at the positive end of the cable. This control circuit may then be used to determine a value for V_l. The other control circuit could then be used to polarize the cable to the voltage at $x = 0$, which allows the current supplied by the first circuit to tend to zero. The boundary condition $(dV/dx)_l = 0$ will then be satisfied, and the cable will be stable since fluctuations at the positive end will be controlled by the clamp circuit at this end rather than being reinforced by inappropriate responses of the control circuit at the negative end. So far as we are aware such an experimental arrangement has not been used in physiological experiments, and it is difficult to foresee conditions in which it might prove useful. Nevertheless, the example serves to illustrate a general point: in assessing the stability of a nonlinear cable, it is essential to include all the physical conditions imposed on the system by the experimental techniques employed.

Short-circuit terminations. These cases are far more laborious to treat than is the open-circuit case since the boundary conditions for evaluating dV/dx are variable and depend on both the current–voltage relation and on the particular value of V_0. At $x = l$, the voltage across the membrane will be zero (notice that this will be absolute zero, not some arbitrary zero such as the resting potential V_A used in most of this chapter), but dV/dx will be a non-zero quantity which, for a given current–voltage relation, will vary with V_0. If solutions for such cases are required it is probably more convenient to use approximate numerical integrations of the cable differential equations (see p. 435) than to attempt to obtain analytical solutions.

Multi-dimensional systems. When current may flow in more than one spatial dimension there are no analogues of the Cole theorem. $V(x)$ and $I(V)$ curves are only obtainable by numerical integration of the appropriate differential equations (see Chapter 5). In the cases of two- and three-dimensional systems, some of the effects we have discussed in this chapter are greatly increased. Thus the 'linearization' of the current–voltage relation is so large that very little information indeed may be obtained on the membrane current density (see Fig. 12.2). Similarly the excitation threshold can be very large and excitation may fail to occur even when the membrane current–voltage relation has a very extensive negative chord conductance region.

Part IV: Applications to excitation and conduction

The liminal length for excitation

In 1937 Rushton introduced the idea that it is necessary to excite a certain minimum length of nerve or muscle in order to initiate a propagated disturbance. Some of the results obtained in this chapter may be used to derive this length and to relate it to the resting and active conductances of excitable cells. As in previous sections, the cable is assumed to approach steady state conditions. The possible application of the results to non-steady state conditions during excitation by brief currents will be discussed in the following section.

As we have already seen, the condition for initiating a propagated action potential is that the inward current generated by areas of membrane whose potential lies above the inward current threshold should exceed the outward current generated by other areas of membrane. The minimum length of cable that must lie above the uniform threshold in order to achieve this condition may be obtained for a membrane obeying eqn (12.12) from Fig. 12.11 or Fig. 12.12. In this case it is 1·5 space constants. In general the value of this length, which following Rushton we will call the liminal length X_{LL}, may be obtained analytically for current–voltage relations that may be fitted by a polynomial up to fifth degree, and numerical integration of the cable equation (see p. 435) allows us to obtain X_{LL} for any current–voltage relations, whatever the function used to describe it.

Nevertheless, it may give some insight into the factors determining X_{LL} if we consider an approximate but much simpler method of derivation that is readily applied to current–voltage relations obtained experimentally, whether or not they are well fitted by analytical functions.

Fig. 12.17 shows eqn (12.12) in the region of the inward-current threshold together with linear approximations for the resting conductance g_r, the slope conductance at threshold g_1, and a slope g_a based on obtaining an approximate linear fit to the current–voltage relation between the uniform V and cable (V_C) thresholds.

As we have shown previously, the appropriate solution for a cable obeying a linear $I(V)$ relation with a negative slope is

$$\Delta V = \Delta V_0 \cos(X\lambda/\lambda_B) \tag{12.18a}$$

$$= (V_C - V_B)\cos(X\lambda/\lambda_B). \tag{12.18b}$$

since the initial value of ΔV in the present case is $V_C - V_B$. Now, if the current–voltage relation is well simulated by a linear segment then the first quarter cycle of the solution given by eqn (12.18b) should correspond to the $V(X)$ relation between $X = 0$ and $X = X_{LL}$ given in Figs. 12.11 and 12.12.

Fig. 12.18 shows a comparison between the curves given by eqn (12.18b) and the full nonlinear solution. It can be seen that the value of X_{LL} given by using the linear segment represented by g_1 is an overestimate of the true value ($X_{LL} = 1·721$ compared to 1·5). This deviation is in the right direction since the current–voltage relation becomes steeper than g_1 when V_c is approached. If we use the slope $g_a = 1·2$ instead of g_1 we obtain a closer estimate of 1·43.

To a first approximation, therefore, X_{LL} is equal to the length of a quarter

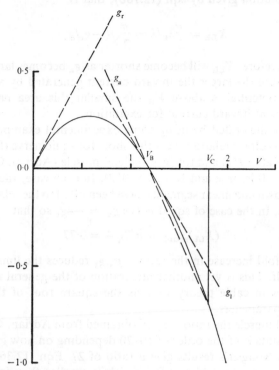

FIG. 12.17. Linear approximations for eqn (12.12) in the region of V_A to V_C. The line given by the slope g_r is used to represent the resting membrane conductance to calculate the resting space constant, $\lambda = \sqrt{(1/g_r r_a)}$. The lines given by g_1 and g_a are used to calculate 'active' space constants as described in text.

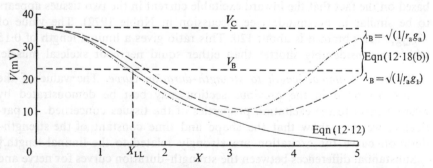

FIG. 12.18. Comparison between solutions for voltage as a function of X given by the linear eqn (12.18b) and that given by the non-linear eqns (12.7) and (12.12). The best approximation to the non-linear solution for the region $0 < X < X_{LL}$ is given by using g_a to calculate the 'active' space constant λ_B.

cycle of the solution given by eqn (12.18b), that is,

$$X_{LL} = \frac{\pi}{2}\lambda_B/\lambda = \frac{\pi}{2}\sqrt{(-g_r/g_a)}. \tag{12.36}$$

In general, therefore, X_{LL} will become shorter as g_a becomes larger. This is to be expected since the larger the inward current generated by areas of membrane whose potential is above V_B the smaller this area needs to be to generate sufficient inward current for excitation.

This may be illustrated by using some experimental examples. Fig. 12.19 shows current–voltage relations near threshold for squid nerve (from Hodgkin Huxley, and Katz 1952) and for frog skeletal muscle (Adrian, Chandler, and Hodgkin 1970; Ildefonse and Rougier 1972). In each case, the values of V_B and V_C are shown and linear segments have been fitted to the relations between these voltages. In the case of squid nerve $g_a = -4g_r$ so that

$$(X_{LL})_{squid} = \pi/2\sqrt{4} = 0.77.$$

Thus, a four-fold increase in the ratio $-g_a/g_r$ reduces the liminal length by about one half. This is yet another illustration of the general result that the distance scales in cable theory vary as the square root of the membrane conductance parameters.

For skeletal muscle the ratio $-g_a/g_r$ obtained from Adrian, Chandler, and Hodgkin's results is of the order of 10–20 depending on how g_r is measured. Ildefonse and Rougier's results give a ratio of 27. Eqn (12.36) then gives a liminal length between 0.35 and 0.5 which is smaller than in squid nerve.

It is more difficult to obtain reliable estimates for the g_a/g_r ratios in other excitable tissues. In the case of cardiac muscle an estimate for Purkinje fibres may be made (see Noble 1972) using Mobley and Page's (1972) estimate of g_r and a value for g_a similar to squid. The latter assumption is based on the fact that the inward excitable current in the two tissues appears to be similar in magnitude (see discussion in Noble 1972). The value of $-g_a/g_r$ then obtained is about 120. This ratio gives a liminal length of 0.15 which is considerably shorter than either squid nerve or skeletal muscle.

Relation of liminal length to strength–duration curve. The value of the results described in the previous section may best be demonstrated by relating them to the excitation properties of the tissues concerned. In particular, we may show that the shape and time constant of the strength–duration curve for excitation are strongly related to the liminal length.[†]

Substantial differences between the strength–duration curves for nerve and muscle fibres have been known to exist for a long time. Davis (1923) and Grundfest (1932) showed that the time constant of the strength–duration curve depends on the geometry of the stimulating current. Excitation using

---

† Katz (1939, p. 105) clearly states that the time constant should depend on X_{LL}. He also discusses early work on the effects of electrode size (see Katz 1939, pp. 77–9).

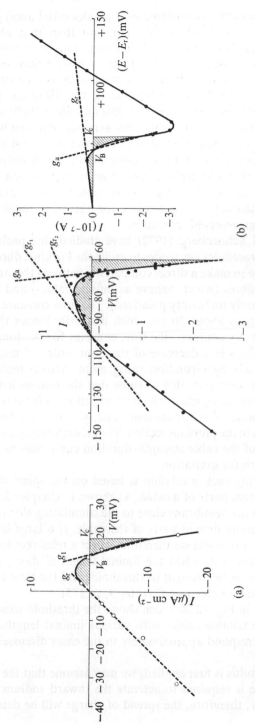

FIG. 12.19. (a) Ionic current–voltage relation for squid nerve obtained by Hodgkin, Huxley, and Katz (1952). The points are replotted from Fig. 10 of Hodgkin et al. V_0 was calculated as the voltage for which the hatched areas become equal. In this case $g_a = -4g_r$. (b) Ionic current–voltage relations for skeletal muscle obtained by Adrian, Chandler, and Hodgkin (1970) (left) and Ildefonse and Rougier (1972) (right). From Adrian et al., g_a is 10–20 times larger than g_r. The upper value is given by using $(g_r)_1$. Ildefonse and Rougier's results give a ratio of 28. Their estimate of g_r corresponds to $(g_r)_2$ in Adrian et al.'s results.

large electrodes (when excitation is initiated over a substantial area) gives a strength–duration curve with a longer time constant than that obtained using very small electrodes. The reduction in time constants on exciting with small electrodes is much greater (about ten-fold) in muscle than in nerve (usually less than a two-fold change). In the case of squid nerve, the strength–duration time constants may be calculated using the Hodgkin–Huxley equations (Cooley, Dodge, and Cohen 1965; Noble and Stein 1966; see also Chapter 9). These calculations agree with the experimental results. The calculated time constant for uniform excitation is about 3 ms and that for excitation of a cable at one point is a little more than 2 ms, a decrease of less than 50 per cent. Noble and Stein (1966) gave an explanation of this decrease in terms of the spread of charge on the cable but were unable to suggest an adequate quantitative explanation for the very much larger reduction in time constant observed in muscle.

Recently Fozzard and Schoenberg (1972) have studied this problem in cardiac muscle. Using intracellular electrodes to excite the Purkinje fibres from sheep heart they were able to make a direct comparison between the strength–duration curves in long fibres (which behave as infinite cables) and in very short fibres (which are nearly uniformly polarized). The time constant for the uniformly polarized fibres is about 30 ms which is a little longer than the membrane time constant. By contrast, the time constant for the long fibres was only about 4 ms, which is a decrease of the same order of magnitude observed in skeletal muscle by Grundfest using more indirect techniques. Fozzard and Schoenberg were also able to show that the liminal length, as measured by inserting a recording electrode at short distances from the point of excitation, is of the order of 0·2 space constants, which is similar to the theoretical estimate given in the previous section. They therefore suggested that the short time constant of the cable strength–duration curve may be related to the short liminal length for excitation.

The reason for expecting such a relation is based on the speed at which charge is applied to different parts of a cable. As shown in Chapter 3, charge is applied very rapidly to the membrane close to the stimulating electrode but spreads more slowly to more distant parts of the cable. If a large length of cable must be brought to threshold we therefore expect a relatively long time constant for excitation whereas, when the liminal length is short, a much shorter length of cable must be brought to threshold, and the time constant for excitation will be much briefer (see Chapter 9, p. 275).

This idea is illustrated in Fig. 12.20 which shows the threshold steady state charge distribution for excitable cables with various liminal lengths. These have been chosen to correspond approximately to the cases discussed in the previous section.

When the current stimulus is first applied, we may assume that the cable is linear since a finite time is required to activate the inward sodium current (see Chapter 8). Initially, therefore, the spread of charge will be determined

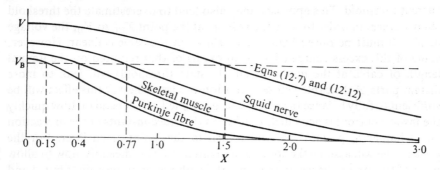

FIG. 12.20. Diagram illustrating threshold $V(X)$ relations for cardiac Purkinje fibre ($X_{LL} = 0.15$), skeletal muscle ($X_{LL} = 0.4$), squid nerve ($X_{LL} = 0.77$), and a cable whose membrane current–voltage relation is given by eqns (12.7) and (12.12) ($X_{LL} = 1.5$). The relations were drawn assuming a sinusoidal solution for voltages above V_B. The curves were then continued below V_B to join the linear cable solution, $V \propto \exp(-X)$.

by the linear equations. The charge will then redistribute itself on the cable so that, when the stimulus current is terminated, the charge near the electrode will fall and that at a distance from the electrode will rise. During this time activation of the inward current will occur and the cable will become non-linear. The charge will therefore redistribute itself towards the nonlinear steady state distribution.

An adequate stimulus will therefore be one that applies sufficient charge to ensure that the cable is at or beyond threshold when the charge redistributes towards the nonlinear steady state distribution. This amount of charge is given by

$$Q_\infty = C \int_0^\infty V_\infty \, dx,$$

which may be calculated from the nonlinear steady state distribution. From Fig. 12.20, it is clear that the shorter the liminal length, the smaller the value of Q_∞. For a given current strength, this means that a shorter stimulus duration is required in order to apply an amount of charge adequate for excitation.

Although such a stimulus will be adequate, it will in general be larger than threshold. The reason for this is that the activation of sufficient inward current for propagation may occur *before* the applied charge has had time to re-distribute towards the steady state distribution. Since, at all times before the steady state, the charge will be still more concentrated on areas of the cable near the electrode, an amount of charge smaller than Q_∞ may then be sufficient to activate a threshold quantity of inward sodium current.

An alternative approach is to calculate the amount of charge necessary at each stimulus strength to raise the liminal length of cable above the inward current threshold. This approach may also tend to overestimate the threshold change since, in order to raise the voltage at the point X_{LL} to V_B, the voltage at $X = 0$ must be raised higher than V_C when the cable is linear. However, some of this excess charge (that is, in excess of that required on the liminal length of cable at the nonlinear steady state threshold) will leak to more distant parts of the cable before excitation occurs. Whether this effect will be sufficient to fully balance the excess charge applied will depend on how quickly the inward current is activated, that is, on the rate constants of the m reaction (see Chapter 8, p. 245). We will return to this point again later. For the present, we will assume the speed of activation to be sufficiently slow to allow most of the excess charge on the liminal length region to be redistributed and explore the consequences of using the liminal length as a parameter for calculating the threshold charge.

Fozzard and Schoenberg (1972) have used exactly this approach to calculate the strength–duration curves for various assumed values of liminal lengths. Their procedure was to use the equation for the response of a linear cable to square current pulses (eqn 3.24) to calculate the time taken for various currents to raise the point X_{LL} to the inward current threshold. Their results are shown in Figs. 12.21 and 12.22. Fig. 12.21 shows the computed strength–duration curves and Fig. 12.22 shows the ratio of cable strength–duration time constant to uniform strength–duration time constants. It can be seen that the strength–duration curve becomes much steeper as the liminal length is decreased and the time constant (τ_{S-D}) becomes much shorter. Thus for a liminal length of 0·15 the time constant is only about one-fifth of that for the strength–duration curve using uniformly applied currents.

As already noted, the main deficiency in this approach is that it does not

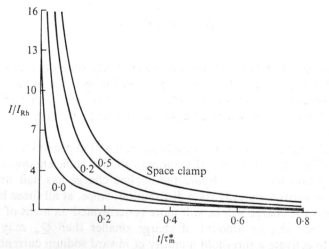

FIG. 12.21. Strength–duration curves computed for cables whose liminal lengths are 0, 0·2, and 0·5 space constants. The relation for a uniformly polarized membrane ('space clamp') is also shown (Fozzard and Schoenberg 1972).

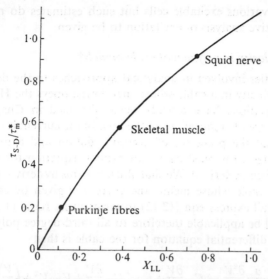

FIG. 12.22. Ratio of the strength–duration curve time constant for point-stimulated fibre $\tau_{\text{S-D}}$ to the strength–duration time constant for the uniformly charged fibre τ_m^* as a function of the liminal length for excitation. The two time constants are equal when X_{LL} is about 0·85 (Fozzard and Schoenberg 1972).

take explicit account of the speed of inward current activation. Implicitly, it is assumed that this process occurs at a rate that is slow enough to allow the excess charge on the liminal length region to redistribute itself to other regions. This assumption is clearly arbitrary but it is not entirely unreasonable, since the time constant for charge redistribution is of the same order of magnitude as that for activation of the inward current, that is, much less than 1 ms (see Chapter 9, p. 269).

Nevertheless, in view of the arbitrariness involved it is unlikely that Fig. 12.21 may be used to give more than a fairly approximate analysis of the strength–duration curve for cable excitation. This is clear from the fact that the time constant of the strength–duration curve decreases as the sodium activation rate is increased by increasing temperatures (see Chapter 11). This effect would not be predicted by the liminal length approach described here, since the absolute magnitudes of the resting and sodium conductances are not very sensitive to temperature. The liminal length, as defined by eqn (12.36) therefore should not change. The effect of temperature on the strength–duration time constant implies that, if the sodium activation process is sufficiently fast, it should be possible to initiate propagation by exciting a length of cable shorter than the liminal length when very brief strong shocks are applied. The liminal-length parameter should therefore be used with some caution in excitation theory. As we have shown in this section, estimates of liminal length based on experimentally observed current–voltage relations may give considerable insight into the differences between the excitation properties of various excitable cells but such estimates do not yet allow a fully quantitative analysis of excitation to be given.

Conduction velocity in polynomial cable models†

The difficulties involved in analytical approaches to the determination of conduction velocity in a cable whose membrane obeys the Hodgkin–Huxley or similar equations have already been discussed in Chapter 10. These difficulties are greatly reduced when the membrane current is represented by a polynomial and the processes of accommodation and activation time are neglected. In fact, for most cases, an explicit equation for the conduction velocity may then be derived. We shall illustrate this by deriving the conduction velocity for a cable whose membrane current is given by eqn (12.12) with $R = 1$. We shall express eqn (12.12) in terms of its roots (V_B and V_D). The derivation will be applicable therefore to all third-degree polynomials.

The partial differential equation for the cable is then

$$\frac{a}{2R_i}\frac{\partial^2 V}{\partial x^2} = C_m\frac{\partial V}{\partial t}+i_i = C_m\frac{\partial V}{\partial t}+V-V^2+\left(\frac{V}{2}\right)^3$$

$$= C_m\frac{\partial V}{\partial t}+V\left(1-\frac{V}{V_B}\right)\left(1-\frac{V}{V_D}\right). \tag{12.37}$$

Using the relation for the case of uniform propagation (see Chapter 10)‡

$$\frac{\partial^2 V}{\partial x^2} = \frac{1}{\theta^2}\frac{\partial^2 V}{\partial t^2} \tag{10.5}$$

we obtain

$$\frac{d^2 V}{dt^2} = K\left\{\frac{dV}{dt}+\frac{V(1-V/V_B)(1-V/V_D)}{C_m}\right\}, \tag{12.38}$$

where $K = \theta^2 2R_i C_m/a$.

For the sake of simplicity we will allow $2R_i C_m/a = 1$ and $C_m = 1$. Other cases can easily be obtained by appropriate scaling (see p. 428). Then

$$\frac{d^2 V}{dt^2} = \theta^2\left[\frac{dV}{dt}+V\left(1-\frac{V}{V_B}\right)\left(1-\frac{V}{V_D}\right)\right]. \tag{12.39}$$

---

† We are very grateful to P. J. Hunter and P. A. McNaughton for suggesting the analysis described in this section. FitzHugh (1969) has used the same approach to derive the conduction velocity for a cable whose membrane current is given by the cubic equation used in models of the van der Pol type (see Chapter 11) and attributes the method to Huxley. The analysis is not restricted to the cubic case. Solutions can be obtained for some odd-degree polynomials of higher than third degree (Hunter, McNaughton, and Noble 1975). In each case, the polynomial describing the trajectory in the $\dot{V}(V)$ plane is a polynomial of degree $(n+1)/2$, where n is the degree of the polynomial describing $i_i(V)$.

‡ For a discussion of conduction in fibres with non-uniform properties see Khodorov and Timin (1975).

Let $dV/dt = \dot{V}$, then eqn (12.39) may be written

$$\dot{V}\frac{d\dot{V}}{dV} = \theta^2\left\{\dot{V}+V\left(1-\frac{V}{V_B}\right)\left(1-\frac{V}{V_D}\right)\right\}$$

or

$$\frac{d\dot{V}}{dV} = \theta^2\left[1+\left\{V\left(1-\frac{V}{V_B}\right)\left(1-\frac{V}{V_D}\right)\Big/\dot{V}\right\}\right]. \tag{12.40}$$

As in Chapter 11, we will make use of the phase-plane diagram (plotting $\dot{V}$ against V) to illustrate the properties of this equation (see Fig. 12.23).

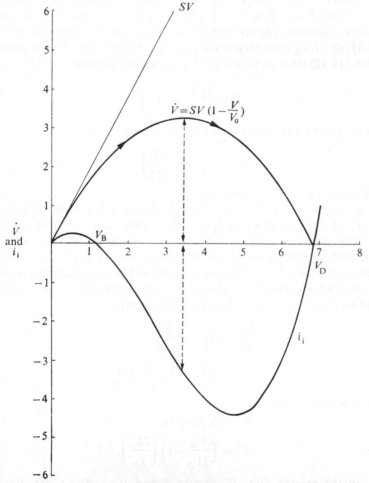

FIG. 12.23. The trajectory $\dot{V}(V)$ for a propagated action potential in a cable whose membrane current is given by eqn (12.12). The initial slope of the trajectory is given by SV, where S is the safety factor for conduction. The interrupted arrows are of equal length to show that, when $\dot{V}$ is maximal, $\dot{V} = -i_i$.

Integrating eqn (12.40) we obtain

$$\dot{V} = \theta^2 \int_0^V \left[1 + \left\{ V\left(1 - \frac{V}{V_B}\right)\left(1 - \frac{V}{V_D}\right) \Big/ \dot{V} \right\} \right] dV. \tag{12.41}$$

This equation and eqn (12.40) may be shown to be satisfied by the solution

$$\dot{V} = \left(\frac{V_D}{2V_B} - 1\right)\left(V - \frac{V^2}{V_D}\right), \tag{12.42}$$

which is plotted in Fig. 12.23. Thus the trajectory is a parabola whose peak value occurs at the voltage $V_D/2$. This solution satisfies the requirements (1) $\dot{V} = 0$ at $V = 0$ and $V = V_D$ and (2) $\dot{V} = -i_i$ when $\dot{V}$ is maximal. The latter requirement arises from the fact that when $\dot{V}$ is maximal, $d^2V/dt^2 = 0$ and $dV/dt + i_i/c_m$ must then equal zero (see eqn (12.38)). This is satisfied by solution (12.42) since at $V = V_D/2$ this equation becomes

$$\dot{V} = \frac{1}{4}\left(\frac{V_D^2}{2V_B} - V_D\right).$$

From the polynomial equation for i_i, we obtain

$$i_i = \frac{1}{4}\left(V_D - \frac{V_D^2}{2V_B}\right),$$

which is equal to $-\dot{V}$, as required.

The factor $(V_D/2V_B - 1)$ in (12.42) and later equations is an important one. It is, in fact, the nearest equivalent to the 'safety factor' defined by Rushton (1937) and others to indicate the extent to which the fibre's ability to be excited and conduct exceeds the minimum. As we shall see, conduction cannot occur unless $V_D > 2V_B$, that is, unless the safety factor S is positive.

To obtain the conduction velocity we may differentiate eqn (12.42) with respect to V and combine with eqn (12.40) to give

$$\frac{d\dot{V}}{dV} = S\left(1 - \frac{2V}{V_D}\right)$$

$$= \theta^2 \left\{ 1 + \frac{(1 - V/V_B)}{S} \right\}, \tag{12.43}$$

which is satisfied when

$$\theta^2 = S^2/(S+1), \tag{12.44}$$

that is,

$$\theta^2 = \left(\frac{V_D}{2V_B} - 1\right)^2\left(\frac{2V_B}{V_D}\right).$$

Since, in the present case, $V_D = 4 + \sqrt{8}$ and $V_B = 8/(4 + \sqrt{8})$, we obtain

$$\theta^2 = \frac{\sqrt{2} + 2 \cdot 25}{\sqrt{2} + 1 \cdot 5}.$$

The time course of the propagated action potential may be obtained by integrating (12.42) with respect to time

$$\int \frac{dV}{V(1-V/V_D)} = St. \tag{12.45}$$

Hence

$$-\ln \frac{(1-V/V_D)}{V} + \ln C = St,$$

from which we obtain

$$V = C \bigg/ \left(\frac{C}{V_D} + \exp(-St)\right), \tag{12.46}$$

C is an arbitrary constant dependent on the definition of the point $t = 0$. If we allow $V = V_D/2$ at $t = 0$, then $C = V_D$ and

$$V = \frac{V_D}{1 + \exp(-St)}. \tag{12.47}$$

This equation is plotted in Fig. 12.24 together with the time course of i_i and i_m for comparison with that shown for a 'complete' action potential in Fig. 10.2. The 'action potential' given by the polynomial model differs from the complete response in having no repolarization phase. The effects on conduction attributable to recovery (or accommodation) processes will be discussed below (see the section *Influence of activation time and accommodation (recovery) on conduction*, p. 429).

The results obtained in this analysis may now be used to note some of the important features of the propagated response that have already been discussed to some extent (and from rather different viewpoints) in Chapters 6 and 10:

1. First, the maximum rate of rise of the propagated action potential must be less than that of the corresponding uniform (or 'membrane') action potential. In the latter case, dV/dt is maximal when i_i is greatest in magnitude, whereas in the propagated case dV/dt becomes maximal at voltages lower than that at which i_i becomes most negative.

2. It is not at all obvious from the time course $V(t)$ or the trajectory $d\dot{V}/dV$ where the 'threshold' potential V_B lies. This is also evident from the fact that V_B only occurs in the parameter S in eqn (12.47), which determines the rate of rise, but not the shape of propagated response. This point serves to emphasize the difficulties involved in analytical approaches that involve assigning a 'voltage threshold' during the propagated response (see Chapter 10. pp. 287).

3. When V is small, that is, at early times, eqn (12.47) approximates to

$$V = V_D/\exp(-St)$$
$$= V_D \exp St, \tag{12.48}$$

which gives the exponential form of the foot of the action potential (see Chapter 6).

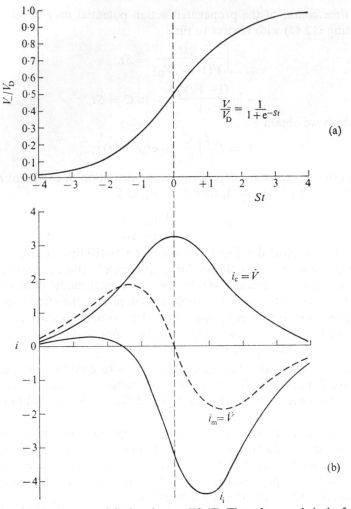

$$\frac{V}{V_D} = \frac{1}{1+e^{-St}}$$

(a)

$i_c = \dot{V}$

$i_m = \dot{V}$

(b)

i_i

FIG. 12.24. (a) 'Action potential' given by eqn (12.47). The voltage scale is the fraction of the peak voltage V_D. The time scale is in units of St, where S is the safety factor.

(b) Time course of i_c, i_i, and i_m. The vertical interrupted line occurs at the point at which $\dot{V}$ is the maximal and $i_m = 0$. This diagram may be compared with that for the complete propagated action potential shown in Fig. 10.2.

4. When $V_B \ll V_D$ (that is, low threshold, large quantity of inward current for excitation), eqn (12.44) becomes

$$\theta^2 \to V_D/2V_B, \tag{12.49}$$

that is, θ is determined simply by the ratio V_D/V_B. When this is the case eqn (12.48) simplifies further to give

$$V = V_D \exp \theta^2 t, \tag{12.50}$$

which is the same result as that given by eqn (6.16) (see Chapter 6 pp. 115–17) when we remember that $a/2R_iC_m = 1$ in the present treatment.

5. The conduction velocity increases as S increases. θ will be zero when S is zero and will approach $\sqrt{S}$ when S is large (see Fig. 12.25). In general, therefore, the conduction velocity increases approximately as the square root of the safety factor. It should be emphasized, however, that the appropriate definition of the safety factor is not entirely independent of the model being used for the membrane ionic current. Thus, it is possible to construct a current–voltage diagram which satisfies the condition $V_D > 2V_B$ but whose shape does not allow the inward current area to exceed the outward current area. The ionic current–voltage relation is then not describable by a cubic equation and higher degree polynomials are required. Hunter, McNaughton, and Noble (1975) have shown that for a certain class of higher degree polynomials the expression $(V_D/2V_B-1)$ is replaced by the general expression:

$$S = \left\{ \frac{V_D^{(n-1)/2}}{\left(\dfrac{n+1}{2}\right)V_B^{(n-1)/2}} - 1 \right\} \qquad (12.51)$$

where n is the degree of the polynomial used for the ionic current. Thus, in the cubic case, where $n = 3$, eqn (12.51) can be seen to reduce to $(V_D/2V_B)-1$, as expected.

In the case of higher degree polynomials eqn (12.51) is appropriate whenever the trajectory $\dot{V}(V)$ is described by a polynomial, $P_2(V)$, of degree $\frac{1}{2}(n+1)$. The conditions on the roots of the ionic current polynomial, $P_1(V)$,

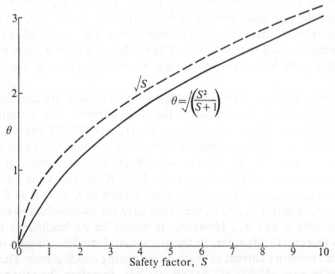

FIG. 12.25. Relation between conduction velocity θ and the safety factor S. The interrupted curve is $\sqrt{S}$.

for this requirement to hold are given in Hunter, McNaughton, and Noble (1975).

6. To simplify the derivation of eqn (12.44) we set $2R_i/a = 1$, $C_m = 1$ and $R = 1$ (see p. 422). A more general equation may be obtained by including these parameters explicitly. From eqns (12.12) and (12.38) it is evident that the polynomial for i_i is divided by the factor RC_m when the equation for d^2V/dt^2 is obtained. Since θ^2 should be replaced by $\theta^2 2R_iC_m/a$, we obtain eqn (12.52) in place of (12.44),

$$\theta^2 = \frac{a}{2R_iC_m}\frac{S^2}{RC_m(S+1)}. \tag{12.52}$$

Hence

$$\theta = \frac{\sqrt{(aR/2R_i)}}{RC_m}\left(\frac{S^2}{S+1}\right)^{\frac{1}{2}}$$

$$= \frac{\lambda_p}{\tau_p}\left(\frac{S^2}{S+1}\right)^{\frac{1}{2}}, \tag{12.53}$$

where $\lambda_p = \sqrt{(aR/2R_i)}$ and $\tau_p = RC_m$.

When S is large this gives

$$\theta = \frac{\lambda_p}{\tau_p}\sqrt{S}. \tag{12.54}$$

Eqn (12.54) bears some resemblance to the conduction velocity equation derived by Rushton (1937, eqn 9) for a cable model with a threshold e.m.f. change. λ_p is similar to Rushton's parameter L; τ_p to his α. The major difference lies in the fact that, in Rushton's equation, θ is proportional to the safety factor, not to its square root. This difference does not arise from differences in the definition of safety factor. In fact, Rushton's definition approaches the one used here when S is large (see Noble 1972, Appendix). It must be related, therefore, to the difference in the shapes of the current–voltage relations used in the models. The model used here is more realistic in not assuming an instantaneous change in ionic current at the threshold potential.

7. Finally, we may draw some further conclusions concerning the dependence of θ on the magnitudes of the ionic currents. The current magnitudes are represented by two factors in eqns (12.52)–(12.53), that is, R and S.

R is a simple scaling factor that determines the resting resistance as well as determining the magnitude of the inward excitatory current. In fact, as already noted earlier in this chapter (p. 382), R has the same value as R_m since the polynomial equations for i_i approximate to $i_i = V/R$ as V becomes small. The parameters λ_p and τ_p therefore have the same values as the passive cable constants λ and τ_m. However, it would be misleading to use these constants in eqn (12.54) since, in the polynomial model, changes in R alter the inward excitatory current as much as the resting conductance. The fact that R appears in eqns (12.52)–(12.54) should not, therefore, be taken to imply that the resting membrane resistance plays a significant role in determining θ.

On the contrary unless S is small, the magnitude of the inward excitatory current is much more important and the presence of R in eqns (12.52)–(12.54) represents this fact. We may conclude therefore that, for the same safety factor (that is, same V_D, V_B, and ratio of excitatory to resting conductance), θ varies as the square root of the conductance to inward current flow.

The factor S depends on i_i in a more complex way. If the excitatory current is increased without changing the resting conductance then the safety factor will increase. Using higher-degree polynomials, it is possible to show that the peak magnitude of i_i is nearly proportional to $\sqrt{(S^2/(S+1))}$ (McNaughton and Noble, unpublished). This also gives the result that θ will vary as the square root of the peak ionic current. This result is intuitively reasonable since, when S is large, the conduction velocity becomes virtually independent of the resting resistance. The relation between θ and the peak value of the excitatory current should not then depend on whether this value is changed without changing the resting resistance (that is, by changing S) or by also changing the resting resistance (that is, by changing R). The dependence of θ on i_i is treated further below (see p. 431).

Influence of activation time and accommodation (recovery) on conduction. The fact that an analytical treatment of conduction is possible when the membrane current is represented by a polynomial is very encouraging, but the results obtained are nevertheless of limited application to experimental situations since the current–voltage relation of the membrane is normally a function of time (see Chapter 8, Fig. 8.12). As we have noted previously, this time dependence is attributable to two major effects: the time taken for inward current to be activated (activation time) and the recovery (or accommodative) processes. Both of these effects reduce the net inward current and, hence, reduce the conduction velocity.

So far as we are aware, it is not yet possible to extend analytical approaches to fully deal with both of these effects, although Hunter, McNaughton, and Noble (1975) have obtained analytical solutions for models including activation time. However, we may give a qualitative account by showing how the conduction velocity versus temperature curve obtained by Huxley (see Chapter 10, Fig. 10.5) may be related to the velocity relations given by simpler approaches. Fig. 12.26 is a diagrammatic attempt to do this. We have not given scales to the ordinates in order to emphasize the point that this discussion is not quantitative.

θ_P represents the conduction velocity that would be derived using the best-fitting polynomial to describe the current-voltage relation when

$$m = m_\infty(V)$$

and n and h are equal to their resting values (see Chapter 8 for definitions of these variables). This relation gives the current occurring at each potential when the activation time is negligible and when no recovery has occurred. As already noted, the value of θ obtained for this case must be larger than any

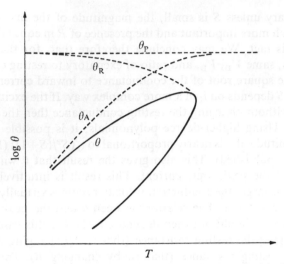

FIG. 12.26. Diagram illustrating relation between the conduction velocities in various cases. θ_P is the conduction velocity for a system with no activation time and no recovery. θ_R is the conduction velocity for a system with instantaneous activation but including recovery. θ_A is the conduction velocity for a system with an activation time but no recovery. θ is the conduction velocity for a system including activation time and recovery. The curve for θ is based on Huxley's curve for squid nerve (see Fig. 10.5). See also Cooley and Dodge (1973).

value obtained experimentally. It will also be relatively independent of temperature, since there are no temperature-dependent rate coefficients to be taken into account and the absolute values of the ionic conductances are not very temperature-dependent.

First consider the effect of introducing the recovery processes. At low temperatures these will be very slow and the conduction velocity for a cable with recovery θ_R will approach θ_P. As the temperature is increased the recovery rate will increase and θ_R will decrease. At a large enough temperature the recovery process will occur sufficiently quickly to reduce the safety factor below zero, and conduction will fail.

Representing the membrane ionic current by a polynomial (usually cubic) plus a first-order recovery term is characteristic of models of the van der Pol type (see Chapter 11). Recently, Noldus (1973) has described the use of a perturbation technique to obtain analyses of conduction in models of this kind. The $\theta(T)$ relation that he has obtained is similar to that for θ_R in Fig. 12.26.

Now consider the effect of including an activation time but no recovery. The ionic current is assumed to activate following a first-order reaction with rate coefficient k. The conduction velocity that would be given by models of this kind is represented in Fig. 12.26 by θ_A. At high temperatures, θ_A will approach θ_P as the activation time (of the order of $1/k$) becomes very small, and eqns (12.52) approximate to eqn (12.12). At lower temperatures, θ_A will be considerably smaller than θ_P, since the time taken to activate the inward current becomes the limiting factor in determining the speed of propagation. The conduction velocities obtained when both activation time and recovery are included will fall below θ_A and θ_R. At low temperatures θ will approach θ_A when the recovery effects are negligible. At high temperatures θ will approach θ_R as the activation time becomes negligible.

Dependence of conduction velocity on magnitude of excitatory membrane conductance. For models not including an activation time the conduction velocity is proportional to the square root of the active membrane conductance g ($= 1/R$) as we showed above (p. 428) for the cubic model. However, this result is not applicable to models which include activation time. The dependence of θ on g is then more complex (Hunter, McNaughton, and Noble 1975). We shall briefly summarize the equations that may be used in various cases and give an intuitive explanation for the results obtained.

Using $P_1(V)$ for the ionic current function (which we initially assume to be time-independent) and $g = 1/R$, eqn (12.38) becomes

$$\frac{\mathrm{d}^2 V}{\mathrm{d}t^2} = K\left\{\frac{\mathrm{d}V}{\mathrm{d}t} + \frac{gP_1(V)}{C_m}\right\}. \tag{12.55}$$

As we have seen in the cubic model discussed above, the *shape* of the trajectory, i.e. the function $P_2(V)$, is dependent only on $P_1(V)$. This result applies generally for all $P_1(V)$ (see Hunter, McNaughton, and Noble 1975). Hence, for any given ionic current function, all action potentials will have the same shape; only the time scale may vary. Let ϕ be a time-scale factor such that $V(\phi t)$ is identical for all cases corresponding to a given $P_1(V)$. Then the parameter ϕ must be a function of K, g, and C_m. Rewriting eqn (12.57) in the terms of ϕt we obtain

$$\frac{\mathrm{d}^2 V}{\mathrm{d}(\phi t)^2} = \frac{K}{\phi}\left\{\frac{\mathrm{d}V}{\mathrm{d}(\phi t)} + \frac{gP_1(V)}{\phi C_m}\right\}. \tag{12.56}$$

The terms $\mathrm{d}^2 V/\mathrm{d}(\phi t)^2$ and $\mathrm{d}V/\mathrm{d}(\phi t)$ are now independent of all parameters except the function $P_1(V)$ which we consider fixed. We may then show that eqn (12.55) is satisfied when

$$K \propto \phi \propto g/c_m$$

and we obtain

$$\theta^2 2R_i C_m/a \propto g/C_m. \tag{12.57}$$

This result is not surprising since the coefficient of the ionic current term in eqn (12.55) is g/C_m and, so far as this term is concerned, a change in g is indistinguishable from a reciprocal change in C_m. Hence the dependence of θ on C_m must include a term reciprocal to that of the dependence of θ on g. In the present case, where θ is proportional to $g^{\frac{1}{2}}$, there must be a term giving θ proportional to $C_m^{-\frac{1}{2}}$.

Although variations in g and $1/C_m$ are indistinguishable so far as the ionic current term is concerned, this is not true for the whole equation, since C_m

also appears in the parameter K. Rearranging eqn (12.57) we obtain[†]

$$\theta \propto (a/2R_iC_m)^{\frac{1}{2}}.(g/C_m)^{\frac{1}{2}}$$

$$\propto (a/2R_i)^{\frac{1}{2}}.g^{\frac{1}{2}}.C_m^{-1},\qquad(12.58)$$

which is the same result as eqn (12.52) when S is constant (that is, the *shape* of the function describing i_i is unchanged).

When the ionic current is assumed to follow a first-order activation process (as in equations of the Hodgkin–Huxley type when $\gamma = 1$ for the m reaction) the conduction velocity becomes proportional to the fourth root of g (see Hunter, McNaughton, and Noble 1975) so that the corresponding equation to eqn (12.58) is

$$\theta \propto (a/2R_iC_m)^{\frac{1}{2}}.(kg/C_m)^{\frac{1}{4}}$$

$$\propto (a/2R_i)^{\frac{1}{2}}.(kg/C_m)^{\frac{1}{4}}.C_m^{-\frac{3}{4}}.\qquad(12.59)$$

where k is the rate coefficient for activation of the conductance. This equation has also been obtained some years ago by Hodgkin (unpublished—see Hodgkin 1975).

The intuitive explanation for this result is that, in models including activation time, when θ is increased, less time is available for activating the ionic current during the action potential. A smaller fraction of the available current is therefore activated and the actual increase in i_i flowing during the action potential is smaller than the increase in g. Only if *both* k and g are increased by the *same* factor does the fractional degree of activation remain constant.

Although eqn (12.59) is more realistic in including the rate of activation of i_i, there remains an important difference between the model and the behaviour of real excitable cells since the sodium activation process is not first-order. As discussed in Chapter 8 (p. 245), there is a delay in the activation process which, in the Hodgkin–Huxley equations, is represented by setting $\gamma = 3$ in the m reaction. Hunter, McNaughton, and Noble (1975) have described a model of this kind and have shown that in this case the equation for θ becomes

$$\theta \propto \left(\frac{a}{2R_i}\right)^{\frac{1}{2}}.(kg)^{\frac{1}{8}}.C_m^{-\frac{5}{8}}.\qquad(12.60)$$

Thus, the conduction velocity depends on an even lower power of g in this case. The result is expected since the decrease in fractional activation with increases in θ is greater when the activation process shows a delay. The result also corresponds to the numerical one obtained by Huxley (1959) and Stein and Pearson (1971) showing that, in the Hodgkin–Huxley model, the curve relating θ to g is very flat at large values of g. There is therefore very

[†] A. F. Huxley (see Hodgkin 1975) has also obtained this result using the same dimensional arguments implied by the present treatment.

little increase in speed to be obtained by further increasing g above a certain value.

Hodgkin (1975) has recently shown that even this small gain is unobtainable in practice. If g is increased by increasing the density of ionic channels then the 'capacitance' attributable to the charge displacement involved in the gating reaction is also increased. Since, in all the cases considered, θ decreases more steeply with C_m than it increases with g, the result is to decrease the conduction velocity at large enough values of channel density. Hodgkin's calculations using eqn (12.59) suggest that the optimal value of conduction velocity is in fact obtained when the ionic current density is similar to that in normal squid nerve. A similar result would be obtained using the more realistic equation, (12.60).

Note on numerical methods

It may be helpful to some readers to have further information on how the equations we have discussed in this chapter may be used. We will discuss two examples. The first illustrates the use of elliptic integral tables. The second concerns the alternative method of numerical integration of the cable differential equation.

Example of use of elliptic integral tables. We will consider the case of the terminated cable showing outward rectification (eqns (12.27)–(12.30)). Numerical results may be obtained in two ways, either by calculating from the elliptic integral solution or from the elliptic function solution. Both elliptic integrals and elliptic functions are tabulated, but we have found it more convenient to use the elliptic integral solution, as the tables of Belyakov *et al.* (1965, pp. 554–72) are particularly detailed and require less elaborate interpolation.

Furthermore, if greater accuracy is required than can be obtained by interpolation (using either simple linear methods or by the technique described by Pearson (1934)), the first elliptic integral may be more readily calculated on a digital computer since it has a rapidly convergent power series expansion:

$$F(\varphi, k) = \sum_{m=0}^{\infty} \binom{-\frac{1}{2}}{m}(-k^2)^m t_{2m}(\varphi)$$

for $0 < \varphi < \pi/2$, $k^2 < 1$,

where

$$\binom{-\frac{1}{2}}{m} = \frac{\Gamma(m+\frac{1}{2})}{\Gamma(\frac{1}{2}).\Gamma(m=1)}$$

and

$$t_{2m}(\varphi) = \frac{(2m-1)}{2m}t_{2(m-1)}(\varphi) - \frac{1}{2^m}\sin^{2m-1}\varphi \cos \varphi.$$

For most values of φ and k, less than five terms are required to obtain 8-figure accuracy.

15

To use the tables, the following procedure is adopted. A value of V_l is selected so that, with given values of p, d, x, and $\sqrt{(R/r_a)}$, V_x may be obtained by finding the value of φ which, for the calculated value of k, results in the first elliptic integral taking the value of

$$\frac{(l-x)}{\sqrt{(R/r_a)}}\sqrt{(pV_l^2+1)}.$$

Once φ is obtained we have

$$V_x = V_l/\cos \varphi.$$

For example, if $p = 1$, $l = 1$, $\sqrt{(R/r_a)} = 1$, $x = 0$ (that is, to find V_0 for a short cable of 'normalized' electrotonic length of 1 when $i_i = (V+V^3)/R$ and $V_l = 1\cdot0$), we have

$$\frac{(l-x)}{\sqrt{(R/r_a)}}\sqrt{(pV_l+1)} = \sqrt{2} = 1\cdot4142,$$

$$k = \sqrt{0\cdot75}.$$

On p. 568 of Belyakov *et al.* (1965) we find that $F(\varphi, \sqrt{0\cdot75})$ is $1\cdot406000$ for $\varphi = 67°$ and $1\cdot435095$ for $\varphi = 68°$, so that an approximate estimate of φ is $67\cdot5°$. Hence $\cos \varphi = 0\cdot3827$ and we obtain the result $V_x = 2\cdot61$. If $V_l = 0\cdot1$, $F(\varphi, \sqrt{0\cdot995}) = 1\cdot005$ when $\varphi = 49°40'$; $\cos \varphi = 0\cdot6472$ and we obtain $V_0 = 0\cdot1545$.

For a linear cable, the ratio of V_l to V_0 in a short cable of this length is $0\cdot6481$ (eqn (4.10)), so that this calculation indicates that for small values of V the nonlinear cable behaves like a linear one, as expected. This may also be checked analytically as follows.

$$\int_b^v \frac{dt}{\sqrt{\{(t^2-b^2)(t^2+a^2)\}}} = g.F(\varphi, k) = g.cn^{-1}(\cos \varphi, k),$$

where cn^{-1} is the inverse Jacobian elliptic function. Therefore

$$\frac{l-x}{\sqrt{(R/r_a)}} = \frac{1}{\sqrt{(1+V_l^2)}} cn^{-1}\left\{\frac{V_l}{V_x}, \left(\frac{1+\frac{1}{2}V_l^2}{1+V_l^2}\right)^{\frac{1}{2}}\right\}.$$

When $V_l \ll 1$, the expressions containing $V_l \to 1$ and, since

$$cn^{-1}(x, 1) = sech^{-1} x,$$

we obtain

$$\frac{l-x}{\sqrt{(R/r_a)}} = sech^{-1}\left(\frac{V_l}{V_x}\right)$$

that is,

$$V_x = V_l \cosh\left\{\frac{(l-x)}{\sqrt{(R/r_a)}}\right\}.$$

Similarly, we obtain

$$V_x = V_0 \frac{\cosh\{(l-x)/\sqrt{(R/r_a)}\}}{\cosh\{l/\sqrt{(R/r_a)}\}},$$

which is identical with eqn (4.10) when R is equated to the linear parameter r_m.

Numerical integration of cable differential equation. This method should be chosen either when the elliptic integral or elliptic function solution is excessively tedious to compute or when the $i_i(V)$ relation is not well fitted by a simple polynomial. Eqn (12.35),

$$d^2V/dx^2 = r_a i_i,$$

is rearranged as a pair of equations:

$$dV/dx = Y \quad \text{and} \quad dY/dx = r_a i_i,$$

which may then be solved using the Runge–Kutta or similar rules for numerical integration. The procedure depends on whether the cable is infinite or terminated.

If the cable is infinite, the procedure is to start the integration at a very small, but increasing, value of V (that is, at some distance away from the current source on the negative x-axis). The integration then proceeds until V equals a chosen value V_0. If a number of values of V_0 are required, the integration is continued until V equals the largest value. The solutions for small values of V_0 are obtained by ignoring the solution beyond the particular V_0 required (cf. Noble and Hall 1963). This simplification arises from the fact (already noted in Chapter 9) that, in the infinite-cable case, there is a unique relation between V and x for any given value of V_∞ so that all the solutions for different values of V_0 may be exactly superimposed by shifting the x-axis.

If the cable is terminated, the procedure is similar except that there is no unique solution. Various values of V_l are chosen to start the integrations and the $V(x)$ curve is computed back to the point $x = 0$ for each value of V_l. This will generate a series of corresponding values of V_0. A particular value of V_0 may be obtained by successive approximation.

13. Mathematical appendix

SOME of the mathematics we have used in this book will be unfamiliar to many physiologists. In this chapter we shall give a brief account of some of these areas. The purpose of the account is not to remove all the difficulties: that would require a textbook in itself. We hope, nevertheless, that it will be helpful to have unfamiliar mathematical methods placed in some relation to better-known ones. In the first place, this may give some degree of confidence in applying the appropriate theory to practical problems. Even a minimum of knowledge about a new function or operation may serve to counter the initial reaction of many biologists when faced with a page of unfamiliar mathematical symbols. Second, for the more mathematically inclined reader, it may be useful to have some references to more extensive treatments. We shall assume that readers are already familiar with the general principles of differentiation, integration, differential equations, and well-used functions such as the exponential and logarithm.

Some properties of linear systems

The first seven chapters of this book are concerned with linear systems. The term 'linear' signifies that the differential equation describing the system contains no products or nonlinear functions of the dependent variable or its derivatives. Thus

$$\frac{d^2V}{dX^2} + V = 0 \tag{13.1}$$

and

$$\frac{d^2V}{dX^2} + \frac{1}{X}\frac{dV}{dX} + V = 1 \tag{13.2}$$

are linear, whereas

$$\frac{d^2V}{dX^2} + V^2 = 0 \tag{13.3}$$

is nonlinear. In particular, linear equations have some useful and very powerful properties, which allow the solution obtained for a single, relatively simple case to be used as a component for the assembly of solutions for more complex cases. Moreover, even in the case of nonlinear systems, it is sometimes possible to reduce a particular problem to a linear form. Within certain restrictions, some of the properties of linear systems may then be used.

The superposition principle

The superposition principle has been used frequently in this book to obtain a solution for an excitation, when that excitation may be represented as the linear sum of simpler forms of excitation for which solutions were already obtained. The principle may be stated in the following general way. Suppose that, in a given linear system, the excitations $i_1(t)$ and $i_2(t)$ produce responses $v_1(t)$ and $v_2(t)$, respectively. Mathematically, this means that, if $i_1(t)$ is substituted into the differential equation (e.g. eqn (13.1)), $v_1(t)$ will satisfy that equation for all t. We can express the relationship between $i_1(t)$ and $v_1(t)$ very simply by using an arrow to relate the excitation to the response:

$$i_1(t) \rightarrow v_1(t)$$

$$i_2(t) \rightarrow v_2(t).$$

The superposition principle states that

$$(A\, i_1(t) + B\, i_2(t)) \rightarrow (A\, v_1(t) + B\, v_2(t)), \tag{13.4}$$

where the coefficients A and B are arbitrary constants. This result may be verified for any linear differential equation by substituting the right-hand expression of eqn (13.4) into the differential equation for v, and rearranging. It may also be verified that the superposition principle is not valid for a nonlinear equation, such as eqn (13.3), where cross-products of the form $v_1(t)v_2(t)$ inevitably arise.

In order to put the superposition principle to use we require simple excitation waveforms whose response is easily obtained. The waveforms most frequently used in this book are the step function and the impulse (delta) function.

The unit step function

As noted in Chapter 2, the step function has been used very frequently in electrophysiology, partly because the even more fundamental impulse function is more difficult to produce to a reasonable degree of accuracy in experimental situations.

Mathematically, the unit step function is often denoted by $u_1(t)$ and is defined as follows:

$$u_1(t) = \begin{cases} 0 & t < 0 \\ 1 & t > 0. \end{cases} \tag{13.5}$$

The function is shown in Fig. 13.1. In some texts, $u_1(t)$ is referred to as the Heaviside unit function $H(t)$.

Knowledge of the response to the unit step function is particularly useful in obtaining the responses to rectangular pulses of finite duration. A rectangular pulse may be represented as the sum of a unit step and an inverted unit step, the latter delayed by the width of the pulse ΔT.

$$\text{Unit rectangular pulse} = u_1(t) - u_1(t - \Delta T). \tag{13.6}$$

If $v_1(t)$ is the response to a unit step, the superposition principle requires

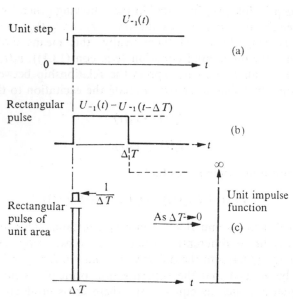

FIG. 13.1. (a) Unit step function. (b) Rectangular pulse formed by two unit step functions. (c) Delta function as a limit of square pulse as $\Delta T \to 0$.

that the rectangular pulse of unit amplitude should produce the response

$$\text{unit rectangular pulse, width } \Delta T \to v_1(t) - v_1(t - \Delta T). \qquad (13.7).$$

The unit impulse function

This function can be thought of as a limiting case of the rectangular pulse. As the rectangular pulse is narrowed to an infinitesimal width ($\Delta T \to 0$), the area of the response will also tend to decrease. It is convenient to compensate for this by increasing the amplitude of the narrowing pulse as $1/\Delta T$, thereby keeping the *area* of the waveform constant (Fig. 13.1). For example, if the amplitude of the pulse is in units of current, a constant area corresponds to a fixed value of applied charge. The infinitely brief pulse of unit area is known as the impulse function, or the Dirac delta function, and is denoted by $u(t_0)$ or $\delta(t)$ (see Pipes and Harvill (1970, p. 166) or Jaeger (1951, p. 37) for further discussion).

The superposition principle allows the unit impulse response to be derived from the unit step response. Scaling expression (13.7) by the amplitude of the pulse ($1/\Delta T$) and letting $\Delta T \to 0$, the unit impulse response $v_0(t)$ becomes

$$v_0(t) = \lim_{\Delta T \to 0} \frac{1}{\Delta T}\{v_1(t) - v_1(t - \Delta T)\} = \frac{dv_1(t)}{dt}. \qquad (13.8)$$

Eqn (13.8) follows from the definition of the derivative; the response to a unit impulse is simply the derivative of the response to a unit step function. In general, if $v(t)$ is the response of a linear, time-independent system to an arbitrary waveform $i(t)$,

$$\frac{di(t)}{dt} \to \frac{dv(t)}{dt}. \tag{13.9}$$

Similar properties may also be demonstrated for higher-order derivatives, as well as integrals of $i(t)$ and $v(t)$.

Although an impulse function is never precisely realizable as an input to a real physical system, it is nevertheless useful as an idealization of very brief pulses. Thus, in Chapter 3 an impulse function was used to simulate the application of a finite amount of charge in a very brief period of time. As shown in Fig. 3.14, the response to an impulse function closely approximates the response to a brief pulse at all but very small times after, and short distances from, the point of excitation. The main virtue of the impulse response is that it is easier to obtain analytically. Furthermore, the results may be used for further analytical investigation, whereas the responses to rectangular pulses of finite duration usually require numerical calculation and the solutions cannot readily be used for further analytical work.

The convolution integral

The impulse function has another important advantage in the analysis of linear systems, arising from the fact that any waveform may be represented as the composite of many impulses occurring at successive instants of time and having appropriate areas (see Fig. 13.2). Using the superposition principle, the response to any waveform may be calculated by summing an infinite number of impulse responses, each suitably scaled and delayed. This calculation is expressed as an integral, the convolution integral.

Let us call the response $v(t)$. Consider first the response at a fixed value of t and the contribution to the response produced by the excitation at some particular previous time u ($u < t$). At time u, the excitation can be represented by a single impulse, area $\phi(u)\,du$, and the rest of the excitation waveform can be ignored. Now if the unit impulse response is denoted by $h(t)$, the impulse at u will give rise to a response that is proportional to $h(t-u)$; the argument $(t-u)$ is simply the delay between u and t. Scaling by the amplitude $\phi(u)\,du$ of the impulse at u, we have

$$\text{impulse at time } u \to \phi(u)\,h(t-u)\,du.$$

$$(\text{area } \phi(u)\,du)$$

The superposition principle allows the contributions from the entire waveform to be added. The summation takes the form of an integral,

$$\int_0^t \phi(u)h(t-u)\,du = v(t). \tag{13.10}$$

As u ranges over all values between 0 and t, the integral incorporates the contributions of all impulses before time t. In real systems there is no point in

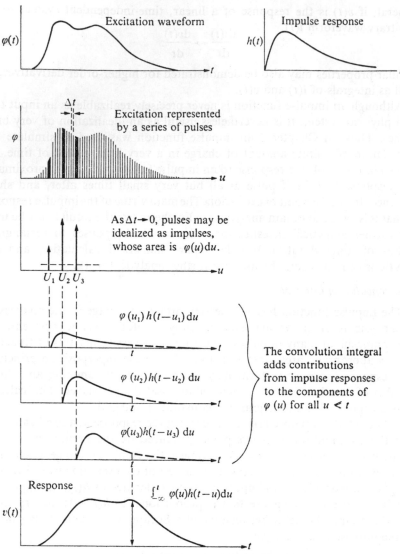

Fig. 13.2. Graphical representation of the convolution integral.

integrating beyond $u = t$. However, for generality, the convolution integral is often written with $+\infty$ as the upper limit of integration. (Whether the upper limit is t or $+\infty$ is immaterial so long as $h(t-u)$ is zero for negative arguments, that is, for $u > t$.) Thus the convolution integral may be stated as

$$\phi(t) \rightarrow v(t) = \int_{-\infty}^{+\infty} \phi(u)h(t-u) \, du. \tag{13.11}$$

It is evident from this brief presentation of the convolution integral that the unit impulse response plays a central role in the analysis of linear systems. One of its transforms (the Laplace transform) is known as the 'transfer function' of the system (see p. 450). The usefulness of the impulse response is even more readily exploited in the Laplace-transform method, which converts the operation of convolving two functions into the simpler operation of multiplying their respective transforms (see p. 451).

Laplace transforms

Throughout this book, the solutions to differential equations arising from various models of current flow are obtained by Laplace-transform methods. These methods are well known in the solution of partial differential equations in many physical problems, in areas such as heat flow and mechanics as well as electrical circuits (see Carslaw and Jaeger (1959) for examples). Nevertheless, it may be helpful to give some introduction in this chapter to the rationale and essential features of the transform method. For more extensive or more rigorous accounts the reader is referred to textbooks by Carslaw and Jaeger (1959), Churchill (1944), or Gardner and Barnes (1942). Milhorn (1964) and Riggs (1970) give many examples of the use of Laplace transforms in the analysis of physiological problems.

What is a transform?

A transform is an operation applied to the terms in an equation, usually with the purpose of making the equation easier to solve. Thus the process of multiplication is simplified, under certain circumstances, by first 'transforming' the variables (say x and y) to their logarithms, $\ln x$ and $\ln y$. The transform of the solution is then obtained simply by addition: $\ln (xy) = \ln x + \ln y$. Finally, the solution itself is obtained by applying an 'inverse transformation', that is, by taking the antilogarithm, to obtain the result xy. In an analogous way, the use of the Laplace transformation, and the corresponding inverse transformation, often simplifies the solution of certain differential equations.

Requirements of a transform for applicability to differential equations

One way of simplifying a differential equation is to transform it into an ordinary algebraic equation which may then be solved by conventional techniques. The Laplace transformation is one of a number of methods which works in this way. Some readers may be familiar with another example, the use of the operator method (see Jaeger 1951, p. 13; Pipes and Harvill 1970, p. 57) which converts linear differential equations with constant coefficients to an algebraic equation (the so-called *characteristic* equation). We shall not specifically discuss the operator method since, within its limited range of application, it is essentially similar to the Laplace-transform method.

The requirements of a useful transform may be stated as follows.

(1) In order to simplify differential equations, the transform should convert the operation of differentiation into some ordinary algebraic operation. This implies that the transform should include the process of integration at some stage.

(2) The transform of a function should be related in rather simple, but general, ways to the transforms of its derivatives and integrals.

(3) The transforms of commonly used functions (e.g. the step function, delta function, and exponential function) should be fairly simple.

(4) There should be an *inverse transformation* corresponding to the transform operation. Once the transform of the solution has been obtained, the inverse transform operation is required to retrieve the solution itself.

These requirements are stated simply from the point of view of applicability to practical problems. From the mathematical standpoint, these requirements may be stated more extensively and in a more rigorous fashion, and for such treatments the reader is referred to the mathematical texts quoted above. The Laplace-transform operation is the simplest integral transform which satisfies the above requirements.

Definition

The Laplace transform of a function $f(t)$ is defined as

$$\mathscr{L}(f(t)) = \bar{F}(s) = \int_0^\infty e^{-st}f(t)\,dt, \tag{13.12}$$

where s is a complex number whose real part is positive. (In some texts (e.g. Carslaw and Jaeger 1959) the variable p is used instead of s.)† This equation clearly satisfies the first requirement, that the transform should involve integration. Since the integral has definite limits, the transformed function will no longer contain the variable t. Hence, in applying the transform operation, differential equations in V and t tend to become algebraic equations in $\bar{V}$ and s. The elimination of differentials in t is of considerable convenience particularly in the solution of partial differential equations (see p. 453).

---

† Different conventions have been used in defining the Laplace-transform operation (Pipes and Harvill 1970, p. 147). Some authors have retained the original formalism introduced by Heaviside:

$$g(p) = p \int_0^\infty e^{-pt}\, h(t)\,dt.$$

Here $g(p)$ is defined as the *p-multiplied transform* of $h(t)$. Clearly, $g(p)/p$ is equivalent to the $F(s)$ which is defined by the unmultiplied notation (eqn (13.12)). Although we have used the more convenient unmultiplied notation throughout this book, the reader is cautioned that some texts use the original Heaviside convention.

We may emphasize the integrative nature of the transform by pointing out that

$$\bar{F}(s)|_{s=0} = \int_0^\infty f(t)\, dt \qquad (13.13)$$

That is, evaluating the transform for $s = 0$ is equivalent to calculating the total area under the function $f(t)$. This relationship is useful in a variety of applications where the integral of a function is to be evaluated and where $\bar{F}(s)$ is already known.

We shall leave the definition of the *inverse transform operation* until a later section. It is sufficient to state here that such an operation exists, and that, in general, it can retrieve the original function after it has undergone Laplace transformation. No information is lost in the successive processes of transformation and inverse transformation. Thus

$$\mathscr{L}^{-1}\{\mathscr{L}(f(t))\} = \mathscr{L}^{-1}\bar{F}(s) = f(t). \qquad (13.14)$$

Here and in the rest of this chapter, we use the notation $\mathscr{L}^{-1}$ for the inverse transform operation and $\mathscr{L}$ for the transform operation itself, in order to emphasize the symmetry of the operations.

Let us now continue with a specific example, eqn (13.1). Applying the transform operation to both sides of the equation, or equivalently, term by term, we get

$$\mathscr{L}\left(\frac{d^2V}{dt^2}\right) + \mathscr{L}(V) = 0. \qquad (13.15)$$

Before proceeding further it is necessary to deal with the Laplace transform of the derivatives of a function.

Transforms of derivatives

We shall now show that the Laplace transform of the derivatives of a function are very simply related to the transform of the function itself—this is condition (2), above. Consider first the transform of dV/dt:

$$\mathscr{L}\left(\frac{dV}{dt}\right) = \int_0^\infty e^{-st}\left(\frac{dV}{dt}\right) dt. \qquad (13.16a)$$

Integrating by parts (see p. 465)

$$\mathscr{L}\left(\frac{dV}{dt}\right) = Ve^{-st}\Big|_0^\infty - \int_0^\infty -se^{-st}V\, dt. \qquad (13.16b)$$

The variable s can be factorized out of the right-hand integral, since it is a constant for the purposes of the integration. Thus

$$\mathscr{L}\left(\frac{dV}{dt}\right) = -V_0 + s\mathscr{L}(V), \qquad (13.16c)$$

where V_0 is the value of V at $t = 0$. Hence, the Laplace transform of the first derivative is simply given by the product of s and $\mathscr{L}(V)$, minus the initial value of V.†

This result introduces an important property of the Laplace-transform method: transformation of the derivative of a variable requires some knowledge of the starting conditions, in this case the initial value of V. In a similar way, the transform of the second derivative requires the knowledge of the initial value of V and dV/dt, and so on. In general, integration by parts shows that

$$\mathscr{L}\left(\frac{d^n V}{dt^n}\right) = s\mathscr{L}\left(\frac{d^{n-1} V}{dt^n}\right) - \left(\frac{d^{n-1} V}{dt^n}\right)_{t=0}. \tag{13.17}$$

This recursion relationship can be reapplied successively to give

$$\mathscr{L}\left(\frac{d^n V}{dt^n}\right) = s^n \mathscr{L}(V) - s^{n-1}(V)_0 - s^{n-2}\left(\frac{dV}{dt}\right)_0 \dots - \left(\frac{d^{n-1} V}{dt^{n-1}}\right)_0. \tag{13.18}$$

Incorporation of boundary conditions in the transform of the solution

It is apparent from the transform of $d^n V/dt^n$ that the transformation of an nth order differential equation will necessitate the inclusion of n values, that of the independent variable itself together with the 1st, 2nd, ... , $(n-1)$th derivatives. These initial values correspond physically to the 'boundary conditions' of the problem. Fortunately, these conditions are known in most applications (e.g. derivation of eqn (3.24) in Chapter 3). Indeed, much of one-dimensional cable theory consists of reapplications of different boundary conditions to the same basic differential equation.

Since the Laplace-transform method directly incorporates boundary conditions into the transformed equation, the solution obtained will be a *particular* solution. General solutions (containing arbitrary constants) are not necessarily obtained.

The over-all method may be outlined in terms of an nth order differential equation

$$\frac{d^n V}{dt^n} + a_1 \frac{d^{n-1} V}{dt^{n-1}} + a_2 \frac{d^{n-2} V}{dt^{n-2}} + \dots a_n V = \phi(t). \tag{13.19}$$

In most cases in this book, $\phi(t)$ corresponds physically to the excitation (e.g. a current waveform) which elicits the response (here written $V(t)$). $\phi(t)$ is often referred to as an 'input function' or 'driving function'.

---

† Eqn (13.16c) leads rather directly to an expression for the steady state value of a function whose Laplace transform is known. Thus

$$V(t = \infty) = \lim_{s \to 0} (s\bar{V}(s)). \tag{13.16d}$$

This may be verified as follows:

$$\lim_{s \to 0} s\mathscr{L}(V) = \lim_{s \to 0} s\bar{V}(s) = V_0 + \lim_{s \to 0} \int_0^\infty e^{-st}\left(\frac{dV}{dt}\right)dt = V_0 + \int_0^\infty \left(\frac{dV}{dt}\right)dt$$

$$= V_0 + V|_0^\infty = V(t = \infty).$$

The method proceeds in the following steps.

1. Transform the differential equation, term by term, applying the above rule for transformation of derivatives (eqn (13.18)). The input function will also be transformed ($\phi(t) \longleftrightarrow \bar{\phi}(s)$):

$$\left\{s^n \bar{V} - s^{n-1}(V)_0 - s^{n-2}\left(\frac{\mathrm{d}V}{\mathrm{d}t}\right)_0 \cdots \left(\frac{\mathrm{d}^{n-1}V}{\mathrm{d}t^{n-1}}\right)_0\right\} +$$

$$+ a_1 \left\{s^{n-1}\bar{V} - s^{n-2}(V)_0 - \cdots \left(\frac{\mathrm{d}^{n-2}V}{\mathrm{d}t^{n-2}}\right)_0\right\} + \ldots a_n \bar{V} = \bar{\phi}(s). \quad (13.20)$$

Note that each derivative gives rise to a polynomial in s.

2. The transformed (or *subsidiary*) equation consists of new terms, containing either $\bar{V}$ or the constants V_0, $(\mathrm{d}V/\mathrm{d}t)_0$, etc. Collect all terms containing $\bar{V}$ on the left-hand side and transfer all the remaining boundary condition terms to the right-hand side. The left-hand side can now be factorized into the product of $\bar{V}$ and a polynomial in s whose coefficients are simply those of the original differential equation.

$$\underbrace{\bar{V}(S^n + a_1 s^{n-1} + a_2 s^{n-2} + \ldots a_n)}_{\text{'characteristic polynomial'}} = \underbrace{\bar{\phi}(s)}_{\substack{\text{input function} \\ \text{polynomial}}} + \underbrace{B(s).}_{\substack{\text{boundary condition} \\ \text{polynomial}}} \quad (13.21)$$

This expression of the transformed equation emphasizes that the boundary conditions are essentially analogous to the input function in their effect. Each boundary condition could be thought of as an impulsive input at $t = 0$, which sets each derivative of V to its appropriate initial value.

3. Solve for $\bar{V}$, dividing both sides by the characteristic polynomial

$$\bar{V} = \frac{\bar{\phi}(s) + B(s)}{s^n + a_1 s^{n-1} + \ldots + a_n}. \quad (13.22)$$

4. Obtain $V(t)$ by performing the inverse transformation operation on $\bar{V}$.

$$V(t) = \mathcal{L}^{-1}(\bar{V}) = \mathcal{L}^{-1}\left\{\frac{\bar{\phi}(s) + B(s)}{s^n + a_1 s^{n-1} + \ldots + a_n}\right\}. \quad (13.23)$$

We may point out once again that the inverse transform operation is performed on an expression that already contains the boundary conditions for the problem.

Methods for obtaining the inverse transform

Up to this point the Laplace-transform method is relatively straightforward. At this stage there are two major alternatives for obtaining the solution.

4a. The inverse transform operation may be performed directly by use of the Bromwich integral:

$$V(t) = \mathcal{L}^{-1}(\bar{V}(s)) = \frac{1}{2\pi j}\oint_c \bar{V}(s)e^{st}\,ds, \qquad (13.24)$$

where $\oint_c$ denotes integration over an appropriate Bromwich contour.

The evaluation of this contour integral will not be discussed here (see Carslaw and Jaeger 1959, Appendix I).

The contour integration procedure has the advantage that the integral may be evaluated numerically (see Bellman, Kaliba, and Lockett 1966) when analytic results (see below) cannot be obtained. Numerical evaluation has been useful in some applications to excitable cells (see Chapter 10, p. 304).

4b. The operation of inverse transformation may be performed on a suitable partial fraction expansion of $\bar{V}(s)$, in which the individual terms are recognizable as members of known transform pairs (some examples are given in Table 13.1). Since very extensive tables of such transform pairs are available, the use of the partial fraction expansion is helpful in most practical situations and will be described in the next section. Roberts and Kaufman (1966) have prepared the most extensive table of transform pairs in a format which conveniently classifies various functions and their transforms. Therefore, we have quoted their classification numbers for transform pairs throughout the text.

Partial fraction expansion

The first step in the partial fraction expansion is to express the denominator of eqn (13.22) (that is, the characteristic polynomial) as the product of n factors:

$$s^n + a_1 s^{n-1} + \ldots + a_n = (s-s_1)(s-s_2)\ldots(s-s_n), \qquad (13.25)$$

here $s_1 \ldots s_n$ denote the n zeroes of the polynomial (that is, the roots of the equation obtained by setting the polynomial equal to zero). If all n roots are distinct, and provided that $\phi(s) + B(s)$ is of order $(n-1)$ or less, the partial fraction expansion is straightforward:

$$\frac{\phi(s)+B(s)}{s^n + a_1 s^{n-1} + \ldots + a_n} = \frac{c_1}{s-s_1} + \frac{c_2}{s-s_2} + \ldots + \frac{c_n}{s-s_n}. \qquad (13.26)$$

Each of the constants $c_1 \ldots c_n$ can be evaluated by clearing fractions and setting equal the respective coefficients of s^n, s^{n-1}, etc.†

† In the case of distinct roots, the general formula for the numerator of the partial fractions may be written (see Hildebrand 1962, p. 67):

$$c_i = \frac{\phi(s_i)+B(s_i)}{\left\{\dfrac{d}{ds}(s^n + a_1 s^{n-1} + \ldots a_n)\right\}_{s=s_i}} \qquad (13.27)$$

Once the constants $c_1 \ldots c_n$ are obtained, their terms may be individually inverse-transformed, using the transform pair

$$\frac{c_i}{s-s_i} \leftrightarrow \begin{cases} c_i \exp(s_i t) & \text{for } t > 0 \\ 0 & \text{for } t < 0. \end{cases} \tag{13.28}$$

(This, and other transform pairs, may be readily verified by using the definition of the Laplace transform, eqn (13.12)).

The solution of the differential equation will consist of the sum of n exponential terms with coefficients $c_1 \ldots c_a$. This result may be familiar from the conventional method for solving linear differential equations with constant coefficients. The zeroes of the characteristic polynomial are identical with the roots of the so-called *characteristic* (or *auxiliary*) *equation* which is generated by the 'operator' method (see Jaeger 1951, p. 13).

Oscillatory solutions

The characteristic polynomial may have complex roots. What is the significance of such complex roots and how are they handled in obtaining the inverse transform?

To answer these questions, we must first point out that physically realizable problems generate characteristic polynomials with purely real coefficients. This implies that complex roots will invariably occur in conjugate pairs. In other words, if there is a root $s_1 = \sigma + j\omega$ there must also be a complex conjugate root, $s_2 = \sigma - j\omega$. (The usual notation for complex conjugation is an asterisk so that, in this case, $s_2 = s_1^*$.)

The significance of complex-conjugate roots may be clarified by a specific example. Consider the fraction

$$\frac{\omega}{s^2 + \omega^2} = \frac{\omega}{(s+j\omega)(s-j\omega)} = \frac{c_1}{s+j\omega} + \frac{c_2}{s-j\omega}. \tag{13.29}$$

In this partial fraction expansion, the constants c_1 and c_2 may be complex quantities. Letting $c_1 = c_1' + c_1''$ and $c_2 = c_2' + c_2''$ and clearing fractions, we find that $c_1 = -c_2 = \tfrac{1}{2}j$, so that

$$\frac{\omega}{s^2 + \omega^2} = \frac{\tfrac{1}{2}j}{s+j\omega} + \frac{-\tfrac{1}{2}j}{s-j\omega}. \tag{13.30}$$

Noting that the definition of the Laplace transform allows s to be complex, the two terms can be transformed by the transform pair for exponentials, eqn (13.28). Thus

$$\frac{\tfrac{1}{2}j}{s+j\omega} + \frac{-\tfrac{1}{2}j}{s-j\omega} \leftrightarrow \begin{cases} \tfrac{1}{2}j(e^{-j\omega t} - e^{j\omega t}) & \text{for } t > 0 \\ 0 & \text{for } t < 0. \end{cases} \tag{13.31}$$

The expression on the right-hand side may be recognized as a sinusoidal function $\sin \omega t$. It is worth emphasizing here that exponential functions and trigonometric (sinusoidal) functions are closely related. Using the standard vector representation of complex numbers (Pipes and Harvill 1970, p. 2)

$$e^{j\theta} = \cos\theta + j\sin\theta \qquad (13.32)$$

and

$$e^{-j\theta} = \cos\theta - j\sin\theta. \qquad (13.33)$$

These two equations may be rearranged to give

$$\cos\theta = \tfrac{1}{2}(e^{j\theta} + e^{-j\theta}) \qquad (13.34)$$

and

$$\sin\theta = \frac{1}{2j}(e^{j\theta} - e^{-j\theta}). \qquad (13.35)$$

In the particular example given here

$$\tfrac{1}{2}j(e^{-j\omega t} - e^{j\omega t}) = \frac{1}{2j}(e^{j\omega t} - e^{-j\omega t}) = \sin\omega t$$

so that the transform pair may be written

$$\frac{\omega}{s^2 + \omega^2} \leftrightarrow \begin{cases} \sin\omega t & t > 0 \\ 0 & t < 0. \end{cases} \qquad (13.36)$$

It may be shown similarly that

$$\frac{s}{s^2 + \omega^2} \leftrightarrow \begin{cases} \cos\omega t & t > 0 \\ 0 & t < 0. \end{cases} \qquad (13.37)$$

This treatment shows how sinusoidal solutions arise from pairs of terms with complex conjugate exponents. In the actual retrieval of solutions from their transformed form, it is unnecessary to expand the fractions $\omega/(s^2 + \omega^2)$ or $s/(s^2 + \omega^2)$ into their respective partial fraction components. If, for example, the characteristic polynomial has roots $s_1 = +j\omega$ and $s_2 = -j\omega$, then the fraction

$$\frac{\phi(s) + B(s)}{(s^2 + \omega^2)(s - s_3)...(s - s_n)}$$

can be expanded as

$$\frac{c_1 s + c_2 \omega}{s^2 + \omega^2} + \frac{c_3}{s - s_3} + \frac{c_4}{s - s_4} + ... \qquad (13.38)$$

It can be shown that the constants c_1 and c_2 must be purely real numbers, so that it is unnecessary to make the assumption that $c_1 = c_1' + c_1''$, etc. By clearing fractions, all of the constants may be evaluated. The expression in eqn (13.38) is then transformed term by term to give the solution

$$c_1 \cos\omega t + c_2 \sin\omega t + c_3 \exp(s_3 t) + c_4 \exp(s_4 t) + ... \qquad (13.39)$$

It may be emphasized at this point that the partial fraction expansion is best performed with an awareness of the types of expressions which are available in tables of transform pairs. In many cases, some rearrangement is necessary to make the transform of the response correspond to a recognizable entry in the table. A number of general theorems are very useful in this respect; some examples are given in Table 13.1.

TABLE 13.1(a)

General relationships of functions and their Laplace transforms

Function	Transform
$f(t)$	$F(s) = \mathcal{L}(f(t)) = \int\limits_0^\infty e^{-st} f(t) dt$
$af(t) + bg(t)$	$a\bar{F}(s) + b\bar{G}(s)$
$\dfrac{df(t)}{dt}$	$sF(s) - f(0)$
$\dfrac{d^n f(t)}{dt}$	$s^n F(s) - \sum\limits_{k=1}^{n} s^{n-k} \dfrac{d^{k-1} f(0)}{dt^{k-1}}$
$\int\limits_0^t f(t)\, dt$	$\dfrac{1}{s} F(s)$
$e^{-at} f(t)$	$F(s+a)$
$\begin{cases} f(t-a) & t > a \\ 0 & t < a \end{cases}$	$e^{-as} F(s)$
$\int\limits_0^t f(u)\, h(t-u)\, du$	$F(s)\, \bar{H}(s)$

TABLE 13.1(b)

Some common transform pairs

Function	Transform
unit step function $v_1(t) = \begin{cases} 1 & t > 0 \\ 0 & t < 0 \end{cases}$	$\dfrac{1}{s}$
delta function $\delta(t)$	1
e^{at}†	$\dfrac{1}{s-a}$
$\sin \omega t$†	$\dfrac{\omega}{s^2 + \omega^2}$
$\cos \omega t$†	$\dfrac{s}{s^2 + \omega^2}$
$e^{-\sigma t} \sin \omega t$†	$\dfrac{\omega}{(s+\sigma)^2 + \omega^2}$
$e^{-\sigma t} \cos \omega t$†	$\dfrac{s+\sigma}{(s+\sigma^2) + \omega^2}$
$e^{b^2 t + ab} \mathrm{erfc}(a/2\sqrt{t} + b\sqrt{t})$	$\dfrac{e^{-a\sqrt{s}}}{s + b\sqrt{s}}$
$e^{-ab}\mathrm{erfc}(a/2\sqrt{t} - b\sqrt{t}) + e^{ab}\mathrm{erfc}(a/2\sqrt{t} + b\sqrt{t})$	$\dfrac{2e^{-\sqrt{s+b}}}{s}$

† In using these transform pairs, it is understood that

$$f(t) = \text{as defined for } t > 0$$
$$= 0 \qquad \text{for } t < 0$$

One very useful theorem states that

$$\mathscr{L}(e^{-at}f(t)) \leftrightarrow F(s+a). \qquad (13.40)$$

This can be derived from the definition of the Laplace transform:

$$\mathscr{L}(e^{-at}f(t)) = \int_0^\infty e^{-st}e^{-at}f(t) = \int_0^\infty e^{-(s+a)t}f(t) = F(s+a).$$

Operationally, this pair is used as follows.

1. Rearrange the transform $\bar{V}(s)$ so that each time s appears, it is written as $(s+a)$.
2. Substitute $s' = s + a$, and find the inverse transform of $V'(s')$, which may be called $V'(t)$.
3. The theorem states that the desired solution is $V(t) = \exp(-at)\,V'(t)$.

This particular manoeuvre has been used at several points in this book, usually with $a = 1$.

Laplace transforms of the step function and delta function

These are the excitation functions most frequently used in this book. The transform of the step function may be obtained very easily since the function is a constant (equal to 1 for the unit step) over the entire range of integration. Thus

$$\mathscr{L}(\text{unit step}) = \int_0^\infty e^{-st}(1)\,dt = \frac{1}{s}. \qquad (13.41)$$

The Laplace transform of the delta function is most easily derived by noting that the delta function is equal to the first derivative of the unit step function. Using eqn (13.16),

$$\mathscr{L}(\text{delta function}) = s\mathscr{L}(\text{step function}) - (\text{step function})_{t=0}. \qquad (13.42a)$$

By convention, $(\text{step function})_{t=0}$ is taken as zero in this derivation. Thus

$$\mathscr{L}(\text{delta function}) = 1. \qquad (13.42b)$$

The fact that the Laplace transform of the delta function is simply unity gives some clue to its usefulness in the analysis of linear systems. If a delta function excitation is applied to a system which is initially at rest (that is, all the initial values of the zeroth through to the $(n-1)$th derivatives are each zero) the response will have the following transform:

$$\bar{H}(s) = \frac{1}{s^n + a_1 s^{n-1} + \ldots a_n}. \qquad (13.43)$$

$\bar{H}(s)$ is simply the reciprocal of the characteristic polynomial and is often referred to as the *transfer function, transmission function,* or *systems function.*

The transform of the response to any excitation $\phi(t)$ can be expressed in terms of the transfer function. Assuming for simplicity that the system is initially at rest (so that $B(s) = 0$), eqn (13.22) can be rewritten as

$$\bar{V}(s) = \bar{H}(s)\bar{\phi}(s). \tag{13.44}$$

This important result gives the transform of the response to $\phi(t)$ in terms of the transfer function of the system. It corresponds directly to the calculation of the response $V(t)$ by means of convolution:

$$V(t) = \int_0^\infty \phi(u)h(t-u)\,\mathrm{d}u. \tag{13.11}$$

In fact it may be demonstrated directly that the convolution of the functions $\phi(t)$ and $h(t)$ is equivalent, in terms of their Laplace transforms, to the multiplication of $\bar{\phi}(s)$ and $\bar{H}(s)$. In many situations it is more convenient to convolve functions by obtaining the inverse transform of the product of Laplace transforms rather than by evaluating the convolution integral directly. This use of Laplace transforms is rather analogous to the use of logarithms in performing a multiplication (see p. 441). As in the case of logarithms in calculating the product of several factors, the multiplication of Laplace transforms is particularly helpful when the response of one system becomes, in turn, the excitation of another system.

Complex frequencies

In an earlier section we discussed the significance of complex roots of the characteristic polynomial $s^n + a_1 s^{n-1} + \ldots a_n$. We indicated that sinusoidal (trigonometric) functions are merely exponentials in a different guise. The close relationship between sinusoids and exponentials is important in understanding the notion of a *complex frequency*. So far we have considered only purely real or purely imaginary exponentials: the purely real exponential,

$\exp(-\sigma t)$ where σ is a real number, and

$\exp(-j\omega t)$ where ω is a real number (that is, where $j\omega$ is purely imaginary).

By combining $\exp(j\omega t)$ with $\exp(-j\omega t)$ we showed that cosine or sine functions arise, and have the *real* argument ωt. ω appears as a frequency, and may be related to the more conventional parameter for frequency f by the relation $\omega = 2\pi f$ (ω has units of radians per second and f has units of hertz).

We are now in a position to consider the meaning of the complex frequency s, where $s = \sigma + j\omega$. Now

$$\exp(-st) = \exp(-\sigma t - j\omega t) = \exp(-\sigma t)\exp(-j\omega t).$$

This equation shows that σ, the real part of s, appears in the exponent of a factor which multiplies the 'oscillatory' exponential $\exp(-j\omega t)$. The factor $\exp(-\sigma t)$ changes only the amplitude; it does not interfere with the frequency of oscillations. The parameter σ, like ω, has units of hertz, and may also be thought of as a rate. If σ is positive, $\exp(-\sigma t)$ will decrease with time, so that σ gives the rate of 'damping' in much the same way that ω determines the rate (frequency) of oscillations. The damping and oscillatory factors then correspond simply to the real and imaginary parts of the complex exponential s. By allowing s to be a complex number, the Laplace transform preserves its applicability to problems of either decremental or oscillatory nature, including those where both decrement and oscillation occur (e.g. in Chapter 11).

A specific example of the general applicability of the results of Laplace-transform analysis is provided by the notion of a space constant. As conventionally derived, the space constant λ is specifically defined as a d.c. quantity, describing the steady state decrement of potential with distance. (λ also appears in the time-dependent part of the solution, but it is nevertheless used as a d.c. quantity, remaining unchanged regardless of the input waveform.) Although the treatment in Chapter 3 does not consider the case of an oscillatory input, the usual analysis can easily be extended to the case of sinusoids (see Eisenberg and Johnson 1970, p. 59).

In principle, the extension simply replaces r_m by z_m, where z_m is explicitly frequency dependent (the effect of c_m is incorporated in z_m, as in eqn (2.5)). r_m is the limiting value of z_m at zero frequency, and r_i is assumed to be purely real at all frequencies. The mathematical treatment is otherwise unchanged. Whereas in the d.c. case

$$V \propto e^{-x/\lambda},\tag{3.25}$$

in the a.c. case

$$V \propto e^{-\gamma x},\tag{13.45}$$

where $\gamma = \sqrt{r_i/z_m}$. γ is called the *propagation constant* and may be written as a complex number $\gamma = \alpha + j\beta$. The real part α is called the *attenuation constant*. α is constant with respect to distance but is frequency-dependent; in the limiting d.c. case $\alpha = 1/\lambda$. β is termed the *phase constant*, since it governs the relative timing of sinusoids at various distances along the cable. β will also be frequency dependent, in general. As Eisenberg and Johnson point out, the concepts of d.c. cable theory can very naturally be extended to the case of sinusoidal inputs (see also Falk and Fatt 1964; Jack and Redman 1971b) since the original analysis does not depend on the assumption that particular parameters should necessarily be real quantities. The generalization to the a.c. case is particularly useful in dealing with the *impedance* of electrical systems.

Relation between Fourier and Laplace transforms

The nature of the complex frequency s provides the key to understanding the similarity between Laplace and Fourier transforms. The Fourier transform, like the Fourier series, is useful in the analysis of periodic (oscillatory) functions. The transform decomposes the periodic function into the sum of an infinite series of sinusoidal functions whose frequency may vary continuously from d.c. to ∞. The Fourier transform is defined formally as follows:

$$\mathscr{F}(\omega) = \int_{-\infty}^{+\infty} e^{-j\omega t} f(t) \, dt, \tag{13.46}$$

where the complex exponential $\exp(-j\omega t)$ is used as the periodic function (rather than $\sin \omega t$ or $\cos \omega t$, as in the respective Fourier sine or cosine series expansions). This definition may be compared to the definition of the Laplace transform:

$$F(s) = \int_{0}^{+\infty} e^{-st} f(t) \, dt. \tag{13.12}$$

There are only two differences between these definitions.

1. The variable s in the Laplace transform corresponds to $j\omega$ in the Fourier transform. This befits the different uses of the transforms; the Laplace transform is applied to the analysis of transients, whereas the Fourier is usually used in describing the ongoing responses to a maintained sinusoidal input (e.g. impedance measurements using a.c. signals).

2. The lower limit of the integration is extended to $-\infty$. Note, however, that if $f(t) = 0$ for all $t < 0$, the Fourier and Laplace transforms become formally identical if s and $j\omega$ are interchanged. The conversion between Fourier and Laplace transforms is useful in certain applications (see Jack and Redman 1971*b*, Appendix).

The use of Laplace transforms in solving partial differential equations

In these notes we have described the use of the Laplace-transform method in solving ordinary differential equations. In general, however, the technique is most useful in solving partial differential equations, since ordinary differential equations may usually be solved fairly easily by the method of characteristic equations or other standard techniques.

The Laplace transform may be used in solving partial differential equations since, for the purposes of integration with respect to one of the variables, the other independent variables may be regarded as constant. Thus in the equation

$$\frac{\partial^2 V}{\partial x^2} + \frac{\partial V}{\partial t} + V = 0,$$

we may transform the terms with respect to t, eliminating t as a variable,

$$\mathscr{L}\left(\frac{\partial^2 V}{\partial x^2}\right) + \mathscr{L}\left(\frac{\partial V}{\partial t}\right) + \mathscr{L}(V) = 0.$$

Denoting $\mathscr{L}(V)$ as $\bar{V}$, as usual, we obtain an *ordinary* differential equation in x

$$\frac{d^2 \bar{V}}{dx^2} + (s+1)\bar{V} = V_0.$$

This equation may now be solved by ordinary methods. Of course, it could also be transformed yet again with respect to x, but this would be an unnecessarily cumbersome method of solving the equation. In general, therefore, the Laplace transform is simply employed to transform a partial differential equation into an ordinary differential equation, and this is the only way in which it has been used in this book.

Bessel functions

In Chapters 5 and 6, solutions of the equation

$$\frac{d^2V}{dR^2}+\frac{1}{R}\frac{dV}{dR}-V = 0 \tag{13.47}$$

are used to describe the cable properties of radially symmetric structures. This equation is called a modified Bessel equation of zero order, and its solutions are given in terms of Bessel functions. Since this differential equation is a particular case of a more general class of equations, we will limit this discussion to those solutions which are referred to in the text. For a more general development of the theory of Bessel functions, the reader is referred to Watson (1966), who also includes useful tables of Bessel functions. A more introductory account is given by Bowman (1958). A brief summary of Bessel functions and their recursion relationships appears in Appendix III of Carslaw and Jaeger (1959).

There are some useful analogies between the properties of functions arising from the two-dimensional model and those arising from one-dimensional cable theory. We hope that these analogies may be helpful to those who are somewhat familiar with the exponential functions obtained in the latter case, but who are unfamiliar with Bessel functions.

In the steady state, the one-dimensional equation corresponding to eqn (13.47) is

$$\frac{d^2V}{dR^2}-V = 0, \tag{13.48}$$

which is eqn (3.13) with $\partial V/\partial t = 0$. Clearly, the two-dimensional equation differs from the one-dimensional equation only in the presence of the term $(1/R)(dV/dR)$.

The general solution to eqn (13.48) may be expressed as the sum of two exponentials:

$$V = A \exp(-X)+B \exp(X), \tag{13.49}$$

where the coefficients of X in the exponents (-1 and $+1$) are given by the roots of the characteristic equation $r^2-1 = 0$. A and B are arbitrary constants. In order to develop the analogy between the solution of eqn (13.48) and that of eqn (13.47), it is helpful to first consider why eqn (13.49) forms a solution to eqn (13.48). This may be done by writing the exponential functions as power series (see p. 446 for the method of obtaining an expansion as the Taylor series):

$$V = A\left(1-X+\frac{X^2}{2!}-\frac{X^3}{3!}...\right)+B\left(1+X+\frac{X^2}{2!}+\frac{X^3}{3!}...\right). \quad (13.50)$$

These series expansions may now be substituted directly in eqn (13.48), and differentiating the series term by term, it may be shown that eqn (13.50) is a valid solution. Another way of viewing the problem, therefore, is to regard eqn (13.50) as those power series which, by definition, satisfy the differential equation (13.48). This point of view may seem less strange when it is realized that the single exponential function $\exp(X)$ is defined as that function which, when differentiated, remains unchanged. Stated in terms of an equation, $\exp(X)$ is the name of the power series which satisfies the equation $dV/dX = V$.

Different ways of expressing the general solution

In this example, $\exp(X)$ or $\exp(-X)$ will each satisfy the differential equation (13.48). Since the equation is second order, and because $\exp(X)$ and $\exp(-X)$ are two independent solutions (one is not a simple multiple of the other) the expression $A \exp(-X)+B \exp(X)$ forms a *general* solution. The term general implies that any *other* function which satisfies eqn (13.48) must be reducible to a linear combination of $\exp(X)$ and $\exp(-X)$. In other words, such a function can be specified by particular values of the coefficients A and B.

A relevant example of other solutions is provided by $\sinh X$ and $\cosh X$, the hyperbolic functions. These can be shown to satisfy eqn (13.48) by virtue of the relations:

$$\frac{d}{dX}(\sinh X) = \cosh X \quad (13.51)$$

and

$$\frac{d}{dX}(\cosh X) = \sinh X. \quad (13.52)$$

From the previous discussion of the general nature of the solution $A \exp(-X)+B \exp(+X)$, it is only to be expected that the hyperbolic functions must be reducible to a linear combination of exponentials. The hyperbolic functions are thus

$$\sinh X = \tfrac{1}{2}\{\exp(X) - \exp(-X)\}, \quad \text{so} \quad B = \tfrac{1}{2}, \quad A = -\tfrac{1}{2}, \quad (13.53)$$
$$\cosh X = \tfrac{1}{2}\{\exp(X) + \exp(-X)\}, \quad \text{so} \quad B = \tfrac{1}{2}, \quad A = +\tfrac{1}{2}. \quad (13.54)$$

This simple example points out that any pair of linearly independent functions will suffice for the expression of the general solution. Although we are accustomed to thinking of exponentials as fundamental functions, so far as the differential equation is concerned we might have started by defining $\cosh X$ as the following power series:

$$\cosh X = 1+\frac{X^2}{2!}+\frac{X^4}{4!}+\frac{X^6}{6!}+... . \quad (13.55)$$

Just as the power series in eqn (13.50) can be shown to satisfy eqn (13.49), so the power series for cosh X can be shown to satisfy the differential equation. A second function would then be required to complete the general solution. Either sinh X, $\exp(X)$, or $\exp(-X)$ would have been possibilities, since, from their power series, it is apparent that none of these are simply multiples of cosh X.

This discussion may now be related to the class of Bessel functions which are solutions to eqn (13.47). Although there are several named power series which satisfy this equation, it is helpful to remember that the general solution may be written as the sum of any pair of such functions (since they are all linearly independent of each other). The choice of *which* pair of functions is not arbitrary in one important sense: a particular functional form may be more convenient for practical reasons. It is often the case that the physical boundary conditions exclude all functions but one. To use the familiar one-dimensional cable solution as an example,

$$V_\infty = V_0 \exp(-X)$$

(cf. eqn (3.25)).

This solution is equivalent to

$$V_\infty = V_0(\cosh X - \sinh X), \tag{13.56}$$

but clearly the former expression is more convenient. By writing the general solution in a form which explicitly includes the appropriate function (e.g. $A \exp(-X) + B \exp(+X)$), the final solution is very immediately expressed in its most convenient form.

The $I_0(R)$ function

One very useful solution to eqn (13.47) is obtained by the method of postulating a power series which satisfies the differential equation (see Bowman, p. 41). This function is called the modified Bessel function of the first kind of zero order $I_0(R)$. Using the argument R,

$$I_0(R) = 1 + \frac{R^2}{2^2} + \frac{R^4}{2^2 4^2} + \frac{R^6}{2^2 4^2 6^2} + \frac{R^8}{2^2 4^2 6^2 8^2} + \cdots . \tag{13.57}$$

The power series expansion shows clearly that $I_0(R) \to 1$ as $R \to 0$. It is the only solution to eqn (13.47) (except for constant multiples of $I_0(R)$) which remains finite at $R = 0$. Because of this characteristic, I_0 is sufficient by itself for cylindrical problems where (as in Chapter 6) the solution and its derivatives remain finite at the axis of the cylinder.

Comparison of the series expansion (13.57) with eqn (13.55) shows that the I_0 and hyperbolic cosine functions are very similar, differing only in the value of the denominators of the series terms. Fig. 13.3 plots the functions for further comparison. Note in particular that

$$\cosh(0) = I_0(0) = 1 \tag{13.58}$$

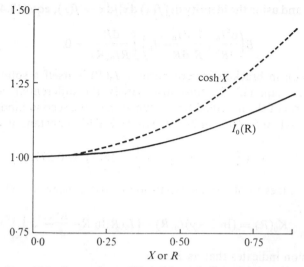

FIG. 13.3. Comparison of I_0 and hyperbolic cosine functions.

and
$$\frac{d}{dX}(\cosh X)_{X=0} = \frac{d}{dR}(I_0(R))_{R=0} = 0. \tag{13.59}$$

Eqn (13.59) corresponds to the physical boundary condition of zero current flow along the X- (or R-) coordinate, which holds for the steady state voltage distributions derived in Chapters 4 and 6.

The derivative of $I_0(R)$ is called $I_1(R)$, the modified Bessel function of the first kind of first order.

$$\frac{d}{dR}I_0(R) = I_1(R) = \frac{R}{2} + \frac{R^3}{2^2 4} + \frac{R^5}{2^2 4^2 6} + \frac{R^7}{2^2 4^2 6^2 8} + \dots. \tag{13.60}$$

This power series expansion is, not surprisingly, similar to that of

$$\frac{d}{dX}\cosh X = \sinh X = \frac{X}{1!} + \frac{X^3}{3!} + \frac{X^5}{5!} + \frac{X^7}{7!} + \dots. \tag{13.61}$$

The function $K_0(R)$

A modified Bessel function of the second kind and of zero order may now be defined as any solution of eqn (13.47) which is not a constant multiple of I_0. It may be shown (Bowman, p. 41) that all such functions may be expressed as

$$AI_0(R) + BI_0(R) \int \frac{dR}{RI_0^2(R)}, \tag{13.62}$$

where it is required that $B \neq 0$. The second term in this equation guarantees that the above function is independent of $I_0(R)$ and may readily be shown to satisfy the differential equation by substitution into eqn (13.47). Making this substitution and using the identity $d\{\int f(x)\,dx\}/dx = f(x)$, eqn (13.47) becomes

$$B\left(\frac{d^2 I_0}{dR^2} + \frac{1}{R}\frac{dI_0}{dR} - I_0\right)\int\frac{dR}{RI_0(R)} = 0. \tag{13.63}$$

The expression in brackets is zero because $I_0(R)$ is itself a solution of eqn (13.47). Thus, eqn (13.62) must also satisfy the differential equation. A particular form of the modified Bessel function of the second kind is given by setting $B = -1$ and $A = (\ln 2 - \gamma)$, where Euler's constant γ is given by

$$\gamma = \lim_{n=-\infty}\left(1 + \tfrac{1}{2} + \tfrac{1}{3} + \tfrac{1}{4}\ldots + \frac{1}{n} - \ln n\right) = 0\cdot 5772\ldots\,.$$

Using these values to obtain an expansion of $K_0(R)$ near $R = 0$,

$$K_0(R) = (\ln 2 - \gamma)I_0(R) - \left(I_0(R)\ln R - \frac{R^2}{4} - \ldots\right). \tag{13.64}$$

This expansion indicates that as $R \to 0$,

$$K_0(R) \to -\ln R \to \infty. \tag{13.65}$$

It is also possible to show (Bowman, eqn 5.45) that the following expansion holds for large values of R:

$$K_0(R) = \left(\frac{\pi}{2R}\right)^{\frac{1}{2}}e^{-R}\left\{1 - \frac{1}{8R} + \frac{1^2 3^2}{2!}\frac{1}{(8R)^2} - \ldots\right\}, \tag{13.66}$$

so that

$$K_0(R) \to \left(\frac{\pi}{2R}\right)^{\frac{1}{2}}e^{-R} \quad \text{as } R \to \infty. \tag{13.67}$$

This equation suggests that $K_0(R)$ is a Bessel function whose behaviour is appropriate for physical problems where the solution decays monotonically (that is, without oscillation) to zero as R becomes large (see Fig. 5.2 for the asymptotic behaviour of $K_0(R)$). Such is the case for the two-dimensional cable treated in Chapter 5.

The derivative of $K_0(R)$ is given by the relation

$$\frac{d}{dR}K_0(R) = -K_1(R),$$

where $K_1(R)$ is the modified Bessel function of the second kind, order one. Fig. 13.4 shows a comparison between $I_0(R)$ and $K_0(R)$.

The Hankel function $H_0^{(1)}(R)$

Some readers may be more familiar with the unmodified Bessel functions of zero order J_0 and Y_0. These are each oscillatory functions which form solutions to the equation

$$\frac{d^2 V}{dR^2} + \frac{1}{R}\frac{dV}{dR} + V = 0, \tag{13.68}$$

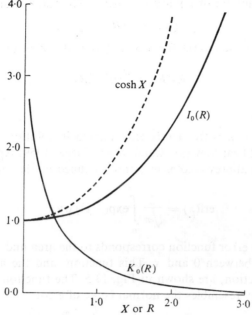

FIG. 13.4. Comparison of I_0 and K_0 functions.

which differs from eqn (13.47) only in the sign of the last term. Given that $J_0(R)$ and $Y_0(R)$ satisfy this equation, it follows that $J_0(jR)$ and $Y_0(jR)$ satisfy

$$\frac{\mathrm{d}^2 V}{\mathrm{d}(jR)^2} + \frac{1}{(jR)}\frac{\mathrm{d}V}{\mathrm{d}(jR)} + V = 0, \tag{13.69}$$

which reduces to eqn (13.47). Thus $J_0(jR)$ and $Y_0(jR)$ are also solutions to the modified Bessel equation.

There is a useful analogy between $J_0(R)$ and $Y_0(R)$ and the circular functions $\sin X$ and $\cos X$ which are solutions to

$$\frac{\mathrm{d}^2 V}{\mathrm{d}X^2} + V = 0. \tag{13.70}$$

In this case, $\cos(jX)$ and $\sin(jX)$ are solutions to

$$\frac{\mathrm{d}^2 V}{\mathrm{d}(jX)^2} + V = 0,$$

which reduces to eqn (13.48). The cosine or sine functions of imaginary argument are rarely used, however, because the exponential and hyperbolic functions are much more familiar.

In the case of the Bessel equation, the relationship between the oscillatory and monotonic functions is sometimes helpful in using existing tables of functions. In particular, some texts (e.g. Jahnke and Embde 1945) tabulate the Hankel function for imaginary arguments. The Hankel function is defined as

$$H_0^{(1)}(R) = J_0(R) + jY_0(R). \tag{13.71}$$

This function is related to the function $K_0(R)$ by the identity

$$K_0(R) = \frac{\pi}{2}jH_0^{(1)}(jR). \tag{13.72}$$

The error function

The *error function* is frequently encountered in solutions to problems in cable theory and heat flow (see Carslaw and Jaeger 1959, Appendix II). The function is often abbreviated as erf, and is defined by the integral

$$\text{erf}(x) = \frac{2}{\sqrt{\pi}} \int_0^x \exp(-y^2)\, dy. \tag{13.73}$$

Graphically, the error function corresponds to the area under the expression $2 \exp(-x^2)/\sqrt{\pi}$ between 0 and x. This function, and the appropriate area beneath the function, are shown in Fig. 13.5. The function has a Gaussian shape and closely resembles the normal curve of error

$$\phi(x) = \frac{1}{\sqrt{2\pi}} \exp\left(\frac{-x^2}{2}\right). \tag{13.74}$$

Note, however, the difference in the coefficient of the exponent and in the scaling of the function itself.

It is useful to point out that, by definition,

$$\frac{d}{dx} \text{erf}(x) = \frac{2}{\sqrt{\pi}} \exp(-x^2). \tag{13.75}$$

This equation states that, as x increases, erf(x) gradually decreases in steepness and attains a steady maximum level. The normalization in eqn (13.73) is appropriate to make the steady level equal to unity. Thus

$$\text{erf}(0) = 0 \tag{13.76}$$

and

$$\text{erf}(\infty) = 1. \tag{13.77}$$

For small values of x, erf(x) can be expressed as a power series:

$$\text{erf}(x) = \frac{2}{\sqrt{\pi}}\left(x - \frac{x^3}{3.1!} + \frac{x^5}{5.2!} - \frac{x^7}{7.3!} + \dots\right)$$

$$= \frac{2}{\sqrt{\pi}} \sum_{n=0}^{\infty} \frac{(-1)^n x^{2n+1}}{(2n+1)n!}. \tag{13.78}$$

Thus erf(x) rises linearly for small values of x with slope $2/\sqrt{\pi}$. Note that the power series expansion contains only odd powers of x, so the error function is necessarily antisymmetric about $x = 0$ (this is to be expected

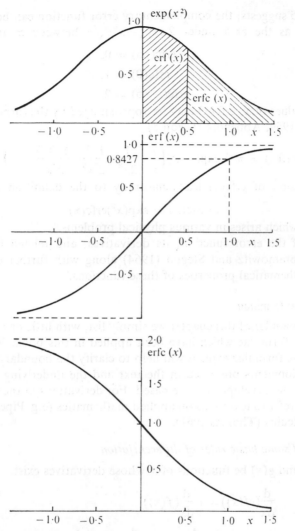

Fɪɢ. 13.5. Error function and complementary error function as errors under Gaussian curve.

from the fact that the error function arises from the integration of an *even* (that is, symmetric) function $\exp(-x^2)$. In other words,

$$\text{erf}(-x) = -\text{erf}(x). \tag{13.79}$$

The *complementary error function* $(\text{erfc}(x))$ may be defined as

$$\text{erfc}(x) = 1 - \text{erf}(x) \tag{13.80}$$

$$= \frac{2}{\sqrt{\pi}} \int_x^\infty \exp(-y^2)\, dy. \tag{13.81}$$

As Fig. 13.5 suggests, the complementary error function can be represented graphically as the area under $2 \exp(-x^2)/\sqrt{\pi}$ between x and ∞. Thus

$$\mathrm{erfc}(\infty) = 0, \qquad (13.82)$$

$$\mathrm{erfc}(0) = 1, \qquad (13.83)$$

and

$$\mathrm{erfc}(-\infty) = 2. \qquad (13.84)$$

For *large* values of x, erfc(x) may be approximated by the formula (Carslaw and Jaeger 1959, Appendix II, eqn 5)

$$\mathrm{erf}(x) = \pi^{-\frac{1}{2}} \exp(-x^2)\left(\frac{1}{x} - \frac{1}{2x^3} + \frac{1.3}{2^2 x^5} - \frac{1.3.5}{2^3 x^7} - \cdots\right). \qquad (13.85)$$

This expansion of erfc(x) leads naturally to the definition of exerfc(x),

$$\mathrm{exerfc}(x) = \exp(x^2)\mathrm{erfc}(x), \qquad (13.86)$$

a function which arises in various physical problems.

Tables of the error function, its derivatives, and related functions are given in Abramowitz and Stegun (1964) along with further discussion of various mathematical properties of these functions.

Mathematical formulae

In the remainder of this chapter we simply list, with little or no derivation, some of the formulae which have been applied in this book. We hope that making these formulae explicit will help to clarify the boundary between the actual developments presented in the text and the underlying mathematics upon which the developments are based. For derivations of the formulae the reader may refer to textbooks on applied mathematics (e.g. Pipes and Harvill 1960) or calculus (Thomas 1961).

Summary of some basic rules of differentiation

Let $f(x)$ and $g(x)$ be functions of x whose derivatives exist.

$$\frac{\mathrm{d}}{\mathrm{d}x}(cf(x)) = c\frac{\mathrm{d}}{\mathrm{d}x}(f(x)). \qquad (13.87)$$

$$\frac{\mathrm{d}}{\mathrm{d}x}(f(x)+g(x)) = \frac{\mathrm{d}}{\mathrm{d}x}(g(x)) + \frac{\mathrm{d}}{\mathrm{d}x}(f(x)). \qquad (13.88)$$

$$\frac{\mathrm{d}}{\mathrm{d}x}(f(x) \cdot g(x)) = f(x)\frac{\mathrm{d}}{\mathrm{d}x}(g(x)) + g(x)\frac{\mathrm{d}}{\mathrm{d}x}(f(x)). \qquad (13.89)$$

$$\frac{\mathrm{d}}{\mathrm{d}x}(f(x))^n = n(f(x))^{n-1}\frac{\mathrm{d}}{\mathrm{d}x}(f(x)). \qquad (13.90)$$

$$\frac{\mathrm{d}}{\mathrm{d}x}e^{f(x)} = e^{f(x)}\frac{\mathrm{d}}{\mathrm{d}x}(f(x)). \qquad (13.91)$$

$$\frac{\mathrm{d}}{\mathrm{d}x}\left(\frac{f(x)}{g(x)}\right) = \left(g(x)\frac{\mathrm{d}}{\mathrm{d}x}(f(x)) - f(x)\frac{\mathrm{d}}{\mathrm{d}x}(g(x))\right)\Big/ g(x)^2. \qquad (13.92)$$

Partial differentiation

In the preceding chapters, solutions are presented to partial differential equations, e.g.

$$\frac{\partial^2 V}{\partial X^2} - \frac{\partial V}{\partial T} - V = 0 \tag{3.13}$$

The operation of partial differentiation is denoted by the ∂ symbol (in place of the ordinary d). $\partial V/\partial T$ is called the first partial derivative with respect to T; $\partial^2 V/\partial X^2$ is called the second partial derivative with respect to X.

Partial differentiation is a rather straightforward extension of the concept of ordinary differentiation. In this particular example, V is a function of two independent variables X and T: this is represented by the expression $V(X, T)$. Partial differentiation with respect to T means that all other variables except T (in this case X) are held constant. V is then differentiated with respect to T as though T were the only independent variable. Mathematically,

$$\frac{\partial V}{\partial T} = \lim_{\Delta T \to 0} \frac{V(X, T+\Delta T) - V(X, T)}{\Delta T}$$

$$= \left(\frac{dV}{dT}\right)_x. \tag{13.93}$$

The rules of partial differentiation are exactly the same as those of ordinary differentiation (e.g. eqns (13.87)–(13.92)) but with the additional feature that the non-participating variables are simply treated as constants.

Differentiation of composite functions: the chain rule

Suppose we have a function $F(x)$ where the variable x is itself a function of t, that is, $x(t)$. Then it is possible to show that, in general (Thomas, p. 77),

$$\frac{dF}{dt} = \frac{dF}{dx}\frac{dx}{dt}. \tag{13.94}$$

This formula is called the *chain rule* because the derivative dF/dt is evaluated in a 'chain' of steps. This equation may also be rearranged to give

$$\frac{dF}{dx} = \frac{(dF/dt)}{(dx/dt)} \tag{13.95}$$

which allows dF/dx to be calculated if dF/dt and dx/dt are already known. For example, this formula was used to compute a trajectory in the phase-plane in Chapter 11, given information as to dI_i/dt and dV/dt. Their ratio gives

$$\frac{dI_i}{dV} = \frac{(dI_i/dt)}{(dV/dt)}. \tag{11.1}$$

The chain rule may also be extended to functions of more than one variable, Suppose $F(x, y)$ where both x and y are functions of t, that is, $x(t)$ and $y(t)$. Then

$$\frac{dF}{dt} = \frac{\partial F}{\partial x}\frac{dx}{dt} + \frac{\partial F}{\partial y}\frac{dy}{dt}, \tag{13.96}$$

where $\partial F/\partial x$ and $\partial F/\partial y$ are partial derivatives as defined above. This rule may be generalized for any number of variables, thus if we consider $F(x, y, z,...)$ the derivative dF/dt will contain one term for each independent variable of F which is itself a function of t.

Use of the differential operator d

In some parts of this book (see Chapter 5, e.g. eqn (5.1)) we have used equations of the form

$$dy = z\,dx \tag{13.97}$$

This equation is not identical in meaning with

$$dy/dx = z, \tag{13.98}$$

since dy/dx means $(d/dx)(y)$, where d/dx is the operation of differentiating with respect to x. The symbol dx in this operator cannot be 'carried across' to the right-hand side to give eqn (13.97) since dx in the operator d/dx has no independent meaning and is certainly not a parameter by which the equation may be multiplied.

However, the symbol d used in (13.97) can be given a meaning (cf. Massey and Kestelman 1959, pp. 76–9): to differentiate with respect to an unspecified (dummy) parameter. Thus, (13.97) may be rewritten

$$dy/d\beta = z(dx/d\beta), \tag{13.99}$$

where β is an unspecified parameter. It may be noted that, using the chain rule, (13.98) may be derived from (13.99). Thus

$$\frac{(dy/d\beta)}{(dx/d\beta)} = \left(\frac{dy}{dx}\right) = z,$$

so that, for all practical purposes (13.97) and (13.98) may be regarded as interchangeable.

This result is convenient since the symbol dy may also be readily compared to the parameter δy used in finite increment approaches (see Massey and Kestelman 1959, p. 80). This approach is sometimes more readily understood by biologists who wish to visualize the processes described in differential equations in graphic terms.

Differentiation of a definite integral

Let $F(x, u)$ be a continuous function of x and u. Consider the function $\phi(u)$, where

$$\phi(u) = \int_{a(u)}^{b(u)} F(x, u)\, \mathrm{d}u. \tag{13.100}$$

It may be shown (see Pipes and Harvill 1970, p. 970) that

$$\frac{\mathrm{d}\phi}{\mathrm{d}u} = \int_a^b \frac{\partial F}{\partial u}\, \mathrm{d}x + F(b, u)\frac{\mathrm{d}b}{\mathrm{d}u} - F(a, u)\frac{\mathrm{d}a}{\mathrm{d}u}. \tag{13.101}$$

This general expression has the following special cases.

1. If a and b are both constants, the last two terms of eqn (13.101) are zero, and

$$\frac{\mathrm{d}\phi}{\mathrm{d}u} = \int_a^b \frac{\partial F}{\partial u}\, \mathrm{d}x. \tag{13.102}$$

2. If $F(x)$ is not a function of u,

$$\frac{\mathrm{d}\phi}{\mathrm{d}u} = F(b)\frac{\mathrm{d}b}{\mathrm{d}u} - F(a)\frac{\mathrm{d}a}{\mathrm{d}u}. \tag{13.103}$$

An example of the use of this formula is given in Chapter 12, where the expression

$$\frac{\mathrm{d}}{\mathrm{d}I} \int_{V_1}^{V_0} i_1\, \mathrm{d}V$$

is evaluated in deriving eqn (12.17).

Integration by parts

This formula is often used in performing integrations where the expression to be integrated can be expressed in the form $f(x)(\mathrm{d}g(x)/\mathrm{d}x)$. The formula states

$$\int f(x) \frac{\mathrm{d}}{\mathrm{d}x}(g(x))\, \mathrm{d}x = f(x)g(x) - \int g(x)\frac{\mathrm{d}}{\mathrm{d}x}(f(x))\, \mathrm{d}x. \tag{13.104}$$

The use of this expression depends on the possibility that the right-hand term is readily integrable.

Power series expansion of functions

Taylor's and Maclaurin's series are two very useful ways of expressing functions. Both are examples of power series. A power series may be defined as an infinite series which has the form

$$S(x) = a_0 + a_1 x + a_2 x^2 + a_3 x^3 + \ldots = \sum_{n=0}^{n=\infty} a_n x^n, \qquad (13.105)$$

where the coefficients a_0, a_1, a_2... are constants independent of x.

The power series is defined for those values of x for which the above summation has a finite value. For such values of x, the series is said to *converge*. The convergence of a series can be predicted from the knowledge of the properties of successive terms of the series (see Pipes and Harvill 1970, p. 831). Some series satisfy the criteria for convergence for values of x within a limited range; the range is termed the *interval* of *convergence*. Outside of the interval of convergence, the series diverges and is generally not useful. In some important cases the interval of convergence extends over all finite values of x. This is true for the exponential, cosine, sine, and hyperbolic functions, as well as for the Bessel function I_0 (but not K_0, which becomes infinite as its argument approaches zero).

An important property of power series can be stated as a theorem (Pipes and Harvill 1970, p. 837): if a function $f(x)$ can be written as either of two power series,

$$f(x) = \sum_{n=0}^{n=\infty} a_n x^n = \sum_{n=0}^{n=\infty} b_n x^n,$$

the coefficients of like powers must be equal, that is to say,

$$a_n = b_n \quad \text{for } n = 0, 1, 2,\ldots$$

This is equivalent to saying that, over the interval of convergence, the function $f(x)$ is *uniquely defined* by its power series.

Power series offer one clear advantage as a means of expressing functions: their definition is a prescription for the numerical calculation of a function's value, for a given argument. Another important use of power series arises from their analytical properties. Within their common interval of convergence, a pair of power series may be added or multiplied together. Their sum or product will necessarily be convergent (Pipes and Harvill 1970, p. 832). Similarly, power series may be differentiated or equated. Since these operations are the only ones implied by a differential equation, it follows that power series can be derived as solutions of differential equations. Indeed, some differential equations are solved by assuming a solution in the form of a power series (see p. 455) and then evaluating the individual coefficients.

Taylor and Maclaurin series. The *Taylor series* gives a prescribed method for obtaining the series expansion of a function. The expansion can be compared to the more familiar operation of fitting an empirical function by an nth degree polynomial. In the case of the Taylor series, the polynomial has an infinite number of terms.

It may be helpful to present the Taylor series in the following explanatory, but non-rigorous, way. Let us imagine that a known function $f(x)$ is to be expanded as a power series over a range of x where the function varies smoothly (that is, where the function and all its derivatives remain finite). We can begin by demanding that, at some point x_0 within the range, the value of the series should match that of the function $f(x_0)$. This demand is readily satisfied if the power series (13.105) is rewritten in the form

$$S(x) = A_0 + A_1(x-x_0) + A_2(x-x_0)^2 + A_3(x-x_0)^3 + \dots . \qquad (13.106)$$

All the terms except the first will be zero at $x = x_0$. Thus, by setting the value of A_0 at $A_0 = f(x_0)$, the function and its expansion will be identical at $x = x_0$.

To make the correspondence between $S(x)$ and $f(x)$ exact, we would require that their respective first derivatives be equal at $x = x_0$. The first derivative of $S(x)$ may be denoted by $S'(x)$, and is given by differentiating eqn (13.103), term by term. We obtain

so that
$$S'(x) = A_1 + 2A_2(x-x_0) + 3A_3(x-x_0)^2 + \dots ,$$
$$S'(x_0) = A_1.$$

Setting $A_1 = f'(x_0) = (d/dx)(f(x_0))$ satisfies the requirement as to the first derivative.

The process of equating derivatives may be continued further:

$$S''(x_0) = f''(x_0) \quad \text{if} \quad A_2 = \frac{f''(x_0)}{2},$$

$$S'''(x_0) = f'''(x_0) \quad \text{if} \quad A_3 = \frac{f'''(x_0)}{3.2},$$

and so on. Taylor's formula simply continues this process to successively higher-order derivatives. The formula may be given therefore in the form

$$f(x) = S(x) = f(x_0) + \frac{f'(x_0)}{1!}(x-x_0) + \frac{f''(x_0)}{2!}(x-x_0)^2 + \dots$$

$$= \sum_{n=0}^{\infty} \frac{f^{(n)}(x_0)}{n!}(x-x_0)^n. \qquad (13.107)$$

Thus, the Taylor expansion gives rise to a power series which will converge if the function has derivatives of all orders (i.e. if $f^{(n)}(x_0)$ is finite for all n). In other words, the Taylor series will be valid over the entire range of x (including x_0) where the function is smoothly varying. Within this interval of convergence the Taylor series will be a unique power-series expansion of the function.

The *Maclaurin series* is a specific case of a Taylor series, where $x_0 = 0$. Thus

$$S(x) = \sum_{n=0}^{n=\infty} \frac{f^{(n)}(0)}{n!} x^n. \qquad (13.108)$$

In the case of Maclaurin series the coefficients $f^{(n)'}(0)/n!$ correspond directly to the constants a_n in the original definition of a power series (13.105). It may be useful at this point to list the Maclaurin series expansions for some commonly used functions:

$$e^x = 1 + \frac{x^1}{1!} + \frac{x^2}{2!} + \frac{x^3}{3!} + \ldots + \frac{x^n}{n!}, \tag{13.109}$$

$$\cos x = 1 - \frac{x^2}{2!} + \frac{x^4}{4!} - \frac{x^6}{6!} + \ldots (-1)^{n-1} \frac{x^{2n-2}}{(2n-2)!}, \tag{13.110}$$

$$\sin x = \frac{x^1}{1!} - \frac{x^3}{3!} + \frac{x^5}{5!} - \ldots (-1)^{n-1} \frac{x^{2n-1}}{(2n-1)!}, \tag{13.111}$$

$$\cosh x = 1 + \frac{x^2}{2!} + \frac{x^4}{4!} + \ldots + \frac{x^{2n}}{(2n)!}, \tag{13.112}$$

$$\sinh x = \frac{x^1}{1!} + \frac{x^3}{3!} + \frac{x^5}{5!} + \ldots \frac{x^{2n+1}}{(2n+1)!}. \tag{13.113}$$

References and author index

Page references to the text are given in square brackets after each bibliographical reference.

ABE, K. and TOMITA, T. (1968). Cable properties of smooth muscle. *J. Physiol.* **196**, 87–100. [11, 23, 117]

ABRAMOWITZ, M. and STEGUN, I. A. (1964). *Handbook of mathematical functions.* National Bureau of Standards, Washington. [395, 462]

ADELMAN, W. J. (1971) (Ed.). *Biophysics and physiology of excitable membranes.* Van Nostrand, New York. [252]

ADRIAN, E. D. (1928). *The Basis of sensation.* Christophers, London. [344]

ADRIAN, R. H. (1969). Rectification in muscle membrane. *Prog. Biophys.* **19**, 339–69. [229, 238, 240]

——, ALMERS, W. (1974). Membrane capacity measurements on frog skeletal muscle in media of low ion content. *J. Physiol.* **237**, 573–605 [103]

——, CHANDLER, W. K., and HODGKIN, A.L. (1966). Voltage clamp experiments in skeletal muscle fibres. *J. Physiol.* **186**, 51–2P. [39, 246]

——, ——, —— (1969). The kinetics of mechanical activation in frog muscle. *J. Physiol.* **204**, 207–30. [121, 124, 126, 128, 130]

——, ——, —— (1970). Voltage clamp experiments in striated muscle fibres. *J. Physiol.* **208**, 607–44. [39, 98, 113, 115, 246, 252, 356, 416, 417]

——, ——, —— (1972). An extension of Cole's theorem and its application to muscle In: *Perspectives in Membrane Biophysics* (Ed. Adelman) 299–309. New York: Gordon & Breach. [380]

——, COSTANTIN, L. L., and PEACHEY, L. D. (1969). Radial spread of contraction in frog muscle fibres. *J. Physiol.* **204**, 231–57. [119, 122, 123, 128, 129, 130, 132]

—— and FREYGANG, W. H. (1962). Potassium conductance of frog muscle membrane under controlled voltage. *J. Physiol.* **163**, 104–14. [39, 75, 238]

—— and PEACHEY, L. D. (1965). The membrane capacity of frog twitch and slow muscle fibres. *J. Physiol.* **181**, 324–36. [105]

——, —— (1973). Reconstruction of the action potential of frog sartorius muscle. *J. Physiol.* **235**, 103–131. [246, 252]

—— and SLAYMAN, C. L. (1966). Membrane potential and conductance during transport of sodium, potassium and rubidium in frog muscle. *J. Physiol.* **184**, 970–1014. [358]

AGIN, D. (1964). Hodgkin–Huxley equations: logarithmic relation between membrane current and frequency of repetitive activity. *Nature, Lond.* **201**, 625–6. [331]

AIDLEY, D. J. (1971). *The physiology of excitable cells.* Cambridge University Press. [3]

ANDERSEN, P., BLACKSTAD, T. W., and LØMO, T. (1966). Location and identification of excitatory synapses on hippocampal pyramidal cells. *Exptl Brain Res.* **1**, 236–48. [174]

——, ECCLES, J. C., and LØYNING, Y. (1963). Recurrent inhibition in the hippocampus with identification of the inhibitory cell and its synapses. *Nature, Lond.* **198**, 540–2. [174]

ANDREOLI, T. E. and WATKINS, M. L. (1973). Chloride transport in porous lipid bilayer membranes. *J. gen. Physiol.* **61**, 809–830. [237]

ANDRONOW, A. A. and CHAIKIN, C. E. (1949) *Theory of oscillations.* Princeton University Press, Princeton, New Jersey. [322]

ARAKI, T. and TERZUOLO, C. A. (1962). Membrane currents in spinal motoneurons associated with the action potential and synaptic activity. *J. Neurophysiol.* **25**, 772–89. [182, 214]

ARMSTRONG, C. M. (1971). Interaction of tetraethylammonium ion derivatives with the potassium channel of giant axons. *J. gen. Physiol.* **58**, 413–37. [240]

—— and HILLE, B. (1972). The inner quaternary ammonium ion receptor in potassium channels of the node of Ranvier. *J. gen. Physiol.* **59**, 388–400. [240]

AUERBACH, A. and BETZ, W. (1971). Does curare affect transmitter release? *J. Physiol.* **213**, 691–705. [66]

BAKER, P. F., BLAUSTEIN, M. P., KEYNES, R. D., MANIL, J., SHAW, T. I., and STEINHARDT, R. A. (1969). The ouabain sensitive fluxes of sodium and potassium in squid giant axons. *J. Physiol.* **200**, 459–96. [361]

——, FOSTER, RACHEL F., GILBERT, D. S., and SHAW, T. I. (1971). Sodium transport by perfused giant axons of *Loligo*. *J. Physiol.* **219**, 487–506. [360]

BALDISSERA, F. and GUSTAFSSON, B. (1971a). Supraspinal control of the discharge evoked by constant current in the alpha-motoneurons. *Brain Res.* **25**, 642–4. [311, 312]

—— —— (1971b). Regulation of repetitive firing in motoneurons by the after-hyperpolarisation conductance. *Brain Res.* **30**, 431–4. [311, 312, 313, 355]

——, ——, and PARMIGGIANI, F. (1973). Adaptation in a simple neurone model compared to that of spinal motoneurones. *Brain Res.* **52**, 382–4. [355]

BARR, L., DEWEY, M. M., and BERGER, W. (1965). Propagation of action potentials and the structure of the nexus in cardiac muscle. *J. gen. Physiol.* **48**, 797. [10]

BARRETT, J. N. and CRILL, W. E. (1971). Specific membrane resistivity of dye-injected cat motoneurons. *Brain Res.* **28**, 556–61. [158]

—— and GRAUBARD, K. (1970). Fluorescent staining of cat motoneurons *in vivo* with beveled micropipettes. *Brain Res.* **18**, 565–8. [144]

BARRY, P. H. and DIAMOND, J. M. (1971). A theory of ion permeation through membranes with fixed neutral sites. *J. membrane Biol.* **4**, 295–330. [240]

BEELER, G. W. and REUTER, H. (1970a). Voltage clamp experiments on ventricular myocardial fibres. *J. Physiol.* **207**, 165–90. [406]

—— —— (1970b). Membrane calcium current in ventricular myocardial fibres. *J. Physiol.* **207**, 191–209. [317]

BELLMAN, R., KALIBA, R. E., and LOCKETT, J. A. (1966). *Numerical inversion of the laplace transform*. American Elsevier, New York. [304, 446]

BELYAKOV, V. M., KRAVTSOVA, P. I., and RAPPOPORT, M. G. (1965). *Tables of elliptic integrals. Part 1.* (Transl. by P. Basu). Pergamon Press, Oxford. [395, 433]

BENNETT, M. V. L. (1966). Physiology of electrotonic junctions. *Ann. N. Y. Acad. Sci.* **137**, 509–39. [224]

—— (1968). Similarities between chemically and electrically mediated transmission. In *Physiological and biochemical aspects of nervous integration*, (ed. F. D. Carlsson) pp. 73–128. Prentice-Hall, Englewood Cliffs, New Jersey. [224]

BERGMAN, C. (1970). Increase of sodium concentration near the inner surface of the nodal membrane. *Pflügers Arch. ges. Physiol.* **317**, 287–302. [357]

——, NONNER, W., and STÄMPFLI, R. (1968). Sustained spontaneous activity of Ranvier nodes induced by combined actions of TEA and lack of calcium. *Pflügers Arch. ges. Physiol.* **302**, 24–37. [343, 355]

—— and STÄMPFLI, R. (1966). Différence de perméabilité des fibres nerveuses myelinisées sensorielles et motrices à l'ion potassium. *Helv. physiol. pharmac. Acta* **24**, 247–98. [301, 343]

BERNSTEIN, J. (1902). Untersuchungen zur Thermodynamik der Bioelektrischen Ströme. I (Erster Teil). *Pflügers Arch. ges. Physiol.* **92**, 521–62. [9, 11]

—— (1912). *Elektrobiologie*. Vieweg, Brunswick, Germany. [11]

BERTHOLD, C-H. (1972). A study on the fixation of large mature feline myelinated ventral lumbar spinal-root fibres. *Acta Soc. Med. upsal.* **73**, Suppl. 9, 1–36. [300]

BEZANILLA, F. and ARMSTRONG, C. M. (1972). Negative conductance caused by entry of sodium and cesium ions into the K channels of squid axons. *J. gen. Physiol.* **60**, 588–608. [240, 259, 260]

——, CAPUTO, C., GONZALEZ-SERRATOS, H., and VENOSA, R. A. (1972). Sodium dependence of the inward spread of activation in isolated twitch muscle fibres of the frog. *J. Physiol.* **223**, 507–23. [130]

BINSTOCK, L. and LECAR, H. (1969). Ammonium ion conductance in the squid giant axon. *J. gen. Physiol.* **53**, 342–60. [259]

BLAUSTEIN, M. P. and GOLDMAN, D. E. (1968). The action of certain polyvalent cations on the voltage-clamped lobster axon. *J. gen. Physiol.* **51**, 279–91 [349]

BLINKS, J. R. (1965). Influence of osmotic strength on cross-section and volume of isolated single muscle fibres. *J. Physiol.* **177**, 42–57 [122]

BONHOEFFER, K. F. (1948). Activation of passive iron as a model for the excitation of nerve. *J. gen. Physiol.* **32**, 69–91. [306, 330]

BOWMAN, F. (1958). *Introduction to Bessel functions.* Dover, New York. [454]

BOYD, I. A. (1965). Differences in the diameter and conduction velocity of motor and fusimotor fibres in nerves to different muscles in the hind limb of the cat. In *Studies in physiology* (Eds. D. R. Curtis and A. K. McIntyre), pp. 7–12. [299]

BROMM, B. and FRANKENHAEUSER, B. (1972). Repetitive discharge of the excitable membrane computed on the basis of voltage clamp data for the node of Ranvier. *Pflügers Arch. ges Physiol.* **332**, 21–7. [342, 343, 348]

BROWN, H. F. and Noble, S. J. (1969). Membrane currents underlying delayed rectification and pacemaker activity in frog atrial muscle. *J. Physiol.* **204**, 717–36. [75, 305]

BROWN, J. E., MULLER, K. J., and MURRAY, G. (1971). Reversal potential for an electrophysiological event generated by conductance changes: mathematical analysis. *Science, N. Y.* **174**, 318. [218]

BROWN, M. C. and STEIN, R. B. (1966). Quantitative studies on the slowly adapting stretch receptor of the crayfish. *Kybernetik* **3**, 175–85. [352]

BURKE, W. (1957). Spontaneous potentials in slow muscle fibres of the frog. *J. Physiol.* **135**, 511–21. [59, 62]

BURKE, R. E. (1968). Group Ia synaptic input to fast and slow twitch motor units of cat triceps surae. *J. Physiol.* **196**, 605–30. [169]

——, FEDINA, L., and LUNDBERG, A. (1971). Spatial synaptic distribution of recurrent and group Ia inhibitory systems in cat spinal motoneurones. *J. Physiol.* **214**, 305–26. [183]

—— and ten BRUGGENCATE, G. (1971). Electrotonic characteristics of alpha motoneurones of varying size. *J. Physiol.* **212**, 1–20. [172]

BURROWS, T. M. O., CAMPBELL, I. A., HOWE, E. J., and YOUNG, J. Z. (1965). Conduction velocity and diameter of nerve fibres of cephalopods. *J. Physiol.* **179**, 39–40P [292]

BYRD, P. W. and FRIEDMAN, M. D. (1954). *Handbook of elliptic integrals.* Springer-Verlag, Berlin. [395, 398]

CALVIN, W. H. (1969a). Dendritic synapses and reversal potentials: theoretical implications of the view from the soma. *Exptl Neurol.* **24**, 248–64. [179]

—— (1969b). Dendritic spikes revisited. *Science, N. Y.* **166**, 637–8. [217]

—— and HELLERSTEIN, D. (1968). Dendritic spikes vs. cable properties. *Science, N.Y.* **163**, 96–7. [217]

—— and SCHWINDT, P. C. (1972). Steps in production of motoneuron spikes during rhythmic firing. *J. Neurophysiol.* **35**, 297–310. [314, 355]

CAPOCELLI, R. H. and RICCIARDI, L. M. (1971). Diffusion approximation and first passage time problem for a model neuron. *Kybernetik,* **8**, 214–23. [372]

CARPENTER, D. O., HOVEY, M. M., and BAK, A. F. (1973). Measurements of intracellular conductivity in *Aplysia* neurons: evidence for organization of water and ions. *Ann. N. Y. Acad. Sci.* **204**, 502–30 [9]

CARSLAW, H. S. and JAEGER, J. C. (1948). *Operational methods in applied mathematics* (2nd edn), p. 359. Oxford University Press. [187]

—— —— (1959). *Conduction of heat in solids* (2nd edn). Oxford University Press. [2, 124, 166, 441, 442, 446, 454, 460]

CHANDLER, W. K., FITZHUGH, R., and COLE, K. S. (1962). Theoretical stability properties of a space-clamped axon. *Biophys. J.* **2**, 105–27. [23, 323]

——, HODGKIN, A. L., and MEVES, H. (1965). The effect of changing the internal solution on sodium inactivation and related phenomena in giant axons. *J. Physiol.* **180**, 821–36. [238]

—— and MEVES, H. (1965). Voltage clamp experiments on internally perfused giant axons. *J. Physiol.* **180**, 788–820. [258]

CHAPMAN, R. A. (1963). 'Pre-pulse' and 'extra impulse' experiments on a type of repetitive crab axon. *J. Physiol.* **168**, 17–18P [243]

CHAPMAN, R. A. (1966). The repetitive responses of isolated axons from the crab, *Carcinus maenas. J. exp. Biol.* **45**, 475–88. [305, 332, 334, 335, 336, 340, 347, 358]

—— (1967). Dependence on temperature of the conduction velocity of the action potential of the squid giant axon. *Nature Lond.* **213**, 1143–4. [289]

CLARK, J. and PLONSEY, R. (1966). A mathematical evaluation of the core conductor model. *Biophys. J.* **6**, 95–112. [3]

—— —— (1968). The extracellular potential field of the single active nerve fiber in a volume conductor. *Biophys. J.* **8**, 842–64. [3, 28]

CLAY, J. R. and GOEL, N. S. (1973). Diffusion models for firing of a neuron with varying threshold. *J. theoret. Biol.* **39**, 633–44. [368]

COLE, K. S. (1928). Electric impedance of suspensions of spheres. *J. gen. Physiol.* **12**, 29. [11]

—— (1941*a*). Impedance of single cells. *Tabul. biol.* **19** (Cellula, Pt 2), 24–7. [9]

—— (1949). Dynamic electrical characteristics of the squid axon membrane. *Archs Sci. physiol.* **3**, 253–8. [7, 12, 241]

—— (1965). Electrodiffusion models for the membrane of squid giant axon. *Physiol. Rev.* **45**, 340–79. [229, 238, 240]

—— (1968). *Membranes, ions and impulses.* University of California Press, Berkeley and Los Angeles. [9, 11, 18, 109, 229, 252, 315]

——, ANTOSIEWICZ, H. A., and RABINOWITZ, P. (1955). Automatic computation of nerve excitation. *J. Soc. ind. appl. Math.* **3**, 153–72. [331]

—— and BAKER, R. F. (1941). Longitudinal impedance of the squid giant axon. *J. gen. Physiol.* **24**, 771–88. [18, 21, 108]

—— and CURTIS, H. J. (1936). Electric impedance of nerve and muscle. *C.S.H. Symp. Quant. Biol.* **4**, 73. [1, 26, 28]

—— —— (1938). Electrical impedance of *Nitella* during activity. *J. gen. Physiol.* **22**, 37–64. [282]

—— —— (1939). Electric impedance of the squid giant axon during activity. *J. gen. Physiol.* **22**, 649–70. [1, 282, 284]

—— —— (1941). Membrane potential of the squid giant axon during current flow. *J. gen. Physiol.* **24**, 551–63. [1, 229, 263]

—— and HODGKIN, A. L. (1939). Membrane and protoplasm resistance in the squid giant axon. *J. gen. Physiol.* **22**, 671–87. [1, 9]

CONNOR, J. A. and STEVENS, C. F. (1971*a*). Inward and delayed outward membrane currents in isolated neural somata under voltage clamp. *J. Physiol.* **213**, 1–19. [305, 315, 344, 345, 356]

CONNOR, J. A. and STEVENS, C. F. (1971*b*). Voltage clamp studies of a transient outward membrane current in gastropod neutral somata. *J. Physiol.* **213**, 21–30. [305, 315, 344].

—— —— (1971*c*). Prediction of repetitive firing behaviour from voltage clamp data on an isolated neurone soma. *J. Physiol.* **213**, 31–53. [305, 315, 344, 345, 346, 347]

CONRADI, S. (1969). On motoneuron synaptology in adult cats. *Acta physiol. scand.* **78**, Suppl, 332. (173)

CONTI, F. and EISENMAN, G. (1965). The steady-state properties of ion exchangers with fixed sites. *Biophys. J.* **5**, 511–30. [229, 238]

COOLEY, J. W. and DODGE, F. A. Jr. (1966). Digital computer solutions for excitation and propagation of the nerve impulse. *Biophys. J.* **6**, 583–99. [252, 270, 273, 278, 302, 331]

——, ——, and COHEN, H. (1965). Digital computer solutions for excitable membrane models, *J. cell. comp. Physiol.* **66**, Suppl. 99–109. [270, 275, 278, 302, 319, 323, 331, 418]

COOLEY, J. W. and DODGE, F. A. (1973). Theoretical limits on the conduction velocity of axons *Abst. Biophys. Soc.* (*17th Annual meeting*) TAM-J4. [430]

COOMBS, J. S., CURTIS, D. R., and ECCLES, J. C. (1957). The interpretation of spike potentials of motoneurones. *J. Physiol.* **139**, 198–231. [214]

——, ECCLES, J. C., and FATT, P. (1955*a*). The electrical properties of the motoneurone membrane. *J. Physiol.* **130**, 291–325. [182, 311]

——, ——, —— (1955*b*). The specific ionic conductance and ionic movements across the motoneuronal membrane that produce the inhibitory post-synaptic potential. *J. Physiol.* **130**, 326–73. [182]

——, ——, —— (1955*c*). Excitatory synaptic action in motoneurones. *J. Physiol.* **130**, 374–95. [182]

——, ——, —— (1955*d*). The inhibitory suppression of reflex discharges from motoneurones. *J. Physiol.* **130**, 396–413. [197]

COPPIN, C. M. L. and JACK, J. J. B. (1972). Internodal length and conduction velocity of cat muscle afferent nerve fibres. *J. Physiol.* **222**, 91–3P. [299, 300]

COSTANTIN, L. L. (1968). The effect of calcium on contraction and conductance thresholds in frog skeletal muscle. *J. Physiol.* **195**, 119–32. [120]

—— (1970). The role of sodium currents in the radial spread of contraction in frog muscle fibres. *J. gen. Physiol.* **55**, 703–15. [100, 120, 408]

—— (1975). Contractile activation in skeletal muscle *Prog. Biophys. Mol. Biol.* **29**, 197–224. [130]

—— and PODOLSKY, R. J. (1967). Depolarization of the internal membrane system in the activation of frog skeletal muscle. *J. gen. Physiol.* **50**, 1101–24. [130]

CRANK, J. (1956). *The mathematics of diffusion.* Clarendon Press, Oxford. [39, 125]

CREMER, M. (1899). Zum Kernleiterproblem. *Z. Biol.* **37**, 550. [1]

—— (1909). Die allgemeine Physiologie der Nerven. In *Handbuch der Physiologie des Menschen* (ed. W. Nagel), Vol. 4, p. 793. Vieweg, Braunschweig. [1]

CUNNINGHAM, W. J. (1958). *Introduction to nonlinear analysis.* McGraw-Hill, New York. [322]

CURTIS, H. J. and COLE, K. S. (1938). Transverse electric impedance of the squid giant axon. *J. gen. Physiol.* **21**, 757–65. [12]

CURTIS, D. R. and ECCLES, J. C. (1959). The time courses of excitatory and inhibitory synaptic actions. *J. Physiol.* **145**, 529–46. [58]

DAVIS, H. (1923). The relationship of the 'chronaxie' of muscle to the size of the stimulating electrode. *J. Physiol.* **57**, 81–2P. [273, 416]

DAVIS, L. and LORENTE DE NO, R. (1947). Contribution to the mathematical theory of electrotonus. *Stud. Rockefeller Inst. med. Res.* **131**, 442–96. [30, 34]

DECK, K. A., KERN, R., and TRAUTWEIN, W. (1964). Voltage clamp technique in mammalian cardiac fibres. *Pflügers Arch. ges. Physiol.* **280**, 50–62. [39, 75]

474   References and author index

DECK, K. A. and TRAUTWEIN, W. (1964). Ionic currents in cardiac excitation. *Pflüger Arch. ges. Physiol.* **280**, 63–80. [399]

DEL CASTILLO, J. and KATZ, B. (1954). Quantal components of the endplate potential. *J. Physiol.* **124**, 560–73. [65]

—— —— (1956). Localization of active spots within the neuromuscular junction of the frog. *J. Physiol.* **132**, 630–49. [59]

—— and MORALES, T. (1967). Extracellular action potentials recorded from the interior of the giant esophagal cell of *Ascaris. J. gen. Physiol.* **50**, 631–45. [118]

DÉLÈZE, J. (1971). The recovery of resting potential and input resistance in sheep heart injured by knife or laser. *J. Physiol.* **208**, 547–62. [67]

DE SOER, C. A. (1969). *Basic circuit theory.* McGraw-Hill, New York. [106]

DIAMOND, J., GRAY, E. G., and YASARGIL, E. M. (1970). The function of the dendrite spine: an hypothesis. In *Excitatory synaptic mechanisms* (ed. P. Andersen and J. K. S. Jansen), pp. 213–22. Universitets forlaget, Osla. [220]

—— and YASARGIL, E. N. (1969). Synaptic function in the fish spinal cord: dendritic integration. *Prog. Brain Res.* **31**, 201–9. [217, 222]

DIAMOND, J. M. and WRIGHT, E. M. (1969). Biological membranes: the physical basis of ion and non-electrolyte selectivity. *A. Rev. Physiol.* **31**, 581–646. [254]

DODGE, F. A. (1972). On the transduction of visual, mechanical and chemical stimuli. *Int. J. Neuroscience* **3**, 5–14. [331, 335, 337]

DONG, E. and REITZ, B. A. (1970). Effect of timing of vagal stimulation on heart-rate in dog. *Circulation Res.* **27**, 635–46. [378]

DROUIN, H. and THE, R. (1969). The effect of reducing extracellular pH on the membrane currents of the Ranvier node. *Pflügers Arch. ges. Physiol.* **313**, 80–8. [259]

DUDEL, J., PEPER, K., RÜDEL, R., and TRAUTWEIN, W. (1966). Excitatory membrane current in heart muscle (Purkinje fibres). *Pflügers Arch. ges. Physiol.* **292**, 255–73. [246]

—— and RÜDEL, R. (1969). Voltage and time dependence of excitatory sodium current in cooled sheep Purkinje fibres. *Pflügers Arch. ges. Physiol.* **315**, 136–58. [246]

DULHUNTY, A. F. and GAGE, P. W. (1973). Electrical properties of toad sartorius muscle fibres in summer and winter. *J. Physiol.* **230**, 619–41. [122]

DURELL, C. V. and ROBSON, A. (1932). *Advanced Algebra.* [395]

DWIGHT, H. B. (1962). *Tables of integrals and other mathematical data.* (4th edn)., New York. [395]

EBASHI, S. and ENDO, M. (1968). Calcium ions and muscle contraction. *Prog. Biophys. mol. Biol.* **18**, 123–83. [98]

——, ——, and OHTSUKI, I. (1969). Control of muscle contraction. *Q. Rev. Biophys.* **2**, 351–84. [99]

ECCLES, J. C. (1957). *The Physiology of nerve cells,* p. 270. Johns Hopkins Press, Baltimore. [214]

—— (1961). Membrane time constants of cat motoneurones and time courses of synaptic action. *Expl Neurol.* **4**, 1–22. [169, 170]

—— (1964). *The Physiology of synapses,* p. 316. Springer-Verlag, Berlin. [53, 192 224]

EHRENSTEIN, G. and GILBERT, D. L. (1966). Slow changes of potassium permeability in the squid giant axon. *Biophys. J.* **6**, 553–66. [356]

—— and LECAR, H. (1972). The mechanism of signal transmission in nerve axons. *A. Rev. Biophys. Bioeng.* **1**, 347–68. [240]

EIDE, E., FEDINA, L. JANSEN, J., LUNDBERG, A., and VYKLICKY, L. (1969a). Properties of Clarke's column neurones. *Acta physiol. scand.* **77**, 125–44. [315, 377]

EIDE, E., FEDINA, L. JANSEN, J., LUNBERG, A., and VYKLICKY, L. (1969*b*). Unitary components in the activation of Clarke's column neurones. *Acta physiol. scand.* **77**, 145–58. [377]

EISENBERG, B. and EISENBERG, R. S. (1968). Selective disruption of the sarcotubular system in frog sartorius muscle. *J. cell. Biol.* **39**, 451–67. [119]

EISENBERG, R. S. (1967). Equivalent circuit of crab muscle fibers as determined by impedance measurements with intracellular electrodes. *J. gen. Physiol.* **50**, 1785–1806. [7, 12, 21, 100, 106, 110, 113]

—— and COSTANTIN, L. L. (1971). The radial variation of potential in the transverse tubular system of skeletal muscle. *J. gen. Physiol.* **58**, 700–1. [407]

—— and ENGEL, E. (1970). The spatial variation of membrane potential near a small source of current in a spherical cell. *J. gen. Physiol.* **55**, 736–57. [132, 133]

—— and GAGE, P. W. (1969). Ionic conductance of the surface and transverse tubular membranes of frog sartorius fibers. *J. gen. Physiol.* **53**, 279–97. [100]

—— and JOHNSON, E. A. (1970). Three-dimensional electrical field problems in physiology. *Prog. Biophys. mol. Biol.* **20**, 1–65. [3, 6, 8, 32, 83, 87, 96, 97, 101, 113, 124, 132, 452]

EISENMAN, G. (1961). In *Symposium on membrane transport and metabolism* (Eds. A. Kleinzeller and A. Kotyk) Academic Press, New York. [254]

—— (1965). Some elementary factors involved in specific ion permeation. *Proc 23rd Int. Congr. Physiol. Sci. (Tokyo)*, p. 489–506. [254]

—— and CONTI, F. (1965). Some implications for biology of recent theoretical and experimental studies of ion permeation in model membranes. *J. gen. Physiol.* **48**, 65–73. [238]

ENGBERG, I. and MARSHALL, K. C. (1971). Mechanism of noradrenaline hyperpolarization in spinal cord interneurones of the cat. *Acta physiol. scand.* **83**, 142–4. [192]

ENGEL, E., BARCILON, V., and EISENBERG, R. S. (1972). The interpretation of current–voltage relations recorded from a spherical cell with a single microelectrode. *Biophys. J.* **12**, 384–403. [133]

ERDELYI, A. (1953) (Ed.). *Higher transcendental functions.* New York. [395]

ERLANGER, J. and BLAIR, E. A. (1938). Comparative observations on motor and sensory fibres with special reference to repetitiousness. *Am. J. Physiol.* **121**, 431–53. [343]

FALK, G. (1968). Predicted delays in the activation of the contractile system. *Biophys. J.* **8**, 608–25. [100, 103, 113, 120, 121, 124]

—— and FATT, P. (1964). Linear electrical properties of striated muscle fibres observed with intracellular electrodes. *Proc. R. Soc.* **B160**, 69–123. [8, 12, 19, 101, 106, 107, 109, 110, 112, 113, 115, 121, 122, 452]

FAMIGLIETTI, E. V. (1970). Dendro-dendritic synapses in the lateral geniculate nucleus of the cat. *Brain Res.* **20**, 181–91. [217]

FATT, P. (1964). An analysis of the transverse electrical impedance of striated muscle. *Proc. R. Soc.* **B159**, 606–51. [19, 109]

—— and GINSBORG, B. L. (1958). The ionic requirements for the production of action potentials in crustacean muscle fibres. *J. Physiol.* **142**, 516–43. [10]

—— and KATZ, B. (1951). An analysis of the end-plate potential recorded with an intracellular electrode. *J. Physiol.* **115**, 320–70. [2, 10, 46, 53, 54, 58, 62, 101, 105]

—— —— (1953). The effect of inhibitory nerve impulses on a crustacean muscle fibre. *J. Physiol.* **121**, 374–89. [10]

FIENBERG, S. E. (1970). A note on the diffusion approximation for single neuron firing problems. *Kybernetik*, **7**, 227–9. [372]

FINEAN, J. B. (1966). The molecular organization of cell membranes. *Prog. Biophys.* **16**, 145. [12]

FitzHugh, R. (1955). Mathematical models of threshold phenomena in the nerve membrane. *Bull. math. Biophys.* **17**, 257–78. [302]

—— (1960). Thresholds and plateaus in the Hodgkin–Huxley nerve equations. *J. gen. Physiol.* **43**, 867–96. [252, 306, 330]

—— (1961). Impulses and physiological states in theoretical models of nerve membrane. *Biophys. J.* **1**, 445–66. [306, 330, 331, 332, 352]

—— (1962). Computation of impulse initiation and saltatory conduction in a myelinated nerve fiber. *Biophys. J.* **2**, 11–21. [298, 301]

—— (1965). A kinetic model of the conductance changes in nerve membrane. *J. cell. comp. Physiol.* **66**, Suppl. 2, 111–17. [241]

—— (1969). Mathematical models of excitation and propagation in nerve. In *Biological engineering* (Ed. H. P. Schwan), pp. 1–85. McGraw-Hill, New York. [286, 306, 315, 330, 422]

—— and Antosiewicz, H. A. (1959). Automatic computation of nerve excitation—detailed corrections and additions. *J. Soc. ind. appl. Math.* **7**, 447–58. [307, 331]

Fozzard, H. A. (1966). Membrane capacity of the cardiac Purkinje fibre. *J. Physiol.* **182**, 255–67. [12, 21, 102, 117]

—— and Schoenberg, M. (1972). Strength–duration curves in cardiac Purkinje fibres: effects of liminal length and charge redistribution. *J. Physiol.* **226**, 593–618. [273, 274, 418, 420, 421]

Frankenhaeuser, B. (1960). Sodium permeability in toad nerve and in squid nerve. *J. Physiol.* **152**, 159–66. [237, 238]

—— (1962). Potassium permeability in myelinated nerve fibres of *Xenopus laevis*. *J. Physiol.* **160**, 54–61. [237]

—— and Hodgkin, A. L. (1956). The after-effects of impulses in the giant nerve fibres of *Loligo*. *J. Physiol.* **131**, 341–76. [340, 357]

—— —— (1957). The action of calcium on the electrical properties of squid axons. *J. Physiol.* **137**, 218–44. [349]

—— and Huxley, A. F. (1964). The action potential in the myelinated nerve fibre of *Xenopus laevis* as computed on the basis of voltage clamp data. *J. Physiol.* **171**, 302–15. [246, 252, 298, 342, 353]

—— and Moore, L. E. (1963a). The effect of temperature on the sodium and potassium permeability changes in myelinated nerve fibres of *Xenopus laevis*. *J. Physiol.* **169**, 431–7. [241]

—— —— (1963b). The specificity of the initial current in myelinated nerve fibres of *Xenopus laevis*. Voltage-clamp experiments. *J. Physiol.* **169**, 438–44. [218, 241]

—— and Vallbo, Å. B. (1965). Accommodation in myelinated nerve fibres of *Xenopus laevis* as computed on the basis of voltage clamp data. *Acta physiol. scand.* **63**, 1–20. [301, 344]

—— and Waltman, B. (1959). Membrane resistance and conduction velocity of large myelinated nerve fibres from *Xenopus laevis*. *J. Physiol.* **148**, 677–82. [356]

Freygang, W. H. Jr. (1965). Tubular ionic movements. *Fedn Proc.* **24**, 1135–40. [121]

——, Rapoport, S. I., and Peachey, L. D. (1967). Some relations between changes in the linear electrical properties of striated muscle fibers and changes in ultra-structure. *J. gen. Physiol.* **50**, 2437–58. [12, 103, 106, 111, 114, 115, 117, 118]

—— and Trautwein, W. (1970). The structural implications of the linear electrical properties of cardiac Purkinje strands. *J. gen. Physiol.* **55**, 524–47. [10, 11, 12, 21, 24, 100, 111, 112, 117]

Fricke, H. (1925). The electrical capacity of suspensions with special reference to blood. *J. gen. Physiol.* **9**, 137–52. [11]

FUJINO, M. YAMAGUCHI, T., and SUZUKI, K. (1961). Glycerol effect and the mechanism linking excitation of the plasma membrane with contraction. *Nature, Lond.* **192**, 1159–61. [100]

FUORTES, M. G. F., FRANK, K., and BECKER, M. C. (1957). Steps in the production of motoneuron spikes. *J. gen. Physiol.* **40**, 735–52. [214]

—— and MANTEGAZZINI, F. (1962). Interpretation of the repetitive firing of nerve cells. *J. gen. Physiol.* **45**, 1163–79. [305, 340]

FURSHPAN, E. J. and POTTER, D. D. (1959). Transmission at the giant motor synapses of the crayfish. *J. Physiol.* **145**, 289–325. [224]

—— —— (1969). Low resistance junctions between cells in embryos and tissue culture. In *Current topics in development biology* (Ed. A. A. Moscona and A. M. Monroy), Vol. 3, p. 95. Academic Press, New York. [11]

GAGE, P. W. and EISENBERG, R. S. (1969a). Capacitance of the surface and transverse tubular membrane of frog sartorius muscle fibers. *J. gen. Physiol.* **53**, 265–78. [100]

—— —— (1969b). Action potentials, after potentials, and excitation contraction coupling in frog sartorius fibers without transverse tubules. *J. gen. Physiol.* **53**, 298–310. [100]

—— and McBURNEY, R. N. (1972). Miniature end-plate currents and potentials generated by quanta of acetylcholine in glycerol-treated toad sartorius fibres. *J. Physiol.* **226**, 79–94. [59]

GARDNER, M. F. and BARNES, J. L. (1942). *Transients in linear systems*, Vol. 1. *Lumped-constant systems*. Wiley, New York. [441]

GASSER, H. S. (1950). Unmedullated fibres originating in dorsal root ganglia. *J. gen. Physiol.* **33**, 651–90. [292, 293]

GEORGE, E. P. (1961). Resistance values in a syncytium. *Aust. J. exp. Biol. med. Sci.* **39**, 267–74. [83]

GERSTEIN, G. L. and MANDELBROT, B. (1964). Random walk models for the spike activity of a single neuron. *Biophys. J.* **4**, 41–68. [367]

GHAUSI, M. S. and KELLY, J. J. (1968). *Introduction to distributed-parameter networks with application to integrative circuits*, p. 326. Holt, Rinehart, and Winston, New York. [149]

GILBERT, D. L. and EHRENSTEIN, G. (1970). Use of a fixed charge model to determine the pK of the negative sites on the external membrane surface. *J. gen. Physiol.* **55**, 822–5. [259]

GILLARY, H. L. (1966). Stimulation of the salt receptor of the blowfly. I. NaCl. *J. gen. Physiol.* **50**, 337–50. [347]

GINSBORG, B. L. (1960). Spontaneous activity in muscle fibres of the chick. *J. Physiol.* **150**, 707–17. [59]

—— (1967). Ion movements in junctional transmission. *Pharmac. Rev.* **19**, 289–316. [53].

GLOBUS, A., LUX, H. D., and SCHUBERT, P. (1968). Soma-dendrite spread of intracellularly injected tritiated glycine in cat spinal motoneurones. *Brain Res.* **11**, 440–5. [144]

—— and SCHEIBEL, A. B. (1967). Synaptic loci on visual cortical neurons of the rabbit: the specific afferent radiation. *Expl. Neurol.* **18**, 116–31. [173]

GLUSS, B. (1967). A model for neuron firing with exponential decay of potentia resulting in diffusion equations for probability density. *Bull. math. Biophys.* **29**, 233–43. [368]

GOEL, N. S., RICHTER-DYN, N. and CLAY, J. R. (1972). Discrete stochastic models for firing of a neuron. *J. theoret. Biol.* **34**, 155–84. [372]

GOLDMAN, D. E. (1943). Potential, impedance and rectification in membranes. *J. gen. Physiol.* **27**, 37–60. [229, 234, 237]

GOLDMAN, D. E. (1964). A molecular structural basis for the excitation properties of axons. *Biophys. J.* **4**, 167–88. [241]

GOLDMAN, L. (1964). The effects of stretch on cable and spike parameters of single nerve fibres; some implications for the theory of impulse propagation. *J. Physiol.* **175**, 425–44 [293]

GOLDMAN, L. and ALBUS, J. S. (1968). Computation of impulse conduction in myelinated fibres; theoretical basis of the velocity-diameter relation. *Biophys. J.* **8**, 596–607. [298, 299, 301]

—— and SCHAUF, C. L. (1972). Inactivation of the sodium current in *Myxicola* giant axons. Evidence for coupling to the activation process. *J. gen. Physiol.* **59**, 659–75. [241, 333]

—— —— (1973). Quantitative description of sodium and potassium currents and computed action potentials in *Myxicola* giant axons. *J. gen. Physiol.* **61**, 361–84. [241, 333]

GONZALEZ-SERRATOS, H. (1966). Inward spread of contraction during a twitch. *J. Physiol.* **185**, 20–1P. [120]

—— (1971). Inward spread of activation in vertebrate muscle fibres. *J. Physiol.* **212**, 777–99. [120]

GORMAN, A. L. F. and MARMOR, M. F. (1970). Contributions of the sodium pump and ionic gradients to the membrane potential of a molluscan neurone. *J. Physiol.* **210**, 897–917. [357, 358]

GRANIT, R. (1955). *Receptors and Sensory perception*, p. 369. Yale University Press. [348]

——, KERNELL, D., and LAMARRE, Y. (1966). Algebraic summation in synaptic activation of motoneurones firing within the 'primary range' to injected currents. *J. Physiol.* **187**, 379–99. [314]

GRAY, E. G. (1959). Axo-somatic and axo-dendritic synapses of the cerebral cortex: an electron microscope study. *J. Anat.* **93**, 420–33. [131]

—— (1969). Electron microscopy of excitatory and inhibitory synapses: a brief review. *Prog. Brain. Res.* **31**, 141–55. [223]

GREENHILL, A. G. (1892). *The applications of elliptic functions*, p. 375. Macmillan, London. [395]

GRUNDFEST, H. (1932). Excitability of the single fibre nerve-muscle complex. *J. Physiol.* **76**, 95–115. [273, 416]

—— (1966). Comparative electrobiology of excitable membranes. In *Advances in comparative physiology and biochemistry* (Ed. O. E. Loewenstein), pp. 1–116. Academic Press, New York. [229]

GRUNER, L. (1965). The steady-state characteristics of nonuniform RC distributed networks and lossless lines. *IEEE Trans. Commun. Technol.* **12**, 241–7 [149]

GUTTMAN, R. (1971). The effect of temperature on the function of excitable membranes. In *Biophysics and physiology of excitable membranes* (Ed. W. J. Adelman, Jr) Van Nostrand, New York. [349]

—— and BARNHILL, R. (1970). Oscillation and repetitive firing in squid axons. Comparison of experiments with computations. *J. gen. Physiol.* **55**, 104–18. [331, 332, 333, 349]

HAAS, H. G., KERN, R., EINWÄCHTER, H. M., and TARR, M. (1971). Kinetics of Na inactivation in frog atria. *Pflügers Arch. ges. Physiol.* **323**, 141–57. [356]

HAGIWARA, S. and OOMURA, Y. (1958). The critical depolarization for the spike in the squid giant axon. *Japan. J. Physiol.* **8**, 234. [332]

HALL, A. E. and NOBLE, D. (1963). Transient responses of Purkinje fibre to non-uniform currents. *Nature, Lond.* **199**, 1294–5. [266, 270, 385, 386, 411]

HANCOCK, H. (1917). *Elliptic integrals*. Wiley, New York. [395]

HARDING, B. N. (1971). Dendro-dendrite synapses, including reciprocal synapses, in the ventrolateral nucleus of the monkey thalamus. *Brain Res.* **34**, 181–5. [217]

HARDY, G. H. (1916). *The integration of functions of a single variable.* Cambridge tracts in math., Vol. 2 (2nd edn.). Cambridge University Press. [395]

HARMON, L. D. and LEWIS, E. R. (1966). Neural modelling. *Physiol. Rev.* **46**, 513–91. [305]

HARRIS, E. J. and SJODIN, R. A. (1961). Kinetics of exchange and net movement of frog muscle potassium. *J. Physiol.* **155**, 221–45. [10]

HARTLINE, H. K. and GRAHAM, C. H. (1932). Nerve impulses from single receptors in the eye. *J. cell. comp. Physiol.* **1**, 277–95. [347]

HAUSWIRTH, O., MCALLISTER, E. NOBLE, D., and TSIEN, R. W. (1969). Reconstruction of the actions of adrenaline and calcium on cardiac pacemaker potentials. *J. Physiol.* **204**, 126–8P. [351]

——, NOBLE D., and TSIEN, R. W. (1968). Adrenaline: mechanism of action on the pacemaker potential in cardiac Purkinje fibres. *Science, N.Y.* **162**, 916–17. [192, 329, 351, 352]

——, ——, —— (1969). The mechanism of oscillatory activity at low membrane potentials in cardiac Purkinje fibres. *J. Physiol.* **200**, 255–65. [325]

HAYDON, D. A. and HLADKY, S. B. (1972). Ion transport across thin lipid membranes: a critical discussion of mechanisms in selected systems. *Q. Rev. Biophys.* **5**, 187–282. [229, 238]

HECHT, H. H., HUTTER, O. F., and LYWOOD, D. (1964). Voltage–current relation of short Purkinje fibres in sodium-deficient solution. *J. Physiol.* **170**, 5P. [75]

HECKMANN, K. (1972). Single-file diffusion. *Biomembranes*, **3**, 127–53. [240]

HEISTRACHER, P. and HUNT, C. C. (1969). The relation of membrane changes to contraction in twitch muscle fibres. *J. Physiol.* **201**, 589–611. [121]

HELLERSTEIN, D. (1968). Passive membrane potentials. A generalization of the theory of electrotonus. *Biophys. J.* **8**, 358–79. [132]

—— (1969). Cable theory and gross potential analysis. *Science N.Y.* **166**, 638–9. [217]

HENČEK, M. and ZACHAR, J. (1965). The electrical constants of single muscle fibres of the crayfish. *Physiologia bohemoslav.* **14**, 297–311. [10, 37]

HERMANN, L. (1879). *Handbuch der Physiologie.* Vogel, Liepzig. [1]

—— (1899). Zur Theorie der Erregungsleitung und der elektrischen Erregung. *Pflügers Arch. ges. Physiol.* **75**, 574. [1]

—— (1905). Beiträge zur Physiologie und Physik des Nerven. *Pflügers Arch. ges. Physiol.* **109**, 95. [1]

HILDEBRAND, F. B. (1962). *Advanced calculus for applications.* Prentice-Hall, Englewood Cliffs, New Jersey. [446]

HILL, A. V. (1936). Excitation and accommodation in nerve. *Proc. R. Soc.* **B119**, 305–55. [272, 344]

—— (1949). The onset of contraction. *Proc. R. Soc.* **B136**, 242–54. [98, 128]

HILLE, B. (1967). The selective inhibition of delayed potassium currents in nerve by tetraethylammonium ion. *J. gen. Physiol.* **50**, 1287–302. [343]

—— (1968). Charges and potentials at the nerve surface: divalent ions and pH. *J. gen. Physiol.* **51**, 221–36. [259, 349]

—— (1970). Ionic channels in nerve membranes. *Prog. Biophys.* **21**, 1–32. [23, 232]

—— (1971). The permeability of the sodium channel to organic cations in myelinated nerve. *J. Gen. Physiol.* **58**, 599–619. [240, 259]

—— (1972). The permeability of the sodium channel to metal cations in myelinated nerve. *J. gen. Physiol.* **59**, 637–658. [240]

—— (1973). Potassium channels in myelinated nerve. Selective permeability to small cations, *J. gen. Physiol.* **61**, 669–686. [240, 259]

HILLE, B. (1974). Ionic selectivity of Na and K channels of nerve membranes. In *Membranes: a series of advances* (Ed. G. Eisenman), Vol. 3. Marcel Dekker, New York. [259]

HÖBER, R. (1910). Eine Methode, die elektrische Leitfahigheit im Innern von Zellen zu messen. *Pflügers Arch. ges. Physiol.* **133**, 237–59. [11]

—— (1912). Ein zweites Verfahren die Leitfahigkeit im Innern von Zellen zu messen. *Pflügers Arch. ges. Physiol.* **148**, 189–221. [11].

HOCHMAN, H. G. and FIENBERG, S. E. (1971). Some renewal process models for single neuron discharge. *J. appl. Prob.* **8**, 802–8. [372]

HODES, R. (1953). Linear relationship between fiber diameter and velocity of conduction in giant axon of squid *J. Neurophysiol.* **16**, 145–54 [292]

HODGKIN, A. L. (1937). Evidence for electrical transmission in nerve. *J. Physiol.* **90**, 183–210 and 211–32. [1, 278, 293]

—— (1939). The relation between conduction velocity and the electrical resistance outside a nerve fibre. *J. Physiol.* **94**, 560–70. [26, 293]

—— (1948). The local electric changes associated with repetitive action in a non-medullated axon. *J. Physiol.* **107**, 165–81. [305, 333, 334, 335, 344, 357]

—— (1964). *The conduction of the nervous impulse.* Liverpool University Press. [252, 298, 335]

—— (1975). The optimum density of sodium channels in an unmyelinated nerve. *Phil. Trans. R. Soc. B.* **270**, 297–300. [432, 433]

—— and HOROWICZ, P. (1959). Movements of Na and K in single muscle fibres. *J. Physiol.* **145**, 405–32. [10]

—— —— (1960). Potassium contractures in single muscle fibres. *J. Physiol.* **153**, 386–403. [98]

—— and HUXLEY, A. F. (1947). Potassium leakage from an active nerve fibre. *J. Physiol.* **106**, 341–67. [340, 357]

—— —— (1952a). Currents carried by sodium and potassium ions through the membrane of the giant axon of *Loligo. J. Physiol.* **116**, 449–72. [2, 21, 241]

—— —— (1952b). The components of membrane conductance in the giant axon of *Loligo. J. Physiol.* **116**, 473–96. [2, 21, 241]

—— —— (1952c). The dual effect of membrane potential on sodium conductance in the giant axon of *Loligo. J. Physiol.* **116**, 497–506. [2, 21, 241]

—— —— (1952d). A quantitative description of membrane current and its application to conduction and excitation in nerve. *J. Physiol.* **117**, 500–44. [2, 21, 117, 241, 245, 252, 278, 335]

——, ——, and KATZ, B. (1952). Measurement of current–voltage relations in the membrane of the giant axon of *Loligo. J. Physiol.* **116**, 424–48. [12, 241, 416, 417]

—— and KATZ, B. (1949a). The effect of sodium ions on the electrical activity of the giant axon of the squid. *J. Physiol.* **108**, 37–77. [234, 237, 287]

—— —— (1949b). The effect of temperature on the electrical activity of the giant axon of the squid. *J. Physiol.* **109**, 240–9. [289, 291]

—— and KEYNES, R. D. (1955). The potassium permeability of a giant nerve fibre. *J. Physiol.* **128**, 61–88. [10, 240]

—— —— (1956). Experiments on the injection of substances into squid giant axons by means of a microsyringe. *J. Physiol.* **131**, 592–616. [361,364, 366]

—— —— (1953). The mobility and diffusion coefficient of potassium in giant axons from *Squid. J. Physiol.* **119**, 513–28. [9]

—— —— (1957). Movements of labelled calcium in squid giant axons. *J. Physiol.* **138**, 253–81. [9]

—— and NAKAJIMA, S. (1972a). The effect of diameter on the electrical constants of frog skeletal muscle fibres. *J. Physiol.* **221**, 105–20. [101, 103, 122]

HODGKIN, A. L. and NAKAJIMA, S. (1972b). Analysis of the membrane capacity in frog muscle. *J. Physiol.* **221**, 121–36. [103, 114, 115, 117, 118]

—— and RUSHTON, W. A. H. (1946). The electrical constants of a crustacean nerve fibre. *Proc. R. Soc.* **B133**, 444–79. [2, 9, 26, 28, 30, 36, 27]

HOFFMAN, B. and CRANEFIELD, P. F. (1958). *Electrophysiology of the heart*. McGraw-Hill, New York. [411]

HOWELL, J. N. (1969). A lesion of the transverse tubules of skeletal muscle. *J. Physiol.* **201**, 515–33. [119]

—— and JENDEN, D. J. (1967). T-tubules of skeletal muscle; morphological alterations which interrupt excitation-contraction coupling. *Fedn Proc.* **26**, 553. [100, 119]

HOYT, R. C. (1963). The squid giant axon. Mathematical models. *Biophys. J.* **3**, 399–431. [241]

—— (1968). Sodium inactivation in nerve fibres. *Biophys. J.* **8**, 1074–97. [241]

HUBBARD, J. I., LLINAS, R., and QUASTEL, D. M. J. (1969). *Electrophysiological analysis of synaptic transmission*. Arnold, London. [66]

HUNTER, P. J., McNAUGHTON, P. A., and NOBLE, D. (1975). Analytical models of propagation in excitable cells. *Prog. Biophys.* **30**, 99–144. [422, 428, 429, 431, 432]

HURSH, J. B. (1939). Conduction velocity and diameter of nerve fibres. *Am. J. Physiol.* **127**, 131–9. [299]

HUTCHINSON, N. A., KOLES, Z. J., and SMITH, R. S. (1970). Conduction velocity in myelinated nerve fibres of *Xenopus laevis*. *J. Physiol.* **208**, 279–89. [299]

HUTTER, O. F. and NOBLE, D. (1960). The chloride conductance of frog skeletal muscle. *J. Physiol.* **151**, 89–102. [237]

—— and WARNER, A. E. (1969). Rectifier properties of the chloride conductance of skeletal muscle at different pH. *J. Physiol.* **200**, 82–3P. [237]

HUXLEY, A. F. (1959a). Ion movements during nerve activity. *Ann. N.Y. Acad. Sci.* **81**, 221–46. [252, 284, 290, 291, 293, 349, 350, 432]

—— (1959b). Can a nerve propagate a subthreshold disturbance? *J. Physiol.* **148**, 80–1P. [290, 302]

—— and STÄMPFLI, R. (1949). Evidence for saltatory conduction in peripheral myelinated nerve fibres. *J. Physiol.* **108**, 315–39. [296, 299]

—— and STRAUB, R. W. (1958). Local activation and interfibrillar structures in striated muscle. *J. Physiol.* **143**, 40–1P. [119]

—— and TAYLOR, R. E. (1958). Local activation of striated muscle fibres. *J. Physiol.* **144**, 426–41. [19, 97, 119]

IANSEK, R. and REDMAN, S. J. (1973). An analysis of the cable properties of spinal motoneurones using a brief intracellular current pulse. *J. Physiol.* **234**, 613–36. [172]

ILDEFONSE, M. and ROUGIER, O. (1972). Voltage-clamp analysis of the early current in frog skeletal muscle fibre using the double sucrose-gap method. *J. Physiol.* **222**, 373–95. [98, 246, 416, 417]

ITO, M. and OSHIMA, T. (1962). Temporal summation of after-hyperpolarization following a motoneurone spike. *Nature, Lond.* **195**, 910–11. [355]

—— —— (1965). Electrical behaviour of the motoneurone membrane during intracellularly applied current steps. *J. Physiol.* **180**, 607–35. [170]

JACK, J. J. B., MILLER, S., PORTER, R., and REDMAN, S. J. (1971). The time course of minimal excitatory post-synaptic potentials evoked in spinal motoneurones by group Ia afferent fibres. *J. Physiol.* **215**, 353–80. [164, 188]

—— and REDMAN, S. J. (1971a). The propagation of transient potentials in some linear cable structures. *J. Physiol.* **215**, 283–320. [47, 70, 170]

—— —— (1971b). An electrical description of the motoneurone, and its application to the analysis of synaptic potentials. *J. Physiol.* **215,** 321–52. [161, 165, 167, 170, 171, 172, 173, 175, 176, 177, 185, 452]

JAEGER, J. C. (1951). *An introduction to applied mathematics.* Clarendon Press, Oxford. [2, 438, 441, 447]

—— (1966). *Introduction to the Laplace transformation.* Methuen, London. [441]

JAHNKE, E. and EMBDE, F. (1945). *Tables of functions with formulas and curves* (4th edn) Dover, New York. [395, 459]

JANSEN, J. K. S., NICOLAYSON, K., and RUDJORD, T. (1966). Discharge pattern of neurons of the dorsal spinocerebellar tract activated by static extension of primary endings of muscle spindles. *J. Neurophysiol.* **29,** 1061–86. [376]

JOHANNESMA, P. I. M. (1968). Diffusion models for the stochastic activity of neurons. In *Neural networks* (Ed. E. R. Caianello), pp. 116–44. Springer-Verlag, Berlin. [368]

—— (1969). Stochastic neural activity. A theoretical investigation. Ph.D. Thesis. Catholic University, Nijmegen, The Netherlands. [367, 368, 369]

JOHNSON, E. A. and LIEBERMAN, M. (1971). Heart: excitation and contraction. *A. Rev. Physiol.* **33,** 479–532. [406]

—— and SOMMER, J. R. (1967). A strand of cardiac muscle: its ultrastructure and the electrophysiological implications of its geometry. *J. Cell. Biol* **33,** 103–29. [10]

JONES, D. S. (1961). *Electrical and mechanical oscillations.* Routledge and Kegan Paul, London. [306, 316, 320, 322, 327]

JONES, E. G. and POWELL, T. P. S. (1969). Morphological variations in the dendritic spines of the neocortex. *J. Cell. Sci.* **5,** 509–19 [131, 218, 219]

JONES, A. W. and TOMITA, T. (1967). The longitudinal tissue resistance of the guinea-pig taenia cell. *J. Physiol.* **191,** 109–10P. [24, 112, 117]

JULIAN, F. J., MOORE, J. W., and GOLDMAN, D. E. (1962). Current–voltage relations in the lobster giant axon membrane under voltage-clamp conditions. *J. gen. Physiol.* **45,** 1217– 38. [246]

KAO, C. Y. and STANFIELD, P. R. (1968). Action of some anions on electrical properties and mechanical threshold of frog twitch muscle. *J. Physiol.* **198,** 291–309. [120, 121]

KATZ, B. (1936). Multiple response to constant current in frog's medullated nerve. *J. Physiol.* **88,** 239–55. [344]

—— (1939). *Electric excitation of nerve.* Oxford University Press. [1, 416]

—— (1948). Electrical properties of the muscle fibre membrane. *Proc. R. Soc.* **B135,** 506–34. [10]

—— (1950). Depolarization of sensory terminals and the initiation of impulses in the muscle spindle. *J. Physiol.* **111,** 261–82. [347, 348]

—— (1966). *Nerve, muscle and synapse.* McGraw-Hill, New York. [3]

—— and MILEDI, R. (1965). The measurement of synaptic delay, and the time course of acetylcholine release at the neuromuscular junction. *Proc. R. Soc.* **B161,** 483–95. [188]

—— and MILEDI, R. (1967). A study of synaptic transmission in the absence of nerve impulses. *J. Physiol.* **192,** 406–36. [218]

KAVALER, F. (1959). Membrane depolarization as a cause of tension development in mammalian ventricular muscle. *Am. J. Physiol.* **197,** 968–970. [10]

KELLERTH, J-O. (1968). Aspects on the relative significance of pre- and postsynaptic inhibition in the spinal cord. In *Structure and function of inhibitory neuronal mechanisms* (Eds. C. von Euler, S. Skoglund, and S. Söderberg, pp. 197–212. Pergamon Press, Oxford. [183]

KELVIN, LORD (WILLIAM THOMPSON) (1855). On the theory of the electric telegraph. *Proc. R. Soc.* **7,** 382–99. [1]

—— (1856). On the theory of the electric telegraph. *Phil. Mag.* **11** (4), 146–60. [1]

KELVIN, LORD (WILLIAM THOMPSON) (1872). *Papers on electrostatics and magnetism.* Macmillan, London. [1]

KERKUT, G. A. and THOMAS, R. C. (1965). An electrogenic sodium pump in snail nerve cells. *Comp. Biochem. Physiol.* 14, 167–83. [358]

KERNELL, D. (1965a). Adaptation and the relation between discharge frequency and current strength of cat lumbosacral motoneurones stimulated by long-lasting injected currents. *Acta physiol. scand.* 65, 65–73. [305, 352]

—— (1965b). High-frequency repetitive firing of cat lumbosacral motoneurones stimulated by long-lasting injected currents. *Acta physiol. scand.* 65, 74–86. [305]

—— (1965c). The limits of firing frequency in cat lumbosacral motoneurones possessing different time course of afterhyperpolarization. *Acta physiol. scand.* 65, 87–100. [305]

—— (1968). The repetitive impulse discharge of a simple neurone model compared to that of spinal motoneurones. *Brain Res.* 11, 685–7. [311, 312, 355]

—— (1969). Synaptic conductance changes and the repetitive impulse discharge of spinal motoneurones. *Brain Res.* 15, 291–4. [311, 312, 314]

—— (1970). Cell properties of importance for the transfer of signals in nervous pathways. In *Excitatory synaptic mechanisms* (Eds. P. Andersen and J. K. S. Jansen), pp. 269–73. Universitets forlaget, Oslo. [311, 312]

—— (1971). Effects of synapses on dendrites and soma on the repetitive impulse firing of a compartmental neuron model. *Brain Res.* 35, 551–5. [311, 314]

—— (1972). The early phase of adaptation in repetitive impulse discharges of cat spinal motoneurones. *Brain Res.* 41, 184–6. [355]

—— and SJÖHOLM, H. (1972). Motoneurone models based on 'voltage clamp equations' for peripheral nerve. *Acta physiol. scand.* 86, 546–62. [354, 355]

—— —— (1973). Repetitive impulse firing: comparisons between neurone models based on 'voltage clamp equations' and spinal motoneurones. *Acta physiol. scand.* 87, 40–56. [342, 354, 355]

KEYNES, R. D., RITCHIE, J. M., and ROJAS, E. (1971). The binding of tetrodotoxin to nerve membranes. *J. Physiol.* 213, 235–54. [23]

KHODOROV, B. I. and TIMIN, E. N. (1975). Nerve impulse propagation along non-uniform fibres. *Prog. Biophys.* 30, 145–84. [422]

KING, R. W. P. (1965). *Transmission-line theory.* Dover, New York. [106]

KNIGHT, B. W. (1972a). Dynamics of encoding in a population of neurons. *J. gen. Physiol.* 58, 734–66. [367, 368]

—— (1972b). The relationship between the firing rate of a single neuron and the level of activity in a population of neurones. Experimental evidence for resonant enhancement in the population response. *J. gen. Physiol.* 59, 767–78. [368]

KOIKE, H., MANO, N., OKADA, Y., and OSHIMA, T. (1970). Repetitive impulses generated in fast and slow pyramidal tract cells by intracellularly applied current steps. *Expl. Brain Res.* 11, 263–81. [314]

KOSTYUK, P. G., KRISHTAL, D. A., and PIDOPLICHKO, V. I. (1972). Potential-dependent membrane current during the active transport of ions in snail neurones. *J. Physiol.* 226, 373–92. [363]

KUNO, M. and LLINAS, R. (1970a). Enhancement of synaptic transmission by dendrite potentials in chromatolysed motoneurones of the cat. *J. Physiol.* 210, 807–21. [217]

—— —— (1970b). Alterations of synaptic action in chromatolysed motoneurones of the cat. *J. Physiol.* 210, 823–38. [217]

—— and MIYAHARA, J. T. (1968). Factors responsible for multiple discharge of neurons in Clarke's column. *J. Neurophysiol.* 31, 624–48. [315]

—— —— (1969). Non-linear summation of unit synaptic potentials in spinal motoneurones of the cat. *J. Physiol.* 201, 465–477. [193]

KUSANO, K. (1970). Influence of ionic environment on the relationship between pre- and postsynaptic potentials. *J. Neurobiol.* **1**, 435–57. [218]

KUSHMERICK, M. J. and PODOLSKY, R. J. (1969). Ionic mobility in muscle cells. *Science N.Y.* **166**, 1297–8. [9, 10]

LAPICQUE, L. (1907). Recherches quantitatifs sur l'excitation electrique des nerfs traitée comme une polarisation. *J. Physiol. Paris* **9**, 622–35. [11]

LÄUGER, P. (1973). Ion transport through pores: a rate theory analysis. *Biochem. Biophys. Acta* **311**, 423–41 [240]

LEV, A. A. (1964). Determination of activity and activity coefficients of potassium and sodium ions in frog muscle fibres. *Nature, Lond.* **201**, 1132–4. [9]

LEVINE, Y. K. (1972). Physical studies of membrane structure. *Prog. Biophys.* **24**, 1–74. [12]

LING, G. (1962). *A physical theory of the living state: the association-induction hypothesis*. Blaisdell, New York. [9]

LIPPOLD, O. C. J., NICHOLLS, J. G., and REDFEARN, J. W. T. (1960). Electrical and mechanical factors in the adaptation of a mammalian muscle spindle. *J. Physiol.* **153**, 209–17. [352]

LLINAS, R. (1964). Mechanisms of supraspinal actions upon spinal cord activities. Differences between reticular and cerebellar inhibitory actions upon alpha extensor motoneurons. *J. Neurophysiol.* **27**, 1117–26. [218]

LLINAS, R. and NICHOLSON, C. (1971). Electrophysiological properties of dendrites and somata in alligator Purkinje cells. *J. Neurophysiol.* **34**, 532–51. [302]

——, NICHOLSON, C., FREEMAN, J. A., and HILLMAN, D. E. (1968). Dendritic spikes and their inhibition in alligator Purkinje cells. *Science N.Y.* **160**, 1132–5. [217]

——, ——, ——, —— (1969). Reply to technical comment by W. H. Calvin and D. Hellerstein, 'Dendritic spikes vs. cable properties'. *Science N.Y.* **168**, 96–7. [162]

—— ——, and PRECHT, W. (1969). Preferred centripetal conduction of dendritic spikes in alligator Purkinje cells, *Science. N.Y.* **168**, 184–7. [217]

—— and TERZUOLO, C. A. (1965). Mechanisms of supraspinal actions upon spinal cord activities. Reticular inhibitory mechanisms upon flexor motoneurons. *J. Neurophysiol.* **28**, 413–22. [183]

LOEWENSTEIN, W. R. and MENDELSON, M. (1965). Components of receptor adaptation in a Pacinian corpuscle. *J. Physiol.* **177**, 377–97. [352]

LORENTE DE NÓ, R. (1938). Synaptic stimulation of motoneurons as a local process. *J. Neurophysiol.* **1**, 195–206. [134]

—— (1947). *A study of nerve physiology*. Rockefeller Institute, New York. [2, 3, 293]

—— and CONDOURIS, C. A. (1959). Decremental conduction in peripheral nerve. Integration of stimuli in the neuron. *Proc. natn. Acad. Sci. U.S.A.* **45**, 592–617. [216, 302]

LUIKOV, A. V. (1968). *Analytical heat diffusion theory*. New York. [2]

LUND, R. D. (1969). Synaptic patterns of the superficial layers of the superior colliculus. *J. comp. Neurol.* **135**, 179–208. [217]

LUSSIER, J. J. and RUSHTON, W. A. H. (1952). The excitability of a single fibre in a nerve trunk. *J. Physiol.* **117**, 87–108. [300]

LÜTTGAU, H. (1961). Weitere Untersuchungen über den passiven Ionentransport durch die erregbare Membran des Ranvierknotens. *Pflügers Arch. ges. Physiol.* **273**, 302–10. [259]

LUX, H. D. (1967). Eigenschaften eines neuron-modells mit dendriten begrenzter länge. *Pflügers Arch. ges. Physiol.* **297**, 238–55. [70, 107, 170, 171]

—— and POLLEN, D. A. (1966). Electrical constants of neurons in the motor cortex of the cat. *J. Neurophysiol.* **29**, 207–20. [169, 171]

—— SCHUBERT, P., and KREUTZBERG, G. W. (1970). Direct matching of morphological and electrophysiological data in cat spinal motoneurons. In *Excitatory synaptic mechanisms* (Eds. P. Andersen and J. K. S. Jansen), pp. 189–98. Universitets forlaget, Oslo. [170]

MACGILLIVRAY, A. D. and HARE, D. (1969). Applicability of Goldman's constant field assumption to biological systems. *J. theor. Biol.* **26**, 113–26. [234, 236]

MACGREGOR, R. J. (1968). A model for responses to activation by axodendritic synapses. *Biophys. J.* **8**, 305–18. [193, 194]

MACNICHOL, E. F., JR. (1958). Subthreshold excitatory processes in the eye of *Limulus*. *Exptl. Cell. Res.* 10, Suppl. 5, 411. [348]

MARMONT, G. (1949). Studies on the axon membrane. 1. A new method. *J. cell. comp. Physiol.* **34**, 351–82. [7, 241]

MARTIN, A. R. (1955). A further study of the statistical composition of the end-plate potential. *J. Physiol.* **130**, 114–22. [65]

—— (1966). Quantal nature of synaptic transmission. *Physiol. Rev.* **46**, 51–66. [65]

MASSEY, H. S. W. and KESTELMAN, H. (1959). *Ancillary Mathematics*, London: Pitman [464]

MATTHEWS, B. H. C. (1931). The response of a single end organ. *J. Physiol.* **71**, 64–110. [347]

MATTHEWS, P. B. C. and STEIN, R. B. (1969). The regularity of primary and secondary muscle spindle afferent discharges. *J. Physiol.* **202**, 59–82. [377]

MAURO, A., CONTI, F., DODGE, F. A., and SCHOR, R. (1970). Subthreshold behaviour and phenomenonological impedance of the squid giant axon. *J. gen. Physiol.* **55**, 497–523. [23, 324]

MCALLISTER, R. E., NOBLE, D., and TSIEN, R. W. (1975). Reconstruction of the electrical activity of cardiac Purkinje fibres. *J. Physiol.* **251**, 1–59. [249, 252, 328]

MCCALLISTER, L. P. and HADEK, R. (1970). Transmission electron microscopy and stereo ultrastructure of the T system in frog skeletal muscle. *J. ultrastruct. Res.* **33**, 360–8. [103]

MEVES, H. (1961). Die Nachpotentiale isolierter mackhaltiger Nervenfasern des Frosches bei tetanischer Reizung. *Pflügers. Arch.* **272**, 336–59. [356]

MILHORN, H. T. (1964). *The Application of control theory to physiological systems.* Saunders, Philadelphia. [441]

MINORSKY, N. (1947). *Introduction to non-linear mechanics.* J. W. Edwards, Ann Arbor. [306]

MOBLEY, B. A. and PAGE, E. (1972). The surface area of sheep cardiac Purkinje fibres. *J. Physiol.* **220**, 547–63. [21, 102, 416]

MONNIER, A. M. (1934). *L'excitation électrique des tissus.* Hermon et Cie, Paris. [1]

MOORE, G. P., PERKEL, D. H., and SEGUNDO, J. P. (1966). Statistical analysis and functional interpretation of neuronal spike data. *A. Rev. Physiol.* **28**, 493–522. [373]

MOORE, J. W. (1958). Temperature and drug effects on squid axon membrane ion conductance. *Fedn. Proc.* **17**, 113. [241]

—— (1963). Operational amplifiers. In *Physical techniques in biological research* (Ed. W. L. Nastuk). p. 77–97. Academic Press, New York and London. [6]

——, NARAHASHI, T., and SHAW, T. I. (1967). An upper limit to the number of sodium channels in nerve membrane? *J. Physiol.* **188**, 99–105. [23]

MÜLLER-MOHNSSEN, H. and BALK, O. (1960). Stationäre elektrische eigenschaften des Ranvierschen Schnürrings während der Rb-depolarisation. *Pflügers Arch. ges. Physiol.* **289**, R8. [259]

MULLINS, L. J. (1968). A single channel or a dual channel mechanism for nerve excitation. *J. gen. Physiol.* **52**, 550–3. [241]

MULLINS, L. J. and BRINLEY, F. J. (1969). Potassium fluxes in dialyzed squid axons. *J. gen. Physiol.* **53**, 704–40. [361]

NAKAJIMA, S. and HODGKIN, A. L. (1970). Effect of diameter on the electrical constants of frog skeletal muscle fibres. *Nature, Lond.* **227**, 1053–5. [103]

—— and KUSANO, K. (1966). Behaviour of delayed current under voltage clamp in the supramedullary neurons of Puffer. *J. gen. Physiol.* **49**, 613–28. [356]

——, NAKAJIMA, Y. and PEACHEY, L. D. (1968). Speed of repolarization and morphology of glycerol-treated muscle fibres. *J. Physiol.* **200**, 115–6P. [119]

—— and ONODERA, K. (1969a). Membrane properties of the stretch receptor neurones of crayfish with particular reference to mechanisms of sensory adaptation. *J. Physiol.* **200**, 161–85. [340, 352, 356, 358, 360]

—— —— (1969b). Adaptation of the generator potential in the crayfish stretch receptors under constant length and constant tension. *J. Physiol.* **200**, 187–204. [352]

—— and TAKAHASHI, K. (1966). Post-tetanic hyperpolarization and electrogenic Na pump in stretch receptor neurone of crayfish. *J. Physiol.* **187**, 105–27. [358]

NAKAMURA, Y., NAKAJIMA, S., and GRUNDFEST, H. (1965). Analysis of spike electrogenesis and depolarizing K inactivation in electroplaques of *Electrophorus electricus*. *J. gen. Physiol.* **49**, 321–49. [356]

NASONOV, D. N. and ROSENTAL, D. L. (1952). *Usp. Sovrem. Biol.* **34**, 161. [286]

NEHER, E. (1971). Two fast transient current components during voltage clamp on snail neurons. *J. gen. Physiol.* **58**, 36–53. [344]

—— and LUX, H. D. (1971). Properties of somatic membrane patches of snail neurons under voltage clamp. *Pflügers Arch. ges. Physiol.* **322**, 35–8. [344]

NELSON, P. G. and LUX, H. D. (1970). Some electrical measurements of motoneuron parameters. *Biophys. J.* **10**, 55–73. [168, 169, 170, 171]

NERNST, W. (1889). Zur Kinetik der in Losung befindlichen Körper: Theorie der Diffusion. *Z. phys. Chem.* **2**, 613–37. [11]

NICHOLSON, C. and LLINAS, R. (1971). Field potentials in the alligator cerebellum and theory of their relationship to Purkinje cell dendritic spikes. *J. Neurophysiol.* **34**, 509–31. [3, 217]

NOBLE, D. (1962a). A modification of the Hodgkin–Huxley equations applicable to Purkinje fibre action and pacemaker potentials. *J. Physiol.* **160**, 317–52. [250, 252]

—— (1962b). The voltage dependence of the cardiac membrane conductance. *Biophys. J.* **2**, 381–93. [91, 96, 120, 385, 386]

—— (1965). Electrical properties of cardiac muscle attributable to inward going (anomalous) rectification. *J. cell. comp. Physiol.* **66**, Suppl., 127–35. [357]

—— (1966). Applications of Hodgkin–Huxley equations to excitable tissues. *Physiol. Rev.* **46**, 1–50. [247, 248, 252, 264, 265, 282]

—— (1972). The relation of Rushton's liminal length for excitation to the resting and active conductances of excitable cells. *J. Physiol.* **226**, 573–91. [416]

—— and HALL, A. E. (1963). The conditions for initiating 'all-or-nothing' repolarization in cardiac muscle: *Biophys. J.* **3**, 261–74. [266, 385]

—— and STEIN, R. B. (1966). The threshold conditions for initiation of action potentials by excitable cells. *J. Physiol.* **187**, 129–62. [46, 252, 263, 266, 267, 270, 271, 272, 273, 274, 275, 278, 331, 418]

—— and TSIEN, R. W. (1968). The kinetics and rectifier properties of the slow potassium current in cardiac Purkinje fibres. *J. Physiol.* **195**, 185–214. [238, 249, 317, 328, 351]

—— —— (1969a). Outward membrane currents activated in the plateau range of potentials in cardiac Purkinje fibres. *J. Physiol.* **200**, 205–31. [249, 317, 319, 399]

—— —— (1969b). Reconstruction of the repolarization process in cardiac Purkinje fibres based on voltage clamp measurements of the membrane current. *J. Physiol.* **200**, 233–54. [250]

—— —— (1972). The repolarization process in heart cells. In *Electrical phenomena in the heart* (Ed. W. C. de Mello), pp. 133–61. Academic Press, New York. [250, 269, 428]

NOLDUS, E. (1973). A perturbation method for the analysis of impulse propagation in a mathematical neuron model. *J. theor. Biol.* **38**, 383–95. [430]

NORMAN, R. S. (1972). Cable theory for finite length dendritic cylinders with initial and boundary conditions. *Biophys. J.* **12**, 25–45. [167]

OFFNER, F., WEINBERG, A. and YOUNG, C. (1940). Nerve conduction theory: some mathematical consequences of Bernstein's model. *Bull. math. Biophys.* **2**, 89–103. [2, 284]

OSAKI, S. and VASUDERAN, R. (1972). On a model of neuronal spike trains. *Math. Biosci.* **14**, 337–41 [372]

PAGE, E., POWER, B., FOZZARD, H. A., and MEDOFF, D. A. (1969). Sarcolemmal evaginations with knob-like or stalked projections in Purkinje fibres of the sheep's heart. *J. ultrastruct. Res.* **28**, 288–300.

PAGE, S. (1965). A comparison of the fine structure of frog slow and twitch muscle fibers. *J. Cell. Biol.* **26**, 477–97. [123]

PAINTAL, A. S. (1966). The influence of diameter of medullated nerve fibres of cats on the rising and falling phases of the spike and its recovery. *J. Physiol.* **184**, 791–811. [301]

—— (1967). A comparison of the nerve impulses of mammalian non-medullated nerve fibres with those of the smallest diameter medullated fibres. *J. Physiol.* **193**, 523–33. [295]

PALTI, Y. and ADELMAN, W. J., JR. (1969). Measurement of axonal membrane conductances and capacity by means of a varying potential control voltage clamp. *J. Membrane Biol.* **1**, 431–58. [18, 317]

PASSOW, H. (1969). Passive ion permeability of the erythrocyte membrane. *Prog. Biophys.* **19**, 425–67. [236]

PEACHEY, L. D. (1965). The sarcoplasmic reticulum and transverse tubules of the frog's sartorius. *J. Cell Biol.* **25**, 209–31. [99, 101, 123, 128]

—— (1968). Muscle. *A. Rev. Physiol.* **30**, 401–40. [101, 102]

—— and ADRIAN, R. H. (1973). Electrical properties of the transverse tubular system . In *Structure and function of muscle* (ed. G. Bourne), Vol. 3. Academic Press, New York. [100, 101, 103, 115, 123]

PEARSON, K. (1934). *Tables of the complete and incomplete elliptic integrals*, reissued from Tome II of Legendre's *Traité des fonctions elliptiques*. London C.U.P. [395, 433]

PEARSON, K. G., STEIN, R. B., and MALHOTRA, S. K. (1970). Properties of action potentials from insect motor nerve fibres. *J. exp. Biol.* **53**, 299–316 [292, 293, 295]

PEDERSEN, C. J. (1970). Crystalline salt complexes of macrocyclic polyethers. *J. Am. chem. Soc.* **92**, 386. [260]

PERKEL, D. H. (1964). A digital-computer model of nerve cell functioning. The Rand Corporation, Memorandum RM-4132-NIH. [374]

—— and BULLOCK, T. H. (1968). Neural coding. *Neurosci. Res. Program Bull.* **6**, 221–348. [378]

——, SCHULMAN, J. H., BULLOCK, T. H., MOORE, G. P., and SEGUNDO, J. P. (1964). Pacemaker neurons: effects of regularly spaced synaptic input. *Science N.Y.* **145**, 61–3. [378]

PESKOFF, A. and EISENBERG, R. S. (1973). Interpretation of some microelectrode measurements of electrical properties of cells. *Ann. Rev. Biophys: Bioeng.* **2**, 65–79 [133]

PETERS, A. and KAISERMAN-ABRAMOF, I. R. (1970). The small pyramidal neuron of the rat cerebral cortex. The perikaryon, dendrites and spines. *Am. J. Anat.* **127**, 321–56. [131]

PHILLIPS, C. S. G. and WILLIAMS, R. J. P. (1965). *Inorganic chemistry.* Clarendon Press, Oxford. [255, 256]

PICKARD, W. F. (1966). On the propagation of the nervous impulse down medullated and unmedullated fibres. *J. theor. Biol.* **11**, 30–45. [296]

—— (1969). Estimating the velocity of propagation along myelinated and un-myelinated fibres. *Math. Biosci.* **5**, 305–19. [287]

—— (1971*a*). Electrotonus on a cell of finite dimensions. *Math. Biosci.* **10**, 201–13. [132]

—— (1971*b*). The spatial variation of plasmalemma potential in a spherical cell polarized by a small current source. *Math. Biosci.* **10**, 307–28. [132]

——, LETTVIN, J. Y., MOORE, J. W., TAKATA, M., POOLER, J., and BERNSYEIN, T. (1964). Calcium ions do not pass the membrane of the giant axon. *Proc. natn Acad. Sci. U.S.A.* **52**, 1173–83. [259]

PIPES, L. A. and HARVILL, L. R. (1970). *Applied mathematics for engineers and physicists* (3rd edn.). McGraw-Hill, New York. [438, 442, 462, 465, 466]

PLONSEY, R. (1964). Volume conductor fields of action currents. *Biophys. J.* **4**, 317–28. [3]

PORTER, R. and MUIR, R. B. (1971). The meaning for motoneurones of the temporal pattern of natural activity in pyramidal tract neurones of conscious monkeys. *Brain Res.* **34**, 127–42. [378]

PRICE, J. L. and POWELL, T. P. S. (1970*a*). The morphology of the granule cells of the olfactory bulb. *J. Cell Sci.* **7**, 91–123. [223]

—— —— (1970*b*). The synaptology of the granule cells of the olfactory bulb. *J. Cell Sci.* **1**, 125–55. [223]

PUGSLEY, I. D. (1966). Contribution of the transverse tubular system to the membrane capacitance of striated muscle of the toad (*Bufo marinus*). *Aust. J. exp. Biol. med. Sci.* **44**, 9–22. [103, 113]

PUMPHREY, R. J. and YOUNG, J. Z. (1938). The rates of conduction of nerve fibres of various diameters in cephalopods. *J. exp. Biol.* **15**, 453. [292]

PURPURA, D. (1967). Comparative physiology of dendrites. In *The Neurosciences: a study program* (eds. G. C. Quarton, T. Melnechuk, and F. O. Schmitt), pp. 372–93. The Rockefeller University Press. [217, 218]

RALL, W. (1953). Electrotonic theory for a spherical neurone. *Proc. Univ. Otago med. School* **31**, 14–15. [132, 134]

—— (1955). A statistical theory of monosynaptic input-output relations. *J. cell. comp. Physiol.* **46**, 373–411. [134]

—— (1959*a*). Branching dendrite trees and motoneuron membrane resistivity. *Expl Neurol.* **1**, 491–527. [67, 131, 134, 135, 136, 144, 145, 161, 187].

—— (1959*b*). Dendrite current distribution and whole neuron properties. Naval Medical Research Institute Research Rep. NM 10 05 00.01.02. [67, 134, 139, 140, 145]

—— (1960). Membrane potential transients and membrane time constant of motoneurons. *Expl Neurol.* **2**, 503–32. [131, 161, 163, 167, 168, 169, 171]

—— (1962*a*). Theory of physiological properties of dendrites. *Ann. N.Y. Acad. Sci.* **96**, 1071–92. [64, 131, 134, 135, 149, 153, 157, 164, 167, 197]

—— (1962*b*). Electrophysiology of a dendrite neuron model. *Biophys. J.* **2**, 145–67. [64, 164]

—— (1964). Theoretical significance of dendritic trees for neuronal input–output relations. In *Neural theory and modelling* (Ed. R. F. Reiss), pp. 73–9. Stanford University Press. [64, 158, 164, 173, 193, 194, 197, 199, 200, 215, 314]

RALL, W. (1967). Distinguishing theoretical synaptic potentials computed for different soma-dendritic distributions of synaptic input. *J. Neurophysiol.* **30**, 1138–68. [48, 64, 161, 164, 173, 176, 185, 193]

—— (1969a). Time constants and electrotonic length of membrane cylinders and neurons. *Biophys. J.* **9**, 1483–1508. [132, 158, 164, 165, 167, 171, 172, 178]

—— (1969b). Distributions of potential in cylindrical coordinates and time constants for a membrane cylinder. *Biophys. J.* **9**, 1509–41. [8, 132, 133]

—— (1970). Cable properties of dendrites and effects of synaptic location. In *Excitatory synaptic mechanisms* (Eds. P. Andersen and J. K. S. Jansen), pp. 175–87. Universitets forlarget, Oslo. [158, 217, 220]

——, BURKE, R. E., SMITH, T. G., NELSON, P. G., and FRANK, K. (1967). Dendrite location of synapses and possible mechanisms for the monosynaptic EPSP in motoneurons. *J. Neurophysiol.* **30**, 1169–93. [194, 198, 224]

—— and RINZEL, J. (1973). Branch input resistance and steady attenuation for input to one branch of a dendritic neurone model. *Biophys. J.* **13**, 648–88. [146]

—— and SHEPHERD, G. M. (1968). Theoretical reconstruction of field potentials and dendrodendritic synaptic interactions in olfactory bulb. *J. Neurophysiol.* **31**, 884–915. [3, 217, 223]

——, ——, REESE, T. S., and BRIGHTMAN, M. W. (1966). Dendrodendritic synaptic pathway for inhibition in the olfactory bulb. *Expl Neurol.* **14**, 44–56. [217]

RALSTON, H. J., III (1968). The fine structure of neurons in the dorsal horn of the cat spinal cord. *J. comp. Physiol.* **132**, 275–302. [217, 223]

—— (1971). Evidence for presynaptic dendrites, and a proposal for their mechanism of action. *Nature. Lond.* **230**, 585–7. [217, 218]

RAMON-MOLINER, E. (1962). An attempt at classifying nerve cells on the basis of their dendritic patterns. *J. comp. Neurol.* **119**, 211–27. [131]

RANG, H. P. and RITCHIE, J. M. (1968). On the electrogenic sodium pump in mammalian non-myelinated nerve fibres and its activation by various external cations. *J. Physiol.* **196**, 183–221. [358]

RASHBASS, C. and RUSHTON, W. A. H. (1949). The relation of structure to the spread of excitation in the frog's sciatic trunk. *J. Physiol.* **110**, 110–35. [9]

RASHEVSKY, N. (1931). On the theory of nervous conduction. *J. gen. Physiol.* **14**, 517–28. [1]

—— (1938). *Mathematical biophysics*. University of Chicago Press.

REDMAN, S. J. (1973). The attenuation of passively propagating dendritic potentials in a motoneurone cable model. *J. Physiol.* **234**, 637–64. [191, 192]

RESCIGNO, A., STEIN, R. B., PURPLE, R. L., and POPPELE, R. E. (1970). A neuronal model for the discharge patterns produced by cyclic inputs. *Bull. math. Biophys.* **32**, 337–53. [368]

REUTER, H. and BEELER, G. W. (1969). Calcium current and activation of contraction in ventricular myocardial fibres. *Science N. Y.* **163**, 399–401. [246]

RICHER, I. (1966). The switch-line: a simple lumped transmission line that can support unattenuated propagation. *IEEE Trans. Commun. Technol.* **CT-113**, 388–92. [286]

RIGGS, D. S. (1970). *Control theory and physiological feedback mechanisms*. Williams and Wilkins, Baltimore. [441]

RINGHAM, G. L. (1971). Origin of nerve impulse in slowly adapting stretch receptor of crayfish. *J. Neurophysiol.* **34**, 773–84. [340]

ROBERTS, G. E. and KAUFMAN, H. (1966). *Tables of Laplace transforms*. Saunders, Philadelphia. [30, 46, 71, 72, 79, 81, 88, 90, 95, 162, 446]

ROBERTSON, J. D. (1960). The molecular structure and contact relationship of cell membranes. *Prog. Biophys.* **10**, 343–418. [12]

RODGERS, D. (1972). Ultrastructural identification of degenerating boutons of monosynaptic pathways in the lumbosacral segments in the cat after spinal hemisection. *Expl Brain Res.* **14**, 293–311. [173]

ROSENBERG, H. (1937*a*). Electrotonus and excitation in nerve. *Proc. R. Soc.* **B124**, 308. [1]

—— (1937*b*). The physico-chemical basis of electrotonus. *Trans. Faraday Soc.* **33**, 1028. [1]

ROSENBLUETH, A., WIENER, N., PITTS, W., and GARCIA-RAMOS, J. (1948). An account of the spike potential of axons. *J. cell. comp. Physiol.* **32**, 275–317. [287]

ROSENFALCK, P. (1969). Intra- and extracellular potential fields of active nerve and muscle fibres. *Acta physiol. scand.*, Suppl., **321**, 1–168. [3]

ROUGIER, O., VASSORT, G., and STÄMPFLI, R. (1968). Voltage clamp experiments on frog atrial heart muscle fibres with the sucrose gap technique. *Pflügers Arch ges. Physiol.* **301**, 91–108. [75]

——, ——, GARNIER, D., GARGOUIL, Y. M., and CORABOEUF, E. (1969). Existence and role of a slow inward current during the frog atrial action potential. *Pflügers Arch. ges. Physiol.* **308**, 91–110. [249]

ROY, B. K. and SMITH, D. R. (1969). Analysis of the exponential decay model of the neuron showing frequency threshold effects. *Bull. math. Biophys.* **31**, 341–57. [372]

RUDJORD, T. and ROMMETVEDT, H. J. (1970). Muscle spindle receptors. Effects of geometrical structure on passive current spread in group Ia terminals. *Kybernetik* **7**, 72–7. [149]

RUSHTON, W. A. H. (1934). A physical analysis of the relation between threshold and interpolar length in the electric excitation of medullated nerve. *J. Physiol.* **82**, 332–52. [1]

—— (1937). Initiation of the propagated disturbance. *Proc. R. Soc.* **B124**, 210. [1, 269, 275, 284, 308, 403, 413]

—— (1951). A theory of the effects of fibre size in medullated nerve. *J. Physiol.* **115**, 101–22. [296, 297]

SABAH, N. H. and LIEBOVIC, K. N. (1969). Subthreshold oscillatory responses of the Hodgkin-Huxley cable model for the squid giant axon. *Biophys. J.* **9**, 1206–22. [23, 304, 324]

SANDERS, F. K. and WHITTERIDGE, D. (1946). Conduction velocity and myelin thickness in regenerating nerve fibres. *J. Physiol.* **105**, 152–74. [299]

SATO, M. (1952). Repetitive responses of the nerve fiber, as determined by recovery process and accommodation *Japan. J. Physiol.* **2**, 277–89. [344]

SCHEIBEL, M. E. and SCHEIBEL, A. B. (1968). On the nature of dendritic spines—report of a workshop. *Commun. Behav. Biol.* **1A**, 231–65. [131, 218]

SCHLÖGL, R. (1954). Elektrodiffusion in freier Lösung und geladenen Membranen. *Z. phys. Chem.* **1**, 305–39. [238]

SCHMIDT, H. and STÄMPFLI, R. (1964). Nachweis unterschliedlicher elektrophysiolo-gischer Eigenschaften motorischer und sensibler Nervenfasern des Frosches. *Helv. physiol. pharmac. Acta* **22**, C143–5. [343]

SCHNEIDER, M. F. (1970). Linear electrical properties of the transverse tubules and surface membrane of skeletal muscle fibers. *J. gen. Physiol.* **56**, 640–71. [12, 103, 106, 107, 111, 113, 114, 121, 129]

—— and CHANDLER, W. K. (1973). Voltage dependent charge movement in skeletal muscle: a possible step in excitation-contraction coupling. *Nature, Lond.* **242**, 244–6. [121, 130]

SCHWAN, H. P. (1954). Electrical properties of muscle tissue at low frequencies. *Z. Naturf.* **96**, 245. [12]

SCHWAN, H. P. (1965). Biological impedance determinations. *J. cell. comp. Physiol.* **66**, Suppl. 2, 5–12. [12, 18]

SCHWINDT, P. C. and CALVIN, W. H. (1972). Membrane-potential trajectories between spikes underlying motoneuron firing rates. *J. Neurophysiol.* **35**, 311–15. [314, 355]

SCOTT, A. C. (1963). Neuristor propagation on a tunnel diode loaded transmission line. *Proc. IEEE* **51**, 240. [286]

—— (1964). Analysis of a myelinated nerve model. *Bull. math. Biophys.*, **26**, 247–54. [286]

—— (1971). Effect of the series inductance of a nerve axon upon its conduction velocity. *Math. Biosci.* **11**, 277–90. [25]

—— (1972). Transmission line equivalent for an unmyelinated nerve axon. *Math. Biosci.* **13**, 47–54. [25]

SEGUNDO, J. P., PERKEL, I. D. H., WYMAN, H., HEGSTAD, H., and MOORE, G. P. (1968). Input–output relations in computer-simulated nerve cells. Influence of the statistical properties, strength, number and inter-dependence of excitatory pre-synaptic terminals. *Kybernetik* **4**, 157–71. [374, 375]

SHAPIRO, B. I. and LENHERR, F. K. (1972). Hodgkin–Huxley axon. Increased modulation and linearity of response to constant current stimulus. *Biophys. J.* **12**, 1145–58. [336, 342]

SHAW, S. R. (1972). Decremental conduction of the visual signal in barnacle lateral eye. *J. Physiol.* **220**, 145–75. [36]

SHIBA, H. (1971). Heaviside's "Bessel cable" as an electric model for flat simple epithelial cells with low resistance junctional membranes *J. theoret. Biol.* **25**, 113–26. [83]

SIGGINS, G. R., OLIVER, A. P., HOFFER, B. J., and BLOOM, F. E. (1971). Cyclic adenosine monophosphate and norepinephrine: effects on transmembrane properties of cerebellar Purkinje cells. *Science N.Y.* **171**, 192–4. [192]

SLOPER, J. J. (1971). Dendro-dendritic synapses in the primate motor cortex. *Brain Res.* **34**, 186–92. [217]

SMITH, D. S. (1966). The organisation and function of the sarcoplasmic reticulum and T-system of muscle cells. *Prog. Biophys. molec. Biol.* **16**, 107. [10]

SMITH, T. G., WUERKER, R. B., and FRANK, K. (1967). Membrane impedance changes during synaptic transmission in cat spinal motoneurons. *J. Neurophysiol.* **30**, 1072–96. [184]

SOKOLOVE, P. G. (1972). Computer simulation of after-inhibition in crayfish slowly adapting stretch receptor neuron. *Biophys. J.* **12**, 1429–51. [361, 364, 366]

—— and COOKE, I. M. (1971). Inhibition of impulse activity in a sensory neuron by an electrogenic pump. *J. gen. Physiol.* **57**, 125–63. [352, 358, 359, 360, 366]

SOMMER, J. R. and JOHNSON, E. A. (1968). Cardiac muscle: a comparative study of Purkinje fibres and ventricular fibres. *J. Cell Biol.* **36**, 497–526. [11, 21, 102]

SPERELAKIS, N. (1972). Electrical properties of embryonic heart cells. In *Electrical phenomena in the heart*, (Ed. de Mello), pp. 1–61. Academic Press, New York. [11]

STEIN, R. B. (1965). A theoretical analysis of neuronal variability. *Biophys. J.* **5**, 173–94. [367, 374]

—— (1967a). The frequency of nerve action potentials generated by applied currents. *Proc. R. Soc.* **B167**, 64–86. [307, 308, 331, 332, 333, 334, 335, 336, 338, 340, 342, 348, 352]

—— (1967b). Some models of neuronal variability. *Biophys. J.* **7**, 37–68. [367, 370, 371, 372, 373]

—— (1967c). The information capacity of nerve cells using a frequency code. *Biophys. J.* **7**, 797–826. [378]

STEIN, R. B. (1970). The role of spike trains in transmitting and distorting sensory signals. In *The neurosciences: second study program*, (Ed. F. O. Schmitt), pp. 597–604. Rockefeller University Press. [368]

—— and FRENCH, A. S. (1970). Models for the transmission of information by nerve cells. In *Excitatory synaptic mechanisms*, (Eds. P. Andersen and J. K. S. Jansen), pp. 147–257. Universitets forlaget, Oslo. [368]

—— and MATTHEWS, P. B. C. (1965). Differences in variability of discharge frequency between primary and secondary muscle spindle afferent endings of the cat. *Nature, Lond.* **208**, 1217–18. [377]

—— and PEARSON, K. G. (1971). Predicted amplitude and form of action potentials recorded from unmyelinated nerve fibres. *J. theoret. Biol.* **32**, 539–58. [296]

STEN-KNUDSEN, O. (1954). The ineffectiveness of the 'window-field' in the initiation of muscle contraction. *J. Physiol.* **125**, 396–404. [98]

STEVENS, C. F. (1966). *Neurophysiology: a primer*, Wiley, New York. [53]

STRETTON, A. O. W. and KRAVITZ, E. A. (1968). Neuronal geometry: determination with a technique of intracellular dye injection. *Science N.Y.* **162**, 132–4. [144]

SUGI, H. and ŌCHI, R. (1967). The mode of transverse spread of contraction initiated by local activation in single frog fibres. *J. gen. Physiol.* **50**, 2167–76. [119]

SUGIYAMA, H., MOORE, G. P., and PERKEL, D. H. (1970). Solutions for a stochastic model of neuronal spike production. *Math. Biosci.* **8**, 323–41. [367, 368, 369]

SZABO, G., EISENMAN, G., and CIANI, S. (1969). The effects of the macrotetralide actin antibiotics on the electrical properties of phospholipid bilayer membranes. *J. membrane Biol.* **1**, 346–82. [260]

TAKEUCHI, A. and TAKEUCHI, N. (1959). Active phase of frog's end-plate potential. *J. Neurophysiol.* **22**, 395–411. [53, 66, 182]

—— —— (1960). On the permeability of endplate membrane during the action of transmitter. *J. Physiol.* **154**, 52–67. [60]

TAKEUCHI, N. (1963). Some properties of conductance changes at the endplate membrane during the action of acetylcholine. *J. Physiol.* **167**, 128–40. [60]

TARR, M. and SPERELAKIS, N. (1964). Weak electrotonic interaction between contiguous cardiac cells. *Am. J. Physiol.* **207**, 691–700. [10]

TASAKI, I. (1939). The electrosaltatory transmission of the nerve impulse and the effect of narcosis upon the nerve fibre. *Am. J. Physiol.* **127**, 211–27. [293]

—— (1950). The threshold conditions on electrical excitation of the nerve fiber. Part II. *Cytologia*, **15**, 219–36. [344]

—— (1953). *Nervous transmission*. Thomas, Springfield. [293, 296, 299]

—— (1959). Conduction of the nervous impulse. In *Handbook of physiology*, Section I. Neurophysiology, pp. 75–121. American Physiological Society, Washington. [296]

—— and HAGIWARA, S. (1957). Capacity of muscle fiber membrane. *Am. J. Physiol.* **188**, 423–9. [106, 111, 117]

TAYLOR, R. E. (1965). Impedance of the squid axon membrane. *J. cell. comp. Physiol.* **66**, Suppl. 2, 21–6. [12]

TEORELL, T. (1935). An attempt to formulate a quantitative theory of membrane permeability. *Proc. Soc. exp. Biol. Med.* **33**, 282–5. [238]

—— (1946). Application of 'square wave analysis' to bioelectric studies. *Acta physiol. scand.* **12**, 235–54. [21]

—— (1951). Zur quantitativen Behandelung der Membranpermeabilität. *Z. Elektrochem.* **55**, 460–9. [238]

—— (1953). Transport processes and electrical phenomena in ionic membranes. *Prog. Biophys.* **3**, 305–69. [238]

TERZUOLO, C. A. and WASHIZU, Y. (1962). Relation between stimulus strength, generator potential and impulse frequency in stretch receptor of *Crustacea. J. Neurophysiol.* **25**, 56–66. [315, 347, 348]

THOMAS, G. B. (1960). *Calculus and analytical geometry.* Addison Wesley, Reading, Massachusetts. [462]

THOMAS, R. C. (1969). Membrane current and intracellular sodium changes in a snail neurone during extrusion of injected sodium. *J. Physiol.* **201**, 495–514. [358, 364]

—— (1972). Electrogenic sodium pump in nerve and muscle cells. *Physiol. Rev.* **52**, 563–94. [358, 361]

TILLE, J. (1965). A new interpretation of the dynamic changes of the potassium conductance in the squid giant axon. *Biophys. J.* **5**, 163–71. [241]

—— (1966). Electrotonic interaction between muscle fibres in the rabbit ventricle. *J. gen. Physiol.* **50**, 189–202. [11]

TOMITA, T. (1966). Membrane capacity and resistance in mammalian smooth muscle. *J. theor. Biol.* **2**, 216–27. [11, 83]

—— (1967). Current spread in the smooth muscle of the guinea pig vas deferens. *J. Physiol.* **189**, 163–76. [11, 25, 97]

—— and WRIGHT, E. B. (1965). A study of the crustacean axon repetitive response: I. The effect of membrane potential and resistance. *J. cell. comp. Physiol.* **65**, 195–209. [305]

TRAUTWEIN, W., DUDEL, J., and PEPER, K. (1965). Stationary S-shaped current-voltage relation and hysteresis in heart muscle fibres. Excitatory phenomena in the Na-free bathing solutions. *J. cell. comp. Physiol.* **66**, Suppl. 2, 79–90. [408]

—— and KASSEBAUM, O. F. (1961). On the mechanism of spontaneous impulse generation in the pacemaker of the heart. *J. gen. Physiol.* **45**, 317–30. [305]

—— and SCHMIDT, R. F. (1960). Zur Membranwirkung des Adrenalins an der Herzmuskelfaser. *Pflügers Arch. ges. Physiol.* **271**, 715–26. [351]

TROSHIN, A. S. (1966). *Problems of cell permeability* (Tr. M. G. Hell; Ed. W. F. Widdas). Pergamon Press, Oxford. [9]

TSIEN, R. W. and NOBLE, D. (1969). A transition state theory approach to the kinetics of conductance changes in excitable membranes. *J. Membrane Biol.* **1**, 248–73. [241, 242]

UCHIZONO, K. (1968). Inhibitory and excitatory synapses in vertebrate and in-vertebrate animals. In *Structure and function of inhibitory neuronal mechanisms*, (Eds. C. von Euler, S. Skoglund, and U. Soderberg), pp. 33–60. Pergamon Press, Oxford. [223]

VALDIOSERA, R., CLAUSEN, C., and EISENBERG, R. S. (1973). Impedance of frog skeletal muscle fibres. *Abstracts, Biophysical Society (17th annual meeting)* TPM-C5. [115]

—— —— —— (1974a). Measurement of the impedance of frog skeletal muscle fibres. *Biophys. J.* **14**, 295–315. [114]

—— —— —— (1974b). Circuit models of the passive electrical properties of frog skeletal muscle fibers. *J. gen. Physiol.* **63**, 432–59. [114]

—— —— —— (1974c). Impedance of frog skeletal muscle fibres in various solutions. *J. gen. Physiol.* **63**, 460–91. [114]

VALDERDE, F. (1967). Apical dendrite spines of the visual cortex and light deprivation in the mouse. *Expl Brain Res.* **3**, 337–52. [131]

VAN DER POL, B. (1926). On relaxation oscillations. *Phil. Mag.* **2**, 978–92. [306, 325]

VAN DER POL, B. and VAN DER MARK, J. (1928). The heartbeat considered as a relaxation oscillation, and an electrical model of the heart. *Phil. Mag.*, Suppl. 6, 763–75. [306, 328]

VAN VALKENBURG, M. E. (1964). *Network Analysis* (2nd edn) Prentice-Hall, Englewood Cliffs, New Jersey. [100]

VITEK, M. and TRAUTWEIN, W. (1971). Slow inward current and action potentials in cardiac Purkinje fibres. *Pflügers Arch. ges. Physiol.* **323**, 204–18. [317]

WACHTEL, H. and KANDEL, E. R. (1971). Conversion of synaptic excitation to inhibition at a dual chemical synapse. *J. Neurophysiol.* **34**, 56–68. [216]

WALLØE, L. (1968). Transfer of signals through a second order sensory neuron. Thesis, University of Oslo. [376, 377, 378]

—— (1970). On the transmission of information through sensory neurons. *Biophys. J.* **10**, 745–63. [378]

——, JANSEN, J. K. S., and NYGAARD, K. (1969). A computer simulated model of a second order sensory neuron. *Kybernetik* **6**, 130–40. [376]

WALTMAN, B. (1966). Electrical properties and fine structure of the ampullary canals of Lorenzini. *Acta physiol. scand.* **66**, Suppl., 264. [82]

WATSON, G. N. (1966). *A treatise on the theory of Bessel functions* (2nd edn). Cambridge University Press. [85, 87, 454]

WAXMAN, S. G. and BENNETT, M. V. L. (1972). Relative conduction velocities of small myelinated and non-myelinated fibres in the central nervous system. *Nature new Biol* **238**, 217–19. [301]

WEBER, H. (1873). Über die stationaren Strömungen der Elektricität in Cylindren. *Bochardt's J. Math.* **76**, 1. [1]

—— (1884). Nachtrag zu Seite 150. *Pflügers Arch. ges. Physiol.* **33**, 162. [1]

WEIDMANN, S. (1952). The electrical constants of Purkinje fibres. *J. Physiol.* **118**, 348–60. [10, 11, 25, 67, 73, 82, 101, 392]

—— (1955). The effect of the cardiac membrane potential on the rapid availability of the sodium carrying system. *J. Physiol.* **127**, 213–24. [246]

—— (1966). The diffusion of radiopotassium across intercalated disks of mammalian cardiac muscle. *J. Physiol.* **187**, 323–42. [10, 11, 102]

—— (1967). Cardiac electrophysiology in the light of recent morphological findings. *Harvey Lect.* **61**, 1–15. [10]

—— (1969). Electrical coupling between myocardial cells. *Prog. Brain Res.* **31**, 275–81. [10]

—— (1970). Electrical constants of trabecular muscle from mammalian heart. *J. Physiol.* **210**, 1041–54. [117]

WEIGHT, F. F. and PADJEN, A. (1973). Slow synaptic inhibition: evidence for synaptic inactivation of sodium conductance in sympathetic ganglion cells. *Brain Res.* **55**, 219–24. [192]

—— and VOTAVA, J. (1970). Slow synaptic excitation in sympathetic ganglion cells: evidence for synaptic inactivation of potassium conductance. *Science N.Y.* **170**, 755–8. [192, 218]

WEINBERG, A. M. (1941). Weber's theory of the Kernleiter. *Bull. math. Biophys.* **3**, 39. [2]

WHITTAKER, E. T. and WATSON, G. N. (1944). *A course of modern analysis.* [395]

WILLIAMS, E. J., JOHNSTON, R. J., and DAINTY, J. (1964). Electrical resistance and capacitance of membranes of *Nitella translucens. J. exp. Bot.* **15**, 1. [107]

WILLIAMS, R. J. P. (1952). The stability of the complexes of the group IIa metal ions. *J. chem. Soc.* 3770–8. [254]

—— (1970). The biochemistry of sodium, potassium, magnesium and calcium. *Q. Rev. chem. Soc.* **24**, 331–65. [254, 255, 257]

WILSON, V. J. and BURGESS, R. R. (1962). Disinhibition in the cat spinal cord. *J. Neurophysiol.* **25**, 392–404. [218]

WOLBARSHT, M. L. (1960). Electrical characteristics of insect mechanoreceptors. *J. gen. Physiol.* **44**, 105–22. [348]

WONG, M. T. T. (1970). Somato-dendritic and dendro-dendritic synapses in the squirrel monkey lateral geniculate nucleus. *Brain Res.* **20**, 135–9. [217]

WOODBURY, J. W. (1961). Voltage and time-dependent membrane conductance changes in cardiac cells. In *Biophysics of physiological and pharmacological actions*, (Ed. A. Shanes) New York: A.A.A.S. [411]

—— (1971). Eyring rate theory model of the current voltage relationships of ion channels in excitable membranes. In *Chemical dynamics: Papers in honor of Henry Eyring* (Ed. J. Hirschfelder), pp. 601–17. John Wiley, New York. [229, 233]

—— and CRILL, W. E. (1961). On the problem of impulse conduction in the atrium. [84, 96]. In *Nervous inhibition* (Ed. Lord Florey), pp. 124–35. Plenum Press, New York. [10, 11]

—— and GORDON, A. M. (1965). The electrical equivalent circuit of heart muscle. *J. cell. comp. Physiol.* **66**, Suppl. 2, 35–42. [10, 11, 25]

—— and MILES, P. R. (1973). Anion conductance of frog muscle membranes: one channel, two kinds of pH dependence. *J. gen. Physiol.* **62**, 324–53. [260]

WOODHULL, A. M. (1973). Ionic blockage of sodium channels in nerve. *J. gen. Physiol.* **61**, 687–708. [259]

WRIGHT, E. B. and TOMITA, T. (1965). A study of the crustacean axon repetitive response: II. The effect of cations, sodium, calcium (magnesium), potassium and hydrogen (pH) in the external medium. *J. cell. comp. Physiol.* **65**, 211–28. [331, 352]

ZUCKER, R. S. (1969). Field potentials generated by dendritic spikes and synaptic potentials. *Science N.Y.* **165**, 409–13. [217]

Supplementary bibliography

General

EISENBERG, R. S. and MATHIAS, R. T. (1980). Structural analysis of electrical properties. *Crit. Rev. Bioeng.* **4**, 203-32.

JACK, J. J. B. (1976). Electrophysiological properties of peripheral nerve. In: *The Peripheral Nerve*, 740-818. (Ed. D. N. London.) Chapman and Hall, London.

—— (1979). An introduction to linear cable theory. In: *The Neurosciences: Fourth Study Program*, 423-37. (Ed. F. O. Schmitt and F. G. Worden.) M.I.T. Press, Cambridge, MA.

MACGREGOR, R. J. and LEWIS, E. R. (1977). *Neural modeling: Electrical signal processing in the nervous system.* Plenum, New York.

RALL, W. (1977). Core conductor theory and cable properties of neurons. In: *Handbook of Physiology* (Sect. 1). *The Nervous System. I. Cellular Biology of Neurons*, 39-97. (Ed. E. R. Kandel.) American Physiological Society, Bethesda, MD.

Chapter 2: Linear electrical properties of excitable cells

CARPENTER, D. O., HOVEY, M. M., and BAK, A. F. (1975). Resistivity of axoplasm. II. Internal resistivity of giant axons of squid and *Myxicola*. *J. gen. Physiol.* **66**, 139-48.

COLE, K. S. (1975). Resistivity of axoplasm. I. Resistivity of extruded squid axoplasm. *J. gen. Physiol.* **66**, 133-8.

—— (1976). Electrical properties of the squid axon sheath. *Biophys. J.* **16**, 137-42.

DEMELLO, W. C. (1976). Influence of the sodium pump on intercellular communication in heart fibres: effect of intracellular injection of sodium ion on electrical coupling. *J. Physiol.* **263**, 171-98.

FISHMAN, H. M. and MOORE, L. E. (1977). Asymmetry currents and admittance in squid axons. *Biophys. J.* **19**, 177-83.

FOSTER, K. R., BIDINGER, J. M., and CARPENTER, D. O. (1976). The electrical resistivity of cytoplasm. *Biophys. J.* **16**, 991-1002.

HUANG, W.-T. and LEVITT, D. G. (1977). Theoretical calculation of the dielectric constant of a bilayer membrane. *Biophys. J.* **17**, 111-28.

LEVIN, D. N. (1981a). Surface capacity of electrically syncytial tissues. *Biophys. J.* **35**, 127-46.

—— (1981b). Unit membrane parameters of electrically syncytial tissues. *Biophys. J.* **35**, 147-65.

MOBLEY, B. A., LEUNG, J., and EISENBERG, R. S. (1975). Longitudinal impedance of single frog muscle fibers. *J. gen. Physiol.* **65**, 97-117.

OHBA, M., SAKAMOTO, Y., TOKUNO, H., and TOMITA, T. (1976). Impedance components in longitudinal direction in guinea pig taenia coli. *J. Physiol.* **256**, 527-40.

POLLACK, G. H. (1976). Intercellular coupling in the atrioventricular node and other tissues of the rabbit heart. *J. Physiol.* **255**, 275-98.

POUSSART, D., MOORE, L. E., and FISHMAN, H. M. Ion movements and kinetics in squid axon. I. Complex admittance. *Ann. N.Y. Acad. Sci.* **303**, 355-79.

STRANDBERG, M. W. P. (1977). The representation of membrane admittance. *Biophys. J.* **20**, 279-83.

TAKASHIMA, S. and YANTORNO, R. (1977). Investigation of voltage-dependent membrane capacity of squid giant axons. *Ann. N.Y. Acad. Sci.* **303**, 306-21.

TAYLOR, R. E. (1977). Electrical impedance of excitable membranes. *Ann. N.Y. Acad. Sci.* **303**, 298–305.
WEINGART, R. (1977). The actions of ouabain on intercellular coupling and conduction velocity in mammalian ventricular muscle. *J. Physiol.* **264**, 341–65.

Chapter 3: Linear cable theory

Application to biological cables

ANAGNOSTOPOULOS, T., TEULON, J., and EDELMAN, A. (1980). Conductive properties of the proximal tubule in *Necturus* kidney. *J. gen. Physiol.* **75**, 553–87.
BLAKE, I. O. (1980). A.C. cable theory and its implications on the measurements of the membrane electric parameters. *J. theor. Biol.* **83**, 595–621.
BYWATER, R. A. R. and TAYLOR G. S. (1980). The passive membrane properties and excitatory junction potentials of the guinea-pig vas deferens. *J. Physiol.* **300**, 303–16.
DOMINGUEZ, G. and FOZZARD, H. A. (1979). Effect of stretch on conduction velocity and cable properties of cardiac Purkinje fibers. *Am. J. Physiol.* **237**, C119–24.
FINK, R. and LÜTTGAU, H. C. (1976). An evaluation of the membrane constants and the potassium conductance in metabolically exhausted muscle fibres. *J. Physiol.* **263**, 215–38.
KOOTSEY, J. M., JOHNSON, E. A., and LIEBERMAN, M. (1977). The cylindrical cell with a time-variant membrane resistance. Measuring passive properties. *Biophys. J.* **17**, 145–54.
LIEBERMAN, M., SAWANOBORI, T., KOOTSEY, J. M., and JOHNSON, E. A. (1975). A synthetic strand of cardiac muscle. Its passive electric properties. *J. gen. Physiol.* **65**, 527–50.
SCOTT, A. C. (1977). *Neurophysics.* Wiley, New York. *Math. Biosci.* **44**, 299.
SINGER, J. J. and WALSH, J. V., Jr. (1980). Passive properties of the membrane of single freshly isolated smooth muscle cells. *Am. J. Physiol.* **239**, C153–61.

Quantitative estimates of synaptic function

ADAMS, W. B. (1976). Upper and lower bounds on the non-linearity of summation of end-plate potentials. *J. theor. Biol.* **63**, 217–24.
GARDNER, D. (1980). Time integral of synaptic conductance. *J. Physiol.* **304**, 181–91.
GINSBORG, B. L., McLACHLAN, E. M., MARTIN, A. R., and SEARL, J. W. The computation of simulated endplate potentials. *Proc. R. Soc. Lond. B.* **213**, 233–42.
McLACHLAN, E. M. and MARTIN, A. R. (1981). Non-linear summation of end-plate potentials in the frog and mouse. *J. Physiol.* **311**, 307–24.
MARTIN, A. R. (1976). The effect of membrane capacitance on the non-linear summation of synaptic potentials. *J. theor. Biol.* **59**, 179–87.
MIYAMOTO, M. D. (1978). Estimates of magnitude and nonlinear summation of evoked potentials at motor end-plate. *J. Neurophysiol.* **41**, 589–99.
QUASTEL, D. M. J. (1979). Correction of end-plate potentials and currents for non-linear summation. *Can. J. Physiol. Pharmacol.* **57**, 702–9.
STEVENS, C. F. (1976). A comment on Martin's relation. *Biophys. J.* **16**, 891–5.
WERNIG, A. (1975). Estimates of statistical release parameters from crayfish and frog neuromuscular junctions. *J. Physiol.* **244**, 107–221.
WILLIAMS, J. D. and BOWEN, J. M. (1977). Integral correction of end-plate potentials: a useful alternative to peak amplitude correction. *J. theor. Biol.* **67**, 313–17.

Other problems

GRECO, E. C., CLARK, J. W., and HARMAN, T. L. (1977). Volume-conductor fields of the isolated axon. *Math. Biosci.* **33**, 235–56.
KOIDE, F. T. (1975). Electrotonus in medullated nerve. *Math. Biosci.* **25**, 363–73.

Chapter 4: Properties of finite cables

HERNANDEZ-NICAISE, M.-L., MACKIE, G. O., and MEECH, R. W. (1980). Giant smooth muscle cells of Beroë. Ultrastructure, innervation and electrical properties. *J. gen. Physiol.* **75**, 79–105.

HUDSPETH, A. J., POO, M. M., and STUART, A. E. (1977). Passive signal propagation and membrane properties in median photoreceptors of the giant barnacle. *J. Physiol.* **272**, 25–43.

ONODERA, K. and TAKEUCHI, A. (1975). Ionic mechanism of the excitatory synaptic membrane of the crayfish neuromuscular junction. *J. Physiol.* **252**, 295–318.

ORTEGA BLAKE, I. (1980). A.C. cable theory and its implications on the measurements of the membrane electrical parameters. *J. theor. Biol.* **83**, 595–621.

Current spread in cardiac muscle as modelled by finite longitudinal or radial cables

ATTWELL, D. and COHEN, I. (1977). The voltage clamp of multicellular preparations. *Prog. Biophys. mol. Biol.* **31**, 201–45.

BEELER, G. W. and MCGUIGAN, J. A. S. (1978). Voltage clamping of multicellular myocardial preparations: capabilities and limitations of existing methods. *Prog. Biophys. mol. Biol.* **34**, 219–54.

COLATSKY, T. J. and TSIEN, R. W. (1979). Electrical properties associated with wide intercellular clefts in rabbit Purkinje fibers. *J. Physiol.* **290**, 227–52.

JAKOBSSON, E., BARR, L., and CONNOR, J. A. (1975). An equivalent circuit for small atrial trabeculae of frog. *Biophys. J.* **15**, 1069–85.

KASS, R. S., SIEGELBAUM, S. A., and TSIEN, R. W. (1979). Three micro-electrode voltage clamp experiments in calf cardiac Purkinje fibres: is slow inward current adequately measured? *J. Physiol.* **290**, 201–25.

MANN, J. E., FOLEY, E., and SPERELAKIS, N. (1977). Resistance and potential profiles in the cleft between two myocardial cells: electrical analog and computer simulations. *J. theor. Biol.* **68**, 1–15.

MOORE, J. W., RAMON, F., and JOYNER, R. W. (1975). Axon voltage-clamp simulations. II. Double sucrose gap method. *Biophys. J.* **15**, 25–35.

RAMÓN, F., ANDERSON, N., JOYNER, R. W., and MOORE, J. W. (1975). Axon voltage clamp simulations. IV. A multicellular preparation. *Biophys. J.* **15**, 55–69.

RUFFNER, J. A., SPERELAKIS, N., and MANN, J. E. (1980). Application of the Hodgkin–Huxley equations to an electric field model for interaction between excitable cells. *J. theor. Biol.* **87**, 129–52.

SCHOENBERG, M., DOMINGUEZ, G., and FOZZARD, H. A. (1975). Effect of diameter on membrane capacity and conductance of sheep cardiac Purkinje fibers. *J. gen. Physiol.* **65**, 441–58.

—— and FOZZARD, H. A. (1979). The influence of intercellular clefts on the electrical properties of sheep cardiac Purkinje fibers. *Biophys. J.* **25**, 217–34.

Chapter 5: Current flow in multidimensional systems

Approaches to three-dimensional problems

BARR, L. and JAKOBSSON, E. (1976). The spread of current in electrical syncytia. In: *Physiology of Smooth Muscle* (Eds. E. Bülbring and M. Shuba) 41–8. Raven Press, New York.

EISENBERG, R. A., BARCILON, V., and MATHIAS, R. T. (1979). Electrical properties of spherical syncytia. *Biophys. J.* **25**, 151–80.

MATHIAS, R. T., EBIHARA, L., LIEBERMAN, M., and JOHNSON, E. A. (1981). Linear electrical properties of passive and active currents in spherical heart cell clusters. *Biophys. J.* **36**, 221–42.

PESKOFF, A. (1979). Electric potential in three-dimensional electrical syncytial tissues.

Bull. math. Biol. **41,** 163–81.

—— and EISENBERG, R. A. (1975). The time-dependent potential in a spherical cell using matched asymptotic expansions. *J. math. Biol.* **2,** 277–300.

—— EISENBERG, R. S., and COLE, J. D. (1976). Matched asymptotic expansions of the Green's function for the electrical potential in an infinite cylindrical cell. *J. appl. Math.* **30,** 222–39.

—— and RAMIREZ, D. M. (1975). Potential induced in a spherical cell by an intracellular point source and an extracellular point sink. *J. math. Biol.* **2,** 301–16.

PURVES, R. D. (1975). Current flow and potential in a three-dimensional syncytium. *J. theor. Biol.* **60,** 147–62.

—— (1976). Microelectrodes in spherical cells. *J. theor. Biol.* **63,** 225–8.

Current spread in two-dimensional arrays of retinal photoreceptors

ATTWELL, D. and WILSON, M. (1980). Behaviour of the rod network in the tiger salamander retina mediated by membrane properties of individual rods. *J. Physiol.* **309,** 287–315.

COPENHAGEN, D. R. and OWEN, W. G. (1980). Current-voltage relations in the rod photoreceptor network of the turtle retina. *J. Physiol.* **308,** 159–84.

DETWILER, P. B. and HODGKIN, A. L. (1979). Electric coupling between cones in turtle retina. *J. Physiol.* **291,** 75–100.

—— —— and MCNAUGHTON, P. A. (1980). Temporal and spatial characteristics of the voltage response of rods in the retina of the snapping turtle. *J. Physiol.* **300,** 213–50.

GOLD, G. H. (1979). Photoreceptor coupling in retina of the toad, *Bufo marinus*. II. Physiology. *J. Neurophys.* **42,** 311–28.

LAMB, T. D. (1976). Spatial properties of horizontal cell responses in the turtle retina. *J. Physiol.* **263,** 239–55.

—— and SIMON, E. J. (1976). The relation between intercellular coupling and electrical noise in turtle photoreceptors. *J. Physiol.* **263,** 257–86.

SCHWARTZ, E. A. (1976). Electrical properties of the rod syncytium in the retina of the turtle. *J. Physiol.* **257,** 379–406.

Electrical properties of some other syncytial tissues

CHAPMAN, R. A. and FRY, C. H. (1978). An analysis of the cable properties of frog ventricular myocardium. *J. Physiol.* **283,** 263–82.

EISENBERG, R. S. and RAE, J. L. (1976). Current-voltage relationships in the crystalline lens. *J. Physiol.* **262,** 285–300.

GRAF, J. (1978). Three-dimensional cable analysis of current spread in the liver. *J. Physiol.* **284,** 122–4.

HAYHOE, H. N. (1974). Shunting effect on resistance measurements in visceral smooth muscle tissue. *J. theor. Biol.* **46,** 295–305.

HOLMAN, M. E., TAYLOR, G. S., and TOMITA, T. (1977). Some properties of the smooth muscle of mouse vas deferens. *J. Physiol.* **266,** 751–64.

JOSEPHSON, R. K. and SCHWAB, W. E. (1979). Electrical properties of an excitable epithelium. *J. gen. Physiol.* **74,** 213–36.

MAIHIAS, R. T., RAE, J. L., and EISENBERG, R. S. (1981). The lens as a nonuniform spherical syncytium. *Biophys. J.* **34,** 61–83.

Chapter 6: Spread of excitation in muscle

Passive properties of skeletal muscle

CHANDLER, W. K. and SCHNEIDER, M. F. (1976). Time course of potential spread along a skeletal muscle fiber under voltage clamp. *J. gen. Physiol.* **67,** 165–84.

DULHUNTY, A. F. and FRANZINI-ARMSTRONG, C. (1977). The passive electrical properties of frog skeletal muscle fibres at different sarcomere lengths. *J. Physiol.* **266**, 687–711.

EISENBERG, R. S. and MATHIAS, R. T. (1980). Structural analysis of electrical properties of cells and tissues. CRC Crit. Rev. Bioeng. (*Chem. Rubb. Co.*), 203–32.

—— —— and RAE, J. S. (1977). Measurement, modelling and analysis of the linear electrical properties of cells. *Ann. N.Y. Acad. Sci.* **303**, 342–54.

GILAI, A. (1976). Electromechanical coupling in tubular muscle fibers. II. Resistance and capacitance of one transverse tubule. *J. gen. Physiol.* **67**, 343–67.

MATHIAS, R. T. (1978). An analysis of the electrical properties of a skeletal muscle fiber containing a helicoidal T system. *Biophys. J.* **23**, 277–84.

—— EISENBERG, R. S., and VALDIOSERA, R. (1977). Electrical properties of frog skeletal muscle fibers interpreted with a mesh model of the tubular system. *Biophys. J.* **17**, 57–94.

NEVILLE, M. C. (with Appendix by Mathias, R. T.) (1979). The extracellular compartments of frog skeletal muscle. *J. Physiol.* **288**, 45–70.

PESKOFF, A. (1979). Electric potential in cylindrical syncytia and muscle fibers. *Bull. math. Biol.* **41**, 183–92.

SCHNEIDER, M. F. and CHANDLER, W. K. (1976). Effects of membrane potential on the capacitance of skeletal muscle fibers. *J. gen. Physiol.* **67**, 125–63.

Voltage-dependent ionic currents

BEATY, G. N. and STEFANI, E. (1976). Calcium dependent electrical activity in twitch muscle fibres of the frog. *Prog. R. Soc. B.* **194**, 141–50.

GILLY, W. F. and HUI, C. S. (1980). Membrane electrical properties of frog slow muscles fibres. *J. Physiol.* **301**, 157–73.

GONZALEZ-SERRATOS, H. (1975). Graded activation of myofibrils and the effect of diameter on tension development during contractures in isolated skeletal muscle fibres. *J. Physiol.* **253**, 321–9.

HILLE, B. and CAMPBELL, D. T. (1976). An improved vaseline gap voltage clamp for skeletal muscle fibres. *J. gen. Physiol.* **67**, 265–93.

JAIMOVICH, E., VENOSA, R. A., SHRAGER, P., and HOROWICZ, P. (1976). Density and distribution of tetrodotoxin receptors in normal and detubulated frog sartorius muscle. *J. gen. Physiol.* **67**, 399–416.

KIRSCH, G. E., NICHOLS, R. A., and NAKAJIMA, S. (1977). Delayed rectification in transverse tubules. Origin of the late after-potential in frog skeletal muscle. *J. gen. Physiol.* **70**, 1–21.

MANDRINO, M. (1977). Voltage-clamp experiments on frog single skeletal muscle fibres: evidence for a tubular sodium current. *J. Physiol.* **269**, 605–26.

NAKAJIMA, S., NAKAJIMA, Y., and BASTIAN, J. (1975). Effects of sudden changes in external sodium concentration on twitch tension in isolated muscle fibers. *J. gen. Physiol.* **65**, 459–82.

SANCHEZ, J. A. and STEFANI, E. (1978). Inward calcium current in twitch muscle fibres of the frog. *J. Physiol.* **283**, 197–209.

STANFIELD, P. R. (1977). A calcium-dependent inward current in frog skeletal muscle fibres. *Pflügers Arch.* **368**, 267–70.

Charge movement possibly associated with excitation-contraction coupling

ADRIAN, R. H. (1978). Charge movement in the membrane of striated muscle. *Ann. Rev. Biophys. Bioeng.* **7**, 85–112.

—— and ALMERS, W. (1976). The voltage dependence of membrane capacity. *J. Physiol.* **254**, 317–38.

—— —— (1976). Charge movement in the membrane of striated muscle. *J. Physiol.* **254**, 339–60.

—— CHANDLER, W. K., and RAKOWSKI, R. F. (1976). Charge movement and mechanical repriming in skeletal muscle. *J. Physiol.* **254**, 361–88.

—— and PERES, A. (1979). Charge movement and membrane capacity in frog muscle. *J. Physiol.* **289**, 83–97.

—— and RAKOWSKI, R. F. (1978). Reactivation of membrane charge movement and delayed potassium conductance in skeletal muscle fibres. *J. Physiol.* **278**, 533–57.

ALMERS, W. (1978). Gating currents and charge movements in excitable membranes: *Rev. physiol. biochem. Pharmacol.* **82**, 96–190.

CHANDLER, W. K., RAKOWSKI, R. F., and SCHNEIDER, M. F. (1976). A non-linear voltage dependent charge movement in frog skeletal muscle. *J. Physiol.* **254**, 245–84.

—— —— —— (1976). Effects of glycerol treatment and maintained depolarization on charge movement in skeletal muscle. *J. Physiol.* **254**, 285–316.

GILLY, W. F. and HUI, C. S. (1980). Voltage-dependent charge movement in frog slow muscle fibres. *J. Physiol.* **301**, 175–90.

HOROWICZ, P. and SCHNEIDER, M. F. (1981). Membrane charge movement in contracting and non-contracting skeletal muscle fibres. *J. Physiol.* **314**, 565–93.

KOVÁCS, L., RÍOS, E., and SCHNEIDER, M. F. (1979). Calcium transients and intramembrane charge movement in skeletal muscle fibres. *Nature* **279**, 391–6.

—— and SCHNEIDER, M. F. (1978). Contractile activation by voltage clamp depolarization of cut skeletal muscle fibres. *J. Physiol.* **277**, 483–506.

MATHIAS, R. T., LEVIS, R. A., and EISENBERG, R. S. (1980). Electrical models of excitation–contraction coupling and charge movement in skeletal muscle. *J. gen. Physiol.* **76**, 1–31.

RAKOWSKI, R. F. (1978). Reprimed charge movement in skeletal muscle fibres. *J. Physiol.* **281**, 339–58.

SCHNEIDER, M. F. (1981). Membrane charge movement and depolarization–contraction coupling. *Ann. Rev. Physiol.* **43**, 507–17.

Chapter 7: Mathematical models of the nerve cell

BROCKMAN, W. H. (1981). Ladder network prediction of transients in linear spatially inhomogeneous cables. *J. theor. Biol.* **92**, 469–78.

BROWN, T. H., FRICKE, R. A., and PERKEL, D. H. (1981). Passive electrical constants in three classes of hippocampal neurons. *J. Neurophysiol.* **46**, 812–27.

—— PERKEL, D. H., NORRIS, J. C., and PEACOCK, J. H. (1981). Electrotonic structure and specific membrane properties of mouse dorsal root ganglion neurons. *J. Neurophysiol.* **45**, 1–15.

BUTZ, E. G. and COWAN, J. D. (1974). Transient potentials in dendritic systems of arbitrary geometry. *Biophys. J.* **14**, 661–89.

CARLEN, P. L. and DURAND, D. (1981). Modelling the postsynaptic location and magnitude of tonic conductance changes resulting from neurotransmitters or drugs. *Neurosci.* **6**, 839–46.

CARNEVALE, N. T. and JOHNSTON, D. (1982). Electrophysiological characterization of remote chemical synapses. *J. Neurophysiol.* **47**, 606–21.

CHRISTENSEN, B. N. and TEUBL, W. P. (1979a). Estimates of cable parameters in lamprey spinal cord neurones. *J. Physiol.* **297**, 299–318.

—— —— (1979b). Localization of synaptic input on dendrites of a lamprey spinal cord neurone from physiological measurements of membrane properties. *J. Physiol.* **297**, 319–33.

EDWARDS, F. R., HIRST, G. D. S., and SILINSKY, E. M. (1976). Interaction between inhibitory and excitatory synaptic potentials at a peripheral neurone. *J. Physiol.* **259**, 647–63.

GRAUBARD, K. (1975). Voltage attenuation within Aplysia neurons: the effect of branching pattern. *Brain Res.* **88**, 325–32.

HORWITZ, B. (1981). Neuronal plasticity: how changes in dendritic architecture can affect the spread of postsynaptic potentials. *Brain Res.* **224,** 412–18.

—— (1981). An analytical method for investigating transient potentials in neurons with branching dendritic trees. *Biophys. J.* **36,** 155–92.

JAFFE, R. A. and SAMPSON, S. R. (1976). Analysis of passive and active electrophysiologic properties of neurons in mammalian nodose ganglia maintained in vitro. *J. Neurophysiol.* **39,** 802–15.

JOYNER, R. W., WESTERFIELD, M., MOORE, J. W. and STOCKBRIDGE, N. (1978). A numerical method to model excitable cells. *Biophys. J.* **22,** 155–70.

KELLER, D. J. and LAL, S. (1976). Membrane voltage changes in a compartmental chain model of a neurone. *Biol. Cybern.* **24,** 211–17.

JOHNSTON, D. (1981). Passive cable properties of hippocampal CA3 pyramidal neurons. *Cell molec. Neurobiol.* **1,** 41–55.

LEVINE, D. S. and WOODY, C. D. (1978). Effects of active versus passive dendritic membranes on the transfer properties of a simulated neuron. *Biol. Cybern.* **31,** 63–70.

LLINAS, R. and NICHOLSON, C. (1976). Reversal properties of climbing fiber potential in cat Purkinje cells: an example of a distributed synapse. *J. Neurophysiol.* **39,** 311–23.

PELLIONICZ, A. and LLINAS, R. (1977). A computer model of cerebellar Purkinje cells. *Neuroscience* **2,** 37–48.

PERKEL, D. H. and MULLONEY, B. (1978a). Electrotonic properties of neurons: Steady-state compartmental model. *J. Neurophysiol.* **41,** 621–39.

—— —— (1978b). Calibrating compartmental models of neurons. *Am. J. Physiol.* **235,** R39–8.

—— —— and BUDELLI, R. W. (1981). Quantitative methods for predicting neuronal behavior. *Neuroscience* **6,** 823–37.

POGGIO, T. and TORRE, V. (1977). A Volterra representation for some neuron models. *Biol. Cybern.* **27,** 113–24.

RALL, W. (1978). Dendritic spines and synaptic potency. In: *Studies in Neurophysiology* (Ed. R. Porter), 203–9. Cambridge University Press.

—— (1981). Functional aspects of neuronal geometry. In: *Neurones without impulses* (Eds. A. Roberts and B. M. H. Bush), 223–54. Soc. Exp. Biol., Seminar Series 6. Cambridge University Press.

RAMON, F., MOORE, J. W., JOYNER, R. W., and WESTERFIELD, M. (1976). Squid giant axons. A model for the neuron soma? *Biophys. J.* **16,** 953–64.

REDMAN, S. J. (1976). A quantitative approach to integrative function of dendrites. *Int. Rev. Physiol. Neurophysiol. II,* **10,** 1–35.

SATO, S. and TSUKAHARA, N. (1976). Some properties of the theoretical membrane transients in Rall's neuron model. *J. theor. Biol.* **63,** 151–63.

SHEPHERD, G. M. and BRAYTON, R. K. (1979). Computer simulation of a dendro dendritic synaptic cricuit for self and lateral inhibition in the olfactory bulb. *Brain Res.* **175,** 377–82.

STRAIN, G. M. and BROCKMAN, W. D. (1975). A modified cable model for neuron processes with non-constant diameters. *J. theor. Biol.* **51,** 475–94.

TSUKAHARA, N., MURAKAMI, F., and HULTBORN, H. (1975). Electrical constants of neurons of the red nucleus. *Exp. Brain Res.* **23,** 49–64.

TURNER, D. A. and SCHWARTZKROIN, P. (1980). Steady-state electrotonic analysis of intracellularly stained hippocampal neurons. *J. Neurophysiol.* **44,** 184–99.

Chapter 8: Nonlinear properties of excitable membranes

ARMSTRONG, C. M. (1975). Evidence for ionic pores in excitable membranes. *Biophys. J.* **15,** 932–33.

ATTWELL, D. and EISNER, D. A. (1978). Discrete membrane surface charge distributions. Effect of fluctuations near individual channels. *Biophys. J.* **24**, 869–75.

—— —— and COHEN, I. (1979). Voltage clamp and tracer flux data: effects of a restricted extracellular space. *Q. Rev. Biophys.* **12**, 213–61.

—— and JACK, J. J. B. (1978). The interpretation of membrane current–voltage relations: a Nernst–Planck analysis. *Prog. Biophys. mol. Biol.* **34**, 81–107.

BEGENISICH, T. (1975). Magnitude and location of surface charges on *myxicola* giant axons. *J. gen. Physiol.* **66**, 47–65.

—— (1979). Conditioning hyperpolarization-induced delays in the potassium channels of myelinated nerve. *Biophys. J.* **27**, 257–65.

—— and CAHALAN, M. D. (1980). Sodium channel permeation in squid axons I. Reversal potential experiments. *J. Physiol.* **307**, 217–42, 243–57.

—— and STEVENS, C. F. (1975). How many conductance states do potassium channels have? *Biophys. J.* **15**, 843–6.

BULLOCK, J. D. and SCHAUF, C. L. (1978). Combined voltage-clamp and dialysis of *Myxicola* axons: behaviour of membrane asymmetry currents. *J. Physiol.* **278**, 309–24.

CAHALAN, M. and BEGENISICH, T. (1975). Sodium-channel selectivity. Dependence on internal permeant ion concentration. *J. gen. Physiol.* **68**, 111–25.

CAMPBELL, D. T. (1976). Ionic selectivity of the sodium channel of frog skeletal muscle. *J. gen. Physiol.* **67**, 295–307.

CHIU, S. Y., RITCHIE, J. M., ROGART, R. B., and STAGG, D. (1979). A quantitative description of membrane currents in rabbit myelinated nerve. *J. Physiol.* **292**, 149–66.

D'ARRIGO, J. S. (1978). Screening of membrane surface charges by divalent cations: an atomic representation. *Am. J. Physiol.* **235**, C109–17.

EASTON, D. M. (1978). Exponentiated exponential model of Na^+ and K^+ conductance charges in squid giant axon. *Biophys. J.* **22**, 15–28.

EBERT, G. A. and GOLDMAN, L. (1976). The permeability of the sodium channel in *Myxicola* to the alkali cations. *J. gen. Physiol.* **68**, 327–58.

EDMONDS, D. T. (1980). Membrane ion channels and ionic hydration energies. *Proc. R. Soc. B.* **211**, 51–62.

—— (1981). A calculation of the current voltage characteristic of a voltage-controlled model membrane ion channel. *Proc. R. Soc. B.* **214**, 125–36.

FINKELSTEIN, A. and ANDERSEN, O. S. (1981). The gramicidin A channel: A review of its permeability characteristics with special reference to the single-file aspect of transport. *J. membr. Biol.* **59**, 155–71.

FREHLAND, E. and FAULHABER, K. H. (1980). Nonequilibrium ion transport through pores. The influence of barrier structures on current fluctuations, transient phenomena and admittance. *Biophys. Struct. Mech.* **7**, 1–16.

FRENCH, R. J. and WELLS, J. B. (1977). Sodium ions as blocking agents and charge carriers in the potassium channel of the squid giant axon. *J. gen. Physiol.* **70**, 707–24.

GILLESPIE, J. I. and MEVES, H. (1980). The time course of sodium inactivation in squid giant axons. *J. Physiol.* **299**, 289–307.

GOLDMAN, L. (1975). Quantitative description of the sodium conductance of the giant axon of *Myxicola* in terms of a generalized second-order variable. *Biophys. J.* **15**, 119–36.

HAGIWARA, S., MIYAZAKI, S., KRASNE, S. R., and CIANI, S. (1977). Anomalous permeabilities of the egg cell membrane of a starfish in K^+ −Tl+ mixtures. *J. gen. Physiol.* **70**, 269–81.

—— —— and ROSENTHAL, N. P. (1976). Potassium current and the effect of cesium on this current during anomalous rectification of the egg cell membrane of a starfish. *J. gen. Physiol.* **67**, 621–38.

HILLE, B. (1975). Ionic selectivity, saturation and block in sodium channels. A four barrier model. *J. gen. Physiol.* **66**, 535–60.

—— and SCHWARZ, W. (1978). Potassium channels as multi-ion single-file pores. *J. gen. Physiol.* **72**, 409–42.

HORN, R., PATLAK, J. and STEVENS, C. F. (1981). The effect of tetramethylammonium on single sodium channel currents. *Biophys. J.* **36**, 321–27.

—— and STEVENS, C. F. (1980). Relation between structure and function of ion channels. *Comm. Mol. Cell. Biophys.* **1**, 57–68.

KEYNES, R. D. and ROJAS, E. (1976). The temporal and steady-state relationships between activation of the sodium conductance and movement of the gating particles in the squid giant axon. *J. Physiol.* **255**, 157–89.

KÖHLER, H.-H. (1977). A single-file model for potassium transport in squid giant axon. Simulation of potassium currents at normal ionic concentrations. *Biophys. J.* **19**, 125–40.

KOLATA, G. B. (1976). Water structure and ion binding: a role in cell physiology? *Science* **192**, 1220–2.

KRAMER, L. (1976). The problem of nonstationary ion fluxes in excitable membranes. *Biophys. Struct. Mech.* **2**, 233–42.

LANDOWNE, D. (1977). Sodium efflux from voltage-clamped squid giant axons. *J. Physiol.* **266**, 43–68.

LÄUGER, P. (1980). Kinetic properties of ion carriers and channels. *J. membr. Biol.* **57**, 163–78.

LEUCHTAG, H. R. and SWIHART, J. C. (1977). Steady-state electrodiffusion. Scaling, exact solution for ions of one charge, and the phase plane. *Biophys. J.* **17**, 27–46.

LEVITAN, E. and PALTI, Y. (1975). Dipole moment, enthalpy and entropy changes of Hodgkin–Huxley type kinetic units. *Biophys. J.* **15**, 239–51.

LEVITT, D. G. (1982). Comparison of Nernst–Planck and reaction-rate models for multiply occupied channels. *Biophys. J.* **37**, 575–87.

—— (1978). Electrostatic calculations for an ion channel. I. Energy and potential profiles and interactions between ions. *Biophys. J.* **22**, 209–19.

—— (1978). Electrostatic calculations for an ion channel. II. Kinetic behavior of the gramicidin A channel. *Biophys. J.* **22**, 221–48.

MCILROY, D. K. (1975). Electric field distributions in neuronal membranes. *Math. Biosci.* **26**, 191–206.

MCQUARRIE, D. A. and MULÁS, P. (1977). Asymmetric charge distributions in planar bilayer systems. *Biophys. J.* **17**, 103–9.

MEVES, H. (1978). Inactivation of the sodium permeability in squid giant nerve fibres. *Prog. Biophys. mol. Biol.* **33**, 207–30.

—— and VOGEL, W. (1977a). Slow recovery of sodium current and 'gating' current from inactivation. *J. Physiol.* **267**, 395–410.

—— —— (1977b). Inactivation of the asymmetrical displacement current in giant axons of *Loligo forbesi*. *J. Physiol.* **267**, 377–93.

MIYAZAKI, S.-I., OHMORI, M., and SASAKI, S. (1975). Potassium rectifications of the starfish oocyte membrane and their changes during oocyte maturation. *J. Physiol.* **246**, 55–78.

MULLINS, L. J. (1975). Ion selectivity of carriers and channels. *Biophys. J.* **15**, 921–39.

—— (1979). The generation of electric currents in cardiac fibers by Na/Ca exchange. *Am. J. Physiol.* **236**, C103–10.

NEMOTO, I., MIYAZAKI, S., SAITO, M., and UTSONOMIYA, T. (1975). Behavior of solutions of the Hodgkin–Huxley equations and its relation to properties of mechanoreceptors. *Biophys. J.* **15**, 469–79.

NONNER, W. (1980). Relations between the inactivation of sodium channels and the immobilization of gating charge in frog myelinated nerve. *J. Physiol.* **299**, 573–603.

PICKARD, W. F. (1976). Generalizations of the Goldman–Hodgkin–Huxley equation. *Math. Biosci.* **30**, 99–111.

—— (1976). A physical model for the passage of ions through an ion-specific channel—II. The potassium-like channel. *Math. Biosci.* **32**, 51–61.

—— and LETTVIN, J. Y. (1976). A physical model for the passage of ions through an ion-specific channel—I. The sodium-like channel. *Math. Biosci.* **32**, 37–50.

REUTER, H. and SCHOLZ, H. (1977). A study of the ion selectivity and the kinetic properties of the calcium-dependent slow inward current in mammalian cardiac muscle. *J. Physiol.* **264**, 17–48.

—— and STEVENS, C. F. (1980). Ion conductance and ion selectivity of potassium channels in snail neurones. *J. membr. Biol.* **57**, 103–18.

SCHWARTZ, T. L. and KADO, R. T. (1977). Permeability, phase-boundary potential and conductance in a cholinergic channel without constant field. *Biophys. J.* **18**, 323–49.

SHINAGAWA, Y. (1978). Analytical solution of the Poisson–Boltzmann equation for membrane potential. *J. theor. Biol.* **72**, 603–10.

—— (1979). Note on the validity of constant field assumptions in relation to the exact solution of Nernst–Planck equations. *J. theor. Biol.* **81**, 333–40.

—— (1980). Invalidity of the Henderson diffusion equation shown by the exact solution of the Nernst–Planck equations. *J. theor. Biol.* **83**, 359–64.

SIMPSON, I., ROSE, B. R., and LOEWENSTEIN, W. R. (1977). Size limit of molecules permeating the junctional membrane channels. *Science* **195**, 294–6.

TAYLOR, R. E. and BEZANILLA, F. (1979). Comments on the measurement of gating currents in the frequency domain. *Biophys. J.* **25**, 338–40.

TREDGOLD, R. H. (1979). On the potential variation in the gramicidin channel. *Biophys. J.* **25**, 373–8.

TSIEN, R. Y. (1978). A virial expansion for discrete charges buried in a membrane. *Biophys. J.* **24**, 561–7.

URBAN, B. W., HLADKY, S. B., and HAYDON, D. A. (1980). Ion movements in gramicidin pores. An example of single-file transport. *Biochim. Biophys. Acta.* **602**, 331–54.

VAIDHYANATHAN (1979). Nernst–Planck Analog Equations and Stationary State Membrane Electric Potentials. *Bull. math. Biol.* **41**, 365–85.

WERBLIN, F. S. (1975). Anomalous rectification in horizontal cells. *J. Physiol.* **244**, 639–57.

Chapters 9 and 10: Nonlinear cable theory: excitation and conduction

ARMON, C. and PALTI, Y. (1980). The weighted reversal potential—a tool to study refractoriness and regenerativity in nerves. *J. theor. Biol.* **83**, 505–15.

BELL, J. (1981). Some threshold results for models of myelinated nerve. *Math. Biosci.* **54**, 181–90.

BOSTOCK, H. and SEARS, T. A. (1978). The internodal axon membrane: electrical excitability and continuous conduction in segmental demyelination. *J. Physiol.* **280**, 273–301.

BOYD, I. A. and KALU, K. U. (1979). Scaling factor relating conduction velocity and diameter for myelinated afferent nerve fibres in the cat hind limb. *J. Physiol.* **289**, 277–97.

CARPENTER, C. A. (1977). A geometric approach to singular perturbation problems with applications to nerve impulse equations. *J. differ. Equn.* **23**, 335–67.

CASTEN, R., COHEN, H., and LAGERSTROM, P. (1975). Perturbation analysis of an approximation to Hodgkin–Huxley theory. *Q. appl. Math.* **32**, 365–402.

CLAY, J. R. (1976). A stochastic analysis of the graded excitatory response of nerve membrane. *J. theor. Biol.* **59**, 141–58.

CLERC, L. (1976). Directional differences of impulse spread in trabecular muscle from mammalian heart. *J. Physiol.* **255**, 335–46.

COHEN, I., ATTWELL, D., and STRICHARTZ, G. (1981). The dependence of the maximum rate of rise of the action potential upstroke on membrane properties. *Proc. R. Soc. B.* **214**, 85–98.

COLDING-JORGENSEN, M. (1976). Can membrane excitation be described without a membrane capacity? *J. theor. Biol.* **57**, 373–83.

DOMINGUEZ, G. and FOZZARD, H. A. (1979). Effect of stretch on conduction velocity and cable properties of cardiac Purkinje fibres. *Am. J. Physiol.* **237**, C119–24.

DONATI, F. and KUNOV, H. (1976). A model for studying velocity variation in unmyelinated axons. *IEEE Trans. biomed. Eng.* **BME 23**, 23–8.

DUCLAUX, R., MEI, N., and RANIERI, F. (1976). Conduction velocity along the afferent vagal dendrites: a new type of fibre. *J. Physiol.* **260**, 487–95.

EASTON, D. M. and SWENBERG, C. E. (1975). Temperature and impulse velocity in giant axon of squid *Loligo pealei*. *Am. J. Physiol.* **229**, 1249–53.

EVANS, J. W. and FEROE, J. (1977). Local stability theory of the nerve impulse. *Math. Biosci.* **37**, 23–50.

FEROE, J. A. (1978). Local existence of the nerve impulse. *Math. Biosci.* **38**, 259–77.

—— (1978). Temporal stability of solitary impulse solutions of a nerve equation. *Biophys. J.* **21**, 103–10.

FITZHUGH, R. (1976). Anodal excitation in the Hodgkin–Huxley nerve model. *Biophys. J.* **16**, 209.

GROSSMAN, Y., PARNAS, I., and SPIRA, M. E. (1979). Differential conduction block in branches of a bifurcating axon. *J. Physiol.* **295**, 283–305.

—— —— —— (1979). Ionic mechanisms involved in differential conduction of action potentials at high frequency in a branching axon. *J. Physiol.* **295**, 307–22.

HASTINGS, S. P. (1976). On travelling wave solutions of the Hodgkin–Huxley equations. *Arch. rational mech. Anal.* **60**, 229–57.

JACK, J. J. B. (1975). Physiology of peripheral nerve fibres in relation to their size. *Br. J. Anaesth.* **47**, 173–82.

KHRAMOV, M. N. and KRINSKII, V. I. (1977). Steady speeds of spread of the stable and unstable impulses, dependence on ionic currents of the membrane. *Biophysics* **22**, 529–36.

MCNEAL, D. R. (1976). Analysis of a model for excitation of myelinated nerve. *IEEE Trans. biomed. Eng.* **BME 23**, 329–37.

MADSEN, E. L. (1977). Theory for a test of the electric cable model of the myelinated axon and saltatory conduction. *J. theor. Biol.* **67**, 203–12.

MARKS, W. B. and LOEB, G. E. (1976). Action currents, internodal potentials, and extracellular records of myelinated mammalian nerve fibers derived from node potentials. *Biophys. J.* **16**, 655–68.

MATSUMOTO, G. and TASAKI, I. (1977). A study of conduction velocity in non-myelinated nerve fibers. *Biophys. J.* **20**, 1–13.

MILLER, R. N. (1979). A simple model of delay, block and one way conduction in Purkinje fibers. *J. math. Biol.* **7**, 385–98.

MOORE, J. W., JOYNER, R. W., BRILL, M. H., WAXMAN, S. G., and NAJAR-JOA, M. (1978). Simulations of conduction in uniform myelinated fibers. Relative sensitivity to changes in nodal and internodal parameters. *Biophys. J.* **21**, 147–60.

PARNAS, I. and SEGEV, I. (1979). A mathematical model for conduction of action potentials along bifurcating axons. *J. Physiol.* **295**, 323–43.

RAMON, F., ANDERSON, N. C., JOYNER, R. W., and MOORE, J. W. (1976). A model for propagation of action potentials in smooth muscle. *J. theor. Biol.* **59**, 381–408.

RINZEL, J. (1975). Neutrally stable travelling wave solutions of nerve conduction equations. *J. math. Biol.* **2**, 205–17.

—— (1975). Spatial stability of travelling wave solutions of a nerve conduction equation. *Biophys. J.* **15**, 975–88.

RITCHIE, J. M. (1983). On the relation between fibre diameter and conduction velocity in myelinated nerve. *Proc. R. Soc. B* (in press).

SELEKTOV, L. Y. and KHODOROV, B. I. (1980). Conduction of a nerve impulse along a myelinized fibre with variation in the parameters of the internodal portion. Mathematical model. *Biophysics* **24**, 930–7.

—— —— (1980). Conduction of a nerve impulse along myelinized fibre on variation of the properties of the membrane of the node of Ranvier. Mathematical model. *Biophysics* **25**, 319–24.

SMITH, D. O. (1980). Mechanisms of action potential propagation failure at sites of axon branching in the crayfish. *J. Physiol.* **301**, 243–59.

—— (1980). Morphological aspects of the safety factor for action potential propagation at axon branch points in the crayfish. *J. Physiol.* **301**, 261–9.

ZYKOV, V. S. and MOROZOVA, O. L. (1980). Speed of spread of excitation in a two-dimensional excitable medium. *Biophysics* **24**, 739–44.

Chapter 11: Repetitive activity in excitable cells

Part 1: Constant current

ADELMAN, W. J. and FITZHUGH, R. (1975). Solutions of the Hodgkin–Huxley equations modified for potassium accumulation in a periaxonal space. *Fed. Proc.* **34**, 1322–9.

ATTWELL, D., COHEN, I., and EISNER, D. (1979). Membrane potential and ion concentration stability conditions for a cell with a restricted extracellular space. *Proc. R. Soc. B.* **206**, 145–61.

BAKER, C. L. (1978). Nonlinear systems analysis of computer models of repetitive firing. *Biol. Cybern.* **29**, 115–23.

—— and HARTLINE, D. K. (1978). Nonlinear systems analysis of repetitive firing behavior in the crayfish stretch receptor. *Biol. Cybern.* **29**, 105–13.

BALDISSERA, F., GUSTAFSSON, B., and PARMIGGIANI, F. (1976). A model for refractoriness accumulation and secondary range firing in spinal motoneurones. *Biol. Cybern.* **24**, 61–5.

—— and PARMIGGIANI, F. (1979). After hyperpolarization conductance time-course and repetitive firing in a motoneurone model with early inactivation of the slow potassium conductance system. *Biol. Cybern.* **34**, 233–40.

BARBI, M., CARELLI, V., FREDIANI, C., and PETRACCHI, D. (1975). The self-inhibited leaky integrator: transfer functions and steady state relations. *Biol. Cybern.* **20**, 51–9.

BRILLINGER, D. R. and SEGUNDO, J. P. (1979). Empirical examination of the threshold model of neuron firing. *Biol. Cybern.* **35**, 213–20.

BRUCKSTEIN, A. M. and ZEEVI, Y. Y. (1979). Analysis of 'integrate-to-threshold' neural coding schemes. *Biol. Cybern.* **34**, 63–79.

CALVIN, W. H. (1975). Generation of spike trains on CNS neurons. *Brain Res.* **84**, 1–22.

—— (1978). Setting the pace and pattern of discharge: Do CNS neurons vary their sensitivity to external inputs via their repetitive firing processes? *Fed. Proc.* **37**, 2165–70.

CARPENTER, G. A. (1977). Periodic solutions of nerve impulse equations. *J. Math. anal. Appl.* **58**, 152–73.

—— (1979). Bursting phenomena in excitable membranes. *SIAM J. appl. Math.* **36**, 334–72.

CLUSIN, W. T. and BENNETT, M. V. L. (1979). The oscillatory response of skate electroreceptors to small voltage stimuli. *J. gen. Physiol.* **73**, 685–702.

—— —— (1979). The ionic basis of oscillatory responses of skate electroreceptors. *J. gen. Physiol.* **73**, 709–23.

COLDING-JØRGENSEN, M. (1976). A description of adaptation in excitable membranes. *J. theor. Biol.* **63**, 61–87.

CONNOR, J. A., WALTER, D., and McKOWN, R. (1977). Neural repetitive firing. Modifications of the Hodgkin–Huxley axon suggested by experimental results from crustacean axons. *Biophys. J.* **18**, 81–102.

FOHLMEISTER, J. F. (1975). Adaptation and accommodation in the squid axon. *Biol. Cybern.* **18**, 49–60.

—— (1979). A theoretical study of neural adaptation and transient responses due to inhibitory feedback. *Bull. math. Biol.* **41**, 257–82.

—— (1980). Electrical processes involved in the encoding of nerve impulses. *Biol. Cybern.* **36**, 103–8.

—— POPPELE, R. E., and PURPLE, R. L. (1972). Repetitive firing: a quantitative study of feedback in model encoders. *J. gen. Physiol.* **67**, 815–48.

—— —— —— (1977). Repetitive firing: quantitative analysis of encoder behavior of slowly adapting sketch receptor of crayfish and eccentric cell of *Limulus*. *J. gen. Physiol.* **69**, 849–77.

FROLOV, A. A. and PETUKHOVA, V. M. (1978). Mathematical model of the adaptation of a neurone allowing for the influence of calcium on the permeability of the membrane for potassium ions. *Biophysics* **23**, 681–8.

—— —— (1979). Mathematical model of the adaptation of a neurone allowing for the influence of calcium on the permeability of the membrane for potassium ions. *Biophysics* **23**, 681–8.

GOLA, M., CHAGNEUX, H., and ARGÉMI, J. (1982). An asymmetrical kinetic model for veratridine interactions with sodium channels in molluscan neurons. *Bull. math. Biol.* **44**, 231–58.

GUEVARA, M. R. and GLASS, L. (1982). Phase locking, period.doubling bifurcations and chaos in a mathematical model of a periodically driven oscillator: a theory for the entrainment of biological oscillators and the generation of cardiac dysrhythmias. *J. math. Biol.* **14**, 1–23.

GUSTAFSSON, B., LINDSTRÖM, S., and ZANGGER, P. (1978). Firing behaviour of dorsal spinocerebellar tract neurones. *J. Physiol.* **275**, 321–43.

HADELEV, K. P., VAN DER HEIDEN, U., and SCHUMACHER, K. (1976). Generation of the nervous impulse and periodic oscillations. *Biol. Cybern.* **23**, 211–18.

HINDMARSH, J. L. and ROSE, R. M. (1982). A model of the nerve impulse using two first-order differential equations. *Nature* **296**, 162–4.

HOLDEN, A. V. (1980). Autorhythmicity and entrainment in excitable membranes. *Biol. Cybern.* **38**, 1–8.

KRYLOV, B. V. and MAKOVSKII, V. S. (1979). Ionic mechanism of analog-code conversion in nerve-fiber membrane. *Doklady Biophys.* **244**, 3–6.

LINKENS, D. A. (1979). Modulation analysis of forced non-linear oscillators for biological modelling. *J. theor. Biol.* **77**, 235–51.

MORRIS, C. and LECAR, H. (1981). Voltage oscillations in the barnacle giant muscle fiber. *Biophys. J.* **35**, 193–213.

NOBLE, D. (1983). Ionic mechanisms of rhythmic firing. *Symp. Soc. Exp. Biol.* (in press).

PARTRIDGE, L. D. and CONNOR, J. A. (1978). A mechanism for minimizing temperature effects on repetitive firing frequency. *Am. J. Physiol.* **234**, C155–61.

—— and STEVENS, C. F. (1976). A mechanism for spike frequency adaptation. *J. Physiol.* **256**, 315–32.

PINSKER, H. M. and BELL, J. (1981). Phase plane description of endogenous neuronal oscillators in *Aplysia*. *Biol. Cybern.* **39**, 211-21.

PLANT, R. E. (1977). Crustacean cardiac pacemaker model—an analysis of the singular approximation. *Math. Biosc.* **36**, 149-71.

—— (1978). The effects of calcium^{++} on bursting neurons. A modelling study. *Biophys. J.* **21**, 217-38.

—— (1981). Bifurcation and resonance in a model for bursting nerve cells. *J. math. Biol.* **11**, 15-32.

—— and KIM, M. (1976). Mathematical description of a bursting pacemaker neuron by a modification of the Hodgkin-Huxley equations. *Biophys. J.* **15**, 227-44.

POKROVSKII, A. N. (1979). Effect of conductivity of the synapses on the conditions of appearance of spikes. *Biophysics* **23**, 662-7.

RAMÓN, F. and MOORE, J. W. (1978). Propagation of action potentials in squid giant axons. Repetitive firing at regions of membrane inhomogeneities. *J. gen. Physiol.* **73**, 595-603.

REKAA, S. and SKAUGEN, E. (1981). Firing behavior in a nerve membrane model with long-term changes of a potassium conductance component. *Math. Biosci.* **55**, 65-87.

RINZEL, J. (1978). Repetitive activity and Hopf bifurcation under point-stimulation for a simple FitzHugh-Nagumo nerve conduction model. *J. math. Biol.* **5**, 363-82.

—— and MILLER, R. N. (1980). Numerical calculation of stable and unstable periodic solutions to the Hodgkin-Huxley equations. *Math. Biosci.* **49**, 27-59.

SCHARSTEIN, H. (1979). Input-output relationship of the leaky-integrator neuron model. *J. math. Biol.* **8**, 403-20.

SCRIVEN, D. R. L. (1981). Modelling repetitive firing and bursting in a small un-myelinated nerve fiber. *Biophys. J.* **35**, 715-30.

SKAUGEN, E. (1975). Repetitive firing behaviour in nerve cell models based upon a simplified form of the Hodgkin-Huxley equations. *Math. Biosc.* **26**, 119-55.

—— (1980). Firing behaviour in stochastic nerve membrane models with different pore densities. *Acta physiol. scand.* **108**, 49-60.

—— and WALLØE, I. (1979). Firing behaviour in a stochastic nerve membrane model based upon the Hodgkin-Huxley equations. *Acta physiol. scand.* **107**, 343-63.

TRAUB, R. D. (1977). Motoneurons of different geometry and the size principle. *Biol. Cybern.* **25**, 163-76.

—— (1977). Repetitive firing of Renshaw spinal interneurons. *Biol. Cybern.* **27**, 71-6.

—— and LLINAS, R. (1979). Hippocampal pyramidal cells: significance of dendritic ionic conductances for neuronal function and epileptogenesis. *J. Neurophysiol.* **42**, 476-96.

ZEEVI, Y. Y. and BRUCKSTEIN, A. M. (1981). Adaptive neural encoder model with self-inhibition and threshold control. *Biol. Cybern.* **40**, 79-92.

Part 2: Varying current

BROWN, T. (1976). Neuron's firing time. *Biol. Cybern.* **21**, 97-101.

BRYANT, H. L. and SEGUNDO, J. P. (1976). Spike initiation by transmembrane current: a white-noise analysis. *J. Physiol.* **260**, 279-314.

COPE, D. K. and TUCKWELL, H. C. (1979). Firing rates of neurons with random excitation and inhibition. *J. theor. Biol.* **80**, 1-14.

FIENBERG, S. E. (1974). Stochastic models for single neuron firing trains: a survey. *Biometrics* **30**, 399-427.

FOHLMEISTER, J. F. (1979). Excitation parameters of the repetitive firing mechanism from a statistical evaluation of nerve impulse trains. *Biol. Cybern.* **34**, 227-32.

GESTRI, G., MASTEBROEK, H. A. K., and ZAAGMAN, W. H. (1980). Stochastic constancy, variability and adaptation of spike generation: performance of a giant neuron in the visual system of the fly. *Biol. Cybern.* **38**, 31-40.

GIGLMAYR, J. (1979). Modulation of point processes as a model of neuronal impulse generation. *Math. Biosc.* **46**, 139–49.

GRECO, E. C. and CLARK, J. W. (1976). A mathematical model of the vagally driven SA nodal pacemaker. *IEEE Trans. biomed. Eng.*, *BME.* **23**, 192–9.

GUTTMAN, R., FELDMAN, L., and JAKOBSSON, E. (1980). Frequency entrainment of squid axon membrane. *J. membr. Biol.* **56**, 9–18.

—— GRISELL, R., and FELDMAN, L. (1977). Strength–frequency relationship for white noise stimulation of squid axons. *Math. Biosci.* **33**, 335–43.

HOCKMAN, H. G. (1980). Models for neural discharge which lead to Markov renewal equations for probability density. *Math. Biosci.* **51**, 125–39.

HOLDEN, A. V. (1976). Models of the stochastic activity of neurones. *Lecture notes in Biomathematics* v. 12. Springer N.Y. 368 pp.

KEENER, J. P., HOPPENSTEADT, F. C., and RINZEL, J. (1981). Integrate-and-fire models of nerve membrane response to oscillatory input. *SIAM J. Appl. Math.* **41**, 503–17.

KOHN, A. F., FREITAS DA ROCHA, A., and SEGUNDO, J. P. (1981). Presynaptic irregularity and pacemaker inhibition. *Biol. Cybern.* **41**, 5–18.

KOSTYUKOV, A. I., IVANOV, Y. N., and KRYZHANOVSKY, M. V. (1981). Probability of neuronal spike initiation as a curve-crossing problem for Gaussian stochastic processes. *Biol. Cybern.* **39**, 157–63.

LEVINE, M. W. and SHEFNER, J. M. (1977). A model for the variability of interspike intervals during sustained firing of a retinal neuron. *Biophys. J.* **19**, 241–52.

LOSEV, I. S. (1978). Limiting behaviour of a model of the impulse activity of a neurone for high frequencies of the input flows and small contributions of one synapse (non-diffusion approximation). *Biophys.* **23**, 507–14.

—— (1980). Effect of non-linear summation of postsynaptic potentials on mean potential and distribution of the inter-impulse intervals of a neurone. *Biophys.* **25**, 324–9.

MUSHA, T., KOSUGI, Y., MATSUMOTO, G., and SUZUKI, M. (1981). Modulation of the time relation of action potential impulses propagating along an axon. *IEEE* Trans. biomed. Eng., BME. **28**, 616–23.

POGGIO, T. and TORRE, V. (1977). A Volterra representation for some neuron models. *Biol. Cybern.* **27**, 113–24.

RICCIARDI, L. M. (1976). Diffusion approximation for a multi-input model neuron. *Biol. Cybern.* **24**, 237–40.

—— and SACERDOTE, L. (1979). The Ornstein–Uhlenbeck process as a model for neuronal activity. 1. Mean and variance of the firing time. *Biol. Cybern.* **35**, 1–9.

RUBIO, J. E. and HOLDEN, A. V. (1975). The response of a model neurone to a white noise input. *Biol. Cybern.* **19**, 191–5.

SAMPATH, G. and SRINIVASAN, S. K. (1977). Stochastic models for spike trains of single neurons. *Lecture notes in Biomathematics* v. 16. Springer, New York, pp. 188.

SATO, S. (1978). On the moments of the firing interval of the diffusion approximated model neuron. *Math. Biosci.* **39**, 53–70.

SCLABASSI, R. J. (1976). Neuronal models, spike trains, and the inverse problem. *Math. Biosci.* **32**, 203–19.

SMITH, C. E. (1979). A comment on a retinal neuron model. *Biophys. J.* **25**, 385–6.

SRINIVASAN, S. K. (1977). A stochastic model of neuronal firing. *Math. Biosci.* **33**, 167–75.

—— and SAMPATH, G. (1975). A neuron model with pre-synaptic deletion and post-synaptic accumulation, decay, and threshold behaviour. *Biol. Cybern.* **19**, 69–74.

—— —— (1976). On a stochastic model for the firing sequence of a neuron. *Math. Biosci.* **30**, 305–23.

TUCKWELL, H. C. (1975). Determination of the inter-spike times of neurons receiving randomly arriving post-synaptic potentials. *Biol. Cybern.* **18**, 225–37.

—— (1979). Synaptic transmission in a model for stochastic neural activity. *J. theor. Biol.* **77**, 65-81.

—— and COPE, D. K. (1980). Accuracy of neuronal interspike times calculated from a diffusion approximation. *J. theor. Biol.* **83**, 377-87.

—— and RICHTER, W. (1978). Neuronal interspike time distributions and the estimation of neurophysiological and neuroanatomical parameters. *J. theor. Biol.* **71**, 167-83.

WAN, F. Y. M. and TUCKWELL, H. C. (1979). The response of a spatially distributed neuron to white noise current injection. *Biol. Cybern.* **33**, 39-55.

WHITE, B. S. and ELLIAS, S. (1979). A stochastic model for neuronal spike generation. *SIAM J. Appl. Math.* **37**, 206-33.

WINFREE, A. T. (1977). Phase control of neural pacemakers. *Science* **197**, 761-3.

YANG, G. L. and CHEN, T. C. (1978). On statistical methods in neuronal spike-train analysis. *Math. Biosci.* **38**, 1-34.

Chapter 12: Non-linear cable theory: analytical approaches using polynomial models

ADRIAN, R. H. (1975). Conduction velocity and gating current in the squid giant axon. *Proc. R. Soc. B.* **189**, 81-6.

ANDERSON, N. and ARTHURS, A. M. (1978). Complementary variational principles for the steady state finite cable model of nerve membranes. *Bull. Math. Biol.* **40**, 735-42.

BELL, J. (1981). Modeling parallel, unmyelinated axons: pulse trapping and ephaptic transmission. *SIAM J. appl. Math.* **41**, 168-80.

BELL, J. and COOK, L. P. (1978). On the solutions of a nerve conduction equation. *SIAM J. appl. Math.* **35**, 678-88.

BUTRIMAS, P. and GUTMAN, A. (1978). Theoretical analysis of an experiment with voltage clamping in the motoneurone. Proof of the N-shape pattern of the steady voltage–current characteristic of the dendrite membrane. *Biophys.* **23**, 897-904.

CARPENTER, G. (1977). Periodic solutions of nerve impulse equations. *J. math. Anal. Appl.* **58**, 152-73.

—— (1977). A geometric approach to singular perturbation problems with applications to nerve impulse equations. *J. diff. Eqns.* **23**, 335-67.

CASTEN, R., COHEN, H., and LAGERSTROM, P. (1975). Perturbation analysis of an approximation to Hodgkin-Huxley theory. *Quart. appl. Math.* **32**, 365-402.

DONATI, F. and KUNOV, H. (1976). A model for studying velocity variations in un-myelinated axons. *IEEE Trans. biomed. Eng. BME.* **23**, 23-8.

EVANS, J. W. and FEROE, J. (1977). Local stability theory of the nerve impulse. *Math. Biosci.* **37**, 23-50.

HASTINGS, S. (1976). On travelling wave solutions of the Hodgkin-Huxley equations. *Arch. rational mech. Anal.* **60**, 229-57.

KOOTSEY, J. M. (1977). The steady-state finite cable: numerical method for non-linear membrane. *J. theor. Biol.* **64**, 413-20.

MAGINU, K. (1978). Stability of periodic travelling wave solutions of a nerve conduction equation. *J. math. Biol.* **6**, 49-57.

—— (1980). Existence and stability of periodic travelling wave solutions to Nagumo's nerve equation. *J. math. Biol.* **10**, 133-53.

MILLER, R. N. and RINZEL, J. (1981). The dependence of impulse propagation speed on firing frequency, dispersion, for the Hodgkin-Huxley model. *Biophys. J.* **34**, 227-59.

MIURA, R. M. (1982). Accurate computation of the stable solitary wave for the Fitzhugh-Nagumo equations. *J. math. Biol.* **13**, 247-69.

PICKARD, W. F. (1974). Electrotonus on a nonlinear dendrite. *Math. Biosci.* **20**, 75-84.

—— (1977). An exact solution to a problem of nonlinear electrotonus. *Math. Biosci.* **33**, 1-4.

RINZEL, J. and MILLER, R. N. (1980). Numerical calculation of stable and unstable periodic solutions to the Hodgkin–Huxley equations. *Math. Biosci.* **49**, 27–59.

RISSMAN, P. (1977). The leading edge approximation to the nerve axon problem. *Bull. math. Biol.* **39**, 43–58.

WALTON, M. K. and FOZZARD, H. A. (1983a). Experimental study of the conducted action potential in cardiac Purkinje strands. *Biophys. J.* (In press.)

—— —— (1983b). The conducted action potential: models and comparison with experiments. *Biophys. J.* (In press.)

Subject index